# Membranes and Ion Transport

## Volume 1

# Membranes
# and
# Ion Transport

## Volume 1

*Edited by*

E. Edward Bittar

*Department of Physiology,*
*The University of Wisconsin,*
*Madison, Wisconsin, U.S.A.*

WILEY-INTERSCIENCE
a division of John Wiley & Sons Ltd.
LONDON    NEW YORK    SYDNEY    TORONTO

Library of Congress Catalog card No. 71-110649

ISBN 0 471 07707 0

Printed in Great Britain
By Photolithography
Unwin Brothers Limited
Woking and London

This work is dedicated to
the late
**EDWARD J. CONWAY**
1894–1968

# Preface

Thirty years have now elapsed since the publication of the celebrated paper by Boyle and Conway. During these years much has happened in the field of membrane metabolism and ion transport. The following work is an attempt to deal with the notable advances that have already been made and to treat the subject both systematically and critically. Membrane transport may be said to be an integral part of the life sciences but it does not yet seem to cover a comprehensive field. To undertake a general survey of it is therefore to invite criticism, in particular at a time when research is being conducted more intensively than ever. Written primarily for the novice and professed student of the field, the work aims at providing an intelligible view of the wide scope and importance of the subject, and of the different lines of research that may lead to a more unified discipline.

The work is divided into three books. Volume 1 contains a section on the structure, chemistry and behaviour of both artificial and natural membranes, and a section on the theoretical aspects of transport. A useful glossary has been appended at the beginning of this volume. Volume 2 deals with ion movements in symmetrical cells and sub-cellular organelles. And Volume 3 begins with an account of ion movements in asymmetrical cells, as well as water movements in cells in general, and ends with a collection of chapters dealing with ion regulatory mechanisms and cellular interaction and continuity.

The field of ion transport owes a great deal to the late Professor E. J. Conway. It therefore seemed more than fitting to dedicate this work to him.

*Madison, Wisconsin*　　　　　　　　　　　　　　　　E. EDWARD BITTAR
*April*, 1970

# Contributors to volume 1

SJOERD L. BONTING — Department of Biochemistry, University of Nijmegen, Nijmegen, The Netherlands.

P. C. CALDWELL — Department of Zoology, University of Bristol, Bristol, England.

S. R. CAPLAN — Biophysical Laboratory, Harvard Medical School, Boston, Massachusetts, U.S.A.

D. CHAPMAN — Department of Chemistry, Sheffield University, Sheffield, England.

HALVOR N. CHRISTENSEN — Department of Biological Chemistry, The University of Michigan, Ann Arbor, Michigan, U.S.A.

A. ESSIG — Renal Service, New England Medical Center Hospitals and Tufts University School of Medicine, Boston, Massachusetts, U.S.A.

D. A. HAYDON — Laboratory of Biophysical Chemistry and Colloid Science, University of Cambridge, Cambridge, England.

R. P. KERNAN — Department of Physiology, University College, Dublin, Ireland.

IRVING M. KLOTZ — Biochemistry Division, Department of Chemistry, Northwestern University, Evanston, Illinois, U.S.A.

S. K. MALHOTRA — Biological Sciences Electron Microscopy Laboratory, University of Alberta, Edmonton, Canada.

PETER MITCHELL — Glynn Research Laboratories, Bodmin, Cornwall, England.

RUSSELL PATERSON — Department of Chemistry, The University, Glasgow, Scotland.

H. ROTTENBERG — Biophysical Laboratory, Harvard Medical School, Boston, Massachusetts, U.S.A

# Glossary of symbols

| | |
|---|---|
| $A$ | affinity of a chemical reaction (kcal mole$^{-1}$) |
| $c_i$ | molar concentration of species i |
| $D_{ii}$ | isotopic diffusion coefficient |
| $E$ | electromotive force of a galvanic cell |
| F | Faraday constant [kcal volt$^{-1}$ (or coulombs) gm equiv$^{-1}$] |
| $f$ | isotopic flux ratio |
| $G$ | Gibbs free energy [kcal (or joules) mole$^{-1}$] |
| $I$ | electric current density |
| $J$ | generalized flow (moles cm$^{-2}$ sec$^{-1}$) |
| $J_{ch}$ | rate of chemical reaction |
| $J_q$ | flow of heat |
| $J_s$ | flow of entropy |
| $K$ | rate constants for a chemical reaction |
| $L$ | phenomenological conductance coefficient [mole$^2$ kcal$^{-1}$ (or joule$^{-1}$) cm$^{-2}$ sec$^{-1}$] |
| $n_i$ | number of moles of component i |
| $m_i$ | molecular mass of species i |
| O | subscript denoting oxidation |
| o | subscript denoting initial value; superscript denoting standard value |
| $P$ | { pressure; subscript also denoting phosphorylation <br> { permeability coefficient |
| $Q$ | heat |
| $q$ | degree of coupling |
| $R$ | phenomenological resistance coefficient (kcal cm$^2$ sec mole$^{-2}$) |
| R | gas constant [kcal (or joules) deg$^{-1}$ mole$^{-1}$] |
| $R^x$ | $\int_o^{\Delta x} (r_{00} - r_{iK})\, dx$ the exchange resistance of test species including interactions with component isotopes |
| $r$ | subscript denoting reaction |
| $r_{ik}$ | local frictional coefficient between trace and abundant isotope species |
| $S$ | entropy |
| $T$ | absolute temperature |
| $t_i$ | transport number of species i |

| | |
|---|---|
| $U$ | internal energy |
| $V$ | volume |
| $x$ | flow coordinate |
| $X$ | thermodynamic force [kcal (or joules) mole$^{-1}$cm$^{-1}$] |
| $X_{exch}$ | the exchange force in the model for a sodium pump |
| $z_i$ | valency of ion i, including sign |
| $Zx$ | the force ratio |
| $\alpha_i$ | stoichiometric coefficients for dissociation of a salt |
| $\delta$ | thickness of a membrane |
| $\Delta$ | difference ($\Delta\mu_i = \mu_i{}^{in} - \mu_i{}^{ex}$, $\Delta\psi = \psi^{in} - \psi^{ex}$, $\Delta P_i = [P_i]_o - [P_i]$) |
| $\eta$ | efficiency of energy conversion |
| $\kappa$ | specific conductivity |
| $\mu$ | chemical potential (kcal mole$^{-1}$) |
| $\tilde{\mu}$ | electrochemical potential (kcal mole$^{-1}$) |
| $v$ | frequency of carrier transport |
| $v_i$ | stoichiometric coefficient of species i in a chemical reaction |
| $d\xi$ | degree of advancement of a chemical reaction |
| $\rho_i$ | specific activity of tracer isotope i |
| $\sigma$ | rate of production of entropy/unit volume |
| $\Phi$ | dissipation function |
| $\psi$ | internal electrical potential of a phase (volt) |
| [ ] | indicates concentration (the distinction between concentration and activity is neglected) |
| $-$ | indicates average value |

# Contents

# I

# Membranes: Biogenesis, Structure and Behaviour

# CHAPTER 1

# Organization and biogenesis of cellular membranes

## S. K. Malhotra

*Biological Sciences Electron Microscopy Laboratory,*
*University of Alberta,*
*Edmonton, Canada*

## I. ORGANIZATION

The functions of cellular membranes during the life of a cell are largely served by the diverse chemical constituents that make up the membrane and by the interactions occurring between these constituents. Lipids and proteins constitute the greater part of the membranes and form the plasma membrane which acts as the permeability barrier of the cell. Carbohydrates make up a small fraction, up to approximately 7 percent, of the total membrane contents and occur mostly attached to lipids (glycolipids) and proteins (glycoproteins). The carbohydrate complement is generally considered to be on the outside of the plasma membrane and to be part of the cell coat (Maddy, 1966; Rambourg and Leblond, 1967; Revel and Ito, 1967), and thus exposed to the extracellular environment. Alterations in the chemical composition of the extracellular material, for example the concentration of $Na^+$ and $Ca^{2+}$, could also influence the molecular arrangement in the plasma membrane (see Lehninger, 1968 for his model of the greater neuronal membrane). There is little direct information on the molecular organization of lipids and proteins or on the lipid–protein interactions in cellular membranes. Since the lipids and proteins of membranes can be completely separated by organic solvents and detergents, their association must be through weak interactions such as electrostatic interactions, hydrogen bonding, London–van der Waals forces and hydrophobic interactions rather than by covalent bonding (Maddy, 1966; Wallach and Gordon, 1968). One can expect major

1

advances in our understanding of the organization of membranes as a result of the current physicochemical and biological emphasis on the subject. This chapter contains introductory remarks on some of such recent considerations relevant to membrane structure and function.

Recent investigations on the structure of natural membranes have largely been aimed at examining the classical Danielli–Davson model (Danielli and Davson, 1935; Danielli, 1967) proposed for the plasma membrane. In this model, dealing with the relationship between the lipid and protein, the lipid complement is assumed to be arranged essentially in a thin continuous layer, possibly bimolecular, and the protein in an extended sheet of insoluble film on either side of the lipid leaflet (figure 1a). The protein is thought to be bound through electrostatic interactions with the polar groups of the lipid. If the protein chains are on the surface of a phospholipid layer, interactions between the non-polar groups of the lipid and protein are considered unlikely. Since the side chains of the proteins (4 Å long) are short as compared to the polar ends (at least 8·5 Å long) of the phospholipid molecules, penetration of the protein side chains between the polar groups of the lipid molecules to effect hydrophobic bonding would expose the fatty acid chains to water. Such interactions may occur when cholesterol is present, because its polar end is much less bulky (Haydon and Taylor, 1963). The extended conformation of the protein contiguous with lipid was suggested, because it was thought that proteins lost their secondary structure at interfaces. Additional protein was considered to be in globular conformation and adsorbed upon the unrolled protein layer. This paucimolecular model evolved as a result of various well-known physicochemical considerations (see Brown and Danielli, 1964; Malhotra, 1969), for example, penetration

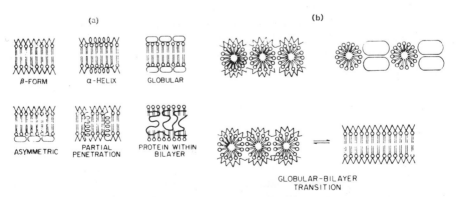

**Figure 1.** Some membrane models based on (a) a bimolecular lipid leaflet arrangement, and (b) globular arrangements (From Lehninger, 1968).

of lipid soluble substances into cells, known electrical properties of some of the cell membranes, extent of lipid in red cell ghosts relative to the surface area of the cell, and tension at the surface of cells. The principal features of the Danielli–Davson model were found to be consistent with the unit membrane concept (Robertson, 1961, 1966 a, b) that emerged from studies by electron microscopy. The basis of the unit membrane concept is the well-known tripartite structure, 60 to 110 Å in width (Sjöstrand, 1963 a; Yamamoto, 1963), that is typically seen in electron micrographs as a transparent layer bordered by a dense layer on either side. The middle transparent layer is assumed to represent the location of the fatty acid chains of the bimolecular lipid leaflet and the dense layers the polar groups of the lipid and the protein molecules. Such a tripartite structure is discerned in all natural membranes and this resemblance has been interpreted as indicative of a basic similarity in their molecular organization. The evidence in favour of the lamellar arrangement of the lipid and the extended form of the protein envisaged in the unit membrane structure is largely obtained from model systems (phospholipid/water mixtures) discussed by Stoeckenius and coworkers (1960), Robertson (1966 b), Korn (1968), Malhotra and Van Harreveld (1968), Riemersma (1968) and Malhotra (1969). This interpretation is supported by the x-ray diffraction data obtained from a variety of plasma membranes (e.g. myelin sheath and red cell membranes) and intracellular membranes (e.g. microsomes and mitochondria, see Finean and coworkers, 1968). The x-ray diffraction patterns are best interpreted as indicative of a lamellar structure which corresponds to the width of a bimolecular lipid leaflet. Moreover, there are similarities in the physical properties of black membranes reconstituted from lipids and those of natural membranes (Mueller and Rudin, 1963, 1968; Thompson and Huang, 1966; Cass and Finkelstein, 1967; Howard and Burton, 1968; Tien and Diana, 1968; Tosteson and coworkers, 1968). The thickness of such reconstituted membranes, their tripartite appearance in electron micrographs (Henn and coworkers, 1967), and their electrical resistance are consistent with the presence of lipid as a bimolecular leaflet in these membranes. In phospholipid/water systems, the phospholipids (which make up the greater part of the lipid complement in natural membranes, see Ashworth and Green, 1966; Korn, 1966; Malhotra and Van Harreveld, 1968; Rouser and coworkers, 1968) form bimolecular leaflets spontaneously, and the lipid will be so oriented that the more polar portions are directed towards the aqueous phase and the hydrophobic portions form a non-polar phase (Danielli, 1967). Until recently the bimolecular leaflet arrangement was the only configuration known to be stable in lipid/water systems and this concept dominated considerations of the membrane structure. It also provided a readily accepted interpretation of the tripartite structure of the membranes commonly discerned in electron micrographs, because in model systems, consisting of phospholipid and water fixed by $OsO_4$ and examined in the electron microscope, the electron density corresponded to the location of the polar ends of the bimolecular lipid leaflet (Finean, 1959; Stoeckenius, 1962; Stoeckenius and Mahr, 1965; Robertson, 1966 b and Riemersma, 1968). Addition of a basic protein to the phospholipid/water systems resulted in increased width of the dense lines in electron micrographs which indicates that the protein was adsorbed on to the lipid leaflets and reacted with $OsO_4$ (Stoeckenius and coworkers, 1960). Whether the contrast seen in natural membranes after fixation by $OsO_4$ is due to the selective accumulation of osmium at the polar groups of the lipid molecules and protein is still not certain (Korn, 1967, 1968). The tripartite unit membrane structure of the inner mitochondrial membranes and the entire lamellar structure of the myelin sheath can be demonstrated in $OsO_4$-fixed material after almost all of the lipid has been extracted (Fleischer and coworkers, 1967; Napolitano and coworkers, 1967). Zahler and coworkers (1968) have reported that the typical tripartite unit membrane structure is preserved in electron micrographs of membranes of mitochondria from which the insoluble (approximately 45%) 'structural' protein had been extracted by acid treatment. The residual fraction, which is membranous and shows the tripartite structure, contains almost all of the phospholipid. These findings emphasize the caution necessary in giving molecular interpretations to the structure of natural membranes visualized in electron micrographs. Moreover, when $OsO_4$ reacts with unsaturated lipids, it binds with the double bonds of unsaturated fatty acids, which can be cross-linked through ester formations. In addition, osmium is also precipitated as $OsO_2$. The possibility remains that $OsO_2$ may be formed during fixation and localized at the interfaces between lipoprotein structures and water. If this were the case, $OsO_2$ would mark only the presence of an aqueous/organic interface rather than specify the sites of polar groups of lipids at interfaces (Korn, 1968). Furthermore, fixation by $OsO_4$ (and other chemical agents used for fixation) may introduce gross alterations in the organization of membrane components. Lenard and Singer (1968) observed that fixation of red cell membranes, bovine serum albumin and apomyoglobin by $OsO_4$ (either alone or preceded by glutaraldehyde) and $KMnO_4$ obliterated the greater part of the helical conformation of the proteins.

Recently, it has been learnt that lipid/water systems exhibit several configurations other than the familiar bimolecular lamellar leaflet. One of these phases is a hexagonal phase which consists of a hexagonal array of water-filled cylinders in a lipid matrix (Luzzati and Husson, 1962; Luzzati and coworkers, 1966). This hexagonal phase exists in a phospholipid/water system at an elevated temperature (37 °C) and a low water content (3%), and can be preserved in electron micrographs by fixation with vapours of

$OsO_4$ (Stoeckenius, 1962). In soap/water systems there is another hexagonal phase in which hydrophobic fatty acid chains are confined to the cylinders arranged in a hexagonal array and the water is localized between the cylinders, that is, the reverse of the one in the above-mentioned hexagonal phase in the phospholipid/water system (Luzzati and Husson, 1962; Stoeckenius, 1962). Moreover, negatively stained preparations of various mixtures of lipids show several structures (lamellar, tubular, hexagonal and helical structures) which could be interpreted satisfactorily by assuming that the lipid is arranged in the form of globular micelles (40 to 50 Å in diameter) and that these micelles aggregate to form specific assemblies (Lucy and Glauert, 1964; Lucy, 1968). It was therefore suggested that in the natural membranes lipid molecules are organized as globular micelles as well as a bimolecular leaflet (figure 1b) and the extent of these configurations is likely to vary not only from one membrane to another, but also within any particular membrane, depending upon its functions. Electron micrographs of the natural membranes showing particulate substructure suggestive of the presence of spherical micelles in a variety of membranes prepared by different techniques have been published (Sjöstrand, 1963 b; Sjöstrand and Elfvin, 1964; Blasie and coworkers, 1965; Nilsson, 1965; Branton, 1966; Malhotra, 1966, 1969; Fernández-Morán, 1967; Malhotra and Eakin, 1967). However, such appearances can be artifacts produced in more than one way (Robertson, 1966 a, b; Stoeckenius, 1966). Moreover, the particulate substructure seen in electron micrographs may not represent lipid micelles, as for example, in the outer segments of the retina, the 40 Å ordered arrangement of spherical particles seems to represent the rhodopsin molecules (Blasie and coworkers, 1966). Freeze-etched preparations of chloroplast membranes show distinct globular subunits of two categories, one 175 Å in diameter and the other 110 Å arranged in a regular array (Branton and Park, 1967; Park and Pheifhofer, 1968). These subunits are considered to be located on the inner surface of the membrane as the frozen chloroplast membrane seems to fracture so as to split into two halves (figure 2). The larger subunits (quantasomes) are considered to be part of the photosynthetic system. The smaller subunits are believed to be part of the matrix in which the larger subunits are embedded.

In addition to the above considerations of the globular micelles as possible arrangements for lipids in the natural membranes, there are other physiological factors which also suggest that certain functions of the membranes require a modification of the simple bimolecular lipid leaflet (see Brandt and Freeman, 1967; Lucy, 1968). For example, adsorption of certain proteins or of macrocyclic compounds, such as valinomycin, by black membranes reconstituted from lipids leads to a decrease in the electrical resistance of the membranes and the lowered resistance is within the range of natural membranes (see Cass and Finkelstein, 1967; Mueller and Rudin, 1968;

Tien and Diana, 1968; Tosteson and coworkers, 1968). A modification of
the bimolecular lipid leaflet may also be responsible for the properties of
low resistance membranes through which a large variety of cells are known
to be electrically coupled (see Loewenstein, 1966; Bullivant and Loewenstein,
1968; Malhotra and Van Harreveld, 1968).

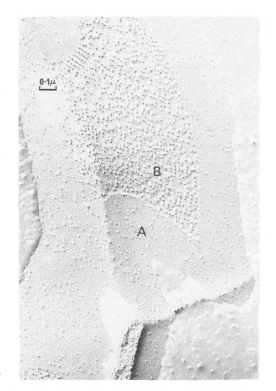

**Figure 2.** Spinach chloroplast membrane prepared by freeze-etching. The face A
is the outside of the membrane, whereas the fracture face B shows the particulate
structures believed to be the quantasomes (see text). (Courtesy of Dr. R. B. Park).

Also, studies on protein conformation in isolated membranes do not
support the view that the bulk of the membrane protein has a β-extended
conformation, as postulated by Danielli and Davson and by Robertson for
their membrane models. The results of more recent studies by the application
of techniques of infrared spectroscopy, optical rotatory dispersion, and
circular dichroism have been interpreted as indicating that the bulk of the
protein is in α-helical and random coil conformation; little protein in the

β-conformation could be detected. In this context, the possibility of reversible conformational changes in proteins should be borne in mind, as even the heat denaturation of enzymes is sometimes reversible (Dawson, 1968). For instance, the nucleotide pyrophosphatase of *Proteus vulgaris* is completely inactivated at 70 °C but it is reactivated on cooling to 37 °C (Swartz and coworkers, 1958). Serum albumin is unfolded in salt-free acid or strong urea solutions but the protein undergoes a conformational change on removal of the acid or urea and becomes water soluble at the isoelectric point (Dawson, 1968). These observations on protein conformation of membranes were made on a variety of plasma membranes as well as intracellular membranes, and the optical activity of the membranes from such diverse sources is similar (Lenard and Singer, 1966; Wallach and Zahler, 1966, 1968; Urry and coworkers, 1967; Chapman and Wallach, 1968). This similarity may reside in the presence of 'structural' proteins in all membranes (Lenard and Singer, 1966). Moreover, these results on optical activity have been interpreted to emphasize hydrophobic interactions between membrane proteins and lipids rather than electrostatic interactions envisaged in the Danielli–Davson model. On the basis of such conclusions, models have been proposed for the structure of membranes in which the helical portions of the protein are considered to be in the interior of the membrane where they interact with lipid through hydrophobic interactions (Lenard and Singer, 1966; Wallach and Zahler, 1966). Green has also found that the lipid and

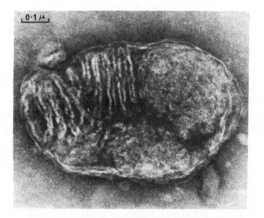

**Figure 3.** A fairly well preserved mitochondrion stained negatively by PTA (phosphotungstate). Mitochondrial fraction was isolated from *Neurospora crassa* (Malhotra and Eakin, 1967). The electron-transparent, more or less parallel disposed structures, are probably mitochondrial cristae. Stalked particles are not evident in such preparations of intact mitochondria, but can be demonstrated in swollen and disrupted mitochondria (see figure 4).

protein interact in membranes predominantly through hydrophobic interactions. Furthermore, he has proposed that the lipids and proteins are
organized in subunits which aggregate to form an intact membrane. Such
lipoprotein subunits are considered to be the basic building blocks of cellular
membranes (see Green and Perdue, 1966).

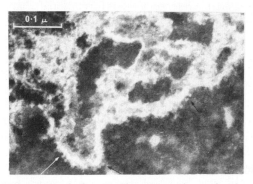

**Figure 4.** Negatively stained (by PTA) preparation of mitochondria isolated
from rat liver showing stalked particles (arrows).

The best available morphological evidence in favour of the subunit
hypothesis has been obtained from negatively stained preparations of
isolated mitochondrial fractions. In electron micrographs of such preparations mitochondria are often swollen and disrupted and frequently they show
regularly disposed particles ($\sim 100$ Å in diameter) attached by a short stalk
($\sim 50$ Å long) to one side of the inner mitochondrial membrane (figures 3
and 4, see Fernández-Morán and coworkers, 1964; Racker and coworkers,
1965; Malhotra and Eakin, 1967; Racker and Hornstman, 1967). These
particles are characteristic of the inner mitochondrial membrane as the
negatively stained preparations of the outer mitochondrial membrane and
other cellular membranes do not show similar stalked particles (Benedetti
and Emmelot, 1968; Emmelot and coworkers, 1968; Prezbindowski and
coworkers, 1968). It has been claimed that the inner mitochondrial membrane is subdivided into base-pieces (110 Å × 40 Å), each with one stalked
particle attached to it (Fernández-Morán and coworkers, 1964). A tripartite
subunit, formed by the spherical head, a stalk and a base-piece, has been
considered to represent a typical structure of the inner mitochondrial
membrane (Green and Perdue, 1966; Fernández-Morán, 1967). Since no
such tripartite subunits have been isolated so far, their morphological entity
remains hypothetical (Stoeckenius, 1966). Moreover, the stalked particles
have not been seen in sections, with one or two possible exceptions, and

it is not certain how the stalked particles are related to the tripartite unit membrane. In view of the finding of ATPase activity in the biochemical fraction corresponding to the stalked particles (Racker and coworkers, 1965; Stoeckenius, 1966; Racker and Horstman, 1967) there is little doubt that the stalked particles are a manifestation of the highly ordered organization of the inner mitochondrial membrane. The precise *in vivo* organization is of course uncertain. X-ray diffraction patterns of the inner mitochondrial membranes do not show any remarkable physically pronounced substructure (Finean and coworkers, 1968; Thompson and coworkers, 1968) as has been observed by electron microscopy. Moreover, the x-ray diffraction patterns of both the inner and the outer mitochondrial membranes are essentially alike; they resemble the patterns obtained from red cell ghosts and myelin sheath, which indicate the existence of a lamellar structure corresponding to the bimolecular lipid leaflet (Finean and coworkers, 1968; Worthington and Blaurock, 1968). Until further evidence becomes available, the differences in the outer and inner mitochondrial membranes discerned in electron micrographs of negatively stained preparations should be considered as exaggerated manifestations of their *in vivo* differences (Thompson and coworkers, 1968).

In view of the great diversity in the chemical constituents and in the lipid to protein ratios in different membranes, it is conceivable that different cellular membranes are not built upon one basic molecular pattern; their functions may necessitate several alternative configurations, each configuration with its own energy level (Maddy, 1967). Moreover, in organized tissues the plasma membrane is not always uniform in its functions along the entire perimeter of the cell. For example, the organization of the plasma membranes at the site where adjacent cells are connected with each other through low resistance pathways (intercellular communication; figures 5 and 6) must be different from elsewhere where the plasma membranes have relatively higher resistance (see Bennett, 1966; Revel and Karnovsky, 1967; Weidmann, 1967; Barr and coworkers, 1968; Bullivant and Loewenstein, 1968; Malhotra and Van Harreveld, 1968; Malhotra, 1969). There is also likelihood of functional compartmentalization within the membranes of the rough surfaced endoplasmic reticulum (Siekevitz and Palade, 1966). The possibility also exists that certain membranes may be organized as a mosaic of functionally discrete complexes, which make the membrane biochemically heterogeneous. For example, Wallach (1967) has reported isolation of two distinct types of vesicles from the plasma membrane fraction of *Ehrlich* ascites carcinoma cells: one type of vesicle showed concentration of ATPase activity, and the other contained agglutinating agent. In contrast, it is likely that in a membrane, such as the myelin sheath which serves essentially as an insulator, the constitutive components are uniformily distributed. Clearly, advances in

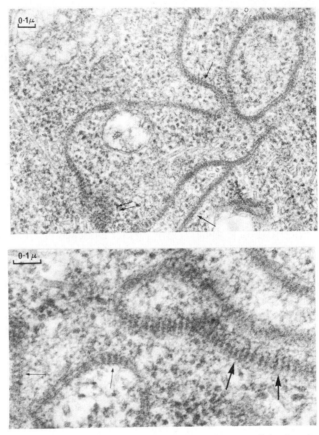

Figure 5

Figure 6

**Figures 5 and 6.** Section through salivary gland of *Drosophila* showing regions of cell contact between the epithelial cells. Single thin arrows indicate where the intercellular space appears to be traversed by an array of bridges connecting apposing unit membranes. Thick arrows show regularly disposed cross bridges but the tripartite unit membranes are not evident in this micrograph. Double arrows indicate a region of cell contact where a regular hexagonal pattern has been demonstrated by Bullivant and Loewenstein (1968). Such contact regions are considered to be the sites through which ions and molecules pass across cells. Fixed in glutaraldehyde and osmium tetroxide (both in phosphate buffer, pH 7·1–7·2), embedded in Araldite. Thin sections were stained in uranyl acetate and lead citrate before examination in the electron microscope.

cytochemical techniques at the electron microscope level could help to map out possible differences in the general pattern of the organization of various cellular membranes (figure 7). Such techniques could provide valuable aid in following the formation of cellular membranes as they are assembled from their constitutive components.

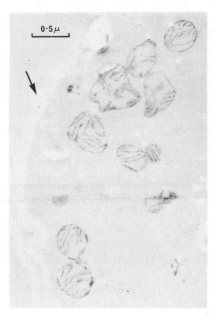

**Figure 7.** Wild type *Neurospora crassa* mycelia in log phase treated to show location of succinic dehydrogenase by reduction of ferricyanide according to the technique of Ogawa and coworkers (1968). After reaction the mycelia were fixed in glutaraldehyde (without post fixation in osmium tetroxide), dehydrated in ethanol and embedded in Epon. Thin sections were examined (without additional staining with heavy metal salts) in a Philips E. M. 200 electron microscope operated at 40 kV with 20 μ objective aperture. The sites of the reaction product indicate that the reaction is specific for the inner mitochondrial membranes. With improved resolution it may be possible to characterize further the sites of such enzymatic reactions. The arrow indicates the cell wall. (From unpublished work by Koke and Malhotra).

## II.  BIOGENESIS

Structures that resemble the intact membrane can be produced *in vitro* from components of disrupted membranes. However, such attempts have not so far been successful in reconstituting membranes with the same properties as the natural membrane from which the constituents are isolated. There is no definite evidence in favour of a *de novo* origin of cellular membranes, though there have been reports suggestive of *de novo* origin of mitochondria, that is, from parts of cells that have no mitochondrial precursors (see Berger, 1964; Lehninger, 1964; Marchant and Smith, 1968; Roodyn and Wilkie, 1968; Wallace and coworkers, 1968) but these reports require substantiation by suitable biochemical characterization. In intact cells, membranes are believed to grow only on existing membranes.

An understanding of the way a membrane is assembled from its constitutive components could provide valuable information about the molecular
organization of the membrane, such as the configuration of lipids and proteins and the lipid–protein interactions. For example, it is implied in the
bimolecular lipid leaflet model that the first step in the formation of the
membrane would be the assembly of the lipid complement; the protein
contiguous with the lipid is likely to be adsorbed later in a pattern probably
determined by the arrangement of the lipid; other components could be
added on in subsequent steps. Such a process of membrane formation
implies that a membrane is produced in more than one step, and consequently
the composition, morphology, and specific biochemical activity will change
during the assembly of the membrane. This scheme emphasizes that there is
a primary membrane structure and that constitutive enzymes and other
components are added later on to this basic membrane (Siekevitz and
coworkers, 1967). The other possible process that could be involved in the
formation of membranes is a single-step process. According to this scheme
a membrane is formed from a simultaneous assembly of newly synthesized
or pre-existing components. This mechanism implies that a membrane has a
relatively constant composition, morphology and specific biochemical
activity at any one time. Such a scheme is consistent with the suggestion that
a membrane is made up of discrete (lipoprotein) subunits held together to
form a continuum membrane (Siekevitz and coworkers, 1967). In this case,
the association of lipids with the protein molecules could be determined by
the structure of the protein (Korn, 1968). If a membrane is a mosaic composed of several functionally distinct patches (units), each with its own
characteristic chemical composition, it is conceivable that each type of unit
is assembled in a single-step but different types of units could be formed at
different times (Dallner and coworkers, 1966 a, b).

The detailed mechanism involved in the formation of a cellular membrane
from its constitutive components is nòt understood, and the data available
so far do not justify a generalization of any one scheme for the formation of
all cellular membranes. Most of the findings reported in the literature seem
to be at the moment best interpreted in favour of the multiple-step formation.
However, the manner of assembly of chloroplast membranes in a mutant of
the unicellular alga *Chlamydomonas* is such that it suggests that a single-step
process may be operative in the assembly of at least these membranes (Ohad
and coworkers, 1967 a, b: summarized by Siekevitz and coworkers, 1967).
The apparent differences in the mode of formation of different membranes
may be due to differences in the rate of assembly of constitutive components
into fully functional membranes (Siekevitz and coworkers, 1967). In the
wild type *Chlamydomonas* the cells can synthesize chlorophyll in the dark as
well as in the light. In a dark-grown mutant, the alga does not have the

ability to synthesize chlorophyll in the dark but can do so in the light, as the higher plants do upon photoactivation. This defect in chlorophyll synthesis is manifested in a defect in membrane formation. When it is placed in the dark the mutant continues to grow and multiply and progressively loses its initial chlorophyll contents, which occurs by simple dilution through cell division. After 6 to 8 generations a minute amount of the original chlorophyll content is left and the alga appears yellow. Concomitant with this change in the chlorophyll content, the disks (membranes) of the chloroplast are gradually disorganized and markedly decreased in extent: only a few vesicular elements remain. This is the only detectable change at the morphological level in the plastid. Upon resumption of chlorophyll synthesis in the light, the formation of the membrane resumes and the organized disks appear within 6–8 hours. There is a distinct connection between the synthesis of chlorophyll and the formation of the chloroplast disk membranes, because even in the dark the plastid can synthesize enzymes located in its matrix as well as those expected to be associated with the disk membranes (for example, cytochrome $f$ and ferredoxin). Moreover, the plastid of the dark grown mutant can synthesize lipids, such as galactolipids and sulpholipids which are considered to be specific for the chloroplast disks. Ohad and coworkers (1967 b) have proposed that chlorophyll molecules may be needed to complex with certain specific proteins before the latter can be released from the plastid ribosomes where they are synthesized. Therefore, it appears that as soon as the synthesis of chlorophyll begins, constitutive proteins and lipids already present in the plastid aggregate to produce membranes. Whether the lipids and proteins are initially assembled as lipoprotein subunits before forming the membrane continuum is not yet known. The pre-existing membranes (for example, vesicular elements) in the plastid may provide centres for growth, if the new membranes are formed only on the existing membranes.

In the above instance, the synthesis of membrane proteins appears to continue even in the absence of membrane formation. Therefore the genes that control the synthesis of these proteins (and hence the production of membrane) are active though the membrane assembly is not apparent. In contrast, the genes that control the synthesis of proteins of the membranes of the endoplasmic reticulum in liver cells of the rat appear to be activated only when membrane formation commences (Siekevitz and coworkers, 1967). In liver cells at birth there is an abundance of rough-surfaced endoplasmic reticulum (with associated ribosomes) but little smooth-surfaced endoplasmic reticulum (free of ribosomes). The latter increases in extent after birth. The membranes of the rough endoplasmic reticulum seem to serve as precursors of the smooth endoplasmic reticulum, as the new membranes are assembled in association with the rough endoplasmic reticulum and subsequently transferred or converted to the smooth endoplasmic reticulum by a process not yet understood (Dallner and coworkers, 1966, a, b).

In the case of membrane organization, findings concerning rates of synthesis and turn over of the proteins and the lipids of the endoplasmic reticulum in rat liver cells are of interest (Siekevitz and coworkers, 1967; Omura and coworkers, 1967). These authors studied the electron transport enzymes and phosphatase activity and found that these enzymes are newly synthesized at birth, because the increases in enzymatic activities could be prevented by injecting actinomycin D and puromycin at birth. The enzymatic activities appear at different times and increase at different rates during development of the rat. Moreover, the total lipids turn over at a rate different from the rate of the total membrane proteins. These findings are difficult to reconcile with the hypothesis that cellular membranes are aggregates of the discrete lipoprotein subunits: if this were the case, the membrane lipid and protein would be expected to turn over at the same rate. The fact that the synthesis of proteins may not be geared to a specific lipid composition is further supported by the finding that alterations in the lipid composition of the endoplasmic reticulum produced by dietary means do not influence the pattern of synthesis of the enzymes associated with the endoplasmic reticulum of the rat liver cells. Also, it has been reported that specific phospholipids (such as cardiolipin) extracted from mitochondria can replace phospholipids in microsomal fraction and the enzyme activity of the microsomal fraction is thereby restored (Siekevitz and coworkers, 1967). The protein synthesis in the liver cells can be impaired by administration of carbon tetrachloride, yet synthesis of phospholipid has been reported to increase and formation of the membranes of the endoplasmic reticulum to be stimulated (Meldolesi and coworkers, 1968). These newly formed membranes produced with $CCl_4$ are presumably deficient in normal biochemical functions. However, these findings are of interest inasmuch as they indicate that the synthesis of membrane proteins is not geared to specific lipid composition of the membrane. (Such an interpretation does not necessarily rule out specific functions for the specific lipid contents of various cellular membranes.) Also, in mitochondria of *Neurospora* the phospholipid to protein ratios can be varied by altering the concentration of precursor molecules in the culture medium without altering the growth rate (Luck, 1965). In a choline-requiring strain, changes in the level of available choline (phospholipid precursor) in the culture medium are manifested in the density of mitochondria. It is possible to grow the fungus with a level of choline which is adequate to support maximum growth. A level of choline in the culture medium ten times higher provides maximum incorporation into phospholipids without affecting the growth rate. Mitochondria from cultures grown with low levels of choline were found to be of higher density and larger (0·6 to 0·7 $\mu$ against

0·2 to 0·25 μ in diameter) than those produced with higher levels of choline.

Membranes are not known to have loose ends and they are continuous structures *in vivo*. As already mentioned, their growth involves accretion of new constitutive components into the existing membranes. A prime example of this is the increase of the plasma membrane of the Schwann cell and oligodendrocyte leading to the formation of the myelin sheath surrounding axons. It is also now well established that mitochondria grow and multiply by division or fragmentation (Luck, 1965; Hawley and Wagner, 1967). Where and how the new material is accumulated by the existing membranes is not certain. It is also not clearly understood how the newly synthesized components are transferred from their site of synthesis to their final destination (which may not be the same as the site of synthesis) and how the assembly of specific constitutive components into functional membranes is regulated. Some of these aspects of membrane biogenesis can be expected to be resolved from study of such organelle systems as mitochondria.

The study of respiratory-deficient mutants of yeast and *Neurospora* has been of fundamental importance in contributing to our knowledge of the genetic control of mitochondrial biogenesis. It is clear from such studies that cytoplasmic genetic determinants as well as nuclear genes are involved in the biogenesis of these organelles (see Haldar and coworkers, 1966, 1967; Woodward and Munkres, 1967; Linnane and coworkers, 1968; Roodyn and Wilkie, 1968). Isolated mitochondria are capable of energy linked incorporation of labeled amino acids into the insoluble ('structural') protein in membrane fractions (Haldar, Freeman and Work, 1966; Roodyn, 1966; Kadenbach, 1967 a, b; Neupert and coworkers, 1967; Beattie, 1968). Whether a similar incorporation of amino acids into the structural protein occurs *in vivo* during biogenesis of mitochondria is not certain (Beattie, 1968). In this respect it is of interest that the presence of a messenger RNA associated with the membranes of the endoplasmic reticulum has been reported in *HeLa* cells. This membrane associated *m*-RNA is distinct from the *m*-RNA of free polysomes and is considered to be possibly of mitochondrial origin (Attardi and Attardi, 1967). This report is compatible with the discovery of structural proteins having similar characteristics (amino acid composition, electrophoretic pattern and immunological behavior) in different membrane fractions (mitochondria, microsomes, nuclear membrane) as well as free in the ground cytoplasm in *Neurospora* (Woodward and Munkres, 1967). Moreover, one of the cytoplasmically inherited (poky) mutant of *Neurospora* carries a single amino acid substitution (tryp→cys) in its mitochondrial structural protein. Structural proteins isolated from other membrane fractions of this mutant carried the same amino acid substitution. These considerations make it highly likely that the primary structure of mitochondrial structural protein is coded by mitochondrial DNA. Woodward and Munkres have proposed that mitochondrial DNA could indirectly control the function and regulation of those proteins, which complex with structural protein, even though their primary structure is specified by the nuclear genes.

In view of the existence of a membrane associated *m*-RNA, possibly of mitochondrial origin, it is quite likely that a fraction of the *m*-RNA of mitochondrial origin remains inside the mitochondria which could direct the synthesis of structural protein of this organelle. The demonstration of ribosomes and their physical characterization in mitochondria of *Neurospora* adds to the components necessary for autonomous protein synthesis in mitochondria (Rifkin and coworkers, 1967). By using methods to separate outer and inner mitochondrial membranes it has been reported that the bulk of amino acids incorporated by isolated rat liver mitochondria is found in the inner membrane fraction (Neupert and coworkers, 1967). In the yeast *Saccharomyces cerevisiae* mitochondria seem to have the capacity to synthesize the insoluble enzymes (cytochromes $a$, $a_3$, $b$ and $c_1$), perhaps in addition to their structural protein. This is indicated by the action of chloramphenicol, which can inhibit the synthesis of these proteins, but does not inhibit the synthesis of soluble proteins, like cytochrome $c$, malate dehydrogenase and fumarase, which seem to be synthesized outside mito-chondrial control, and presumably at the cytoplasmic ribosomes. The incorporation of amino acids by the cytoplasmic ribosomes can be inhibited by cycloheximide, which does not inhibit amino acid incorporation by isolated mitochondria (Clark-Walker and Linnane, 1967; Lamb and co-workers, 1968). The picture that emerges from this and other similar investigations on animal cells (Halder and coworkers, 1966, 1967; Kadenbach, 1967 a, b; Beattie, 1968) is that the mitochondrial soluble proteins, in general, are synthesized outside the mitochondria and then transferred into the mitochondria where they are presumably integrated into mitochondrial structure. Mitochondria can also synthesize lipids but it is not certain whether all the mitochondrial lipids are synthesized within the mitochondrion (Kaiser and Bygrave, 1968; Roodyn and Wilkie, 1968). There is the possi-bility that a part of the mitochondrial lipid complement is synthesized in association with the endoplasmic reticulum and subsequently transferred to the mitochondria (Stein and Stein, 1969). It is conceivable that the lipid required for the formation of the outer mitochondrial membrane is synthe-sized outside the mitochondrion, as the formation of the outer mitochondrial membrane has been reported to be under the control of non-mitochondrial systems (Clark-Walker and Linnane, 1967).

The mechanism by which lipid and protein synthesized in association with the endoplasmic reticulum are transferred to the mitochondria is not known. The resemblance in the chemical composition of the outer mitochondrial membrane and the smooth endoplasmic reticulum (Parsons and coworkers, 1967; Sottocasa and coworkers, 1967) suggests that the endoplasmic reticulum may have a role to play in this transfer of newly synthesized materials to the mitochondria. In *Neurospora* (and perhaps in other such

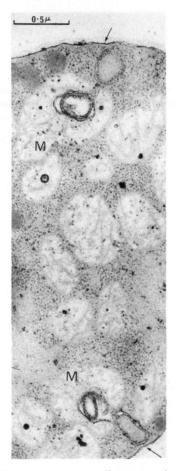

**Figure 8.** Wild type *Neurospora crassa* (from germinating conidia), showing mesosome-like structures within two mitochondria (M). The mesosomes can be distinguished from the mitochondrial membranes because of their differential staining properties. By examination of serial sections it has been shown that the mitochondrial mesosomes are continuous with the plasma membrane shown at arrows (see figures in Malhotra, 1968, 1969). Fixed in glutaraldehyde and then in osmium tetroxide (both in phosphate buffer, pH 7·1–7·2) and embedded in Araldite, sections were routinely stained in uranyl acetate and lead citrate before examination in the electron microscope.

B

microorganisms) this function of transfer may be served by another membrane system, which morphologically resembles the bacterial mesosomes (Malhotra, 1968, 1969). In *Neurospora* the mesosome-like structures have also been observed within mitochondria (figure 8) and are in direct continuity

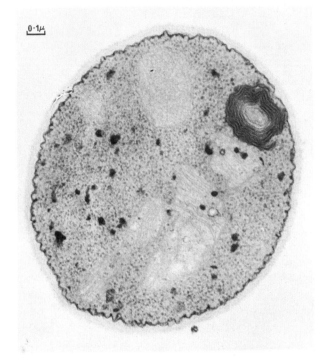

**Figure 9.** Wild type *Neurospora crassa* (from germinating conidia), showing a lamellar membranous structure resembling bacterial mesosomes which can be shown to be continuous with the plasma membrane, shown in this micrograph as a dense line underneath the amorphous cell wall. Prepared by the procedure detailed in the caption to figure 8.

with the plasma membrane (figure 9), as are the mesosomes in bacteria. The mitochondrial mesosomes were first observed in the poky mutant of *Neurospora* which carries a cytoplasmically inherited mutation that has a pleiotropic effect on the synthesis of the respiratory enzyme and other mitochondrial systems (Hardesty and Mitchell, 1963; Woodward and Munkres, 1966; Malhotra and Eakin, 1967; Eakin and Mitchell, 1969). It was thought that the presence of mesosomes in this mutant may be related to some biochemical deficiency of its mitochondria (Malhotra, 1968). However, similar mesosomes have been since then observed in mitochondria of the wild type *Neurospora* (figure 8; Beck and Greenawalt, 1968; Malhotra, 1969). Earlier Yotsuyanagi (1966) found such membranous structures in mitochondria of a yeast, and thought that they were formed from the membranes of mitochondria, since connection with the plasma membrane was not seen. Thus it would seem that the presence of mesosomes in mitochondria of these microorganisms may be of greater general importance than previously thought. The bacterial mesosomes are considered to serve as an anchorage for the DNA during replication of the bacterial genome (Salton, 1967; Ryter, 1968). They are also thought to play a role in oxidation–reduction (Salton, 1967; Ryter, 1968). Yotsuyanagi compared his mitochondrial mesosomes with bacterial mesosomes and considered that his yeast mesosomes serve the same functions as bacterial mesosomes. Whether the mesosomes in mitochondria of *Neurospora* are similar to the bacterial mesosomes in their functioning is being investigated. However, the finding of these mesosomes in mitochondria is consistent with the speculation that during the course of evolution mitochondria evolved from microorganisms which parasitized the host cell (Lehninger, 1964).

## III. CONCLUSIONS

Our concept of the molecular organization of cellular membranes is presently in a state of flux. Recent investigations by a variety of biochemical and biophysical techniques have been highly critical of membrane models based on a simple bimolecular lipid leaflet arrangement. While it is likely that the structure of cellular membranes involves some modifications of the simple bimolecular leaflet concept, not enough is known to specify these modifications in molecular terms (Maddy, 1966). Alternative models based on lipoprotein subunits as basic building blocks for the membranes have been proposed but definite evidence in support of such models is still lacking (Korn, 1968). An understanding of the process of formation of a membrane from its constitutive components could help to resolve some of the questions of membrane organization. A study of the biogenesis of membranes in well characterized mutants of microorganisms, such as *Neurospora* and *Saccharomyces* could aid in elucidating some of the fundamental aspects of membrane organization.

### Acknowledgement

The author's research was supported by grants from the National Research Council of Canada. The technical assistance provided by Mrs. S. Prasad, Mr. V. Prier and Mr. M. Kobalcik is greatly appreciated.

## REFERENCES

Ashworth, L. A. E. and C. Green (1966) *Science,* **151,** 210
Attardi, B. and G. Attardi (1967) *Proc. Natl. Acad. Sci. U.S.,* **58,** 1051
Barr, L., W. Berger and M. M. Dewey (1968) *J. Gen. Physiol.,* **51,** 347
Beattie, D. S. (1968) *J. Biol. Chem.,* **15,** 4027
Beck, D. P. and J. W. Greenawalt (1968) *J. Cell Biol.,* **39,** 11 A (abstract)
Benedetti, E. L. and P. Emmelot (1968) In A. J. Dalton and F. Haguenau (Eds.) *The Membranes,* Academic Press, New York, p. 33
Bennett, M. V. L. (1966) *Ann. N.Y. Acad. Sci.,* **137,** 509
Berger, E. R. (1964) *J. Ultrastruct. Res.,* **11,** 90
Blasie, J. K., M. M. Dewey and C. R. Worthington (1966) *J. Histochem. Cytochem.,* **14,** 789 (abstract)
Blasie, J. K., M. M. Dewey, A. E. Blaurock and C. R. Worthington (1965) *J. Mol. Biol.,* **14,** 143
Brandt, P. W. and A. R. Freeman (1967) *Science,* **155,** 582
Branton, D. (1966) *Proc. Natl. Acad. Sci. U.S.,* **55,** 1048
Branton, D. and R. B. Park (1967) *J. Ultrastruct. Res.,* **19,** 283
Brown, F. and J. F. Danielli (1954) In G. H. Bourne (Ed.) *Cytology and Cell Physiology,* Academic Press, New York, p. 239
Bullivant, S. and W. R. Loewenstein (1968) *J. Cell Biol.,* **37,** 621
Cass, A. and A. Finkelstein (1967) *J. Gen. Physiol.,* **50** 1765
Chapman, D. and D. F. H. Wallach (1968) In D. Chapman (Ed.) *Biological Membranes: Physical Fact and Function,* Academic Press, London, p. 125
Clark-Walker, G. and A. Linnane (1967) *J. Cell Biol.,* **34,** 1
Dallner, G., P. Siekevitz and G. E. Palade (1966 a) *J. Cell Biol.,* **30,** 73
Dallner, G., P. Siekevitz and G. E. Palade (1966 b) *J. Cell Biol.,* **30,** 97
Danielli, J. F. (1967) In K. B. Warren (Ed.) *Formation and Fate of Cell Organelles,* Academic Press, London, p. 239
Danielli, J. F. and H. Davson (1935) *J. Cellular Comp. Physiol.,* **5,** 495
Dawson, R. M. C. (1968) In D. Chapman (Ed.) *Biological Membranes: Physical Fact and Function,* Academic Press, London, p. 203
Eakin, R. T. and H. K. Mitchell (1969) *Arch. Biochem. Biophys.,* **134,** 160
Emmelot, P., A. Viser and E. L. Benedetti (1968) *Biochim. Biophys. Acta.,* **150,** 364
Fernández-Morán, H. (1967) In G. C. Quarton, T. Melnechuk and F. O. Schmitt (Eds.) *The Neurosciences,* Rockefeller University Press, New York, p. 281
Fernández-Morán, H., T. Oda, P. V. Blair and E. E. Green (1964) *J. Cell Biol.,* **22,** 63
Finean, J. B. (1959) *J. Biophys. Biochem. Cytol.,* **6,** 123
Finean, J. B., R. Coleman, S. Knutton, A. R. Limbrick and J. E. Thompson (1968) *J. Gen. Physiol.,* **51,** 19
Fleischer, S., B. Fleischer and W. Stoeckenius (1967) *J. Cell Biol.,* **32,** 193
Green, D. E. and J. F. Perdue (1966) *Ann. N.Y. Acad. Sci.,* **137,** 667
Haldar, D., K. Freeman and T. S. Work (1966) *Nature,* **211,** 9
Haldar, D., K. B. Freeman and T. S. Work (1967) *Biochem. J.,* **102,** 684
Hardesty, B. A. and H. K. Mitchell (1963) *Arch. Biochem. Biophys.,* **100,** 330
Hawley, E. S. and R. P. Wagner (1967) *J. Cell Biol.,* **35,** 489
Haydon, D. A. and Janet Taylor (1963) *J. Theoret. Biol.,* **4,** 281
Henn, F. A., G. L. Decker, J. W. Greenawalt and T. E. Thompson (1967) *J. Mol. Biol.,* **24,** 51
Howard, R. E. and R. M. Burton (1968) *J. Amer. Oil Chemists' Soc.,* **45,** 202
Kadenbach, B. (1967 a) *Biochim. Biophys. Acta.,* **134,** 430
Kadenbach, B. (1967 b) *Biochim. Biophys. Acta.,* **138,** 651
Kaiser, W. and F. L. Bygrave (1968) *Eur. J. Biochem.,* **4,** 582
Korn, E. D. (1966) *Science,* **153,** 1491
Korn, E. D. (1967) *J. Cell Biol.,* **34,** 627
Korn, E. D. (1968) *J. Gen. Physiol.,* **52,** 257

Lamb, A. J., G. D. Clark-Walker and A. W. Linnane (1968) *Biochim. Biophys. Acta.*, **161**, 415
Lehninger, A. L. (1964) *The Mitochondrion*, W. A. Benjamin, New York
Lehninger, A. L. (1968) *Proc. Natl. Acad. Sci. U.S.*, **60**, 1069
Lenard, J. and S. J. Singer (1966) *Proc. Natl. Acad. Sci. U.S.*, **56**, 1828
Lenard, J, and S. J. Singer (1968) *J. Cell Biol.*, **37**, 117
Linnane, A. W., J. Lamb, C. Christodoulou and H. B. Lukins (1968) *Proc. Natl. Acad. Sci. U.S.*, **59**, 1288
Loewenstein, W. R. (1966) *Ann. N.Y. Acad. Sci.*, **137**, 441
Luck, D. J. L. (1965) *Amer. Naturalist*, **99**, 241
Lucy, J. A. (1968) *Brit. Med. Bull.*, **24**, 127
Lucy, J. A. and A. M. Glauert (1964) *J. Mol. Biol.*, **8**, 727
Luzzati, V. and F. Husson (1962) *J. Cell Biol.*, **12**, 207
Luzzati, V., F. Reiss-Husson, E. Rivas and T. Gulik-Krzywicki (1966) *Ann. N.Y. Acad. Sci.*, **137**, 409
Maddy, A. H. (1966) *Intern. Rev. Cytol.*, **20**, 1
Maddy, A. H. (1967) In K. B. Warren (Ed.) *Formation and Fate of Cell Organelles*, Academic Press, London, p. 255
Malhotra, S. K. (1966) *J. Ultrastruct. Res.*, **15**, 14
Malhotra, S. K. (1968) *Nature*, **219**, 1267
Malhotra, S. K. (1969) *Prog. Biophys. Mol. Biol.*, **20**, 67
Malhotra, S. K. and R. T. Eakin (1967) *J. Cell Sci.*, **2**, 205
Malhotra, S. K. and A. Van Harreveld (1968) In E. E. Bittar and N. Bittar (Eds.) *The Biological Basis of Medicine*, **1**, Academic Press, London, p. 3
Marchant, R. and D. G. Smith (1968) *Biol. Rev.*, **43**, 459
Meldolesi, J., L. Vincenzi, P. Bassan and M. T. Morini (1968) *Lab. Invest.*, **19**, 315
Mueller, P. and D. O. Rudin (1963) *J. Theoret. Biol.*, **4**, 268
Mueller, P. and D. O. Rudin (1968) *J. Theoret. Biol.*, **18**, 222
Napolitano, L., F. Lebaron and J. Scaletti (1967) *J. Cell Biol.*, **34**, 817
Neupert, W., D. Brdiczka and Th. Bucher (1967) *Biochem. Biophys. Res. Commun.*, **27**, 488
Nilsson, S. E. G. (1965) *J. Ultrastruct. Res.*, **12**, 207
Ogawa, K., T. Saito and H. Mayahara (1968) *J. Histochem. Cytochem.*, **16**, 49
Ohad, I., P. Siekevitz and G. E. Palade (1967 a) *J. Cell Biol.*, **35**, 521
Ohad, I., P. Siekevitz and G. E. Palade (1967 b) *J. Cell Biol.*, **35**, 553
Omura, T., P. Siekevitz and G. E. Palade (1967) *J. Biol. Chem.*, **242**, 2389
Park, R. B. and A. O. A. Pheifhofer (1968) *Proc. Natl. Acad. Sci. U.S.*, **60**, 337
Parsons, D. F., G. R. Williams, W. Thompson, D. F. Wilson and B. Chance (1967) In J. M. Tager, S. Pappa, E. Quagliarello and E. C. Slater (Eds.) *Round Table Discussion on Mitochondrial Structure and Compartmentation*, Adriatica Editrice, Bari
Prezbindowski, K. S., F. F. Sun and F. L. Crane (1968) *Exptl. Cell Res.*, **5**, 241
Racker, E. and L. L. Horstman (1967) *J. Biol. Chem.*, **242**, 2547
Racker, E., D. D. Tyler, R. W. Estabrook, T. E. Conover, D. F. Parsons and B. Chance (1965) In T. E. King, H. S. Mason and M. Morrison (Eds.) *Oxidases and Related Redox Systems*, Academic Press, New York, p. 1077
Rambourg, A. and C. P. Leblond (1967) *J. Cell Biol.*, **32**, 27
Revel, J. P. and S. Ito (1967) In B. D. Davis and L. Warren (Eds.) *The Specificity of Cell Surfaces*, Prentice-Hall, New Jersey, p. 211
Revel, J. P. and M. J. Karnovsky (1967) *J. Cell Biol.*, **33**, C7
Riemersma, J. C. (1968) *Biochim. Biophys. Acta.*, **152**, 718
Rifkin, M. R., D. D. Wood and D. J. L. Luck (1967) *Proc. Natl. Acad. Sci. U.S.*, **58**, 1025
Robertson, J. D. (1961) In *Biophysics of Physiological and Pharmacological Actions*, American Association for the Advancement of Science, Washington, D.C., p. 63
Robertson, J. D. (1966 a) *Ann. N.Y. Acad. Sci.*, **137**, 421
Robertson, J. D. (1966 b) In G. E. W. Wolstenholme and M. O'Connor (Eds.) *Principles of Biomolecular Organization*, Little, Brown and Company, Boston, p. 357

Roodyn, D. B. (1966) In J. M. Tager, S. Papa, E. Quagliariello and E. C. Slater (Eds.) *Regulation of Metabolic Processes in Mitochondria*, Elsevier, Amsterdam, p. 383

Roodyn, D. B. and D. Wilkie (1968) *The Biogenesis of Mitochondria*, Methuen, London

Rouser, G., G. J. Nelson, S. Fleischer and G. Simon (1968) In D. Chapman (Ed.) *Biological Membranes: Physical Fact and Function*, Academic Press, London, p. 5.

Ryter, A. (1968) *Bacterial Rev.*, **32**, 39

Salton, M. R. J. (1967) *Ann. Rev. Microbiol.*, **21**, 417

Siekevitz, P. and G. E. Palade (1966) *J. Cell Biol.*, **30**, 519

Siekevitz, P., G. E. Palade, G. Dallner, I. Ohad and T. Omura (1967) In H. J. Vogel, J. O. Lampen and V. Bryson (Eds.) *Organizational Biosynthesis*, Academic Press, London. p. 331

Sjöstrand, F. S. (1963 a) *J. Ultrastruct. Res.*, **9**, 561

Sjöstrand, F. S. (1963 b) *J. Ultrastruct. Res.*, **9**, 340

Sjöstrand, F. S. and L. G. Elfvin (1964) *J. Ultrastruct. Res.*, **10**, 263

Sottocasa, G. L., B. Kuylenstierna, L. Ernster and A. Bergstrand (1967) *J. Cell Biol.*, **32**, 415

Stein, O. and Y. Stein (1969) *J. Cell Biol.*, **40**, 461

Stoeckenius, W. (1962) *J. Cell Biol.*, **12**, 221

Stoeckenius, W. (1966) In G. E. W. Wolstenholme and M. O'Connor (Eds.) *Principles of Biomolecular Organization*, Little, Brown and Company, Boston, p. 418,

Stoeckenius, W. and S. C. Mahr (1965) *Lab. Invest.*, **14**, 1196

Stoeckenius, W., J. H. Schulman and L. M. Prince (1960) *Kolloid-Z.*, **169**, 170

Swartz, M. N., N. O. Kaplan and M. F. Lamborg (1958) *J. Biol. Chem.*, **232**, 1051

Thompson, J. E., R. Coleman and J. B. Finean (1968) *Biochim. Biophys. Acta.*, **150**, 405

Thompson, T. E. and C. Huang (1966) *Ann. N.Y. Acad. Sci.*, **137**, 740

Tien, T. H. and A. L. Diana (1968) *Chem. Phys. Lipids*, **2**, 55

Tosteson, D. C., T. E. Andreoli, M. Tieffenberg and P. Cook (1968) *J. Gen. Physiol.*, **51**, 373

Urry, D. W., M. Mednieks and E. Bejnarowicz (1967) *Proc. Natl. Acad. Sci. U.S.*, **57**, 1043

Wallace, P. G., M. Huang and A. W. Linnane (1968) *J. Cell Biol.*, **37**, 207

Wallach, D. F. H. (1967) In B. D. Davis and L. Warren (Eds.), *The Specificity of Cell Surfaces*, Prentice-Hall, New Jersey, p. 129

Wallach, D. F. H. and A. Gordon (1968) *Federation Proc.*, **27**, 1263

Wallach, D. F. H. and P. H. Zahler (1966) *Proc. Natl. Acad. Sci. U.S.*, **56**, 1552

Wallach, D. F. H., and P. H. Zahler (1968) *Biochim. Biophys. Acta.*, **150**, 186

Weidmann, S. (1967) *Harvey Lectures* 1965–66, Academic Press, New York, p. 1

Woodward, D. O. and K. D. Munkres (1966) *Proc. Natl. Acad. Sci. U.S.*, **55**, 872

Woodward, D. O. and K. D. Munktres (1967) In H. J. Vogel, J. O. Lampen, and V. Bryson (Eds.), *Organizational Biosynthesis*, Academic Press, London, p. 489

Worthington, C. R. and A. E. Blaurock (1968) *Nature*, **218**, 87

Yamamoto, T. (1963) *J. Cell Biol.*, **17**, 413

Yotsuyanagi, M. Y. (1966) *Compt. Rend. Acad. Sci., Paris*, **262**, 1348

Zahler, W. L., A. Saito and S. Fleischer (1968) *Biochem. Biophys. Res. Commun.*, **32**, 512

# CHAPTER 2

# The chemical and physical characteristics of biological membranes

*D. Chapman*

*Department of Chemistry,*
*Sheffield University,*
*Sheffield, England.*

## I.  INTRODUCTION

The idea of a cell membrane existed for many years almost entirely at a conceptual level; physiological experiments suggested a barrier to the exchange of material between the cell and its surroundings. It was called a membrane because the barrier was thought to be a thin layer completely enclosing the cell contents. The membrane appeared to have definite mechanical and physical properties and, even in the absence of any direct analyses, suggestions were made as to its structure and organization.

Ponder (1961) pointed out that the definition of the term 'cell membrane' is a matter of contention. Cell biologists use the term 'cell membrane' in at least three different senses:

i) *The anatomical sense.* In this sense the cell membrane is the external limiting region of the cell, visible occasionally as a darkly staining region in the light microscope and with more certainty in the electron microscope as a layer (or pair of layers).

ii) *The biochemical sense.* In this case the cell membrane is a 'fraction' of the cell prepared by the techniques of selective disintegration of the whole cell, followed by differential centrifugation. A preparation is obtained which can be analysed chemically and which can be compared by electron microscopy with the 'cell membrane' seen in the whole cell.

iii) *The physiological sense.* In this case it is a hypothetical structure devised to explain certain data on permeability and which also explains other data on the distribution of metabolites and other molecules between the cell and the fluid in which the cell is immersed.

It is generally assumed (Ponder, 1961; Dervichian, 1955) that these three definitions of cell membrane refer to the same entity. The 'cell membrane' of the biochemists may, however, include many substances which are not present in the 'cell membrane' of the anatomist and may have lost many components during the separation and washing procedures.

### A.  Types of Animal-Cell Membranes

#### 1. *Plasma Membranes*

The plasma membrane interposes a boundary between the cell and its environment by the active transport of ions and nutrients, and it creates and maintains the interior environment of the cell. In certain tissues, e.g. of the intestinal mucosa and kidney, substances are transposed across cellular barriers in the interests of maintaining the internal environment of the whole organism. In the nervous system, the electrochemical potentials to which the transport of $Na^+$ and $K^+$ ions give rise are used for the transmission and processing of information. Schwann cells, in the peripheral nervous system, and oligodendrogliocytes, in the central nervous system, insulate the axons of neurones from their neighbours by myelin formation and thereby bring about a speeding up of impulse transmission by means of saltatory conduction.

Electron microscope data suggest that plasma membranes are often coated with a surface coat which takes a variety of forms.

Certain types of cells—epithelial cells, cardiac and smooth-muscle cells— make specialized contacts characterized by modifications of the plasma membrane. One type of contact, the desmosome or macula adhaerens, appears to provide anchorage and involves a somewhat wider than normal intercellular space (220–250 Å) filled with a band of lightly staining material. The tight junction involves membrane-to-membrane apposition without intervening space.

## 2. Cytoplasmic Membranes

The cytoplasmic membranes (cytomembranes, endoplasmic reticulum) are usually classified as 'rough' and 'smooth'. The roughness is due to the ribosomes that stud their surface and these are numerous in cells that synthesize protein for export. Both types of membrane form arrays of cisternae or tubules which vary considerably in form and intercommunicate with each other and perhaps with the cell exterior. The Golgi apparatus is a specialized form of smooth membrane, characteristically seen as a somewhat isolated pattern of stacked cisternae, and it is particularly prominent in cells that elaborate secretions rich in carbohydrate.

## 3. Organelle Membranes

Subcellular organelles are the least homogeneous. The membranes, comprising the nuclear envelope, are thought to be homologous with the endoplasmic reticulum; they are studded with pores about $0·05$ $\mu$ in diameter which may facilitate the exchange of macromolecules (messenger ribonucleic acid, repressors) essential for the genetic control of the cell.

The essential structural features of the mitochondrion are the double membrane and the characteristic cristae. As first stressed by Palade (1953) the number of cristae and their conformation varies considerably from one type of mitochondrion to another. However, the elements that appear to be common to all mitochondria are an outer membrane, an inner membrane, cristae, mitochondrial 'sap' or 'matrix', dense granules and DNA.

## II.  CHEMICAL CHARACTERIZATION

### A.  Membrane Isolation

To understand the structure of biological membranes at the molecular level, their various components must be identified and isolated in a pure state. This is by no means easy, particularly because the membrane preparation must lack none of the membrane components themselves. The problem of isolating pure biological membrane material therefore involves establishing a method for destroying cells, isolating the membrane material, demonstrating the materials' precise origin, and, finally, estimating the purity of the preparations.

The biological membrane which has received considerable attention is that of the erythrocyte—the red blood cell. When erythrocytes are subjected to certain osmotic pressures or enzymes, they undergo haemolysis—a process in which the cell membrane structure loosens sufficiently to allow the cell contents to escape without itself being ruptured completely. All that remains are the 'ghosts' of the original cells—a collection of erythrocyte membranes.

Recently many other membrane systems have been examined including other plasma membranes, mitochondria, chloroplast membranes and bacterial membranes.

The success of the isolation of the membrane material depends also on its identification. Electron microscopy is sometimes useful. Surface antigens which attach to specific target chemicals have been used as marker materials. A difficulty in this approach is to ensure that the marker molecule is specific to the membrane. Attempts have been made to overcome this by artificially labelling the membrane with a chemical known to be confined to the membrane and easily detectable during the separation of the membranes from the constituents of the disrupted cells.

To appreciate which materials are contaminants of this membrane fraction is also rather difficult. With erythrocyte ghosts the question has been raised as to whether haemoglobin is a true component of the membrane or whether it is a contaminant. This question is still unresolved and the significance of the part played by haemoglobin in the membrane structure is somewhat obscure.

Recently modern analytical techniques have been increasingly applied to membrane preparations. The membrane materials which have been analysed appear to contain both lipid, protein and carbohydrate material.

## B. Analysis

Recent advances in separation and analytical techniques have now begun to make it possible to determine the detailed lipid composition of animal and plant membranes (Rouser and coworkers, 1968). As a particular membrane can contain various lipid classes the analysis is by no means straightforward. This is further complicated by the fact that, within each lipid class, there is a variety of chain lengths and unsaturation. Added to this is the complication of other groupings which can be present in the fatty acid chains, such as cyclopentene, hydroxyl or methyl groups. The separation and analysis of the protein material of membranes is a more difficult problem but is now receiving a great deal of attention.

The sampling procedures are a vital part of the analysis of the tissues. A lipid composition may not be representative because, with cell membranes, the lipid composition of the different membranes of one cell type can vary, and there can be variations in lipid composition dependent upon the species and also upon environmental factors affecting the species.

The organelles must be isolated and freed from all contaminating matter. The organelle may be characterized by means of electron microscopy to provide information on morphological integrity and this may be supplemented by enzyme assay. Postmortem degradation has to be guarded against and can be prevented by rapidly cooling the organelles to low temperatures ($\sim 4\,°$C). This has been found to preserve lipid composition for periods of weeks and up to months prior to analysis.

The extraction of lipids from biological tissue is generally complicated by a number of factors. This is because

i) some lipids are linked with protein or carbohydrate and these complexes are usually insoluble in the normal lipid solvents such as ether, alcohol and chloroform;

ii) some lipids are only soluble in a limited range of these solvents; and

iii) some non-lipid constituents of tissues are also dissolved by these solvents.

Generally there is no difficulty in extracting triglycerides from tissues such as adipose tissue or oil seeds. The triglycerides are not bound into lipoproteins or lipocarbohydrate complexes and they are easily soluble in the lipid solvents. Triglycerides can usually be extracted with successive uses of a solvent such as acetone which first extracts the water from the tissue and then the triglycerides themselves. Alternatively, the tissue can be dried and extracted with any desired solvent.

Phospholipids, sphingolipids and polar lipids are usually present in a bound form and, in this case, solvents such as methanol, ethanol or acetone are used to break the linkages. Ethanol–ether and chloroform–methanol solvent systems are frequently used. Subsequently the solvent is removed from this extract of the wet tissue by a vacuum evaporation.

After washing this solution the total lipid is obtained by removing the solvent. It may then contain phospholipids, sphingolipids and simple lipids. Some contamination of the lipids can occur. The phospholipids in particular solubilize many non-lipid components such as sugars, free amino acids, sterols, urea and many inorganic substances. These may be removed by special washing techniques or by chromatographic techniques using column or paper chromatography.

A number of techniques are used for the detailed fractionation of the lipids. These include countercurrent distribution methods, paper chromatography methods and column chromatography.

The composition of some common membranes in terms of lipid and protein content are given in table 1. It can be seen that the lipid to protein ratio varies considerably from one type of membrane to another.

**Table 1.** Protein and lipid composition of animal and bacterial membranes

| Origin of membrane | Amino acid | Molar ratio* Phospho-lipid | Cholesterol | Amino ratio (protein/lipid†) |
|---|---|---|---|---|
| Myelin[1] | 264 | 111 | 75 | 0·43 |
| Erythrocyte[2] | 500 | 31 | 31 | 2·5 |
| *Bacillus licheniformis*[3] | 610 | 31 | 0 | 4·8 |
| *Micrococcus lysodeikticus*[3] | 524 | 29 | 0 | 4·3 |
| *Bacillus megaterium*[4] | 520 | 23 | 0 | 5·4 |
| *Streptococcus faecalis*[5] | 441 | 31 | 0 | 3·4 |
| *Mycoplasma laidlawii*[6] | 442 | 25·2 | 2·3 | 4·1 |

[1]O'Brien and Sampson (1965)
[2]Maddy and Malcolm (1965)
[3]Salton and Freer (1965)
[4]Weibull and Bergström (1958)
[5]Shockman and coworkers (1963)
[6]Razin, Morowitz and Terry (1965)

*Data are calculated from the percentage compositions given in the references indicated, using the appropriate molecular weights.

†The approximate area occupied by a monomolecular film assuming an average amino acid occupies 17 Å² (Weibull and Bergström, 1958), a phospholipid molecule 70 Å² (Bear and coworkers, 1941; Van den Heuvel, 1963), and a cholesterol molecule 38 Å² (Van den Heuvel, 1963).

(From Korn, 1966).

## 1. *Lipids*

Analysis of cell membranes has shown that a particular membrane may contain one or more different types of lipid. A recent summary is shown in table 2 of some of the lipid classes which have been observed.

**Table 2.** Composition of animal and bacterial membranes

|  | Myelin | Erythro-cyte | Mito-chondria | Micro-some | *Azoto-bacter agilis* | *Escheri-chia coli* * |
|---|---|---|---|---|---|---|
| Cholesterol | 25 | 25 | 5 | 6 | 0 | 0 |
| Phosphatidylethanolamine | 14 | 20 | 28 | 17 | 100 | 100 |
| Phosphatidylserine | 7 | 11 | 0 | 0 | 0 | 0 |
| Phosphatidylcholine | 11 | 23 | 48 | 64 | 0 | 0 |
| Phosphatidylinositol | 0 | 2 | 8 | 11 | 0 | 0 |
| Phosphatidylglycerol | 0 | 0 | 1 | 2 | 0 | 0 |
| Cardiolipin | 0 | 0 | 11 | 0 | 0 | 0 |
| Sphingomyelin | 6 | 18 | 0 | 0 | 0 | 0 |
| Cerebroside | 21 | 0 | 0 | 0 | 0 | 0 |
| Cerebroside sulphate | 4 | 0 | 0 | 0 | 0 | 0 |
| Ceramide | 1 | 0 | 0 | 0 | 0 | 0 |
| Lysylphosphatidylglycerol | 0 | 0 | 0 | 0 | 0 | 0 |
| Unknown or other | 12 | 2 | 0 | 0 | 0 | 0 |

(From Korn, 1966).
*These values are only approximate, and ignore minor components.

The structures of some of the phosphoglycerides are shown in figure 1.

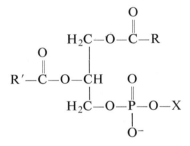

where X = —H                              Phosphatidic acid                    (i)

= $-CH_2-CH_2-\overset{+}{N}(CH_3)_3$        Phosphatidylcholine            (ii)
                                          or lecithin

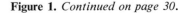

= $-CH_2-CH_2-\overset{+}{N}$         Phosphatidyl(*N*-dimethyl)-       (iii)
                                          ethanolamine

**Figure 1.** *Continued on page 30.*

**Figure 1.** *Continued from previous page.*

$$= -CH_2-CH_2-\overset{+}{N}\overset{CH_3}{\underset{H_2}{\Big\langle}}$$ Phosphatidyl($N$-methyl)-ethanolamine   (iv)

$$= -CH_2-CH_2-\overset{+}{N}H_3$$ Phosphatidylethanolamine   (v)

$$= -CH_2-\overset{\overset{+}{N}H_3}{\underset{}{C}}H-CO_2H$$ Phosphatidylserine   (vi)

$$= -CH-\overset{\overset{+}{N}H_3}{\underset{CH_3}{C}}H-CO_2H$$ Phosphatidylthreonine   (vii)

$$= -CH_2-\underset{OH}{C}H-CH_2OH$$ Phosphatidyl glycerol   (viii)

$$= -CH_2-\underset{OH}{C}H-\underset{O}{C}H_2$$ *o*-Amino acid ester of phosphatidyl glycerol   (ix)

$$\underset{}{C}=O$$
$$H_2N-\underset{R}{C}H$$

$$= -CH_2-\underset{OH}{C}H-CH_2O-PO_3H_2$$ Phosphatidyl glycero-phosphate   (x)

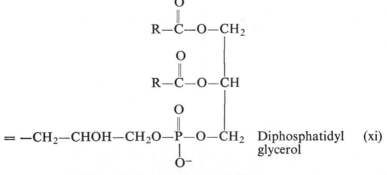

$$= -CH_2-CHOH-CH_2O-\overset{O}{\underset{O^-}{\overset{\|}{P}}}-O-CH_2$$ Diphosphatidyl glycerol   (xi)

**Figure 1.** *Continued on page 31.*

**Figure 1.** *Continued from previous page.*

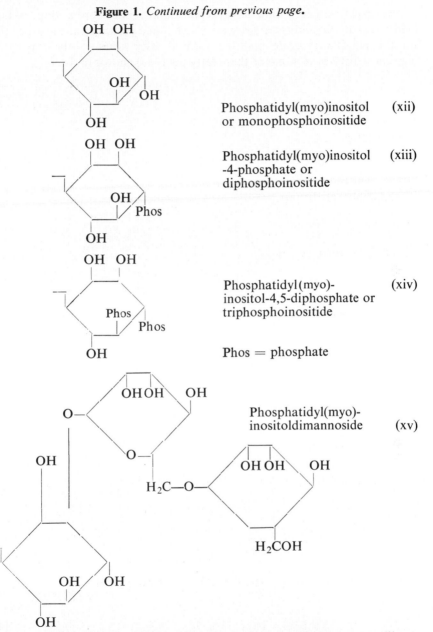

Phosphatidyl(myo)inositol    (xii)
or monophosphoinositide

Phosphatidyl(myo)inositol    (xiii)
-4-phosphate or
diphosphoinositide

Phosphatidyl(myo)-    (xiv)
inositol-4,5-diphosphate or
triphosphoinositide

Phos = phosphate

Phosphatidyl(myo)-
inositoldimannoside    (xv)

If the possibility of internal salt formation does not exist (structures viii to xv), the phosphate anion may be bound to a metal cation.

**Figure 1.** The structures of natural phosphoglycerides.

   Associated with each lipid class present in membranes are a range of fatty acids. Usually the saturated fatty acids are found esterified at the 1-position on the phosphoglyceride and unsaturated fatty acids at the 2-position. Typical structures of some of these fatty acids are shown in figure 2.

$$CH_3(CH_2)_{16}C\diagup^{O}\diagdown_{OH}$$

Stearic acid 18:0
(Octadecanoic acid)

$$CH_3(CH_2)_7CH{=}CH(CH_2)_7C\diagup^{O}\diagdown_{OH}$$

Oleic acid 18:1
(Octadec-9-enoic acid)

$$CH_3(CH_2)_4CH{=}CHCH_2CH{=}CH(CH_2)_7C\diagup^{O}\diagdown_{OH}$$

Linoleic acid 18:2
(Octadeca-9:12-dienoic acid)

$$CH_3CH_2CH{=}CHCH_2CH{=}CHCH_2CH{=}CH(CH_2)_7C\diagup^{O}\diagdown_{OH}$$

Linolenic acid 18:3
(Octadeca-9:12:15-trienoic acid)

$$CH_3(CH_2)_4CH{=}CHCH_2CH{=}CHCH_2CH{=}CHCH_2CH{=}\atop CH(CH_2)_3C\diagup^{O}\diagdown_{OH}$$

Arachidonic acid 20:4
(Eicosa-5:8:11:14-tetraenoic acid)

**Figure 2.** The structures of a range of naturally occurring fatty acids.

Other variations on the fatty acid structure include the presence of a cyclopropane ring, e.g. lactobacillic acid, present in certain bacteria, contains this grouping, and a cyclopentene ring which occurs in hydnocarpic, chaulmoogric and gorlic acids.

The structure of sphingomyelin is

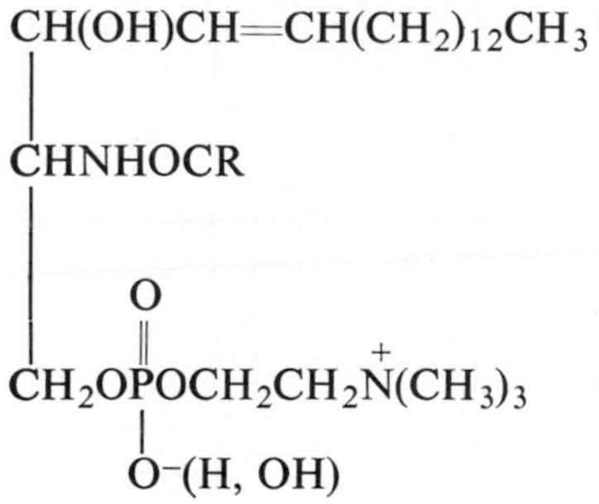

$CH(OH)CH=CH(CH_2)_{12}CH_3$ where R is a fatty alkyl group

|
$CHNHOCR$

|      $O$
|      ‖
$CH_2OPOCH_2CH_2\overset{+}{N}(CH_3)_3$
|
$O^-(H, OH)$

Sphingomyelin

The similarity of the structure of sphingomyelin and the phosphatidylcholines is apparent.

The fatty acid amides of the sphingosines are known. These are called the ceramides. The structure of a ceramide is shown:

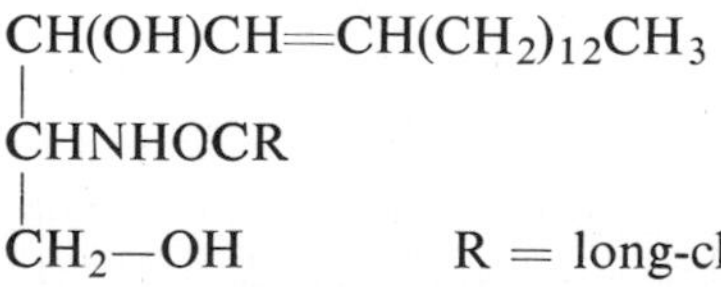

$CH(OH)CH=CH(CH_2)_{12}CH_3$
|
$CHNHOCR$
|
$CH_2-OH$          R = long-chain alkyl group

Ceramide

Cerebrosides contain either sphingosine or dihydrosphingosine, a long-chain fatty acid, and a sugar. They are similar to the sphingomyelins except that, instead of the phosphorylcholine group, they contain a sugar. The sugar is either glucose or galactose in a β-glycosidic linkage on the terminal carbon of the sphingosine. The term is used generally for ceramide-monohexosides. In the past cerebrosides have been variously designated as galactolipids, glycolipids and glycosphingosides.

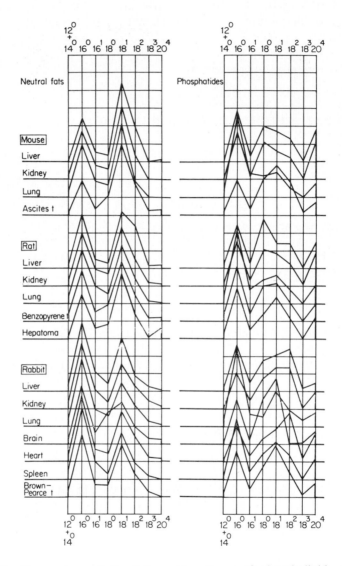

**Figure 3.** Comparison of the fatty acid patterns of phospholipids and non-phospholipid fractions from several tissues of three animal species (Veerkamp and coworkers, 1962). On the abscissa, the fatty acids are indicated by means of the number of carbon atoms and the number of double bonds. On the ordinate, weight percentages of fatty acid methyl esters are plotted. The distance between two horizontal lines corresponds to 10% fatty acid.

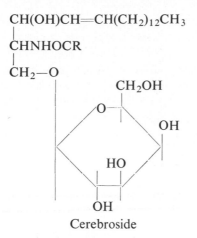

$$CH(OH)CH\!=\!CH(CH_2)_{12}CH_3$$

CHNHOCR

$$CH_2\!-\!O$$

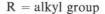

R = alkyl group

Cerebroside

Comparisons of the fatty acid analyses of the non-phospholipid and phospholipid fractions of several tissues from a number of mammals (Veerkamp and coworkers, 1962) show that the neutral lipid fractions possess some degree of animal specificity, i.e. the fatty acid patterns may differ from animal to animal, but resemble one another for different tissues of one animal (figure 3). In the rat the fatty acid composition of neutral lipid fractions from several tissues appeared to be rather similar to that of the

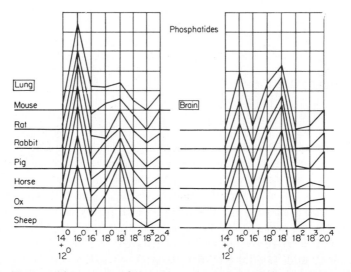

**Figure 4.** Fatty acid patterns of the phospholipid fractions from lung and brain tissues from several mammalian species (Veerkamp and coworkers, 1962).

abdominal depot fat. The phospholipid fractions of the tissues studied were again found to be usually richer in stearic and poorer in palmitic acid when compared with the non-phospholipids. Furthermore, the phospholipids from several tissues of *one* animal species revealed a less characteristic fatty acid pattern than that given by the neutral lipids. By contrast, the fatty acids from phospholipids may exhibit a certain degree of similarity in homologue tissues of different animals. This type of specificity is demonstrated by comparing the fatty acid pattern of phospholipids from lung and brain tissue of a number of species (figure 4).

Sialic acid molecules are also found to be associated with certain membranes, e.g. sialic acids have been found to be associated with erythrocyte and *Ehrlich* ascites carcinoma cells. The general structure of the sialic acids is as shown in figure 5.

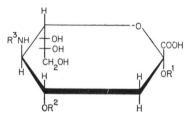

**Figure 5.** Structure of sialic acids.*

*The term neuraminic acid ($C_9H_{17}O_8N$) is reserved for the unsubstituted structure $R^1$, $R^2$ and $R^3$ = H. Naturally occurring neuraminic acids are *N*-acylated (e.g. *N*-acetyl-neuraminic acid, $R^1$ = H, $R^2$ = H, $R^3$ = $CH_2CO$) and some in addition are *O*-substituted ($R^2$ = $CH_3 \cdot CO$). Sialic acid is the group name for all acylated neuraminic acids. Neuraminidase cleaves the glycosidic bond joining the keto group of neuraminic acid to D-galactose or D-galactosamine (usually in *N*-acetyl form) ($R^1$ = sugar).

## 2. *Protein*

During the past few years a number of studies have been made on the isolation and characterization of proteins from erythrocyte membranes. Maddy (1966) has described a method for the solubilisation of the protein of ox erythrocyte by *n*-butanol fractionation. Zwall and van Deenen (1968) have described a method for the solubilization of erythrocyte ghosts using *n*-pentanol. In contrast to the butanol procedure, not only the proteins but also the lipids were recovered in the water layer.

Rosenberg and Guidotti (1968) have also carried out studies on the characterization of the protein component of mammalian erythrocyte membranes. Exhaustive lipid extraction of the membrane produced a glycoprotein residue containing greater than 90% of the total membrane protein. The protein was found to be soluble in formic acid and sodium dodecyl sulphate solutions. The protein of this membrane appears to be a heterogeneous collection of proteins, many in the molecular weight range near 50,000. Amino acid composition of the membrane protein shows a high percentage of amino acids with long non-polar side chains. There is a small excess of amino acids (13·76% as compared with 12·18% basic amino acids). Molecular weights of the major protein components fall in the range normal for most globular proteins. The amino acid composition of lipid-extracted membrane protein is shown in table 3.

**Table 3.** Amino acid composition of lipid-extracted membrane proteins

| Amino acid | Residues per 100 residues |
|---|---|
| Lysine | 5·21 |
| Histidine | 2·44 |
| Arginine | 4·53 |
| Aspartic acid* | 8·49 |
| Threonine | 5·86 |
| Serine | 6·26 |
| Glutamic acid* | 12·15 |
| Proline | 4·26 |
| Glycine | 6·73 |
| Alanine | 8·15 |
| Half-cystine | 1·08 |
| Valine | 7·10 |
| Methionine | 2·02 |
| Isoleucine | 5·29 |
| Leucine | 11·34 |
| Tyrosine | 2·41 |
| Phenylalanine | 4·20 |
| Tryptophan | 2·49 |

*One mole of $NH_3$ was formed for every three aspartic and glutamic acids.
(From Rosenberg and Guidotti, 1968).

## III. SOME PHYSICAL PROPERTIES OF COMPONENTS

### A. Thermotropic Mesomorphism

In addition to the capillary melting point, other phase changes have been shown to occur with phospholipids at lower temperatures. For example, when a pure phospholipid, dimyristoylphosphatidylethanolamine, containing two fully saturated chains is heated from room temperature up to the capillary melting point, a number of thermotropic phase changes occur (i.e. phase changes caused by the effect of heat). This was shown by infra-red spectroscopic techniques (Byrne and Chapman, 1964), by thermal analysis techniques (Chapman and Collin, 1965) and has now been studied by a variety of physical techniques (Chapman and coworkers, 1966; Chapman and Salsbury, 1966).

The main conclusions from these various physical studies are that:

i) Even with the fully saturated phospholipid at room temperature, some molecular motion occurs in the solid. This is evident from the p.m.r. spectra and from the i.r. spectra taken at liquid nitrogen and at room temperature.

ii) When the phospholipid is heated to a higher temperature it reaches a transition point, a marked endothermic change occurs, and the hydrocarbon chains in the lipid 'melt' and exhibit a very high degree of molecular motion. This is evident in the appearance of the i.r. spectrum and also in the narrow n.m.r. line width. The broad diffuse appearance of the i.r. spectrum is consistent with the chains flexing and twisting and with a 'break-up' of the all-planar *trans* configuration of the chains.

The fact that the phase transition is concerned primarily with the hydrocarbon chains of the phospholipid is confirmed by the x-ray data. This shows that the space taken up by the glycerol and polar group remains essentially unchanged when this phase transition occurs.

iii) When phospholipids contain shorter chain lengths, or unsaturated bonds, those marked endothermic phase transitions occur at lower temperatures (Chapman and coworkers, 1967). The temperature at which these transitions occur parallels the behaviour of the melting point of the related fatty acids. The transition temperatures are high for the fully saturated long chain phospholipids, lower when there is a *trans* double bond present in one of the chains, and lower still when there is a *cis* double bond present.

iv) An important conclusion which we obtain from these phase transition studies is that, above the endothermic transition temperature, a given phospholipid will be in a highly mobile condition with its hydrocarbon chains flexing and twisting. This is a *fundamental property* of the phospholipid and we can expect this chain mobility to occur in whatever situation the phospholipid occurs unless, for special reasons, this motion is somehow inhibited. We can envisage that inhibition of chain motion by interaction with other molecules (e.g. water or protein) would provide one of these special reasons. In other circumstances, due to less perfect packing arrangements, we might expect, at a particular temperature, even greater mobility of the chains of the lipid and, indeed, diffusion of the whole lipid molecules.

## B. The Effects of Water

### 1. *Transition Temperatures*

Small amounts of water can have unusual effects upon the phospholipid mesomorphic behaviour. Thus the diacylphosphatidylcholines (lecithins) exhibit additional liquid crystalline forms between the first transition temperature and the capillary point (Chapman and coworkers, 1967). The intermediate liquid crystalline form is found to exhibit x-ray spacings consistent with a cubic phase organization. On the other hand, if all the water is removed from the phospholipid, the lipid no longer exhibits this phase.

When these phospholipids are examined in increasing amounts of water the various physical techniques, such as microscopy, n.m.r. spectroscopy or differential thermal analysis, show that as the amount of water increases, the marked endothermic transition temperature for a given phospholipid falls (Chapman and coworkers, 1967). The transition temperature does not fall indefinitely; it reaches a limiting value independent of the water concentration. We can understand this if we regard the effect of water as leading first to a 'loosening' of the ionic structure of the phospholipid crystals. This, in turn, affects the whole crystal structure and a reduction, up to a certain limit, of the dispersion forces between the hydrocarbon chains. Large amounts of energy are still required to counteract the dispersion forces between the chains and quite high temperatures are still required to cause the chains to melt. The limiting transition temperatures parallel the melting point behaviour of the analogous fatty acids becoming lower with increasing unsaturation. This further reduction of the endothermic transition temperatures by water means that the natural phospholipids extracted from biological membranes usually exhibit this crystalline to liquid crystalline transition many degrees below the biological environmental temperature. At the biological environmental temperature we should expect the phospholipids which contain highly unsaturated chains to be in a highly mobile and fluid condition.

This implies that it is necessary to provide a reason if these unsaturated phospholipids do not occur in a highly mobile condition at this environmental temperature. If they are not in a highly mobile condition in a cell membrane, then we need to have some inhibitory interaction, such as interaction with cholesterol or protein, etc. to explain this. Recent studies have shown that cholesterol can inhibit the chain mobility of some phospholipids (Chapman and Penkett, 1966; Hubbell and McConnell, 1968).

## 2. *Bound Water*

Some of the water added to the phospholipid appears to be *bound* to the lipid, e.g. 1,2-dipalmitoylphosphatidylcholine binds about 20% water (Chapman and coworkers, 1967). This water does not freeze at 0 °C and calorimetric studies made with lipid–water mixtures show that it is only after more than 20% water has been added to the lipid that a peak at 0 °C is

observed. Some d.s.c. traces for lipid–water mixtures of different concentrations are shown in figure 6. This 'bound' water may have considerable
relevance to interactions of anaesthetics, drugs and ions with biological
membranes. If this 'bound' water varies, either in its properties or in its total
amount, dependent upon the type of ion or interacting molecule, this in turn
may alter transport and diffusion properties across the membrane. The
amount of 'bound' water associated with the constituent lipids and proteins
will perhaps provide a limit for the amount of water which can be removed
from biological membranes before they lose their organization.

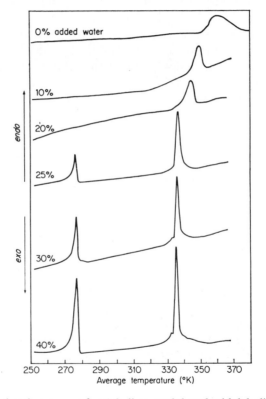

**Figure 6.** D.s.c. heating curves for 1,2-distearoylphosphatidylcholine in increasing
amounts of water.

### 3. *Monolayer Properties*

Monolayer studies of phospholipids have been carried out for a considerable number of years. Usually this work has been performed with natural

phospholipid mixtures and, in the vast majority of cases, with egg yolk phosphatidylcholine. In recent years a few studies have been made with pure synthetic phospholipids (e.g. van Deenen and coworkers, 1962). These show that the fully saturated phospholipids exhibit, at room temperature, mono-layers which are more condensed than are the unsaturated phospholipids containing *cis* hydrocarbon chains, i.e. the saturated lipids occupy less area at low surface pressures than do the unsaturated derivatives. These results can be compared with the d.t.a. results. The type of monolayer observed is related to the degree of flexing and twisting of the hydrocarbon chains of the lipid (Chapman, 1966; Phillips and Chapman, 1968).

### 4. *Phase Polymorphism*

Phospholipids in the presence of water can also form different liquid crystalline phases. In some cases, as the concentration of water varies, transitions from lamellar to hexagonal phases occur. These phases have been fully discussed by Luzzati and Husson (1962).

In our laboratory recent studies on pure 1,2-diacylphosphatidylcholines in water show that there is no lamellar/hexagonal transition over a wide range of concentration. The presence of impurities, such as ions, can, however, have an appreciable effect upon the amount of water taken up by the lipid (Chapman and coworkers, 1967).

The fact that lecithins of different chain length in water adopt a lamellar or bilayer type structure over a wide variation of concentration is interesting. At first sight this may appear to lend some support to the idea that a natural membrane may be built up on this bilayer-type organization. This, however, need not necessarily be the case. The influence of the membrane protein may be quite considerable on the resultant structure, and the organization of the membrane will depend upon the mode of interaction of the lipid and protein. Both the lipid and protein separately take up stable configurations to mini-mize hydrophobic and maximize hydrophilic interactions with the surround-ing water. The important question for membrane organization is whether the combined lipid–protein complex attempting to balance these interactions will adopt an entirely different arrangement. The way in which membranes are synthesized naturally may also be relevant to this point.

## IV. PHYSICAL PROPERTIES OF MEMBRANES

### A. Electron Microscope Studies

The introduction of the electron microscope in the 1930's provided a tool which can form images of structures previously considered to be submicro-scopic. The observation of clearly defined layer structures at the border of cell boundaries provided strong and independent confirmation of the existence of cell membranes.

The earlier electron microscope studies have been extensively reviewed, e.g. Robertson, 1959, 1964 and 1966. We begin with a discussion of the unit membrane concept.

## 1. *The Unit Membrane Concept*

The similarity in appearance of the electron micrographs of many different types of cell membranes were considered by Robertson (1957, 1958, 1959) to be evidence for a common structure at the surface of a wide variety of cells. Using a new fixative procedure involving $KMnO_4$, he was able to observe a similar three-layered unit, approximately 75 Å thick, at the surface of a number of different cell types as well as in many different cellular organelles. This unit appeared as two dense lines about 20 Å wide separated by a lighter space of 35 Å.

Robertson (1957) proposed a model of the membrane which he termed a 'unit membrane'. This concept appeared to clarify and unify a wide body of information including other physical data as well as other electron microscope data. By referring to the biological membrane as a unit, he emphasized not only that all three parts of the triple-layered 75 Å structure seen in the electron microscope were part of one membrane, but also that all membranes had a similarity of molecular arrangement and origin.

The electron microscopic data for other plasma membranes have recently been summarized by Elbers (1964) and this data is consistent with the suggestion of Robertson (1957).

In general with plasma membranes, triple-layered structures are also observed after fixation in $KMnO_4$, and often, but not always, after fixation in $OsO_4$. The endoplasmic reticulum (cytomembranes), the outer, inner and cristae membranes of mitochondria, chloroplast membranes, the two membranes of the nuclear envelope, and the membranes of bacterial protoplasts and spheroplasts are also revealed as triple-layered structures in electron micrographs.

Despite these general similarities there has been some reluctance (Stoeckenius, 1966) to accept this proposal of an identical unit membrane structure for all the membranes in cells and particularly for the mitochondrial membranes.

In general, membranes are thought to be about 75 Å wide. There are considerable variations in dimensions of membranes. The overall widths of triple-layered plasma membranes appear to vary from about 50 Å to perhaps 130 Å (Elbers, 1964). How much of this variation is due to differences in the methods of preparation and how much to fundamental differences in structure is not clear. Perhaps the clearest electron micrographic indication of differences among membranes was obtained by Sjöstrand (1963) who compared adjacent membranes in single-cell sections of mouse kidney and pancreas fixed with $OsO_4$ and $KMnO_4$. The thinnest membranes (mitochondrial and $\alpha$-cytomembranes) were 50 Å to 60 Å, and the thickest membranes (plasma and zymogen granules) were 90 Å to 100 Å. These variations indicate the difficulty of interpreting micrographs in terms of molecular structure.

Several other observations have appeared which seem to disturb the 'unit membrane' picture. Hillier and Hoffman (1953) studied the structure of the membrane of erythrocytes by using shadowing techniques which revealed a mosaic structure. They suggested that the erythrocyte envelope was composed of plaques situated on the outside of a fibrous network joined together by lipids. Recently, Glaeser and coworkers (1966) have studied the membrane structure of $OsO_4$ fixed erythrocytes viewed 'face on' by electron microscopy. This supports the conclusions of Hillier and Hoffman (1953) and shows that the surface of the rat red cell membrane has a 'pebbly' appearance at the level of 400 Å to 500 Å. These authors suggest that these bumps on the surface may be associated with a filamentous structure, the bumps representing the tops of loops of the filaments*.

Also using a shadowing technique, Frey-Wyssling and Steinmann (1948) examined the structural features of the closed flattened sacs which constitute the internal membrane system of the plant chloroplast. These workers had previously noted that the chloroplast membranes failed to show any substantial intrinsic birefringence which would be expected if they contained highly oriented lipid bilayers as in myelin. These membranes show a repeating granular structure which suggests that the membranes may be composed of an array of micellar or globular subunits.

Sjöstrand (1963) observed globular sub-units in one of the opaque layers of mitochondrial membranes and smooth endoplasmic reticulum, and an asymmetry in the electron opacity of the dense lines in the plasma membrane. Subunits, or cross-linkages bridging the gap between the two opaque bands of the triple-layered structure, have been observed by Robertson (1963) who later reinterpreted them as an electron optical artifact derived from a mosaic pattern in the plane of one or both surfaces of the triple layer. In several instances, hexagonal mosaic patterns have been seen on the surfaces of plasma membranes (Benedetti and Emmelot, 1965). Strong evidence for membrane subunits is suggested in a recent paper by Blasie and coworkers (1965). Outer segment membranes of frog retina were isolated and were oriented in ultracentrifugal pellets. Electron microscopic surface views of negatively stained membranes and low-angle x-ray diffraction patterns (see p. 51) from unfixed, unstained pellets showed square array of spherical

---

*See also the studies using the scanning electron microscope, p. 47.

particles. The unit cell size was about 70 Å and the particles had a non-polar core about 40 Å in diameter.

When attempts are made to obtain further information about the interpretation of the electron microscope data of fixed sectioned tissues in molecular terms, we find that there are a number of difficulties.

Korn (1966) has critically discussed this question in some detail. He points out that Robertson (1959) and Fernández-Morán and Finean (1957) have questioned whether the manganese atom is responsible for any of the electron opacity in micrographs of $KMnO_4$-fixed cells. He suggests that, if this is not the case, there is then no way to interpret the dense lines in micrographs of membranes fixed with $KMnO_4$ in molecular terms. This is a serious deficiency because it is in such preparations, as we have already seen, that the triple-layered structure is most reproducibly and distinctly seen. There appears to be only one study in which cells were chemically studied during fixation with $KMnO_4$. Korn and Weisman (1966) found that the lipids of amoebae were essentially unaffected by fixation with $1\%$ $KMnO_4$ for 1 hour at $0\,^{\circ}C$. All the neutral lipids and about half of the phospholipids were extracted from the amoebae during dehydration in ethanol.

Similar triple-layered structures are often observed when $OsO_4$ is used as a fixative and various experiments have been devised so as to ascertain the site of the fixation process. This fixation was originally thought to take place at the sites of the double bonds in the hydrocarbon chain. Wigglesworth (1957) suggested that osmium can cross-link through the ethylenic double bonds and has given evidence to show the occurrence of an insoluble polymeric complex of lipid and osmium which he considers is the basis of cytological fixation. This observation was considered to be consistent with the fact, noted by many authors, that fully saturated phospholipids, such as the phosphatidylethanolamines, do not react easily at room temperature with osmium tetroxide, although suggestions had been made that brominated or hydrogenated phospholipids can take up osmium without fixation occurring. Bahr (1954) showed that many amino acids would also react with osmium tetroxide, and so some authors have considered that the dense lines observed in electron micrographs of cell membranes represent protein, whereas the light central line corresponds either to the hydrocarbon chain region of the lipid, or to a gap produced by removal of lipid in the preparative stages for electron microscopy (Robertson, 1960).

Stoeckenius (1962a,b) studied natural lipids after reaction with osmium tetroxide, and concluded that the dark area seen in electron micrographs corresponds to osmium located at the polar groups of the phospholipid molecules. Finean (1962), after an analysis of x-ray data of fixed and unfixed tissue, also considered that the osmium is located between the polar groups of the bimolecular sheets of lipid. Riemersma (1963) considers, after chromatographic analysis of the intermediates formed during osmium fixation of unsaturated lecithins, that the initial reaction is with the double bonds, but that there is a subsequent migration of osmium derivatives to the polar groups. Despite this work, some authors are still of the opinion that the osmium is located at the double bond of the lipids (Hayes and coworkers, 1963).

Chapman and Fluck (1966), in an attempt to clarify the problem, attempted to fix saturated phospholipids with $OsO_4$, arguing in this case that, as there were no double bonds present, any successful fixation must be at the polar group. This study showed that, whilst fully saturated phosphatidylethanolamine reacts with $OsO_4$, the fully saturated lecithins do not.

Korn (1966) concludes that the dense lines in membranes fixed with osmium tetroxide *reveal nothing* about the molecular orientation of the phospholipid, in the original membrane. He also points to the fact that triple-layered membranes are seen in osmium-fixed mitochondria from which all the lipid has previously been removed by extraction with acetone (Fleischer and colleagues, 1965). [*Escherichia coli* B (van Iterson, 1965) also gives a triple-layered membrane despite the fact that this organism contains essentially no unsaturated fatty acids.] If lipids are not necessary to reveal the triple-layered structure, then its explanation in terms of particular molecular configurations cannot be correct.

We see that there are still many doubts concerning a full acceptance of the 'unit membrane'. Many of the supporting arguments for its acceptance depend upon studies of myelin which may be chemically, metabolically and functionally different from all other membranes.

## 2. *The Negative Staining Technique*

The negative staining technique using phosphotungstate has increased in popularity in recent years. In this method the biological membranes are immersed in a pool of electron-dense material (e.g. sodium phosphotungstate) which dries to form an electron-dense glass (Brenner and Horne, 1959). Membranes appear as regions of electron transparency against a dark background. The specimen is not sectioned.

The possibility of artifacts being introduced with the negative staining technique has been considered. Bangham and Horne (1964) show a number of different phase structures with artificial lipid mixtures, some of which resemble structures seen in negatively stained natural membranes. Whittaker (1966) has commented that the negative staining technique only reveals those structures sufficiently tightly packed to exclude the negative stains. Proteins possessing an α-helix are clearly seen, whilst those with an open structure are not. In some instances the fine structures observed can be generated on the electron microscope grid by interaction with negative stains during drying. There is the further possibility of spontaneous or enzyme-catalysed rearrangements occurring when biological organelles are ruptured.

### 3. *The Freeze-etching Technique*

Recently a new technique has been introduced to electron microscopy (Moor and coworkers, 1961) which attempts to overcome the many problems associated with chemical-fixation methods and the production of artifacts associated with these methods. This is the technique of freeze-etching.

Freeze-etching involves six preparational steps: pretreatment of the object, freezing, chipping of the frozen specimen followed by etching and coating and, finally, the cleaning of the replica. One of the most important processes is to get a clean fracture plane through the frozen specimen. This

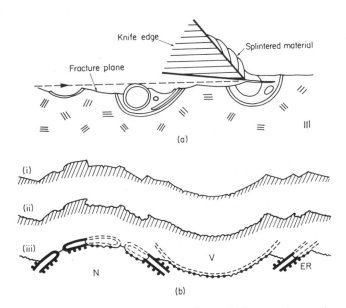

**Figure 7.** (a) Diagram of the cutting procedure which actually consists of a fine splintering of the deep-frozen object (yeast cells). (b) Diagram of how the splintering and the etching reveal the fine structure of a frozen object: (i) cross-section through the fracture plane; (ii) etched fracture plane, showing the fine structure; (iii) the reconstructed structural details of the recorded object. N, nucleus, showing a partially removed envelope. V, totally removed vacuole, rendering possible a surface view of the adjacent cytoplasmic ground substance. ER, endoplasmic reticulum, fractured at a low angle. (From Moor and Mühlethaler, 1963).

can be done by chipping under high vacuum at a controlled object temperature. In this method there is no chemical treatment of the object during the whole procedure until the replica is formed on the fracture plane from the frozen specimen. By means of snap freezing and/or glycerol impregnation, cells have been frozen so as to preserve the life of the organisms.

Freeze-etching depends upon freezing a very small sample very rapidly (the sample is frozen in freon cooled with liquid nitrogen) usually in the presence of glycerol and then putting the rapidly frozen specimen on to the cold stage of a vacuum coating unit. After a very high vacuum has been obtained, say 2 or $3 \times 10^{-6}$ $\tau$, the surface of the specimen is cut with a cold knife. The temperature of the specimen is then raised to $-100\,°C$ and a thin layer of ice is sublimed off. The cut surface is therefore etched by vacuum sublimation (see figure 7). Immediately after this the surface of the specimen is shadowed with a heavy metal and a carbon replica is made. This replica is then examined in the electron microscope.

Using this technique Moor and Mühlethaler (1963) reported the results of a study, using the freeze-etching method, on yeast cells, *Saccharomyces cerevisiae*. They show that the cell wall and cytoplasmic membrane are clearly separated and suggest that the fracture plane either follows the surface of the cell wall or penetrates the whole wall perpendicularly. By combining surface views and cross-fracture views, they were able to obtain a three-dimensional image of the invaginations of the yeast cytoplasmic membrane. The invaginations have an average length of 3000 Å, a width of 200 Å to 300 Å and a depth of 500 Å. The cross-fractured membranes give images the same as those shown by membranes in sections of permanganate-fixed material. Thus the membrane consists of three subunits each about 25 Å thick. They suggest that this is in agreement with the structure of the unit membrane suggested by Robertson (1957, 1958).

## 4. *The Scanning Electron Microscope*

Recently the scanning electron microscope has been applied to the structure of red blood cells and a method of ion etching has been used in combination with this method (Lewis and coworkers, 1968). Dependent upon the intensity of the etching process, the structure can be seen at different levels within the membrane. These results suggest that the membrane is heterogeneous and that there are small etch-resistant circular areas dispersed over the membrane. These appear to be similar in size and feature to those observed by Glaeser and coworkers (1966).

Another feature of the membrane is the presence of indentation or pores particularly prominent in the concave area of the cell (see figure 8).

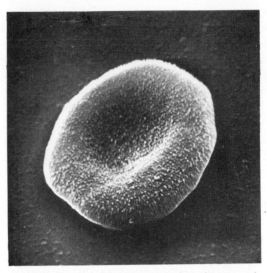

**Figure 8.** Normal red blood cell, as seen by scanning electron microscope after ion etching in argon for 10 minutes at 1 μtorr. The superficial layers of the membrane have been removed leaving the more resistant areas intact, ×6000. (From Lewis and coworkers, 1965)

## B. Electrical Properties

Electrical studies of biological materials have been carried out for many years. They have been carried out over quite a wide frequency ranging from about 5 c/sec up to 30,000 megacycles. This has shown in general three dispersions which appear to be characteristic of separate relaxation mechanisms so that, when the dielectric constant of the biological material is plotted as a function of frequency, three steps are observed which are referred to as α, β and γ dispersions.

Suspensions of erythrocytes in physiological saline solution have been studied in the radio-frequency range. They showed β-dispersion type behaviour and could be explained on the basis of models which assumed a shell surrounding the cell. Thus the existence of a membrane sufficiently low in conductance to form a membrane barrier was proven 30–40 years ago by impedance analysis.

Fricke (1925; 1933) determined the specific capacity and resistance of red cell membranes from the impedance of a suspension of cells. When the alternating current of low frequency is passed through a cell suspension, most of it is carried in the medium and the cells behave as insulators. As the frequency of the current is raised, the impedance of the suspension falls until a frequency is reached at which current passes through the cells. Provided the cells are spherical and the static capacity of the membrane is high compared with the internal resistance, the membrane capacity can be calculated. The specific capacity of the cell suspension, $C$, is given by

$$C = C_o A q \left(1 - \frac{r_1}{r_2}\right)$$

where $C_o$ is the specific capacity of 1 cm$^2$ of membrane, $r_1$ and $r_2$ are the specific resistances of the suspension and of the medium alone, $2q$ is the major axis of the spheroid, and A is a constant. The membrane capacity of spherical cells of several species is about 1 $\mu$F/cm$^2$ and its conductance too small (less than 0·1 mho/cm$^2$) to be noticeable. The effect of various materials on the cell membrane has been studied; thus saponin causes a decrease in the capacity of the membrane indicating a breakdown of the arrangement of the lipid molecules. On the other hand, haemolysis with distilled water does not destroy the membrane, and so the capacity of the membrane remains unchanged although the conductance through the medium and membrane becomes identical. Thus osmotic lysis allows an increase in permeability such that ions exchange readily, whilst the capacity of the membrane which is determined by the bulk of the membrane is unaltered. By assuming a dielectric constant of 3, the impedance measurements have been used to determine the thickness of the lipid part of the membrane. This gives a value of about 40 Å. The use of a value for the dielectric constant of 3 has been criticized on the basis that the membrane contains lipid, protein and water and the dielectric constant of protein is higher than 10. It has been argued that the impedance measurement detects only the lipid part of the membrane, but this only holds if the protein material is considered to be arranged in series with the lipid component. In a parallel arrangement the measurement would integrate over all dielectric constants.

## C. X-ray Diffraction Studies of Membranes

The myelin sheath is considered to be derived from a multiple-folded Schwann cell surface and may be a model system for the study of cell-membrane structure in general.

X-ray diffraction patterns have been obtained from nerve bundles maintained in a physiologically active state in irrigation cells mounted on the x-ray diffraction camera, and from nerve bundles sealed in thin-walled glass capillary tubes containing a physiological solution such as Ringer's solution. In both cases the diffraction patterns are observed to be the same.

When the nerve specimens are examined in a direction perpendicular to the fibre axis, using a symmetrically collimated x-ray beam, a meridionally accentuated ring is observed at 4.7 Å and a faint ring is observed at 9·4 Å.

c

These have been shown to be myelin reflections. Other reflections due to myelin also occur at low angles. These low-angle reflections have been accounted for as diffraction orders from a single fundamental repeating unit varying from 150 Å to 180 Å in the different types of nerve examined. Myelin from different sources gives different low-angle reflections. Peripheral nerve from mammals gives five low-angle reflections showing marked alterations in intensities through the orders and indicating a fundamental repeating unit of about 180 Å. The myelin of central origin, such as brain white matter, spinal cord, or optic nerve, gives only two-angle reflections corresponding to a repeat unit of about 80 Å. (The presence of the short spacing near 4·7 Å suggests that the hydrocarbon chains of the lipids present in the myelin sheath are in a liquid-like condition.) The variation in the intensities of the low-angle diffractions is interpreted to arise from variations in electron density along the axes of the myelin unit. It has been assumed that the fundamental repeating unit consists of two parts having very similar distributions of x-ray scattering power. From the intensities of the odd-order reflections the magnitude of the difference between the two parts has been estimated. The 'difference factor' is appreciable in peripheral nerve myelin but negligible in optic nerve. Diffraction patterns from the structure along the fibre axis appear as complete rings. This has been interpreted to mean that the long axis of the rod-shaped unit cell is oriented radially in the myelin sheath.

Diffraction data has been used to provide information about the molecular organization of myelin. The first ideas about this were based on a polarization–optical analysis of freshly isolated nerve fibres before and after treatment with absolute alcohol. This led to the conclusion that the lipid molecules are oriented radially and the non-lipid material arranged in concentric layers. The layers of oriented lipid molecules alternate with layers of non-lipid material. X-ray data led to the further conclusion that the peripheral nerve consisted of two lipoprotein layers, each of which consists of bimolecular leaflets of 67 Å, sandwiched between protein layers of 25 Å with interposed water layers of about another 25 Å thickness. Further studies by Finean (1960), taking consideration of the contraction of the lipid layers during drying, has led to a more detailed molecular arrangement of the myelin. The changes observed with frog sciatic nerve when dried are that the myelin unit shrinks to about 145–148 Å and, at the same time, three independent, diffraction lipid phases are produced.* The residual myelin unit gives x-ray reflections at about 146 Å and 73 Å. X-ray diffraction studies have also been carried out on various types of nerve at different temperatures.

Additional information about the structure of myelin has been obtained from calculations of the electron density distribution in the membrane.

*See a discussion of calorimetric studies of the effects of drying myelin, p. 60.

Finean (1962) and Finean and Burge (1963) found one-dimensional electron density distribution in a direction perpendicular to the plane of the membrane to show two distinct peaks. These were assumed to represent the phosphate groups of two phospholipid molecules placed tail-to-tail. The separation of the peaks was about 50 Å.

Finean and coworkers (1966) have also studied rat erythrocyte ghosts using low angle x-ray diffraction patterns during a controlled drying of an erythrocyte preparation. The results were compared with those obtained during a parallel study of myelin isolated from guinea-pig brain. The diffraction changes which accompany the dehydration of a sample of erythrocyte ghosts are essentially similar to those described for isolated myelin. A clearly defined point during dehydration at which changes in low-angle x-ray diffraction patterns occur, indicates a modification of membrane structure.

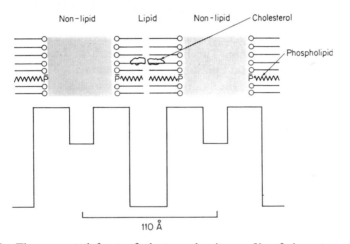

**Figure 9.** The suggested form of electron density profile of the rat erythrocyte membrane with schematic drawings of phospholipid and cholesterol molecules. (From Finean and coworkers, 1966)

The electron density profile of the repeating unit assuming a unit membrane structure is shown in figure 9. The shallow trough in the electron density curve is suggested to correspond to the narrow band of low intensity which separates unit membrane features in the layered system observed in electron micrographs.

These experiments demonstrated that the pattern of the natural hydrated membranes can also be obtained through the rehydration of fully dried samples. Husson and Luzzati (1963) rehydrated frozen-dried preparations of human erythrocyte ghosts and detected a lamellar system with a periodicity of about 170 Å.

With regard to the large angle reflections, the reflection at about 4·6–4·7 Å, observed with membranes and observed with myelin, has been interpreted as representing the repeat distance between parallel phospholipid hydrocarbon chains aligned perpendicular to the membrane. This reflection is also obtained when the phospholipid is in a liquid crystalline condition.

In general the x-ray diffraction studies do not support the idea that these 'membranes' consist of subunits but many other studies need to be carried out to be certain that membranes contain only a bimolecular layer structure.

### D. Spectroscopic Studies

#### 1. *Infra-red Spectroscopic Studies*

Infra-red spectroscopy has been applied to the study of certain membranes, including myelin (Chapman, 1965, Maddy and Malcolm, 1965, 1966) and to the erythrocyte and plasma membranes of *Ehrlich* ascites carcinoma (Wallach and Zahler, 1966). The infra red spectra of human erythrocyte ghost material and the total lipid are shown in figure 10. The broad diffuse bands in the spectrum show the fluidity of the chains in this structure. The main

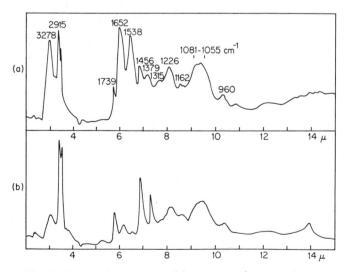

**Figure 10.** (a) The infrared spectrum of human erythrocyte ghost material. (b) The infrared spectrum of the total phospholipid from the ghost material. (From Chapman and coworkers, 1968)

application of the infra red spectroscopic technique has been to obtain information about protein conformation in the membrane. Some studies (Chapman and coworkers, 1968) have pointed out the importance of a band at 720 cm$^{-1}$ correlated with an all planar *trans* configuration of the lipid chains.

The position of certain bands, referred to as amide I and amide II bands in the infra-red spectra of peptide chains, differ, depending upon the conformation of these chains. Thus the amide I band is located at 1652 cm$^{-1}$ and is associated with an $\alpha$-helical and/or random coil conformation of peptide chains. The amide II band at about 1535 cm$^{-1}$ does not allow distinction between the $\alpha$- and $\beta$-conformation. A band at 1630 cm$^{-1}$ is correlated with a $\beta$-conformation. The infra-red spectra of erythrocyte and *Ehrlich* ascites carcinoma membranes do not show a strong band at 1630 cm$^{-1}$, suggesting that there is no extensive $\beta$-structure in these membranes.

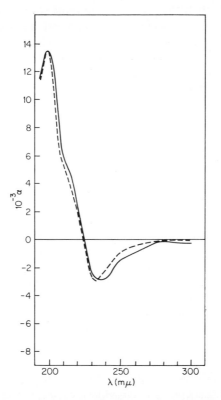

**Figure 11.** Optical rotatory dispersion of plasma membrane in aqueous suspension: ———, plasma membrane; - - - - -, poly-L-glutamic acid (pH 4·25). The values for poly-L-glutamic acid have been reduced by a factor of 4·5. (From Wallach and Zahler, 1966)

Lipid extraction from the membrane abolishes the band at 1740 cm$^{-1}$ due to carbonyl stretching in fatty acid esters. There is also some reduction of the two amide bands due in part to the extraction of sphingomyelin. However, lipid extraction does not produce a detectable transition to a β-conformation of the protein.

### 2. *Optical Rotatory Dispersion Studies*

Optical rotatory dispersion (o.r.d.) studies have also been applied to the study of protein conformation in biological membranes (Wallach and Zahler, 1966). The specific reduction [α] is a function of wavelength for plasma membrane vesicles in an aqueous suspension as shown in figure 11. The optical rotation of α-helical poly-L-glutamic acid, reduced by a factor of 4·5, is included for purposes of comparison.

The o.r.d. spectrum of the plasma membrane shows two small inflections between 300 and 250 mμ, one centred near 290 mμ and one at about 255 mμ. Maximal negative rotation is at 236–237 mμ, whilst it has a zero rotation at 226±1 mμ. A major feature of the o.r.d. spectrum of the membrane in aqueous suspension is the small rotational amplitude throughout the 195–320 mμ region. Similar results are obtained with erythrocyte ghosts and chloroplast lamellae, and have been interpreted to indicate very low helical contents. These are based on the assumption that the only conformations present are α-helix and random coil. The shape of the o.r.d. patterns has been attributed, however, to the presence of other conformational combinations, so that the evidence suggests that there is a mixture of α-helix, random coil and small amounts of β-conformation in the membrane.

The irregular feature of the o.r.d. spectrum in the 250–300 mμ region has been associated with the presence of aromatic side chain chromophores held in an asymmetric array. The third distinctive feature of the o.r.d. feature is the location of the trough at 237 mμ. This is observed with chloroplasts and erythrocyte ghosts. It has been suggested that the anomalous position of this trough is due to hydrophobic reactions between membrane lipids and proteins. Such interactions are suggested by a strong extrinsic Cotton effect arising from lipophilic pigments in chloroplast membranes It is also supported by a blue shift produced by lysolecithin. This alters the structure of the lipid phase but it does not effect full dissociation of the protein from the lipid.

Recent studies have been reported by Lenard and Singer (1966) using optical rotatory dispersion and circular dichroism with erythrocyte membrane preparations in aqueous buffer. They also claim that these show features which strongly suggest the presence of some helical configuration in the membrane protein. The appearance of two minima in the c.d. spectra is thought to be highly significant. The c.d. minima at 208 and 22 m$\mu$, observed with typically $\alpha$-helical polypeptides in protein, are attributed to the $\pi^0$–$\pi^-$ and $n_1$–$\pi^-$ electronic transitions of the peptide group respectively. The presence of the two minima is completely dependent upon the helical arrangement of the peptide groups although their relative magnitudes vary with the polymer and solvent. On the other hand, both c.d. minima in the aqueous membrane suspensions occur at slightly longer wavelengths than are typical in the $\alpha$-helix, as does the minimum in the o.r.d. spectrum. They agree with Wallach and Zahler (1966) who suggest that the Cotton effects observed in the aqueous membrane preparations reflect the presence of polypeptide $\alpha$-helices which are in a special local environment so as to produce the observed red shift. The special environment could include adjacent interacting helices. They conclude that the observed o.r.d. spectrum can be reasonably accounted for as a summation of $\alpha$-helical and random coil contributions. A mixture of $\alpha$-helix and $\beta$-form would not give such an inflection. Closely similar o.r.d. and c.d. spectra are obtained with membrane preparations from diverse sources such as red blood cells and *Bacillus subtilis*. The o.r.d. spectra of membranes of *Mycoplasma laidlawii* (PPLO) also exhibit a pronounced minimum at about 235 m$\mu$.

Lenard and Singer (1966) suggest that the correlation of these o.r.d. and c.d. spectra in various types of membranes show that they reflect a common characteristic feature of the proteins of biological membranes, particularly since they are unusual with simple protein systems. They also suggest that these spectra are associated with structural protein. The idea is that the structural protein might exhibit similar conformational characteristics and interaction with different membranes. If, on the other hand, the helical structures are associated with peripheral rather than the structural protein of the membranes, widely different physical properties might be expected among different membrane types.

### 3. *Nuclear Magnetic Resonance Studies*

The first experiments carried out (Chapman and coworkers, 1967) examined the proton magnetic resonance spectroscopy of membranes, in particular of the erythrocyte membranes or ghosts, when they are dispersed in $D_2O$. These studies have shown that, despite the fact that we are dealing with insoluble materials composed of lipid, protein, water, cholesterol and sugar molecules, useful information can indeed be obtained.

The high resolution proton magnetic resonance (p.m.r.) spectra* of

---

*These spectra are obtained by dispersing the erythrocyte membrane fragments in $D_2O$ using sonication. Studies of sonicated erythrocyte membranes show that the effect of sonication is to break the membrane into small fragments.

erythrocyte membranes is shown in figure 12. It can be seen that peaks associated with some of the functional groups present in the membrane can be clearly observed. Thus there is a peak at $6 \cdot 8\tau$ which arises from the 9 protons present in the $^+N(CH_3)_3$ group of the phosphatidylcholine and sphingomyelins present in the membrane. Peaks are also observed which

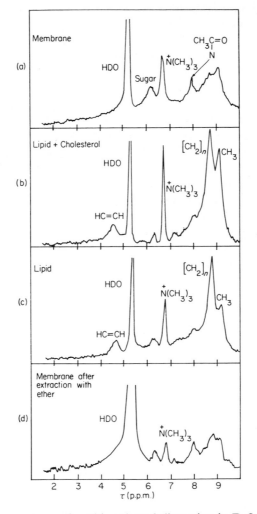

**Figure 12.** P.m.r. spectrum of a 5% sonicated dispersion in $D_2O$: (a) erythrocyte membrane fragments; (b) total equivalent membrane lipid (phospholipid + cholesterol); (c) total equivalent phospholipid; (d) lipid-deficient membrane fragments extracted with ether to remove all cholesterol and about 25% phospholipid (64 scans each). (From Chapman and coworkers, 1968)

have been assigned to the protons present in sugar groupings (6·3 p.p.m.) and also to sialic acid material (7·3 p.p.m.) which is known to be associated with the erythrocyte ghost material. The p.m.r. spectra of ox brain ganglioside and *n*-acetyl neuraminic acid give peaks near these positions and aid the assignment. What is interesting is that the considerable number of protons present in the hydrocarbon chains of the phospholipids do not give rise to a strong peak. A broad hump occurs at 8.8 p.p.m. rather than a sharp peak. No signal is apparent at 4·7 p.p.m. due to the HC=CH protons which are present in the chains.

It is interesting that the $^+N(CH_3)_3$ protons show up so clearly in the n.m.r. spectra of the erythrocyte membrane fragments. This implies that, in the membrane, the protons in this group are able to move quite freely. This suggests that the local viscosity about these choline protons is not high. It is also interesting that the n.m.r. peaks associated with the sugar and sialic acid groups should give rise to high resolution signals, since there have been many conjectures about the presence of sugar and sialic acid on the surface of the membrane. The first tentative conclusion obtained from this high resolution n.m.r. spectrum is that the hydrocarbon chain of the lipid appears to be hindered in its molecular motion, either as a result of being associated with cholesterol or protein, or with both of these membrane constituents. The n.m.r. spectrum is consistent with the membrane being a fairly compact structure of protein and lipid chains with sugar sialic acid groups and the choline groups situated on the outside of this compact structure.

Additional high resolution p.m.r. spectra of the erythrocyte membrane and its constituents reveal further information about membrane construction. Thus, in figure 12 a spectrum of the phospholipid extracted from the membrane is shown at the same concentration as the phospholipid present in the membrane itself, and the large signal associated with the $[CH_2]_n$ protons can be seen. The spectrum of the total lipid, including the cholesterol, is also shown and it can be seen that in this case the inhibition of the chain signal by the cholesterol is not as great as is observed when cholesterol interacts with egg yolk lecithin. This difference in effect may arise because of the much greater unsaturation present in the human red blood cell lipids, and it can be compared with the monolayer results of Demel (1966) which show little condensing effect by cholesterol with phospholipids of this type.

It is suggested that the inhibition of the $[CH_2]_n$ signal in the membrane n.m.r. spectrum is due to a reduction in the lipid chain mobility caused by an interaction between these chains and the protein present in the membrane. Additional support for this idea comes from the n.m.r. spectrum of the erythrocyte membrane after treatment with ether. Ether treatment removes all the cholesterol as well as a small amount of the phospholipid. The effect of this treatment, although drastic, is not understood in detail and does not seem to allow chain $[CH_2]_n$ mobility because no increase in the $[CH_2]_n$ signal at 8·8 p.p.m. is observed. The n.m.r. evidence seems to favour some lipid-chain-protein interaction occurring in the human erythrocyte membrane.

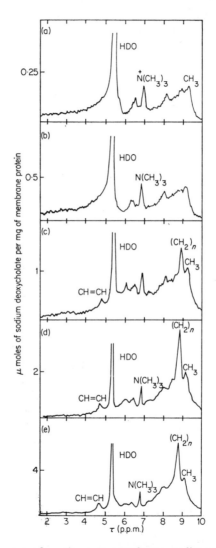

**Figure 13.** P.m.r. spectra of erythrocyte membrane codispersed by sonication in $D_2O$ at 5% (w/v) concentration with increasing concentrations of sodium deoxy-cholate (64 scans). Spectra in(d) and (e) are printed at one-half and one-quarter the sensitivity of (a), (b) and (c). (From Chapman and coworkers, 1968)

Addition of a molecule such as sodium deoxycholate causes a progressive increase in the alkyl chain signal to occur, and it appears that the sodium deoxycholate has weakened the lipid–protein interaction and increased the chain mobility. In fact the whole spectrum of the lipid can be seen, including the protons associated with the double bonds of the alkyl chains at 4·7 p.p.m., as well as other lipid groupings. What is also interesting is that the spectrum of the sodium deoxycholate itself does not show up, implying that the sodium deoxycholate has entered into some complex with the membrane material and, at the same time, releases the lipid. The concentration of sodium deoxycholate required for complete haemolysis of these red blood cells has been shown to be about 1 %. This is just the concentration at which the alkyl chain $[CH_2]_n$ signal at 8·8 p.p.m. markedly shows an increase in intensity. These spectra are shown in figure 13.

Studies on the effects of other molecules, using n.m.r. spectroscopy, have also been carried out and these show interesting interaction effects. The high resolution n.m.r. spectrum of lysolecithin in $D_2O$ shows a number of similar peaks to that of normal lecithin, i.e. it shows a choline signal and it shows a large alkyl $[CH_2]_n$ signal. When lysolecithin is added to the membrane, the p.m.r. spectra show surprisingly that the chain signal from lysolecithin has disappeared, although there has been an increase in the area associated with the choline peak. This shows that the interaction of lysolecithin with the erythrocyte membrane on the one hand, and sodium deoxycholate on the other, are quite different although both these materials are known to cause lysis of red blood cells.

Brown and Shorey (1963) have shown that cell envelopes of *Halobacterium halobium* are stable in high salt concentrations (4–5 M NaCl), but dissolve rapidly as the salt solution in which they are suspended is diluted. In distilled water they give lipoprotein complexes. The intact cells are considered to be bounded by a single membrane, devoid of cell walls. Some evidence has, however, been given recently (Stoeckenius and Rowen, 1967) to suggest the existence of a labile cell wall.

The p.m.r. spectrum of the envelopes in 4 M NaCl in $^2H_2O$ (figure 14) is featureless with only very weak and broad absorption between 8–9 p.p.m. However, when disaggregation of these envelopes occurs by dilution with $^2H_2O$ (to salt concentrations less than 0·8 M), a spectrum with well resolved proton resonance signals is observed.

It was concluded that disaggregation of the envelopes at low ionic strength produces increased freedom for some amino acid protons of the protein. The assignments are consistent with the evidence (Brown, 1965) that only 20 % of the lipoprotein of the envelope is lipid and that, upon dilution, more groups of the envelope protein become available for titration.

The application of n.m.r. spectroscopy to membranes is still in its infancy,

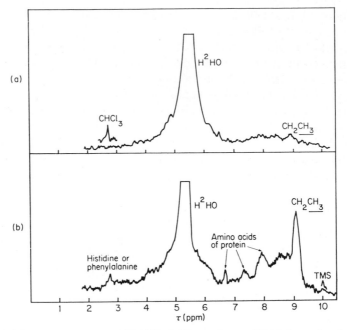

**Figure 14.** The p.m.r. spectra at 33·4 °C of cell envelopes of *Halobacterium halobium:* (a) in 4M NaCl in $^2H_2O$ and (b) disaggregated by dilution in $^2H_2O$ to salt concentrations less than 0·8M NaCl (1024 scans). In (a) $CHCl_3$ was used as marker and in (b) tetramethylsilane (TMS) was used.

but it is already apparent that considerable penetration in our understanding of the organization of membranes and their interaction with various molecules, e.g. hormones, drugs, etc. may occur as a result of studies using this technique both with model and real membrane structures.

## V.   THERMAL STUDIES

Thermal techniques have recently been applied to membrane systems (Ladbrooke and coworkers,* 1968). Myelin isolated from ox brain white matter and the lipids extracted from this material have been examined by thermal analysis techniques including differential thermal analysis and differential scanning calorimetry. The results provide information about the organization of lipids and water in the myelin structure. They show that:

i) On drying myelin a crystallization and precipitation of the lipid and cholesterol takes place. Endothermic transitions associated with the lipid and cholesterol can then be observed (figure 15).

*Biochim. Biophys. Acta* (1968), **164**, 101

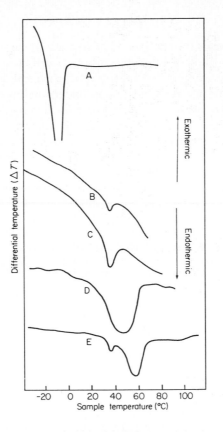

**Figure 15.** Dehydration of myelin. Differential thermal analysis heating curves for samples of myelin previously equilibrated at different relative humidities. Approximate final water contents (wt. % water) are: (a) 30%, (b) 15%, (c) 10%, (d) 5%, (e) 3%.

ii) The total lipid extract in water does not show a detectable endothermic transition but the cholesterol-free lipid does. In the absence of cholesterol, part of the myelin lipid is crystalline at body temperature.

iii) With wet myelin no thermal transitions are detectable. In this case the lipids and cholesterol appear to be organized into a single phase. The presence and organization of the cholesterol in the membrane appear to prevent the lipids from crystallizing.

iv) To maintain the organization of the lipid in the myelin there appears to be a critical amount of water required. This water is unfreezable at 0°C and may correspond to 'bound' water.

# VI.  CONCLUSIONS

Over the last few years the application of new analytical techniques has led to a greater understanding of the composition of cell membranes. A start has also been made attempting to relate physicochemical studies on the components, such as the phospholipids, to an understanding of the structure and function of cell membranes. The alternative and important approach of examining isolated membranes using infra-red and n.m.r. spectroscopy, o.r.d. and c.d. studies, x-ray methods and new electron microscope techniques, also appears to have great potential for determining the manner in which the components are organized into the final functioning membrane structure.

## REFERENCES

Bahr, G. F. (1954) *Expl. Cell Res.*, **7**, 457
Bangham, A. D. and R. W. Horne (1964) *J. Mol. Biol.*, **8**, 660
Bear, S. R., K. J. Palmer and F. O. Schmitt (1941) *J. Cellular Comp. Physiol.*, **17**, 355
Benedetti, E. L. and P. Emmelot (1965) *J. Biophys. Biochem. Cytol.*, **29**, 299
Blasie, J. K., M. M. Dewey, A. E. Blaurock and C. R. Worthington (1965) *J. Mol. Biol.*, **14**, 143
Brenner, S. and R. W. Horne (1959) *Biochim. Biophys. Acta.*, **34**, 103
Brown, A. D. (1965) *J. Mol. Biol.*, **12**, 491
Brown, A. D. and C. D. Shorey (1963) *J. Cell Biol.*, **18**, 681
Byrne, P. and D. Chapman (1964) *Nature*, **202**, 987
Chapman, D. (1965) In *The Structure of Lipids*, Methuen, London
Chapman, D. (1966) *Ann. N.Y. Acad. Sci.*, **137**, 745
Chapman, D., P. Byrne and G. G. Shipley (1966) *Proc. Roy. Soc. (London) Ser. A*, **290**, 115
Chapman, D. and D. T. Collin (1965) *Nature*, **206**, 189
Chapman, D. and D. J. Fluck, (1966) *J. Biophys. Biochem. Cytol.*, **30**, 1
Chapman, D., V. B. Kamat, J. de Gier and S. A. Penkett (1967) *Nature*, **213**, 74
Chapman, D., V. B. Kamat, J. de Gier and S. A. Penkett (1968) *J. Mol. Biol.*, **31**, 101
Chapman, D., V. B. Kamat and R. J. Levene (1968) *Science*, **160**, 314
Chapman, D. and S. A. Penkett (1966) *Nature*, **211**, 1304
Chapman, D. and N. J. Salsbury (1966) *Trans. Faraday Soc.*, **62**, 2607
Chapman, D., R. M. Williams and B. D. Ladbrooke (1967) *Chem. Phys. Lipids*, **1**, 445
Demel, R. A. (1966) Ph.D. Thesis, University of Utrecht, Utrecht
Dervichian, D. G. (1955) In *Exposés actuels: Problems de structures, d'ultrastructures et de functions cellulaires*, Masson, Paris, Chapter 4
Elbers, P. F. (1964) *Recent Prog. Surf. Sci.*, **2**, 443
Fernández-Morán, H. and J. B. Finean (1957) *J. Biophys. Biochem. Cytol.*, **3**, 725
Finean, J. B. (1960) *J. Biophys. Biochem. Cytol.*, **8**, 31
Finean, J. B. (1962) *Circulation*, **26**, 1151
Finean, J. B. and R. E. Burge (1963) *J. Mol. Biol.*, **7**, 672
Finean, J. B., R. Coleman, W. G. Green and A. R. Limbrick (1966) *J. Cell Sci.*, **1**, 287
Fleischer, S., B. Fleischer and W. Stoeckenius (1965) *Federation Proc.*, **24**, 296
Frey-Wyssling, A. and E. Steinmann (1948) *Biochim. Biophys. Acta.*, **2**, 254
Fricke, H. (1925) *J. Gen. Physiol.*, **9**, 137
Fricke, H. (1933) *Cold Spring Harbour Symp. Quant. Biol.*, **1**, 117
Glaeser, R. M., T. Hayes, H. Mel and C. Tobias (1966) *Expl. Cell Res.*, **42**, 467
Hayes, T. L., F. T. Lindgren and J. W. Gofman (1963) *J. Biophys. Biochem. Cytol.*, **19**, 251
Hillier, J. and J. F. Hoffman (1953) *J. Cellular Comp. Physiol.*, **43**, 203
Hubbell, W. L. and H. M. McConnell (1968) *Proc. Natl. Acad. Sci.., U.S.*, **61**, 12

Husson, F. and V. Luzzati (1963) *Nature*, **197**, 822
Korn, E. D. (1966) *Science*, **153**, 1491
Korn, E. D. and R. A. Weisman (1966) *Biochim. Biophys. Acta.*, **116**, 309
Ladbrooke, B. D., R. M. Williams and D. Chapman (1968) *Biochim. Biophys. Acta.*, **150**, 333
Lenard, J, and S. J. Singer (1966) *Proc. Natl. Acad. Sci. U.S.*, **56**, 1828
Lewis, S. M., J. S. Osborn and P. R. Stuart (1965) *Nature*, **220**, 614
Luzzati, V. and F. Husson (1962) *J. Cell. Biol.*, **12**, 207
Maddy, A. H. (1966) *Biochim. Biophys. Acta.*, **117**, 193
Maddy, A. H. and B. R. Malcolm (1965) *Science*, **150**, 1616
Maddy, A. H. and B. R. Malcolm (1966) See Kavanau, J. L. *Science*, **153**, 213
Moor, H. and K. Mühlethaler (1963) *J. Biophys. Biochem. Cytol.*, **17**, 609
Moor, H., K. Mühlethaler, H. Waldner and A. Frey-Wyssling (1961) *J. Biophys. Biochem. Cytol.*, **10**, 1
O'Brien, J. S. and E. L. Sampson (1965) *J. Lipid Res.*, **6**, 537
Palade, G. E. (1953) *J. Histochem. Cytochem.*, **1**, 188
Phillips, M. C. and D. Chapman (1968) *Biochim. Biophys. Acta.*, **163**, 301
Ponder, E. (1961) In J. Brachet and A. E. Mirsky (Eds.), *The Cell*, Vol. 2, Academic Press, New York
Razin, S., H. J. Morowitz and T. T. Terry (1965) *Proc. Natl. Acad. Sci., U.S.*, **54**, 219
Riemersma, J. C. (1963) *J. Histochem. Cytochem.*, **11**, 436
Robertson, J. D. (1957) *J. Biophys. Biochem. Cytol.*, **3**, 1043
Robertson, J. D. (1958) *Intern. Kongr. Elektronen mikroskopie*, Springer, Berlin, **4**, 159
Robertson, J. D. (1959) *Biochem. Soc. Symp.*, **16**, 3
Robertson, J. D. (1960) *Prog. Biophys. Biophys. Chem.*, **10**, 343
Robertson, J. D. (1963) *J. Biophys. Biochem. Cytol.*, **19**, 201
Robertson, J. D. (1964) In M. Locke (Ed.), *Cellular Membranes in Development*, Academic Press, New York and London, p.1
Robertson, J. D. (1966) In G. E. W. Wolstenholme and M. O'Connor (Eds.), *Principles of Biomolecular Organization*, J. and A. Churchill Ltd., London p. 357
Rosenberg, S. A. and G. Guidotti (1968) *J. Biol. Chem.*, **243**, 1985
Rouser, G., G. J. Nelson, S. Fleischer and G. Simon (1968) In D. Chapman (Ed.), *Biological Membranes*, Academic Press, London
Salton, M. R. J. and J. H. Freer (1965) *Biochim. Biophys. Acta.*, **107**, 531
Shockman, G. D., J. J. Kolb, B. Bakay, M. J. Conover and G. Toennies (1963) *J. Bacteriol.*, **85**, 168
Sjöstrand, F. S. (1963) *J. Ultrastruct. Res.*, **9**, 561
Stoeckenius, W. (1962a) *J. Biophys. Biochem. Cytol.*, **12**, 221
Stoeckenius, W. (1962b) *Circulation*, **16**, 1066
Stoeckenius, W. (1966) In G. E. W. Wolstenholme and M. O'Connor (Eds.), *Principles of Biomolecular Organization*, J. A. Churchill Ltd., London p. 418
Stoeckenius, W. and R. Rowen (1967) *J. Cell Biol.*, **34**, 365
van Deenen, L. L. M., U. M. T. Houtsmuller, G. H. de Haas and E. Mulder (1962) *J. Pharm. Pharmaceut.*, **14**, 429
van den Heuvel, F. A. (1963) *J. Amer. Oil Chemists' Soc.*, **40**, 455
van Iterson, W. (1965) *Bacteriol. Rev.*, **29**, 299
Veerkamp, J. H., I. Mulder and L. L. M. van Deenen (1962) *Biochim. Biophys. Acta.*, **57**, 299
Wallach, D. F. H. and P. H. Zahler (1966) *Proc. Natl. Acad. Sci., U.S.*, **56**, 1552
Weibull, C. and L. Bergström (1958) *Biochim. Biophys. Acta.*, **30**, 340
Whittaker, V. P. (1966) In J. N. Tager, F. Papa, E. Quagliariello and E. C. Slater (Eds.), *Regulation of Metabolic Processes in Mitochondria.* Elsevier, Amsterdam p.1
Wigglesworth, V. B. (1957) *Proc. Roy. Soc. (London) Ser. B*, **147**, 185
Zwaal, R. F. A. and L. L. M. van Deenen (1968) *Biochim. Biophys. Acta.*, **150**, 323

# CHAPTER 3

# The organization and permeability of artificial lipid membranes

D. A. Haydon

*Laboratory of Biophysical Chemistry & Colloid Science,*
*University of Cambridge,*
*Cambridge, England*

## I. INTRODUCTION

As is well known, many biological membranes are extremely thin and exhibit a remarkable range of permeability to different molecules and ions. The very low permeability to most ions and polar molecules and the associated higher permeability to non-polar molecules, considered in conjunction with some knowledge of the types of materials present in the membranes, has led inevitably to the idea that those regions of the membrane not directly concerned in passing certain selected species must be effectively continuous sheets of hydrocarbon. Both of the major components of membranes, lipids and proteins, contain hydrocarbon moieties which could form part of these sheets. Regardless of which type of substance is the principal contributor to the sheets, the question arises as to whether the presence of any form of hydrocarbon sheet is consistent with the properties of the membrane.

Further questions arise as to whether lipids or proteins are the most likely to be directly responsible for the highly selective transfer of ions and molecules across the hydrocarbon sheets, and as to what particular structural features a molecule should have in order that it may be highly selective in what it transfers or allows to pass.

It may also be asked what fraction of the area of a membrane need be concerned with, for example, ion transfer in order to account for the observed fluxes. There are, of course, many other questions that might be asked, but for the purposes of this discussion those mentioned above are good examples.

While the direct investigation of the complicated biological membranes is yielding an increasing amount of valuable information on individual types of membrane, the variety of membranes is so great that the large differences between them often tend to obscure those features which they have in common. This, and the sheer difficulty of examining particular components of many membranes makes it very difficult to establish the general physicochemical principles which apply in such structures.

The intensive study of simple systems which has occurred in recent years has, by contrast, supplied a considerable amount of fundamental information regarding the properties of lipids and proteolipids. While these investigations have not established to any extent the structure of a biological membrane, they have given answers to questions such as those posed above and they have given strong indications of the type of answers to be expected to certain other questions. For example, a considerable amount is now known about the properties of bimolecular lipid leaflets and it is possible to say that certain properties of a biological membrane are consistent with this type of structure and that certain others are not. Again, substances are now known that are able to induce ion transfer across lipid membranes with an extraordinary selectivity between not only cations and anions, but even between sodium and potassium. It does not follow, of course, that the biological membrane uses similar substances as carriers, but the possibility exists, and this is valuable knowledge in that it gives us something to look for in a very difficult analytical problem.

The following account is devoted particularly to the description of the properties of the thin lipid membranes studied in the last few years by a number of authors commencing with Mueller and coworkers (1962) and

Taylor (1963). In sections II and III, respectively, the structure and some properties of the purely lipid films are described. Section IV is concerned with the relationship of these films to the lipid lamellae of the smectic mesophase formed by a number of lipids in an aqueous medium, while in section V the modifications in properties of the lipid films which occur in presence of proteins, polypeptides, etc. are described. Limitations of space preclude that this chapter be a comprehensive or critical treatment and no attempt has been made to describe in detail many aspects of the field. The object has been merely to summarize the present state of knowledge on lipid films, giving preference to information likely to be of relevance to biological problems.

## II.  THE STRUCTURE OF OPTICALLY BLACK LIPID FILMS IN AQUEOUS MEDIA

A simple optically black bimolecular lipid film may be formed under aqueous solutions by the extension and drainage of a solution of a water-insoluble polar lipid in a water-insoluble solvent. Stearoyloleylphosphatidyl choline or glycerol monooleate in n-decane are good examples of such a system. The drainage is at first brought about by hydrostatic pressure and capillary suction but as the films become grey or black the London–van der Waals forces are thought to become predominant. A detailed description of the drainage mechanism and of the nature of the London–van der Waals forces in thin films has been given for the closely analogous aqueous soap films by Mysels and coworkers (1959) and by Scheludko (1967). The various experimental techniques for the formation of the lipid films have been described by Bangham (1968) and Tien and Diana (1968).

### A. The Film Thickness

During the drainage of the films the thickness decreases uniformly until a certain critical thickness is reached at which black film is produced. The boundary between the thin black film and the relatively thick coloured film can be seen to be very sharp and in fact corresponds to a decrease in thickness of about 0·1 micron. The black film remains constant in appearance with time (subject to certain conditions which will be mentioned later) and its thickness can be estimated from measurements of its electrical capacitance per unit area, or from its optical reflectance. The film thickness is an important quantity and these two procedures for its determination will be discussed in more detail.

In both approaches the basic measurements of capacitance or reflectance can be made with high precision. The interpretation of both measurements, however, requires a knowledge not only of the structure in depth of the film but also of the effective dielectric constants and refractive indices of the various layers of the film. The general structure is not very seriously in doubt. Thus, it is known that the films are formed by bringing together two mono-layers at hydrocarbon–water interfaces and that, in the simple system under discussion, neither of the two components (polar lipid nor solvent hydro-carbon) can pass into the aqueous phase. It is also known from well-established surface chemical principles that the monolayers in question are oriented so that their polar groups are on the aqueous side of the interface and that their chains remain in the hydrocarbon solvent. The polar groups of the lipid may ionize in the aqueous solution, so producing an ionic double layer, and short range interaction between the water and the ions and dipoles of the lipid polar groups may produce an abnormal structure in the aqueous solution immediately adjacent to the film surface. Figure 1 illustrates the laminated structure of the film.

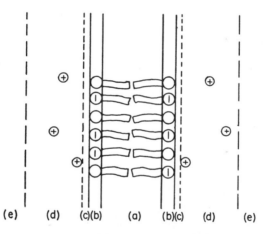

**Figure 1.** The cross-section of a black lipid film in aqueous solution. a: hydrocarbon interior; b: polar groups of the stabilizing lipid; c: perturbed aqueous solution; d: diffuse ionic layer; e: bulk aqueous solution. The space between the lipid chains is filled by solvent hydrocarbon.

To assess accurately the contribution of these various regions to the effective overall dielectric constant or refractive index of the film is quite difficult. For the electrical capacitance there are strong theoretical and experimental reasons to suppose that, usually, the contributions of the ionic double layer, the perturbed aqueous solution and the polar groups are not significant (Hanai and coworkers, 1964, 1965a; Everitt and Haydon, 1968). The problem reduces, therefore, to the assignment of a dielectric constant to the hydrocarbon region. It has been concluded, on the basis of various considerations, that, for films which are formed from alkyl chain polar lipids and n-alkanes, the dielectric constant of this region is given, to a very close approximation, by that of a bulk liquid hydrocarbon of similar composition to that of the film (Hanai and coworkers, 1965c; Haydon, 1968, 1969). With this assumption it is possible to calculate from the electrical capacitance measurements the thickness of the inner or hydrocarbon region of the membrane.

For the interpretation of optical reflectance measurements it is customary to neglect the contribution of the diffuse part of the ionic double layer, but the contributions of the inner region of the double layer, the perturbed aqueous solution and the polar layer groups cannot safely be ignored. In order to obtain the thickness of the film from the reflectance it is, strictly speaking, necessary to know the refractive indices of all the layers individually and the thickness of all but one of them. This information, however, is not normally available and consequently an average refractive index for all the layers is used, which is obtained from a Brewster angle measurement. The precise meaning of a thickness calculated in this way is not very clear, although it is normally considered to correspond to the polar group and

**Table 1.** Thickness of black lipid films

| Film components and aqueous solution | Technique | Thickness (Å) |
|---|---|---|
| Egg yolk phosphatidyl choline + n-tetradecane, chloroform and methanol in 0·1 N NaCl | Reflectance | 72 ± 10[1] |
| Egg yolk phosphatidyl choline + n-decane in 0·1 N NaCl | Reflectance | 77 ± 10[2] |
| Egg yolk phosphatidyl choline + n-decane in 0·1 N NaCl | Capacitance | 48 ± 1[3] |
| Glycerol distearate + n-hexane in 0·1 N NaCl | Reflectance | 50 ± 5[4] |
| Glycerol mono-oleate + n-decane in 0·1 N NaCl | Capacitance | 47 ± 1[5] |

[1]Huang and Thompson (1965); Thompson and Huang (1966); [2]Tien (1967); [3]Hanai and coworkers (1964); [4]Tien and Davidowicz (1966); [5]Taylor and Haydon (1966).

hydrocarbon regions. Some of the problems of the optical measurements have been discussed by Duyvis (1962), Smart and Senior (1966) and Scheludko (1967).*

The results of thickness measurements by electrical and optical methods are complementary in that the first gives the thickness only of the inner or hydrocarbon layer while the second gives the total thickness. The data from the two methods are reasonably consistent. Some values are shown in table 1. It can be seen that the thickness deduced from the capacitance is less than that from the reflectance by an amount which is at least of the right order of magnitude for the thickness of the polar group layers. Provided the average chain length of the polar lipids in the film remains constant, variations in the nature of the aqueous phase, the non-polar solvent, the temperature and the applied electric field do not usually change the film thickness by more than a few percent. Changes in the hydrocarbon chains of the polar lipids can, however, cause large variations in thickness. Thus, introduction of cholesterol into an egg yolk phospholipid film may cause a reduction in thickness of ca. 12 Å (Hanai and coworkers, 1965b). A more systematic

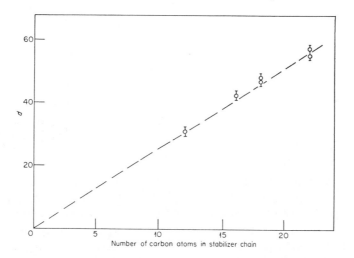

Number of carbon atoms in stabilizer chain

**Figure 2.** The thickness $d$ of the hydrocarbon region of the black films as a function of stabiliser chain length. The dashed line corresponds to a film thickness of twice the chain length of the stabilizer. (Courtesy of the Faraday Society).

*Since this article was written, a more precise analysis of the optical properties of lecithin–aliphatic hydrocarbon films has been published [R. J. Cherry and D. Chapman (1969) *J. Mol. Biol.*, **40**, 19]. In this work an allowance has been made for the birefringence of the film, and the overall thickness of the film was calculated to be $62 \pm 2$ Å.

study of the influence of the chain length of the polar lipid on the thickness of the hydrocarbon region of the films was made by Taylor and Haydon (1966). Films were made from normal chain lipids of $C_{12}$ to $C_{22}$. The results are shown in figure 2. It seems clear that the thickness of the films is directly proportional to the chain length of the stabilizing polar lipid, and also that the thickness is effectively equal to twice the length of the extended chains.

The thickness of the films discussed so far is that thickness resulting primarily from the balance of the London–van der Waals compression and chain (or steric) repulsion forces. A short calculation shows, however, that the application of an applied potential across the film could well give rise to a compression in excess of the London–van der Waals forces, and it is to be expected that, as a consequence, the films may become thinner in electrical fields (Babakov and coworkers, 1966; Haydon and Overbeek, 1966). It is unlikely from theoretical considerations that the effect would be greater than a few percent and, indeed, Babakov and coworkers were unable to detect any thinning of the film by optical reflectance measurement. The electrical capacitance per unit area does, however, increase to a small extent with the applied field. The effect was discernible in the measurements of Hanai and coworkers (1964) and more detailed data for phospholipid films has been published by Rosen and Sutton (1968). The latter authors interpret their capacitance data not in terms of a thickness change but rather by assuming a dipole reorientation. An extensive investigation of glycerol monooleate films in the author's laboratory (Andrews and Haydon, unpublished work) has, however, shown that for these films, at least, there is a significant thickness change and that the dipole reorientation mechanism is not likely. The maximum thickness decrease likely to have occurred under an applied a.c. potential of 400 mV was approximately 13%.

## B. The Film Composition

At the beginning of the film formation process the film is very thick and it usually remains in this state sufficiently long for the polar lipid in the solvent hydrocarbon to attain its equilibrium adsorption at the film surfaces (figure 3a). When fully drained and in the thin optically black state the film is just two lipid molecules in thickness and further thinning under the London–van der Waals forces is evidently prevented by the steric interaction of the lipid chains (figure 3b). The compressive effect of the London–van der Waals forces tends to squeeze out of the film those lipids which are inhibiting the thinning and, as a consequence, the interfaces of the equilibrium thin film are somewhat poorer in polar lipid molecules than were the corresponding interfaces prior to drainage. The composition of the film may be found from the superficial density of the polar lipid and an assumption as to

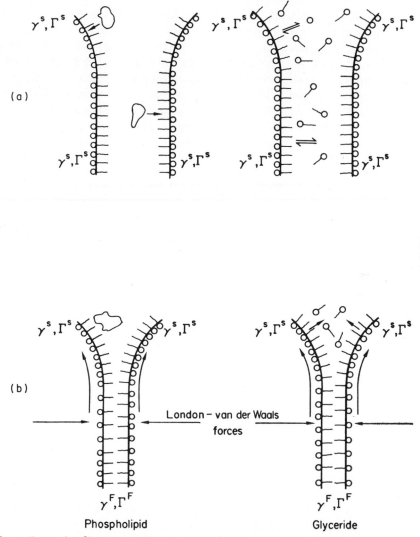

$$\gamma^s, \Gamma^s$$

$$\gamma^s, \Gamma^s$$

(a)

$$\gamma^s, \Gamma^s$$

$$\gamma^s, \Gamma^s$$

$$\gamma^s, \Gamma^s$$

$$\gamma^s, \Gamma^s$$

(b)

$$\gamma^s, \Gamma^s$$

$$\gamma^s, \Gamma^s$$

London – van der Waals
forces

$$\gamma^s, \Gamma^s$$

$$\gamma^s, \Gamma^s$$

$$\gamma^F, \Gamma^F$$

$$\gamma^F, \Gamma^F$$

Phospholipid                     Glyceride

**Figure 3.** a: the film in the thick state prior to drainage. The polar lipid forms equilibrium monolayers at the interfaces. Phospholipid is usually insoluble, at a molecular level, in aliphatic hydrocarbon and exists as large aggregates. The monolayer is thus formed by spreading rather than by adsorption. Glycerides are, by contrast, molecularly soluble and the monolayer is, in this case, formed by adsorption. b: the black film. The polar lipid tends to be squeezed out into the bulk phase interfaces by the action of the London–van der Waals forces. The surface excess of polar lipid in one of the interfaces of the black film $\Gamma^F/2$ is consequently smaller (although only slightly so) than the surface excess in the equilibrium bulk phase interface $\Gamma^S$. The tension per interface of the film $\gamma^F/2$ is also slightly smaller than that of the bulk phase interface $\gamma^S$.

the density of the hydrocarbon core of the film. The former has so far only been estimated for two or three types of film and in each case a quasi-thermodynamic approach has been used. Descriptions of the method have been given by Cook and coworkers (1968) and Haydon (1969). The details differ according to whether the lipids in question are soluble in hydrocarbon and form adsorbed films (such as glycerol monooleate) or are insoluble and form insoluble or spread monolayers (such as some phospholipids). For both types of system an accurate knowledge of the tension in the thin film is necessary. It is not usually possible to obtain this information directly for

**Table 2.** The composition of some black lipid films

| Film components | Number of molecules per cm² film ($\times 10^{-14}$) | Moles decane / Moles polar lipid | Volume decane / Volume polar lipid chains |
|---|---|---|---|
| Glycerol monooleate (10mM) | 5·0 | 1·32 | 0·85 |
| n-decane | 6·6 | | |
| Egg yolk phosphatidyl choline (7mM) | 3·2 | 1·4 | 0·5 |
| n-decane | 4·5 | | |
| Egg yolk phosphatidyl choline (7mM) | 3·0 | | |
| Cholesterol (14·5mM) | 1·2 | | |
| n-decane | (0–1·3) | (<0·3) | (<0·1) |

various technical reasons but it has been shown for one system (Haydon and Taylor, 1968) and is likely to be generally true, that the film tension is only very slightly different from twice the tension of the single equilibrium bulk interface. [This will not be true for very thin films ($\leqslant 30$ Å) but such systems are not very relevant to the present discussion]. It then follows that although the adsorption of lipid at the thin film and bulk phase interfaces must differ, it does so only by a very small amount, and superficial densities of molecules estimated by the application of the Gibbs equation or Langmuir trough techniques to the bulk interface will apply also to the thin film. The results for three types of film are shown in table 2.

Although the polar lipids are quite closely spaced the films evidently contain appreciable amounts of the solvent hydrocarbon, as would be expected from the nature of the system. The results for the films containing cholesterol are, except for the phospholipid figure, less certain than the others, but nevertheless indicate that the solvent hydrocarbon has been largely displaced by the cholesterol.

In the above description of the thickness and composition of thin films only simple systems have been considered and even in these some complications have not been mentioned. If any of the constituents of the solution from which the films are formed are appreciably soluble in water, the thin film may not reach an equilibrium state. The thickness may, in consequence, be perturbed and the determination of the composition will be much more difficult. On the other hand, in films in which the constituents are not significantly water soluble it is often found that numerous very small lenses of solvent form and persist in the black film for long periods (Andrews and Haydon, 1968). Although this phenomenon does not affect the foregoing discussion of the thickness and composition of the thin films, it does tend to complicate both optical measurements and attempts to apply the methods of isotopic analysis to the films (Henn and Thompson, 1968).

## III.  THE PROPERTIES OF BLACK LIPID FILMS IN RELATION TO THEIR STRUCTURE

From the thickness and composition studies and from general observations (for example that the small lenses of hydrocarbon mentioned above move readily in the plane of the film), it is clear that the black lipid films are effectively thin sheets of liquid hydrocarbon sandwiched between the hydrated polar groups of the stabilizing lipid molecules. The tension in the film is usually between 1 and 5 dyne/cm although under some circumstances, such as when complex mixtures of lipids are used or with long ageing of the film, there is evidence of nearly zero tension.

The permeability of the film to a variety of different types of molecule or ion is qualitatively and, for water, even semiquantitatively predictable from the supposition that the rate-limiting stage of the transfer involves the dissolution and diffusion of the molecule in liquid hydrocarbon.

### A. The Permeability to Water

It was noted at an early stage in the study of black lipid films that they were appreciably permeable to water (Mueller and coworkers, 1964). The early measurements suggested that the permeability obtained from tritiated water diffusion experiments was about ten times less than that obtained from osmotic flow (Huang and coworkers, 1964; Hanai and Haydon, 1965). This raised the interesting possibility that a coupling of two unidirectional fluxes was occurring (as has been inferred for diffusion across some cell membranes) and that the results might be explicable in terms of, for example, the single file diffusion model proposed by Hodgkin and Keynes (1955). Alternatively, the water-filled pore model of Pappenheimer and coworkers

(1951) might have been applicable. However, it has now been shown, that there is no real difference between the isotopic and osmotic permeabilities. The discrepancy arose from the presence of unstirred layers of aqueous solution adjacent to the membrane surfaces and from the failure to correct for the fall in concentration of the diffusing species across this region. This boundary layer effect is much more important for isotope diffusion than for osmotic flow. In the former instance it is possible, in unstirred systems, for the boundary layers to constitute almost the total resistance to transfer (Hanai and coworkers, 1966). Mechanical stirring is not usually able to reduce the boundary layer resistance to a negligible value and indirect means have to be used to find the true membrane permeability (Cass and Finkelstein, 1967; Everitt and coworkers, 1969). The permeability as measured by osmotic flow usually requires a correction for boundary layer effects of not more than a few percent. This is a consequence of the density gradients and back diffusion of solute which occur during osmotic flow. Both these effects tend to restore the water concentration at the membrane surfaces to its bulk value, the former by causing natural convection in the boundary layers and the latter simply by interdiffusion of solute and solvent (Everitt and Haydon, 1969).

The osmotic and isotopic diffusion data suggest that there is no interaction between transferring water molecules in the membrane. This conclusion is consistent with the membrane being either an isotropic liquid in which the transferring water molecules dissolve to give an ideal solution, or a structure containing pores of molecular dimensions each of which contains, on average, no more than one water molecule. The former model of the membrane, as a slab of liquid hydrocarbon, permits some semiquantitative predictions of the water permeability to be made. Thus the solubility and diffusion coefficient of water in some liquid hydrocarbons are known (Schatzberg, 1965) and it is a simple matter to calculate the permeability of a slab of hydrocarbon approximately 50 Å in thickness. The difficulties for the black films arise first from the fact that the films so far investigated consist of mixtures of hydrocarbon chains of different lengths and secondly, from the fact that it is debatable to what extent a 50 Å partially oriented assembly of chains may be considered as a slab of bulk liquid. Some results of permeability measurement and of the permeability predicted on the basis of a single hydrocarbon liquid are given in table 3. As can be seen from this the agreement is fair.

The water permeability depends on the lipid composition of the film. The most spectacular results are those of Huang and Thompson (1966) (table 3). Unfortunately, the precise manner in which the lipid varied in these films is not known. The effect of cholesterol in lowering the permeability of a phosphatidyl choline film has been demonstrated by Finkelstein and Cass (1967), but again the precise composition of these films is not known.

**Table 3.** The water permeability of black lipid films

| Film components and aqueous solutions | Film composition (molecules/cm²) | Water permeability (cm/sec) |
|---|---|---|
| Egg phosphatidyl choline<br>Cholesterol<br>n-Decane<br>NaCl, urea or sucrose | $3 \ \times 10^{14}$<br>$1 \cdot 2 \times 10^{14}$<br>$0-1 \cdot 3 \times 10^{14}$ | $2 \cdot 0 \times 10^{-3}$ (20° C)[1,2] |
| Various phosphatidyl choline preparations, n-tetradecane, chloroform, methanol, NaCl | not known | $1 \cdot 73-10 \cdot 4 \times 10^{-3}$ (36 °C)[3] |
| Hypothetical film of 45 Å thickness of<br>n-Hexadecane<br>Squalene | —<br>— | $3 \cdot 85 \times 10^{-3}$ (25 °C)<br>$1 \cdot 7 \ \times 10^{-3}$ (25 °C) |

[1]Hanai and Haydon (1966); [2]Everitt and Haydon (1969); [3]Huang and Thompson (1966).

The temperature dependence of the water permeability can also be predicted from the data on the solubility and diffusion coefficient of water in hydrocarbons. Expressed as an activation energy this is found to be in the region of 11 kcal/mole and is, to a first approximation, independent of temperature. The permeability should therefore increase rapidly with temperature. So far the only detailed study of temperature dependence is that for a lecithin–cholesterol film and, indeed, the permeability increases by approximately 400% between 20° and 40°C, giving an activation energy of 14·6 kcal/mole (Redwood and Haydon, 1969). It was deduced, however, that the increase of temperature also decreased the cholesterol content of the film and hence a permeability increase should have occurred for this reason alone. When a rough correction is applied for this effect from the data of Finkelstein and Cass, the activation energy is lowered to about 11 kcal/mole.

For sodium chloride and urea solutions no significant influence of the concentration of the solutions on the permeability was found (Hanai and Haydon, 1965). In general, however, the nature of the aqueous solution should influence the permeability if only because the adsorption or spreading of the lipid in the film might be affected and the film composition changed.

## B. The Permeability to Non-ionic Substances other than Water

Relatively little work has been carried out on this property of the black films. This is partly, if not entirely, due to technical difficulties. Thus, on the basis of the liquid hydrocarbon slab model the films would not be expected to be very permeable to polar substances. The quantities of material transferred would be small and difficulties of measurement would arise. Substances such as urea and the sugars fall into this category. Urea and sucrose, when used in osmotic experiments have reflection coefficients of effectively unity (Hanai and Haydon, 1965) and indeed, by the use of an isotopic method, Vreeman (1966) concluded that the permeability of a lipid film to urea was only approximately $4 \cdot 2 \times 10^{-6}$ cm/sec. Vreeman also measured permeabilities to glycerol and erythritol by an isotopic method and found values of approximately $4 \cdot 5$ and $0 \cdot 75 \times 10^{-6}$ cm/sec, respectively.

The films would be expected to be much more permeable to lipid-soluble substances and it might be thought that measurements with such substances would be easier. In fact, however, the strong partition of the transferring substance into the film tends to cause complications such as spontaneous thickening or rupture of the film.

The permeability to a number of organic substances has been investigated by Bean and coworkers (1968). The substances chosen were not very lipid soluble, although they permeated much more rapidly than glycerol and erythritol and in some instances, more rapidly than water. The nature of the membrane lipid had very little influence on the permeabilities. The authors concluded that for unionized molecules the relative permeabilities are as would be expected from a mechanism involving solubility and diffusion in a lipid phase.

## C. The Permeability to Ions

An outstanding property of the black lipid films is their very high electrical resistance. The reported values vary considerably, but are usually between $10^7$ and $10^{10}$ ohm cm$^2$. While values at the lower end of this range are quite feasible, some may be leakage rather than true thin film values, and it is desirable that the validity of the measurement be confirmed by the demonstration that the resistance is proportional to the reciprocal of the film area. There is little or no evidence of electronic conduction across the films and it is evidently the ions of the aqueous solutions which carry the current.

A specific resistance of $10^9$ ohm cm$^2$ for a film of thickness 50 Å yields a bulk specific resistance for the material of the film of $2 \times 10^{15}$ ohm cm$^2$. A value of this order of magnitude can be accounted for on the assumption that the resistance is determined by a layer of wet liquid hydrocarbon. There is, nevertheless, good evidence that the transfer of ions across the films is a more complicated process than is the transfer of unionized molecules. Some early results indicated that there may be a saturation of the transfer mechanisms at relatively low aqueous electrolyte concentrations and it was suggested that a major factor in the mechanism was the formation of an ion–lipid complex (Hanai and coworkers, 1965c). Subsequent work has thrown considerably more light on the situation (Pagano and Thompson, 1968). These authors examined conductances, transference numbers and isotope fluxes for egg yolk phosphatidyl choline films in sodium chloride solutions. It was found that whereas the isotopic flux of the sodium corresponded closely with the current carried by the sodium ions, the flux of chloride ions was some 400–1,500 times greater than that needed to account for the current. The isotopic chloride flux also did not increase with concentration above approximately $0.1M$ NaCl. It was inferred that the majority of the chloride ions that crossed the film did so in combination with some carrier molecule which could return across the film only by combining with other chloride ions, so transporting them in the opposite direction. Accordingly, for this mechanism there would be an equal movement of chloride ions in both directions simultaneously and hence no current would pass. The number of carrier molecules required to account for the results was extremely small compared to the number of phospholipid molecules present in the film. In consequence, the carrier molecule could well have been a trace impurity in the phospholipid, which would have been quite undetectable by analytical methods.

There have been several investigations of the electrical resistance of phospholipid films in inorganic electrolytes other than sodium chloride. With one exception, these have not revealed any systems with resistances remarkably lower than for sodium chloride. The exception is for films in potassium iodide solutions (Läuger and coworkers, 1967a) where, in presence of iodine the resistance falls by as much as three orders of magnitude (Läuger and coworkers, 1967b; Rosenberg and Jendrasiak, 1968; Finkelstein and Cass, 1968). It is interesting that molecular iodine is lipid soluble and that it forms complexes with iodide ions to give $I_3^-$ and polyiodides of higher molecular weight. On this and other experimental evidence, Finkelstein and Cass (1968) argue that the molecular iodine acts as a carrier for the iodide ions, and that the current is carried by polyiodides which dissolve and diffuse in the film hydrocarbon.

## IV. A COMPARISON OF THE BLACK LIPID FILM WITH A LEAFLET OF A PHOSPHOLIPID SMECTIC MESOPHASE

The smectic mesophase of a phospholipid consists of a large number of bimolecular lipid leaflets arranged one inside the other such that each forms a closed surface (Bangham, 1968). There seems no obvious reason to suppose that isolated single leaflets should not exist and indeed it appears that these may now be obtainable under certain conditions, albeit still as closed vesicles (Seufert and Stockenius, 1968). These lipid leaflets are obviously more appropriate systems to examine than are the black lipid films if it is desired to know the properties of continuous sheets of phospholipid. Unlike the black lipid films they contain no solvent hydrocarbon and, as a consequence, some of their properties differ in important respects. Perhaps the most obvious difference is in the mechanical states of the two types of membrane.

The black film exhibits a tension characteristic of a liquid and a tendency to assume a planar form. The only exception to this is the spherical (soap bubble) films made by Pagano and Thompson (1968), which nevertheless require an excess pressure inside to maintain their shape. The tension referred to here is merely the tension of a two-sided film of given thickness drawn from a large volume of solution of given composition, and in most cases of interest is nearly equal to twice the tension of the bulk solution interface (Haydon and Taylor, 1968). The isolated pure lipid leaflet does not have an analogous tension, as in extension or contraction material cannot be withdrawn from or returned to an equilibrium bulk phase. Instead, an extension can occur only with a concomitant thinning of the leaflet and, as the amount of thinning that can occur is strictly limited, it is to be expected that after only a small extension the leaflet will rupture. In an isotropic aqueous solution the lipid leaflet shows no sign of being under tension apart from a tendency to form closed vesicles, so removing exposed edges from contact with water.

The different mechanical condition of the black film and the lipid leaflet is, of course, only one manifestation of their different thermodynamic states. The difference in composition, and hence in chemical potential of the lipid, is another. As a consequence, the black film and lipid leaflet may well interact in different ways or, at least, to different extents with other substances such as polypeptides and proteins. It is not easy to make any general comments on this question, although one fairly obvious point might profitably be mentioned. If the product of the interaction of a film or leaflet with another substance is a largely uniform two-dimensional film-like structure, qualitatively similar results may be expected. If this is not so, however, it is probable that the film will rupture with some violence, in which event the system will be unsuitable for discovering the nature of the products of the interaction. The leaflet or assembly of leaflets, on the other hand, may or may not disperse but in any case it should not be very difficult to discover the nature of the products of the interaction. In other words, where interactions with substances which cause *extensive* disruption of the uniform hydrocarbon sheet occur, the pure lipid leaflet is obviously the most suitable system to examine. When, on the other hand, the disruption of the hydrocarbon sheet is relatively minor, such as with some proteins and macrocyclic molecules (section V) or when small sparingly lipid-soluble molecules such as water permeate the membrane, the black film has important advantages in its large controlled area and accurately measurable electrical and permeability properties.

To conclude this section it is appropriate to point out some other respects in which the black film and lipid leaflet differ, or may differ.

The thickness of the hydrocarbon region of the two structures, even when both contain the same polar lipid, is evidently different. As has been described in section II the thickness of the hydrocarbon region of a black film is closely given by twice the length of the extended chains of the stabilizing lipid. For egg phosphatidyl choline this thickness is approximately 47 Å. It has been shown by Small and Bourgès (1966) that the thickness of the hydrocarbon region of the leaflet of pure phospholipid in the presence of excess water is, by contrast only approximately 33 Å. It is to be expected, therefore, that inasmuch as the electrical capacitance of a membrane is inversely proportional to the thickness of its hydrocarbon part, the value for the leaflet should be about 0·56 $\mu F/cm^2$ as compared with 0·39 $\mu F/cm^2$ for the black film. The former value is presumably more appropriate than the latter to compare with a biological membrane.

The compression of a black film in an electric field has been mentioned in section II. This compression probably takes place against the resistance to distortion or reorientation of the chains of the phospholipid. If, in the pure leaflet, the chains are already considerably restricted and distorted, consistent with the smaller thickness of the leaflet, it is probable that the compressibility will be substantially smaller.

Finally, it is interesting to note that although the hydrocarbon region of the pure leaflet of, say, egg phosphatidyl choline is essentially liquid (Chapman, 1968), the average molecular weight of the hydrocarbon moieties is larger than in the black film and therefore its viscosity (speaking in bulk terms) is likely to be higher. There are therefore some reasons, though admittedly crude, for supposing that diffusion in the leaflets may be slower than in the black films and, in consequence, their permeability may be generally somewhat lower.

One of the main difficulties in working with the smectic mesophase is obtaining data for a known area of a single leaflet. With this reservation it appears that for some substances, at least, the permeability of the lipid leaflet is of the same order of magnitude as for the black film. The work so far carried out in this field has been reviewed by Bangham (1968).

## V. THE ASSOCIATION OF POLYPEPTIDES, PROTEINS AND OTHER LARGE MOLECULES WITH BLACK LIPID FILMS AND THE ENHANCEMENT OF ION PERMEABILITY

It soon became evident from studies of films containing only lipids that although these structures alone were able to account qualitatively for some of the properties of biological membranes, they did not seem likely to account for the high and very selective permeability to certain species, especially inorganic ions, which the biological systems possess. As a consequence, the influence on the lipid membranes of a very large number of substances has been examined. These may be subdivided broadly on a size basis into proteins and macromolecules on the one hand, and relatively small molecules such as some polypeptides and macrocyclic substances having molecular weights in the region of 1,000 to 2,000, on the other. The large molecules will be discussed first.

## A. Proteins and macromolecules

It has been established that some, at least, of the common water soluble proteins merely adsorb on to a black lipid film without affecting either the electrical capacitance or conductance of the film. This was shown by Hanai and coworkers (1965b) who examined systems containing bovine serum albumin, egg albumin and insulin. The adsorbed protein does, however, affect the surface potential of the film, and thus may influence the permeability if the film has been made permeable by other means. Tsofina and coworkers (1966) attributed the lack of permeability change under the conditions of the above experiments to the failure of the protein to penetrate the hydrocarbon region of the film which, in turn, they attributed to the close packing of the phospholipid molecules. They therefore devised a means of formation of the black film such that the protein (egg albumin in their experiments) became more strongly adsorbed and perhaps extensively denatured. Quite low resistances were reported (approximately $10^4$ ohm $cm^2$) for the resulting films, but no rectification was observed when the membranes separated solutions of sodium and potassium chloride. The authors describe experiments with potassium oxalate and calcium chloride, however, in which rectification and selective permeability to potassium ions was achieved.

It would perhaps be expected that more interesting results would be obtained from the interaction of membrane transport proteins with lipid films, rather than from globular proteins selected at random. At least one such protein has been tried, a sulphate-binding protein from a transport system of *Salmonella typhimurium* (Pardee and Prestidge, 1966), but evidently without success (Finkelstein and Mueller and Rudin, unpublished results). As Pardee (1968) points out, however, although the protein was known to be one of the proteins involved in sulphate transport and is evidently near the surface of the membrane, it is by no means certain that it is the one that actually facilitates the translocation of the ion across the membrane.

Conductance changes in black lipid films in presence of proteins have been reported by del Castillo and coworkers (1966) and Barfort and coworkers (1968). In both investigations the conductance change was brought about by the simultaneous presence of an antigen and an antibody and it was evidently the interaction of these two substances at the film surface that caused the effect. The first group of authors mention results which suggest that in their experiments the conductance changes might well have occurred in thick film rather than in the bilayer. This, however, was apparently not so in the second investigation and the supposed interaction of antigen and antibody across the black film produced conductance increases of up to × 1,600. Unfortunately, there is little evidence as to the mechanism of the effect.

Lesslauer and coworkers (1968) have described a conductance increase of approximately one thousand times in black phospholipid films in the presence of phospholipase A (molecular weight approximately 14,000). It was necessary, in order to obtain stable films, to use phospholipids which were not attacked by the enzyme and concentrations of the enzyme of $10^{-5}$M were needed to produce the large conductance increases. The capacitance of the black film was not affected by the enzyme and it was established that the conductance increase occurred only over the black areas of the film. Beyond these observations nothing is known of the mechanism.

Probably the most interesting in its effects, although also perhaps the least well characterized of the macromolecules so far discussed, is the so-called Excitability Inducing Molecule (EIM) isolated by Mueller and coworkers (1962), from bacterially desugared egg whites. This substance interacts with a black lipid film and lowers its resistance to a value which depends on the applied potential. When an electrolyte gradient is present across the film and protamine is titrated into the compartment containing the EIM, fully developed action potentials appear (Mueller and Rudin, 1967a). The membrane lipids must evidently contain sphingomyelin, and no other basic protein was found which could replace the protamine. Mueller and Rudin suggest that the EIM produces channels through the film which are selectively permeable to cations, but that on addition of protamine some of these become anion permeable. The opposing membrane potentials which these two types of channel tend to generate, combined with the potential dependence of the permeability, then produces the action potentials. The molecular weight of the EIM is believed to be less than $10^5$ and has been found to consist of two moieties, protein and RNA (Kushnir, 1968). By means of TEAE–cellulose chromatography this author was able to separate the protein and the RNA, but neither of the separated fractions had any membrane activity. On recombination of the moieties activity was partially restored.

## B. Polypeptides and Macrocyclic Substances

Linear polypeptides of up to about eight amino acids which are of uniformly D or L configuration do not appear to have any effect on the

D

black films. Larger molecules of a similar type, such as poly-L-glutamic acid, poly-L-lysine and poly-L-arginine, similarly have no effect. Whether or not this is true for many other polypeptides is difficult to say as these tend to be insoluble in suitable solvents. It has been known for some time, however, that many polypeptides produce permeability changes in bacterial and red cell membranes (see for example Silman and Sela, 1967) and it has been realized recently that certain macrocyclic antibiotics [although by no means all (Tosteson and coworkers, 1968)] have a profound influence on ion transfer across biological membranes (Moore and Pressman, 1964; Lardy and coworkers, 1967; Harris, 1968). From the evidence of these authors it appeared that the macrocyclic substances assisted the ions across a lipid barrier in the membrane and Lev and Buzhinsky (1967) and Mueller and Rudin (1967b) showed that they were indeed very effective in enhancing cation transfer across black lipid films. A large number of macrocyclic substances have been prepared, either from natural sources or by direct synthesis and their structure discussed in relation to their biological activity (see for example Ovchinnikov and coworkers, 1967) but as yet there is a scarcity of systematic data on their interaction with lipid films. The two

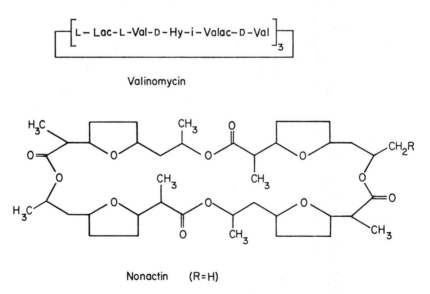

Valinomycin

Nonactin    (R=H)

Monactin    (R = CH$_3$)

**Figure 4.** The cyclic dodecadepsipeptide valinomycin and the macrotetrolides non-actin and monactin. L-Lac-L-Val $=$ L-lactyl-L-valine; D-Hy-i-Valac-D-Val$=$D-α-hydroxyisovaleryl-D-valine.

substances most extensively studied from this point of view are valinomycin and monactin (figure 4).

Valinomycin is a cyclic dodecadepsipeptide and monactin is a macro-tetrolide. Both are non-ionic and have antibiotic properties. They are sparingly soluble in aqueous solutions but may be introduced into black lipid films by this means. They may also be introduced via the lipid solution, but if no antibiotic is present in the aqueous phase the lipid film tends to lose its ion permeability, presumably owing to the diffusion of the antibiotic out of the lipid. The valinomycin and monactin have large effects on the film permeability at very low concentrations (e.g. $10^{-11}$ to $10^{-6}$ mole/l) and in solutions of alkali metal salts produce changes in the film conductance which are proportional to the antibiotic concentration (Gotlib and co-workers, 1968; Tosteson and coworkers, 1968; Eisenman and coworkers, 1968). Thus for phospholipid films in $0.01$M NaCl, $10^{-7}$M monactin gives conductances of approximately $10^{-5}$ ohm$^{-1}$cm$^2$, (Eisenman and coworkers, 1968) and similar results have been found for valinomycin (Tosteson and coworkers, 1968). In contrast to the EIM, the film conductance in these systems is not significantly dependent on the applied voltage.

One of the most remarkable properties of valinomycin and monactin, especially the former, is the very high selectivity in membrane permeability which they induce to the various univalent ions. From the measurement of the transference numbers for the anion and cation of a single electrolyte, it is found for both substances that the cation carries practically all the current. It is also found, from the measurement of biionic potentials, that selectivity between univalent cations can be very high. Thus, for valinomycin, the permeabilities are in the order,

$$H^+ > Rb > K^+ > Cs^+ > Na^+ \approx Li^+$$

The precise ratios of one ion to another are somewhat different in the various reports (owing partly, perhaps, to the different lipids used in each investigation) but it is generally agreed that the selectivity coefficients $K_{AB} = P_A/P_B$) referred to the lithium permeability ($P_B$) as unity are of the order of $K_{Cs/Li} \sim 200$, $K_{K/Li} \sim 400$ and $K_{Rb/Li} \sim 900$. $K_{KN/a}$ is therefore approximately 400 (Mueller and Rudin, 1967b; Lev and Buzhinsky, 1967). Gotlib and co-workers (1968) have reexamined this question in more detail and give corrected data from which they conclude that the permeability ratio of potassium to sodium may be considerably higher. The same authors have also shown that, as one might expect, the conductance ratios for similar concentrations of the various alkali metal chlorides, are in reasonably good agreement with the above mentioned permeability ratios obtained from biionic potentials. For monactin Eisenman and coworkers (1968) give the slightly different sequence,

$$K^+ > Rb^+ > Cs^+ > Na^+ > Li^+$$

and $K_{Na/Li} \sim 2$, $K_{Cs/Li} \sim 3$, $K_{K/Li} \sim 90$ and $K_{Rb/Li} \sim 30$. These values are again independent of whether they are determined from biionic potentials or conductance ratios. $K_{K/Na}$ for monactin is therefore approximately 45.

The mechanism of action of valinomycin and monactin on black lipid films has not been fully established, but enough is known to enable a working hypothesis to be suggested. Pioda and coworkers (1967) have shown that monactin and nonactin form crystalline complexes with potassium, sodium and ammonium thiocyanate, and Kilbourn and coworkers (1967) have determined the structure of the nonactin–KNCS complex. It was found that the 32-membered ring 'can be described as resembling the seam of a tennis ball with the unhydrated $K^+$ ion at the centre of the ball and with the methyl substituents and the methylene groups of the furane rings on the outside.' The thiocyanate anions were distributed in a disordered manner in the cavities of the crystal. The potassium ion in the complex is evidently 8-coordinated, to the four oxygens of the furane rings and to the four keto groups. From an examination of the infrared spectrum, it was found that the structure of the complex was probably similar in chloroform solution to that in the crystal. The potassium ion is thus enclosed within an approximately spherical ball which presents only non-polar groups to the exterior. The free energy of transfer of this complex from an aqueous to a hydrocarbon environment is obviously likely to be considerably lower than that for a normal hydrated potassium ion. The structure of the sodium ion complex has not been worked out, but Kilbourn and coworkers comment that on the evidence available the conformation is likely to be somewhat similar to that for potassium. The selectivity between sodium and potassium would then presumably originate from the closeness of the fit and hence the relative strengths of the chelation for the two ions, and this in turn would be reflected in the concentration of the complex within the lipid film. Whether or not the valinomycin–alkali metal ion forms a complex similar in general character to that of the non-actin complexes is not known. Valinomycin is nevertheless a similar type of molecule (figure 4) and it would not be surprising if it does.

A knowledge of the form of the ion–antibiotic complex unfortunately does not tell us how the ion is transferred across the lipid film. As Kilbourn and coworkers point out, the nonactin may exist as a stack of molecules aligned across the membrane, such that the ion may jump from one molecule to the next. Alternatively, the complex, which has a diameter less than the thickness of the hydrocarbon part of the membrane, may in effect dissolve in the liquid hydrocarbon and diffuse across. Yet another possibility is that the hydrocarbon part of the film may become constricted at the site of the complex and distinction between a 'pore' consisting of a single complex on the one hand, and a diffusing complex on the other, may become impossible. Eisenman and coworkers (1968) have attempted to test the hypothesis that the complex dissolves in the membrane hydrocarbon and diffuses across under an electrochemical potential gradient. They derived the appropriate equations for the model, and then measured the concentration dependence of conductance and the selectivity coefficients for black films of phospholipid in presence of monactin, and also the selective partitioning of the various alkali metal–monactin complexes between bulk aqueous and non-aqueous phases. The film data has already been mentioned, but it is pertinent to remark that the linear relation observed between monactin concentration and film conductance is in satisfactory agreement with the hypothesis. The selectivity obtained from the bulk-phase redistribution experiments was in remarkably good agreement with the selectivity of the film. The existing evidence for monactin and nonactin, assuming them to be not significantly different, therefore supports the dissolution–diffusion mechanism, although it may not be safe to suppose that there is a unique explanation for the rather slender data so far available. The linear dependence of conductance and concentration for valinomycin is, of course, consistent with a dissolution–diffusion mechanism for this substance. The ability of valinomycin to solubilize $^{86}Rb^+$ in a non-aqueous medium in presence of a lipid soluble anion (laurate) has been demonstrated by Pressman (1968) who, from this and other considerations has proposed that valinomycin acts in lipid membranes essentially by complexing with and 'carrying' the ions (Pressman and coworkers, 1967).

The average density of antibiotic molecules and ion complexes in the lipid film or, in other words, the extent of the adsorption of these species from the aqueous solution into the film is not known. The very low concentrations in the aqueous solution, and the similarly low adsorption in the film, necessary to produce the observed conductance changes, make the discovery of this quantity an exceptionally difficult problem. The only information so far available comes from the indirect estimations of Gotlib and coworkers (1968) and Tosteson and coworkers (1968). Gotlib and coworkers measured the increase of capacitance of a black phospholipid film in presence of $10^{-6}$M valinomycin in 0·1M lithium chloride solution. They then assumed a dielectric constant of 80 for the valinomycin and 2 for the interior of the lipid film, and deduced that 0·28 % of the film area was occupied by the antibiotic. Tosteson and coworkers measured the distribution coefficient of valinomycin between n-decane in bulk and 0·1M potassium chloride solution. They then adopted the 'liquid hydrocarbon slab' model of the black film, such as was used to interpret the water permeability (section III A) and calculated simply on the basis of the volume ratio of bulk and film hydrocarbon the number of valinomycin molecules to be expected per unit area of the film. The result was that in $10^{-7}$M valinomycin there were approximately $2 \times 10^{11}$ valinomycin molecules per square centimetre of black film. Depending on the area of a valinomycin molecule in the film, it is readily calculated that this result could just be in agreement with that of Gotlib and coworkers to within an order of magnitude. There are obviously some dubious assumptions in both estimates, but even if the results are only very roughly correct, they emphasize that few molecules are required to substantially modify the ion permeability of a lipid membrane.

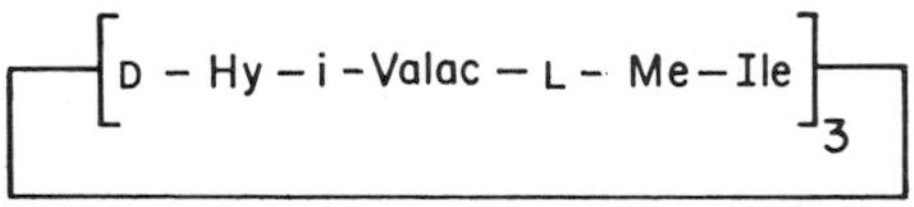

$$\left[ \text{D} - \text{Hy} - \text{i} - \text{Valac} - \text{L} - \text{Me} - \text{Ile} \right]_3$$

Enniatin A

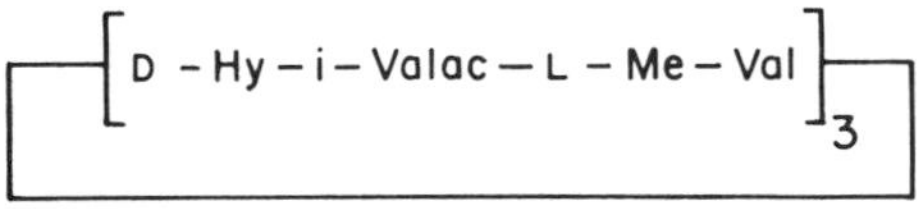

$$\left[ \text{D} - \text{Hy} - \text{i} - \text{Valac} - \text{L} - \text{Me} - \text{Val} \right]_3$$

Enniatin B

**Figure 5.** The cyclic hexadepsipeptides enniatin A and enniatin B.
L-Me-Ile ≡ *N*-methyl-L-isoleucine;
L-Me-Val ≡ *N*-methyl L-valine; D-Hy-i-Valac ≡ D-α-hydroxyisovaleryl-D-valine.

While there are other antibiotics which seem likely to function in a similar manner to monactin, nonactin and valinomycin, they have not been so thoroughly investigated and less can be said about them. Enniatin A and B are cyclic hexadepsipeptides (figure 5) which have been examined in black phospholipid films by Mueller and Rudin (1967b). In solutions of 0·05M electrolyte, enniatin B gave $K_{\text{K/Na}} = 37$. In spite of a smaller ring size, therefore, these molecules still exhibit considerable selectivity. Gramicidins A, B and C (figure 6) are, by contrast, somewhat larger than valinomycin and are evidently not cyclic molecules. They are effectively insoluble in aqueous solutions and in aliphatic hydrocarbons, although they are dispersed in the latter by polar lipids. Like valinomycin and monactin, they are very effective in lowering the conductance of lipid films and do not greatly affect the film capacitance or its frequency dependence (Haydon, 1968).

HCO-L-Val-Gly-L-Ala-D-Leu-L-Ala-D-Val-L-Val-D-Val-L-Tryp-D-Leu-*L-Tryp*-D-
Leu-L-Tryp-D-Leu-L-Tryp-NH · CH₂CH₂OH
Valine Gramicidin A

HCO-L-Val-Gly-L-Ala-D-Leu-L-Ala-D-Val-L-Val-D-Val-L-Tryp-D-Leu-*L-Phe*-D-
Leu-L-Tryp-D-Leu-L-Tryp-NH · CH₂CH₂OH
Valine Gramicidin B

HCO-L-Val-Gly-L-Ala-D-Leu-L-Ala-D-Val-L-Val-D-Val-L-Tryp-D-Leu-*L-Tyr*-D-
Leu-L-Tryp-D-Leu-L-Tryp-NH · CH₂CH₂OH
Valine Gramicidin C

**Figure 6.** The linear peptides gramicidin A, B and C. Each substance is also found with L-isoleucine instead of L-valine at the left-hand end of the chain (see Sarges and Witkop, 1965).

Their selectivity is, however, much less than for valinomycin and monactin. Although they are strongly cation (as opposed to anion) selective, the value of $K_{K/Na}$ is given by Mueller and Rudin (1967b) as only approximately 6. While it seems quite likely that gramicidin may form chelates with cations in a somewhat similar manner to nonactin, nothing is known concerning the mechanism of ion transfer. The so-called crown polyethers of Pederson (1967) have the ability to increase the conductance of black lipid films in electrolyte solutions, although at considerably higher concentrations than the antibiotics discussed so far, and they also exhibit some selectivity between the alkali metal cations (Eisenman and coworkers, 1968). These authors examined the mechanism of action of one of the polyethers (figure 7) in the same way as for monactin, but found, by contrast, very poor evidence in support of a dissolution–diffusion model.

A quite different type of substance, the polyene antibiotics, have also been found to affect the ion permeability of black lipid films. The interaction of these substances with black lipid films, and their tendency to cause instability

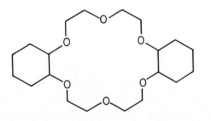

**Figure 7.** The cyclic polyether dicyclohexyl-18-crown-6.

was reported by van Zutphen and coworkers (1966) and subsequent investigations have been carried out by Andreoli and Monahan (1968) and
Finkelstein and Cass (1968). Attention has centred on three substances,
nystatin, amphotericin B and filipin. A probable structure of nystatin is
shown in figure 8. As can be seen, it is a cyclic molecule of molecular weight

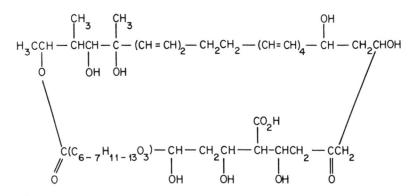

**Figure 8.** The approximate formula for nystatin.

approximately 932 and incorporates conjugated double bonds as well as
one amino and one carboxylic acid group. Each of the three substances has
only been found to influence lipid films when cholesterol was present in the
film-forming solution. Filipin, at $10^{-6}$ to $10^{-5}$M in the aqueous solution,
rapidly caused rupture of the films, and prior to the rupture did not very
significantly influence their electrical resistance or ionic selectivity (Andreoli
and Monahan, 1968). Nystatin and amphotericin B, on the other hand, in
somewhat similar concentrations, did not cause instability of the films and
produced very large changes in conductance and selectivity. The conductances, which were ohmic, varied with a high power of the antibiotic concentration. Andreoli and Monahan found a power of 4·5 while Finkelstein and
Cass found 10. The lipids from which the film were made were evidently
different in the two investigations, however, and the discrepancy may be
explicable on this basis. According to transference number measurements,
the films were rendered preferentially permeable to anions by both nystatin
and amphotericin B. Appreciable cationic conductance was retained,
however, and although the ratio of the transference numbers for the anion
and cation was approximately 10, the actual conductance due to the cations
was nevertheless increased by several orders of magnitude by the antibiotics

(Andreoli and Monahan, 1968). A large decrease in film conductance with increasing temperature (yielding a $Q_{10} \sim 10^4$) has been reported by Finkelstein and Cass (1968). The qualitative explanation of this effect is likely to be quite complicated as several important parameters of the system change with temperature. The decrease in cholesterol content of the film with increasing temperature (Redwood and Haydon, 1969) is likely to be of particular importance. Thus, according to Redwood and Haydon, an increase in temperature of 10 °C can have the same effect on the film capacitance as a three-fold decrease in the ratio of cholesterol to phospholipid in the film-forming solution. If this observation is combined with the results of Andreoli and Monahan concerning the effect of the cholesterol/phospholipid ratio on the film conductance, it is then clear that loss of cholesterol alone could account for the whole effect.

The present state of knowledge of the mechanism of action of the polyene antibiotics indicates that this is probably quite different to that inferred for the valinomycin or monactin type of molecule. The very strong dependence of conductance on antibiotic concentration suggests immediately that aggregates are formed in the lipid film, and both Andreoli and Monahan and Finkelstein and Cass have speculated that these aggregates form pores. Strong support for this idea comes from the fact that the permeability of the film to water, urea, ethylene glycol, glycerol, propionamide and erythritol is increased by nystatin (Finkelstein and Cass, 1968). From the reflection coefficients for erythritol ($\sim 0.79$) and glucose and sucrose ($\sim 1$), these authors deduce that the pore radius is approximately 4 Å.

To conclude this description of the interaction of antibiotics with black lipid films the recent report of Mueller and Rudin (1968) on the cyclopeptide alamethicin must be mentioned. Alamethicin contains some nineteen amino acids but its structure is as yet unknown. It produces a selectivity for cations rather than anions, but between cations the selectivity is poor. The conductance in presence of alamethicin and in symmetrical bathing solutions is strongly dependent on the potential across the film, such that the low conductance state corresponds to small or zero potential difference. In an electrolyte gradient, however, or with appropriate addition of a basic protein, such as a protamine, regions of negative resistance are produced in the current–voltage curves and by suitable combination of both the electrolyte gradient and the protamine, Mueller and Rudin were able to obtain single or rhythmic action potentials similar in form to those given by EIM.

## VI  SOME CONCLUSIONS

In the introduction some examples were given of the ways in which the study of artificial membranes may help towards the understanding of biological membranes. The work on black lipid films, which has been described above, has indicated very clearly many of the adequacies and inadequacies of the simple bimolecular lipid leaflet model. It has been shown that in a purely lipid membrane, the electrical properties and the permeability are determined primarily by the thickness and nature of the hydrocarbon core of the film rather than by the polar groups. Although this conclusion may, in retrospect, never have seemed in doubt, it has often been questioned. The electrical capacitance of a lipid film is not incompatible with the results for cell membranes and the very high electrical resistance of a lipid film suggests a satisfactory explanation of the general impermeability of cell membranes to ions. The appreciable permeability of lipid films to water shows that an extensive lipid barrier is not inconsistent with the magnitude of the water permeability of cells, and the permeability of the lipid films to other non-ionic solutes is consistent with the properties which have often been assumed for the lipid part of a biological membrane.

It has become obvious, however, that a purely lipid membrane is very unlikely to have the relatively high electrical conductance and the very selective ion permeability characteristics of biological membranes. The incorporation of such properties into a lipid film is now known to be possible by the addition of some macrocyclic antibiotics and other large molecules. By this means the lipid films may be endowed with conductances and ion selectivities somewhat similar to those encountered in biological membranes without necessarily changing their capacitance or their water permeability, or by affecting more than a fraction of a percent of their area. The detailed examination of the macrocylic substances has revealed a highly specific mechanism for the transference of selected ions (so far, notably $K^+$) across a lipid barrier, which seems to operate in certain biological membranes as well as in lipid films (Harris, 1968).

In addition to a low conductance and a highly selective ion permeability, it has been shown that a slightly modified lipid film may exhibit rhythmic potential fluctuations or action potentials somewhat similar to those of nerve cell membranes. It must be emphasized that, at present, there is little or no evidence that a normal cell membrane embodies any of the macrocyclic or other substances which give such interesting imitative results in lipid films. It is likely, however, that whatever substances are employed by a cell will be present in extremely small quantities. The discovery of the type of structure which may be used in ion transfer should be of considerable assistance in the discovery of the means actually used.

## REFERENCES

Andreoli, T. E. and M. Monahan (1968) *J. Gen. Physiol.*, **52**, 300
Andrews, D. M. and D. A. Haydon (1968) *J. Mol. Biol.*, **32**, 149
Babakov, A. V., L. N. Ermishkin and E. A. Liberman (1966) *Nature*, **210**, 953
Bangham, A. D. (1968) *Progr. Biophys. Mol. Biol.*, **18**, 29
Barfort, P., E. R. Arquilla and P. O. Vogelhut (1968) *Science*, **160**, 1119
Bean, R. C., W. C. Shepherd and H. Chan (1968) *J. Gen. Physiol.*, **52**, 495
Cass, A. and A. Finkelstein (1967) *J. Gen. Physiol.*, **50**, 1765
del Castillo, J., A. Rodriguez, C. A. Romero and V. Sanchez (1966) *Science*, **153**, 185
Chapman, D. (1968) In D. Chapman (Ed.) *Biological Membranes*, Academic Press, London, p. 125
Cook, G. M. W., W. R. Redwood, A. R. Taylor and D. A. Haydon (1968) *Kolloid-Z.*, **227**, 28
Duyvis, E. M. (1962) Ph.D. dissertation, University of Utrecht, Utrecht
Eisenman, G., S. M. Ciani and G. Szabo (1968) *Federation Proc.*, **27**, 1289
Everitt, C. T. and D. A. Haydon (1968) *J. Theoret. Biol.*, **18**, 371
Everitt, C. T. and D. A .Haydon (1969) *J. Theoret. Biol.*, **22**, 9
Everitt, C. T., W. R. Redwood and D. A. Haydon (1969) *J. Theoret. Biol.*, **22**, 20
Finkelstein, A. and A. Cass (1967) *Nature*, **216**, 717
Finkelstein, A. and A. Cass (1968) *J. Gen. Physiol.*, **52**, 145s
Gotlib, V. A., E. P. Buzhinsky and A. A. Lev (1968) *Biophysics (USSR), English Transl.*, **13**, 562
Hanai, T. and D. A. Haydon (1965) *J. Theoret. Biol.*, **11**, 370
Hanai, T., D. A. Haydon and J. Taylor (1964) *Proc. Roy. Soc. (London), Ser. A*, **281**, 377
Hanai, T., D. A. Haydon and J. Taylor (1965a) *J. Theoret. Biol.*, **9**, 278
Hanai, T., D. A. Haydon and J. Taylor (1965b) *J. Theoret. Biol.*, **9**, 422
Hanai, T., D. A. Haydon and J. Taylor (1965c) *J. Theoret. Biol.*, **9**, 433
Hanai, T., D. A. Haydon and W. R. Redwood (1966) *Ann. N.Y. Acad. Sci.*, **137**, 731
Harris, E. J. (1968) In L. Bolis and B. A. Pethica (Ed.), *Membrane Models and the Formation of Biological Membranes*, North Holland, Amsterdam, p. 247
Haydon D. A. (1968) *J. Amer. Oil. Chemists' Soc.*, **45**, 230
Haydon, D. A. (1969) in D. C. Tosteson (Ed.) *The Molecular Basis of Membrane Function*, Prentice-Hall Inc., New Jersey, p. 111
Haydon, D. A. and J. Th. G. Overbeek (1966) *Discussions Faraday Soc.*, **42**, 66
Haydon, D. A. and J. L. Taylor (1968) *Nature*, **217**, 739
Henn, F. A. and T. E. Thompson (1968) *J. Mol. Biol.*, **31**, 227
Hodgkin, A. L. and R. D. Keynes (1955) *J. Physiol (London)*, **128**, 61
Huang, C. and T. E. Thompson (1965) *J. Mol. Biol.*, **13**, 183
Huang, C. and T. E. Thompson (1966) *J. Mol. Biol.*, **15**, 539
Huang, C., L. Wheeldon and T. E. Thompson (1964) *J. Mol. Biol.*, **8**, 148
Kilbourn, B. T., J. D. Dunitz, L. A. R. Pioda and W. Simon (1967) *J. Mol. Biol.*, **30**, 559
Kushnir, L. D. (1968) *Biochim. Biophys. Acta.*, **150**, 285
Lardy, H. A., S. N. Graven and S. Estrada-O (1967) *Federation Proc.*, **26**, 1355
Läuger, P., W. Lesslauer, E. Marti and J. Richter (1967a) *Biochim. Biophys. Acta.*, **135**, 20
Läuger, P., J. Richter and W. Lesslauer (1967b) *Ber. Bunsenges. Physik. Chem.*, **71**, 906
Lesslauer, W., A. J. Slotboom, N. M. Postema, C. H. de Haas and L. L. M. van Deenen (1968) *Biochim. Biophys. Acta.*, **150**, 306
Lev, A. A. and E. P. Buzhinsky (1967) *Tsitologiya*, **9**, 102
Moore, C. and B. C. Pressman (1964) *Biochem. Biophys. Res. Commun.*, **15**, 562
Mueller, P. and D. O. Rudin (1967a) *Nature*, **213**, 603
Mueller, P, and D. O. Rudin (1967b) *Biochem. Biophys. Res. Commun.*, **26**, 398
Mueller, P. and D. O. Rudin (1968) *Nature*, **217**, 713
Mueller, P., D. O. Rudin, H. T. Tien and W. C. Westcott (1962) *Nature*, **194**, 979

Mueller, P., D. O. Rudin, H. T. Tien and W. C. Westcott (1964) In J. F. Danielli, K. C. A. Pankhurst and A. C. Riddiford (Eds.), *Recent Progress in Surface Science*, 1, Academic Press, New York, p. 379

Mysels, K. J., K. Shinoda and S. Frankel (1959) *Soap Films*, Pergamon, New York

Ovchinnikov, Yu. A., V. T. Ivanov and M. M. Shemyakin (1967) In H. C. Beyerman, A. van de Linde and W. Maasen ven den Brink (Eds.), *Peptides*, North Holland, Amsterdam, p. 173

Pagano, R. and T. E. Thompson (1968) *J. Mol. Biol.*, **38**, 41

Papenheimer, J. R., E. M. Renkin and L. M. Borrero (1951) *Amer. J. Physiol.*, **167**, 13

Pardee, A. B. (1968) *Science*, **162**, 632

Pardee, A. B. and L. S. Prestidge (1966) *Proc. Natl. Acad. Sci. U.S.*, **55**, 189

Pedersen, C. J. (1967) *J. Amer. Chem. Soc.*, **89**, 7017

Pioda, L. A. R., H. A. Wachter, R. E. Dohner and W. Simon (1967) *Helv. Chim. Acta.*, **50**, 1373

Pressman, B. C. (1968) In L. Bolis and B. A. Pethica (Eds.), *Membrane Models and the Formation of Biological Membranes*, North Holland, Amsterdam, p. 258

Pressman, B. C., E. J. Harris, W. S. Jagger and J. H. Johnson (1967) *Proc. Natl. Acad. Sci. U.S.*, **58**, 1949

Redwood, W. R. and D. A. Haydon (1969) *J. Theoret. Biol.*, **22**, 1

Rosen, D. and A. M. Sutton (1968) *Biochim. Biophys. Acta.*, **163**, 226

Rosenberg, B. and G. L. Jendrasiak (1968) *Chem. Phys. Lipids*, **2**, 47

Sarges, R. and B. Witkop (1965) *Biochemistry*, **4**, 2491

Schatzberg, P. (1965) *J. Polymer. Sci.*, c **10**, 87

Scheludko, A. (1967) *Adv. Colloid Interface Sci.*, **1**, 392

Seufert, W. D. and W. Stoeckenius (1968) Abstracts of papers presented at I.U.P.A.B. Symposium on Biophysical Aspects of Permeability, Jerusalem, p. 58

Silman, H. I. and M. Sela (1967) In G. D. Fasman (Ed.), *Poly-α-amino Acids*, Edward Arnold, London and Marcel Dekker, New York, p. 605

Small, D. M. and M. Bourgès (1966) *Mol. Cryst.*, **1**, 541

Smart, C. and W. A. Senior (1966) *Trans. Faraday Soc.*, **62**, 3253

Taylor, J. L. (1963) Ph.D. Dissertation, University of Cambridge, Cambridge

Taylor, J. L. and D. A. Haydon (1966) *Discussions Faraday Soc.* **42**, 51

Thompson, T. E. and C. Huang (1966) *J. Mol. Biol.*, **16**, 576

Tien, H. T. (1967) *J. Theoret. Biol.*, **16**, 97

Tien, H. and E. A. Dawidowicz (1966) *J. Colloid Interface Sci.*, **22**, 438

Tien, H. T. and A. L. Diana (1968) *Chem. Phys. Lipids*, **2**, 55

Tosteson, D. C., T. E. Andreoli, M. Tieffenberg and P. Cook (1968) *J. Gen. Physiol.*, **51**, 373s

Tsofina, L. M., E. A. Liberman and A. V. Babakov (1966) *Nature*, **212**, 681

Vreeman, H. J. (1966) *Koninkl. Ned. Akad. Wetenschap. Proc. Ser. B*, **69**, 564

van Zutphen, H., L. L. M. van Deenen and S. C. Kinsky (1966) *Biochem. Biophys. Res. Commun.*, **22**, 393

CHAPTER 4

# Water : Its fitness as a molecular environment

Irving M. Klotz

*Biochemistry Division, Department of Chemistry,*
*Northwestern University,*
*Evanston, Illinois, U.S.A.*

## I. INTRODUCTION

Over fifty years ago L. J. Henderson (1913) published a collection of essays entitled 'The Fitness of the Environment', a chapter of which was called 'Water'. In it Henderson pointed out how the unusual macroscopic properties of water—its inertness, high specific and latent heats, unusual density–temperature dependence, thermal conductivity, etc.—make it uniquely suited as an environment for living organisms. In the course of the past half-century biology has pushed its horizons by many orders of magnitude in the direction of the molecular constitution and structure of cell constituents and their relationships to biological function. It is striking how apt an expression is Henderson's when it is extended to the new scale of examination. Water is indeed remarkably fit as an environment when viewed at the molecular level as well as at the organism level. Let us examine, therefore,

**Table 1.** Some physicochemical properties of water

| | Liquid at 25°C | | Ice at −10°C | |
| --- | --- | --- | --- | --- |
| | $H_2O$ | $D_2O$ | $H_2O$ | $D_2O$ |
| 1. Dissociation constant | | | | |
|    $K = C_{H^+}C_{OH^-}$ | $1{\cdot}008 \times 10^{-14}$ | $1{\cdot}35 \times 10^{-15}$ | $2{\cdot}40 \times 10^{-23}$ | |
|    $K = C_{H^+}C_{OH^-}/C_{H_2O}$ | $1{\cdot}821 \times 10^{-16}$ | | | |
| 2. Enthalpy of dissociation | $13{\cdot}5$ kcal/mole | $14{\cdot}3$ | $25$–$30$ | |
| 3. Entropy of dissociation | $-18{\cdot}7$ cal/mole deg | $-20{\cdot}4$ | | |
| 4. Ionic concentration $C_{H^+} = C_{OH^-}$ | $1{\cdot}004 \times 10^{-7}$ mole/l | $0{\cdot}44 \times 10^{-7}$ | $1{\cdot}4 \times 10^{-10}$ | $0{\cdot}32 \times 10^{-10}$ |
| 5. Molecular concentration $C_{H_2O}$ | $55{\cdot}34$ mole/l | | $50{\cdot}9$ | |
| 6. Specific conductance $L_{exp.}$ | $5{\cdot}7 \times 10^{-8}$ ohm$^{-1}$cm$^{-1}$ ($1{\cdot}2 \times 10^{-8}$ at 0°) | $2{\cdot}5 \times 10^{-3}$ | $10^{-9}(0{\cdot}3 \times 10^{-8}$ at 0°) | |
| 7. Mobility of $H^+$  $u_{H^+}$ | $3{\cdot}62 \times 10^{-3}$ cm$^2$/volt sec | | $0{\cdot}075$ | $0{\cdot}0115$ |
| 8. Activation energy of mobility | | | $0 \pm 0{\cdot}7$ kcal/mole | $\sim 1$ |
| 9. Mobility of $OH^-$  $u_{OH^-}$ | $1{\cdot}98 \times 10^{-3}$cm$^2$/volt sec | | $5 \times 10^{-2}$ | |
| 10. Rate constant of dissociation $k_D$ | $2{\cdot}5 \times 10^{-5}$sec$^{-1}$ | | $3{\cdot}2 \times 10^{-9}$ | $2{\cdot}7 \times 10^{-11}$ |
| 11. Apparent activation energy of dissociation | $15{\cdot}5$–$16{\cdot}5$ kcal/mole | | $22{\cdot}5$ | $25$ |

| | | | | |
|---|---|---|---|---|
| 12. Rate constant of neutralization, $k_R$ (recombination of $H^+$ and $OH^+$) | $1\cdot4\times10^{11}$ l/mole sec | | $8\cdot6\times10^{12}$ | $1\cdot3\times10^{12}$ |
| 13. Apparent activation energy of recombination | $2\cdot3$ kcal/mole | | | |
| 14. 'Effective' dielectric constant | $78\cdot54$ | $78\cdot25$ | $3\cdot2$–$95$ | |
| 15. Diffusion coefficient | $2\cdot4\times10^{-5}$ cm$^2$/sec | | $8\times10^{-11}(-1\cdot8^{\circ}\mathrm{C})$ | |
| 16. Proton transfer $H_2O + H_3O^+$    Rate    Activation energy | $1\cdot1\times10^{-10}$ l/mole sec $2\cdot6$ kcal/mole | | | |
| 17. Proton transfer: $H_2O + OH^-$    Rate    Activation energy | $6\times10^9$ l/mole sec $5$ kcal/mole | | | |
| 18. Dipole moment | $1\cdot84\times10^{-18}$ e.s.u. (vapour) $2\cdot5\times10^{-18}$ e.s.u. (liquid) | | $3\times10^{-18}$ (ice) | |

Collected from the following sources:
J. H. Wang, C. V. Robinson and I. S. Edelman (1953) *J. Amer. Chem. Soc.*, **75**, 466
H. Gränicher (1958) *Proc. Roy. Soc. Ser. A*, **247**, 453
M. Eigen and L. DeMaeyer (1958) *Proc. Roy. Soc. Ser. A*, **247**, 505
L. Onsager and M. Dupuis (1960) *Rendiconti S.I.F.*, **10**, 294
A. Loewenstein and A. Szöke (1962) *J. Amer. Chem. Soc.*, **84**, 1151
M. Eigen (1963) *Naturwissenschaften*, **50**, 426
W. Luck (1964) *Fortschr. Chem. Forsch.*, **4**, 653
E. Wicke (1966) *Angew. Chem., Inter. Ed. Engl.*, **5**, 107
A. K. Covington, R. A. Robinson and R. G. Bates (1966) *J. Phys. Chem.*, **70**, 3820

the molecular structure and behavior of pure water first, and then of some of its solutions.

## II. WATER

### A. Some Molecular Properties

Shown in table 1 are a number of physicochemical properties of water which are relevant to an understanding of its molecular structure and behavior. Particularly interesting are some comparisons of (liquid) water with ice.

In the very first entry (table 1) we note that the self-dissociation constant of $H_2O$ is much smaller in ice than in water. (This decrease in dissociation is clearly determined by the much higher enthalpy of dissociation). Another measure of the same effect is shown in the fourth row: the concentration of $H^+$ ions in ice is a thousand-fold smaller than in water. With this information at hand it is astonishing to find that the specific electrical conductances of liquid and solid are approximately the same order of magnitude (at $0\,°C$) since the concentration of ionic carriers is obviously so much smaller in ice. Bearing in mind, however, that the electrical conductance, $L$, is determined by the *product* of the *mobility*, $u$, and the *concentration*, $C$, of the ionic carriers,

$$L = 96,500 \; C(u_{H^+} + u_{OH^-}) \tag{1}$$

then it must follow that in ice $H^+$ and $OH^-$ have unusually high mobilities. That this is indeed the case is indicated in table 1, rows 7 and 9. Thus we are led to an alternative bewildering conclusion that the ionic mobility of protons in the rigid matrix of ice is greater than in the mobile liquid state of water. This also means that the structure of ice must permit some unusual behavior.

The rate of dissociation of water molecules is given in row 10 of table 1. The rate constant $k_D$

$$H_2O \underset{k_R}{\overset{k_D}{\rightleftharpoons}} H^+ + OH^- \tag{2}$$

of about $10^{-5}$ sec, implies an apparent mean lifetime for an individual water molecule of about ten hours. Nevertheless 55 M water (the concentration of pure liquid) produces $10^{-7}$M $H^+$ and $OH^-$ in less than $10^{-4}$ sec.

The rate constant, $k_R$, for recombination of $H^+$ and $OH^-$ (row 12 in table 1) is an extremely large number, $1 \cdot 4 \times 10^{11}$. It implies that 1 M $H^+$ and 1 M $OH^-$ would recombine in $10^{-11}$ sec. This is essentially the velocity at which these ions can diffuse toward each other; the recombination is thus a diffusion-controlled reaction, i.e. every time an $H^+$ and an $OH^-$ ion collide they combine to give $H_2O$.

It is thus astonishing to find that $k_R$ in ice, which is $8.6 \times 10^{12}$, is even faster than in water, since the diffusion of ions in a solid solvent would be expected to be much reduced, and certainly not increased, in comparison to that observed in a fluid solvent.

The dipole moment of water, approximately 2–3 Debye units (row 18, table 1), is not exceptional in comparison with other molecules of similar structure. It is slightly higher than that of ethyl ether (1·1 Debye), and lower than that of acetone (2·9 Debye). On the other hand, the molar refraction of liquid water is exceptionally small (table 2). Since this macroscopic quantity is related to the polarizibility, which is directly involved in dispersion interactions, the exceptionally low value of the molar refraction of water may be responsible for some of its unique characteristics as a solvent for biological macromolecules (Hanlon, 1966).

These are a few of the unique physicochemical characteristics of water. It behoves us then to try to interpret some of these properties and related aspects of the behavior of water in terms of its molecular structure. This will now be attempted.

**Table 2.** Molar refractions of liquids

| Substance | Molar refraction (cm³) |
|---|---|
| Water | 3·7 |
| Methanol | 8·2 |
| Formamide | 10·6 |
| Ethylene glycol | 14·5 |
| Dimethylsulfoxide | 20·1 |
| N,N-Dimethylformamide | 20·0 |
| Urea | 13·7 |
| Dimethyl urea | 23·3 |
| Tetramethyl urea | 33·2 |

From S. Hanlon (1966)

## B. Structure of Water

Let us note first some of the dimensions of individual water molecules.

Geometric interatomic parameters of H and O atoms of $H_2O$ differ slightly in the gaseous state from those in the solid and liquid. Infrared spectra of water vapor (Darling and Dennison, 1940) show that the O—H interatomic distance is 0·96 Å and the H—O—H bond angle is 104°31′ (figure 1). These values are slightly different in ice (Peterson and Levy,

1957); the O—H bond distance is 0·99 Å and the bond angle is 109°5′
(figure 2) which is almost exactly equal to a tetrahedral bond angle. The
radius of a water molecule in ice is 1·38 Å (Bernal and Fowler, 1933;
Lonsdale, 1958).

Quantum mechanical considerations (Pople, 1953) show that the two
pairs of free electrons of the oxygen atom in the water molecule lie in orbitals
that, together with the two orbitals connecting bonded hydrogen atoms, are

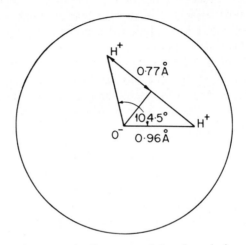

**Figure 1.** O—H interatomic distance and bond angle in gaseous water.

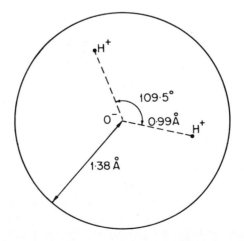

**Figure 2.** O—H interatomic distance and bond angle in ice.

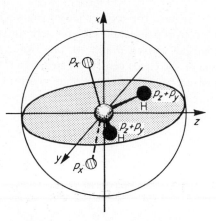

**Figure 3.** Charge distribution and tetrahedral orientation of orbitals around oxygen atom in water.

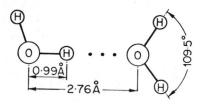

**Figure 4.** Hydrogen bond between two water molecules. Distances and bond angle are for ice.

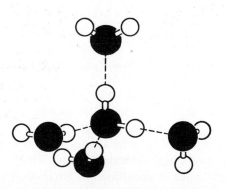

**Figure 5.** Tetrahedral arrangement of water molecules around central water molecule.

arranged tetrahedrally around the oxygen atom (figure 3). This disposition of H atoms and free-electron pairs makes it uniquely suited for 4-coordinated interactions in condensed phases. Each of the two protons can form a hydrogen bond to one of the (two) centres of negative charge on the oxygen atom (figure 4). Since each oxygen can act as acceptor for two hydrogen bonds and as a donor for two, each water molecule is surrounded tetrahedrally* by four other water molecules (figure 5).

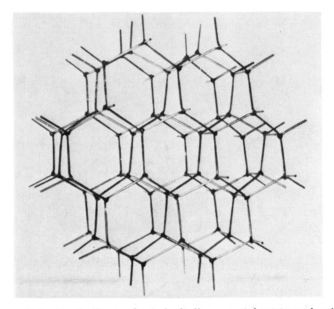

**Figure 6.** Lattice built up of tetrahedrally arranged water molecules.

In ice this tetrahedral orientation of neighboring water molecules extends in all directions, as indicated in figure 6. The normal form of ice is the tridymite-type structure (figure 7) which contains both chair and boat forms of the hexagonal rings formed by the oxygen atoms. A cubic, diamond-type form of ice also exists at low temperatures (about $-100\,°C$) in which the oxygen atoms of the hexagons are arranged only in a chair conformation (figure 7). Even rarer forms of ice exist at very low temperatures (amorphous, below $-140\,°C$) or at very high pressures (Kamb, 1965).

Hydrogen-bonded water molecules in a tetrahedral array also exist in water (Danford and Levy, 1962). The $O—H\cdots O$ bond distance in the liquid

*For a description of some non-tetrahedral bonds see Wicke (1966).

**Figure 7.** Portion of the tridymite lattice of ice showing both chain and boat forms of hexagons.

is somewhat longer than in the solid and varies with temperature (Morgan and Warren, 1938). At room temperature this distance is 2·86 Å (van Eck and coworkers, 1958) (cf. figure 4). Furthermore in the liquid state, defects, distortions and deformations exist which disrupt the highly regular array of the solid state. The structure in any local region in the liquid also fluctuates with time.

No overwhelmingly convincing model for the structure of liquid water has yet appeared. This in part reflects the unsettled status of the theory of the liquid state for even very simple substances. For water there is even a very wide divergence of views about the fraction of O—H's not hydrogen-bonded in the liquid. At one extreme are the uniformistic models (Bernal and Fowler, 1933; Lennard-Jones and Pople, 1951), which view the liquid as differing from the solid only in being a 'looser' lattice with distortions and deformations of the hydrogen-bonds and with the four bonds of every water molecule in the liquid bending and twisting independently. In contrast, since the time of Roentgen, water has also been viewed as a mixture of ice-like clusters, of variable size, and single water molecules (Eucken, 1948, 1949; Frank and Wen, 1957; Samoilov, 1965). Many different estimates have been made of the fraction of non-bonded O—H's in liquid water, some based largely on theoretical considerations and others on physical measurements (for a summary see Luck, 1964). These give results varying from about 70 % down to near zero. In part the discrepancies can be attributed to the different time scales inherent in the method of examining the water, and in part to differences in meaning of 'non-bonded' versus 'distorted' or 'deformed' O—H· · ·O groups. In any event, there is general agreement that the structure of liquid water is *qualitatively* similar to that in ice in having tetrahedral arrays of molecules, but long-range order is not present in the liquid state.

## C. Mobility in Water

The long range order in water and ice provides an unusual matrix for transport of protons and of related mobile defects. As a first step in this direction let us examine proton structure in water.

It was originally suggested by Collie and Tickle (1899) at the end of the last century that $H^+$ in water should be represented as $H_3O^+$ (zero-order hydration). More recently (Eigen, 1964) there has been much evidence from mass spectrometry, infrared spectra and crystal structures of certain hydrates that solvation of $H_3O^+$ leads to the entity $H_9O_4^+$ (primary hydration) (figure 8). Within this $H_9O_4^+$ entity the $H^+$ jumps freely. This entity, further-

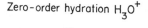

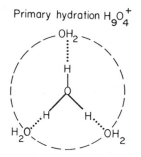

**Figure 8.** Zero-order and primary hydration entities of $H^+$ in water. The $H_9O_4^+$ is further bonded to external water molecules which constitute secondary hydration.

more, is also hydrogen-bonded to additional water molecules all around its periphery, which constitute secondary hydration. Thus it now becomes possible for a proton to 'drift' by concerted jumps from one water molecule to another throughout the bulk water or ice, as shown in figure 9. The effective mobility of a proton moving by the Grotthus mechanism (von

Surplus
proton

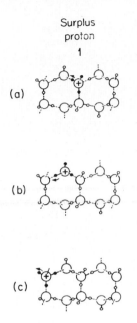

**Figure 9.** Proton transport by concerted jumps along a chain of hydrogen bonded water molecules. The H atoms on oxygen which are $H_3O^+$ momentarily, are shown as ● circles, the other H atoms as ○ and oxygen atoms as O.

Grotthus, 1806; Hückel, 1928) can thus be much greater than would be possible by direct bodily transport of the first proton from its initial position. In ice, in which the $H_2O$ lattice is essentially unperturbed over long distances, the rate of $H^+$ transport is determined by the tunneling of the proton through the potential barrier from one $H_3O^+$ to the adjoining recipient $H_2O$; this takes place in about $1 \times 10^{-13}$ sec. (Eigen, 1964). On the other hand, in liquid water the transport of $H^+$ is limited by the time necessary for structural rearrangement of water molecules to produce a primary hydration structure at the new position of $H^+$ and to form a new secondary structure. These transformations in the liquid are distinctly slower (Eigen, 1964) than the tunneling time of an $H^+$ between adjacent bonded water molecules. Nevertheless, the mobility of an $H^+$ ion is still ten times that of a $Li^+$ in liquid water.

The long-range lattice structure in water and ice also provides a matrix for the transport of other defects. An $OH^-$ ion is in essence a proton deficiency. A schematic representation of a mobile proton deficiency is shown in figure 10.

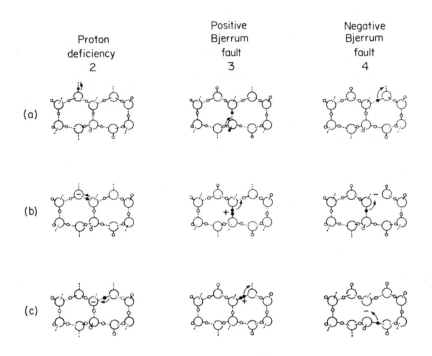

**Figure 10.** Diffusion of defects in ice and water by concerted jumps within the lattice of $H_2O$ molecules.

Besides surplus protons or proton deficiencies, ice or water may have *orientation* defects within the lattice of water molecules. As is shown in figure 10, a so-called 'positive Bjerrum fault (D-defect)', is generated if an O—H shifts around to produce —OH HO— with two protons between two O atoms. The position of this D-defect may be transferred to a neighboring site by a consecutive rotation of the neighboring water molecule (see 3a→ b→c). An alternative orientation defect, shown in part 4 of figure 10 arises from a rotation of a water molecule which leaves an empty site between two O· · ·O atoms. Again the location of this 'negative Bjerrum fault (L-defect)' may be shifted readily to a neighboring site (see 4a→b→c).

In view of the evidence of these concerted pathways for a mobile proton in water, we can now understand some of the unusual molecular properties described above. For example, the unexpectedly high electrical conductance of ice, despite the lower concentration of $H^+$ as compared to water, clearly is a reflection of the high mobility of protons due to the more rapid transport by the concerted chain transfer in the unperturbed lattice of the solid. The same process accounts for the very high rate constant, $k_R$, for recombination of $H^+$ with $OH^-$ in ice. This occurs faster than the diffusion-controlled reaction because the concerted transport offers a pathway totally independent of normal diffusion processes and many orders of magnitude faster. Concerted transport also occurs in liquid water as is evident, for example, from the fact that the electrical mobility of $H^+$ in aqueous solutions is about ten times faster than that of any other cation. Similarly the mobility of $OH^-$ in water is several times greater than that of other anions.

In these molecular properties water is unique as a solvent. It provides a particularly fit environment for $H^+$ and $OH^-$ transfer so necessary in cellular reactions.

## III. HYDRATION

Water is a very versatile solvent: it will dissolve ionic, polar and apolar solutes. Each of these solutes interacts with the solvent to produce a hydrated structure which is the net result of the interplay of the specific character of the solute and the nature of the solvent.

### A. Ionic Hydration

The physical properties of aqueous ionic solutions provide overwhelming evidence for strong hydration of the solute ions. Perhaps most striking are the very large hydration energies (heats of hydration) (Hunt, 1963) for carrying an ion from the gas phase into the liquid (table 3). Even for a singly charged ion the heat of hydration is near 100 kcal/mole; by contrast, for a small apolar molecule such as Ar or $CH_4$, the heat of hydration (from the gaseous phase) is only 3 kcal/mole (Frank and Evans, 1945). It is thus obvious that the primary hydration layer immediately surrounding an ion must be one that interacts very strongly with the electrostatic field of the charged central species (figure 11). Similarly the large negative entropies of hydration (table 3) are consistent with a hydration layer of irrotationally bound water.

Even without a detailed molecular picture the magnitudes of the heats and entropies of hydration of ions can be quantitatively accounted for on electrostatic grounds (Hunt, 1963) particularly if the ionic radii in solution are assumed to be near those of the gaseous ion (Stokes, 1964) rather than the ionic radii in crystals.

Despite the high ion–dipole interaction energy between an ion and its first hydration layer, the exchange lifetimes of bound water molecules are very short on an anthropomorphic time-scale. Such lifetimes vary over an enormous range (table 4). For single-charged cations, such as $Na^+$ or $K^+$, the lifetime is $10^{-9}$ sec, while for $Al^{3+}$ it is about 7·5 sec (Fiat and Connick, 1968). Even a lifetime of $10^{-9}$ sec, however, is very long compared to that of

　　　　　　　　　　　　　　　　　　　　　*Membranes and Ion Transport*

**Tabel 3.** Ionic properties

| Ion | Mobility cm²/volt sec — In water at 25° | In ice | Heats of hydration (kcal/mole) | Entropies of hydration (eu) | Ionic radius in crystal (Å) | van der Waals radius in gaseous ion (Å) |
|---|---|---|---|---|---|---|
| $H^+$ | $36.2 \times 10^{-4}$ | $0.1$ | $-258$ | $-26$ | | |
| $Li^+$ | $4.0$ | $10^{-8}$ | $-121$ | $-28$ | $0.60$ | |
| $Na^+$ | $5.3$ | | $-95$ | $-21$ | $0.95$ | $1.35$ |
| $K^+$ | $7.6$ | | $-75$ | $-12$ | $1.33$ | $1.67$ |
| $NH_4^+$ | $7.6$ | | | | | |
| $Ca^{2+}$ | | | $-337$ | $-50$ | $0.99$ | $1.48$ |
| $Fe^{2+}$ | | | $-456$ | $-65$ | $0.76$ | |
| $Cu^{2+}$ | | | $-500$ | $-62$ | $0.72$ | |
| $Al^{3+}$ | | | $-1109$ | $-111$ | $0.50$ | $1.05$ |
| $OH^-$ | $19.8 \times 10^{-4}$ | | $-111$ $(-135)$ | | | |
| $F^-$ | $5$ | $10^{-8}$ | $-121$ | $-36$ | $1.36$ | $1.91$ |
| $Cl^-$ | $7.9$ | | $-90$ | $-24$ | $1.81$ | $2.25$ |
| $Br^-$ | $8.1$ | | $-82$ | $-20$ | $1.95$ | $2.30$ |
| $I^-$ | $8.0$ | | $-71$ | $-14$ | $2.16$ | $2.55$ |
| $S^{2-}$ | | | $-330$ | $-31$ | $1.84$ | |

Collected from the following sources:
M. Eigen and L. De Maeyer (1958) *Proc. Roy. Soc. Ser. A*, **247**, 505
R. H. Stokes (1964) *J. Amer. Chem. Soc.*, **86**, 979
J. P. Hunt (1963) *Metal Ions in Aqueous Solution*, W. A. Benjamin, New York

a water molecule in a cluster of liquid water molecules (Frank and Wen, 1957) and much longer than the time of a molecular vibration (table 4).

At greater distances than the first hydration layer, the electric field of the central ion is evidently strong enough to interfere with the proper orientation of water molecules into their normal unperturbed structural arrangements. Yet the ion–dipole interaction at such long distances is not strong enough to induce an irrotationally bound conformation similar to the first hydration layer. The net result then is a breakdown in water structure (Frank and Wen, 1957), i.e. a relatively disorganized region is created (figure 11) in which water molecules are more mobile or fluid. Ions that produce these effects are often called 'structure-breakers'. On the other hand, some ions, particularly small ones, such as $F^-$ or multiple-charged ions may actually produce a net increase in 'rigidity' of surrounding water molecules; these are called 'structure-formers'. With this picture one can rationalize many of the properties of ionic solutions (Frank and Wen, 1957; Wicke, 1966) such as heat capacities, volumes, viscosities, dielectric dispersion, infrared spectra, microwave spectra, nuclear magnetic resonance, etc. Direct examination of

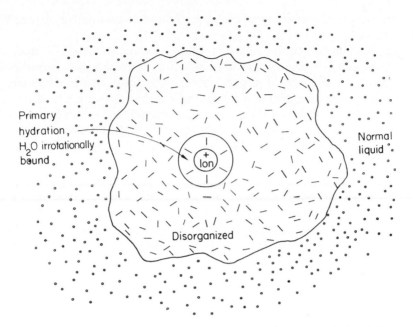

**Figure 11.** Model of structure of water in neighbourhood of simple ion.

the orientation of water molecules in ionic solutions is still not feasible but this may become possible through extension of X-ray diffraction studies of concentrated solutions (Brady and Krause, 1957).

## B. Hydration of Apolar Solutes

### 1. *Crystalline Clathrates*

Some insight into the structure of aqueous solutions of apolar molecules can probably be obtained from an examination of the crystalline hydrates formed by a large number of these molecules. Apolar polyhedral hydrates have actually been known since the experiments of Davy (1811) and of Faraday (1823) with chlorine hydrate. Rare gas hydrates were discovered by Villard (1896) and by deForcrand (1903, 1925) and many others have been prepared during the last hundred years (Pickering, 1893; Hagan, 1962). The molecular structures of these substances, however, have only been worked out recently by X-ray diffraction methods (Claussen, 1951; Pauling and Marsh, 1952; Stackelberg and Müller, 1954; McMullan and Jeffrey, 1959).

On the basis of crystal structure studies, one can classify all of the apolar hydrates into three groups, as shown in table 5. This classification is based in part on the stoichiometry of the hydrates and in part on the structure of the unit cell.

The compounds that are found in Class I hydrates (table 5) include the rare gases such as argon and krypton, inorganic molecules such as phosphine or $SO_2$ and a number of organic molecules, some hydrocarbon-like in nature but some also containing non-hydrocarbon constituents. The compounds in Class I fit into a lattice made up of a unit cell containing forty-six water molecules and eight large cavities. Six of these cavities are tetra-kaidecahedra with an internal diameter of 5·9 Å. Two cavities are dodeca-hedra of 5·2 Å diameter. If only the larger cavities are filled by an inert gas molecule, M, then the formula for the unit cell is $6M \cdot 46H_2O$ or stoichiometrically $M(H_2O)_{7.67}$. If all the cavities are occupied by M the corresponding formulae are $8M \cdot 46H_2O$ and $M(H_2O)_{5.75}$, respectively. In all examples of Class I hydrates the faces of the polyhedra are either pentagons or hexagons.

**Table 4.** Lifetimes of bound water

|  | Sec |
| --- | --- |
| Molecular vibration | $10^{-13}$ |
| Proton jump in H-bond | $10^{-13}$ |
| Water in liquid cluster | $10^{-11}$ |
| Water molecule in first hydration layer | |
| $\quad$ $Br^-$ | $0.7 \times 10^{-11}$ |
| $\quad$ $Li^+$, $Na^+$, $K^+$, $Rb^+$, $Cs^+$ | $10^{-9}$ |
| $\quad$ $Ca^{2+}$, $Sr^{2+}$, $Ba^{2+}$ | $10^{-9}$ |
| $\quad$ $Mn^{2+}$ | $10^{-8}$ |
| $\quad$ $Cu^{2+}$ | $10^{-7}$ |
| $\quad$ $Co^{2+}$, $Mg^{2+}$ | $10^{-6}$ |
| $\quad$ $Ni^{2+}$ | $10^{-5}$ |
| $\quad$ $Fe^{3+}$ | $10^{-5}$ |
| $\quad$ $Be^{2+}$ | $10^{-2}$ |
| $\quad$ $Al^{3+}$ | $7.5 \times 10^{0}$ |
| $\quad$ $Cr^{3+}$ | $1.5 \times 10^{5}$ |

Collected from the following sources:
R. E. Connick and R. E. Poulson (1959) *J. Chem. Phys.*, **30**, 759
H. H. Baldwin and H. Taube (1960) *J. Chem. Phys.*, **33**, 206
M. Eigen (1963) *Pure Appl. Chem.*, **6**, 97
E. Wicke (1966) *Angew. Chem. Intern. Ed. Engl.*, **5**, 107
D. Fiat and R. E. Connick (1958) *J. Amer. Chem. Soc.*, **90**, 608

**Table 5.** Polyhedral hydrates

| Class I | | Class II | Class III |
|---|---|---|---|
| *Guest Molecules* | | | |
| Ar | $CH_4$ | $CHCl_3$ | $(n\text{-}C_4H_9)_4N^+F^-$ |
| Kr | $C_2H_2$ | $CH_3CHCl_2$ | $(n\text{-}C_4H_9)_4N^+{-}O_2CC_6H_5$ |
| $Cl_2$ | $C_2H_4$ | $(CH_3)_2O$ | $[(n\text{-}C_4H_9)_4N^+]_2WO_4{}^{2-}$ |
| $H_2S$ | $C_2H_6$ | $C_3H_8$ | $(i\text{-}C_5H_{11})_4N^+F^-$ |
| $PH_3$ | $CH_3Cl$ | $(CH_3)_3CH$ | $(n\text{-}C_4H_9)_3S^+F^-$ |
| $SO_2$ | $CH_3SH$ | $C_3H_7Br$ | $(n\text{-}C_4H_9)_4P^+Cl^-$ |
| $C_2H_5NH_2$ | $CH_3CHF_2$ | $(CH_3)_2CO$ | $(CH_3)_3N$ $(CH_2)_6N_4$ |
| $(CH_3)_2NH$ | $CH_2CHF$ | | $n\text{-}C_3H_7NH_2$ $(CH_3)_4N^+OH^-$ |
| $\overline{CH_2CH_2O}$ | | | $i\text{-}C_3H_7NH_2$ |
| | | $C_6H_6$ | $(C_2H_5)_2NH$ |
| $\overline{CH_2CH_2CH_2O}$ | | $\overline{CH_2CH_2CH_2CH_2O}$ | $(CH_3)_3C\text{-}NH_2$ |
| | | cyclo-$C_6H_{12}$ | $C_4H_9OH$ |
| | | $\overline{CH_2CH_2OCH_2CH_2O}$ | |
| *Stoichiometry* | | | |
| $M \cdot 5\frac{3}{4} H_2O$ | | $M \cdot 17H_2O$ | $M \cdot (5\text{--}40)H_2O$ |
| $M \cdot 7\frac{2}{3} H_2O$ | | $M \cdot M'_2 \cdot 17H_2O$ | |
| *Unit Cell* | | | |
| $46H_2O$ | | $136H_2O$ | Variable number of $H_2O$ |
| Polyhedra | | Polyhedra | Polyhedra |
| 2 $H_{12}$(5 Å) | | 16 $H_{12}$(5 Å) | $H_8$, $H_{12}$, $H_{14}$, $H_{15}$ |
| 6 $H_{14}$(6 Å) | | 8 $H_{16}$(7 Å) | $H_{16}$, $H_{17}$, $H_{18}$, $H_{60}$, etc. |
| Faces | | Faces | Faces |
| Pentagons | | Pentagons | Quadrilaterals |
| Hexagons | | Hexagons | Hexagons |
| | | | Heptagons |
| $M_2 \cdot M_6 \cdot 46H_2O$ | | $M_8 \cdot M'_{16} \cdot 136H_2O$ | |

($H_n$ symbolizes a polyhedron with $n$ faces)

Class II compounds (table 5) have a unit cell containing one hundred and thirty-six water molecules and twenty-four cavities. Eight of these are hexakaidecahedra with a diameter of 6·9 Å and sixteen are dodecahedral cavities of 4·8 Å diameter. The inert M molecules occupy the eight large holes. Thus the formula for the unit cell is $8M \cdot 136H_2O$ or $M(H_2O)_{17}$ stoichiometrically. In this group of hydrates, mixed hydrates are also common in which a small molecule such as $H_2S$ fills the sixteen smaller cavities in addition to the eight M's in the unit cell. These mixed hydrates thus have the stoichiometric formula $M(H_2S)_2(H_2O)_{17}$.

In both Class I and Class II hydrates each guest molecule M is completely enclosed in one polyhedron with twelve to sixteen faces. Diagrams of such polyhedra are shown in the upper two rows of figure 12.

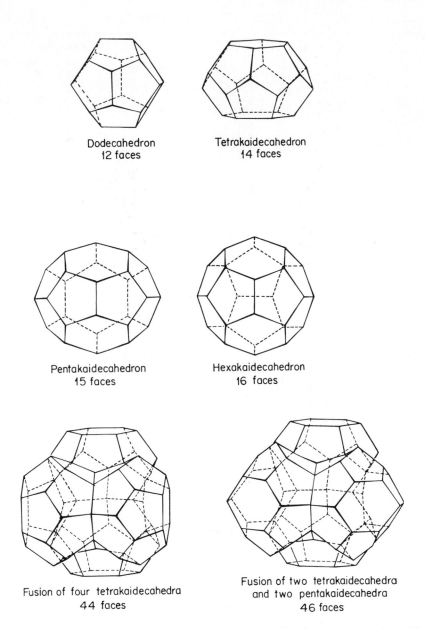

Dodecahedron
12 faces

Tetrakaidecahedron
14 faces

Pentakaidecahedron
15 faces

Hexakaidecahedron
16 faces

Fusion of four tetrakaidecahedra
44 faces

Fusion of two tetrakaidecahedra
and two pentakaidecahedra
46 faces

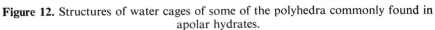

**Figure 12.** Structures of water cages of some of the polyhedra commonly found in apolar hydrates.

The Class III polyhedra, however, are more complicated. They demonstrate that it is not necessary to squeeze the guest molecule into a simple small polyhedron in order to have a stable hydrate. A cage may be obtained composed of portions of a number of the simpler polyhedra. Some examples are shown in the larger polyhedra in the bottom row of figure 12. Thus there is no single unit cell for all the substances in Class III. Each of them can form a specific type of ice-like cage around it.

All of the polyhedra shown in figure 12 are made up of faces containing pentagons or hexagons. Figure 13 (Beurskens and Jeffrey, 1964) shows an even further departure from the relative symmetry of the polyhedra of figure 12. The cages in figure 13 are much more unsymmetrical and complex and are bounded by quadrilateral rhombi as well as by pentagonal and hexagonal faces. The cavities are both large and irregular. The hydrate whose structure is shown in figure 13 has the formula $(n\text{-}C_4H_9)_3 S^+F^- \cdot 23$

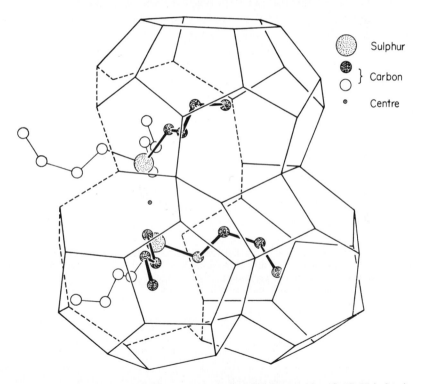

**Figure 13.** Projection of clathrate cavity containing a pair of $(C_4H_9)_3S^+$ ions. Only one-half of the cavity is shown, that into which the shaded carbon chains project. The second half surrounds the open-circle carbon chains. The complete polyhedron is constituted of sixty faces.

$H_2O$. Only half of the complete water cage is shown in figure 13. The remaining half would be a mirror image of that shown. Within a full cage there are two guest molecules.

$$
\begin{array}{lr}
C_4H_9\diagdown & \diagup C_4H_9 \\
C_4H_9-S^+\cdot\cdot\cdot^+S-C_4H_9 \\
C_4H_9\diagup & \diagdown C_4H_9
\end{array}
$$

Particularly striking is the fact that the two triplets of apolar butyl groups of the two molecules face away from each other; they tend to be far apart and imbedded in (three pairs of subsidiary) water cages rather than adjacent to each other and excluding intervening water molecules. The six incomplete polyhedra enclosing each of the alkyl side chains are symmetrically arranged around the centre of symmetry at the midpoint of the $S^+\cdot\cdot\cdot S^+$ line. There are three types of these incomplete compartments with different sizes and shapes made up of the following polygonal faces: (i) nine pentagons and two hexagons; (ii) seven pentagons, two hexagons and one quadrilateral; and (iii) eight pentagons and one hexagon.

In summary, figures 12 and 13 show that a very wide variation is possible with respect to size and shape of the water cages enclosing an apolar moiety. It is thus apparent that water is a remarkably versatile substance in relation to hydrate formation, being capable of forming a large variety of cage-like structures to accommodate itself to a whole gamut of apolar groups.*

## 2. *Apolar Solutes in Aqueous Solution*

Dilute aqueous solutions of non-polar solutes exhibit a number of properties which indicate stabilization of the water structure of the solvent. Table 6, for example, compares the solubilization heat in liquid water with the heat of formation of the clathrate hydrate for a number of non-polar

*Frequently comparisons are made between $D_2O$ and $H_2O$ in their interactions with various solutes. It is of interest to note in passing that $D_2O$ forms polyhedral hydrates with higher melting points than corresponding $H_2O$ hydrates (Worley, Stryker and Klotz, unpublished experiments). This is indicated in table below.

| Guest Molecule | Melting Point (°C) | |
|---|---|---|
| | $H_2O$ Hydrate | $D_2O$ Hydrate |
| Tetrabutylammonium chloride | 14·8 | 16·5 |
| Tetrabutylammonium bromide | 13·1 | 15·0 |
| Tetrabutylammonium acetate | 15·6 | 17·4 |
| Tetrabutylammonium tungstate | 15·2 | 16·2 |

**Table 6.** Solubility and hydrates at 0 °C

| Gas | $-\Delta H°$ cal/mole Gas dissolving in liquid water→solution | $-\Delta H°$ cal/mole Gas dissolving in ice→hydrate |
|---|---|---|
| $CH_4$ | 4621 | 4553 |
| $H_2S$ | 5140 | 5550 |
| $C_2H_6$ | 5560 | 5850 |
| $Cl_2$ | 6180 | 6500 |
| $SO_2$ | 7420 | 7700 |
| $CH_3Br$ | 7390 | 8100 |
| $Br_2$ | 8750 | 8300 |
| $C_3H_8$ | 6860 | 6300 |
| $CH_3I$ | 8500 | 7300 |
| $C_2H_5Cl$ | 8400 | 8700 |

From D. N. Glew (1962) *J. Phys. Chem.*, **66**, 605

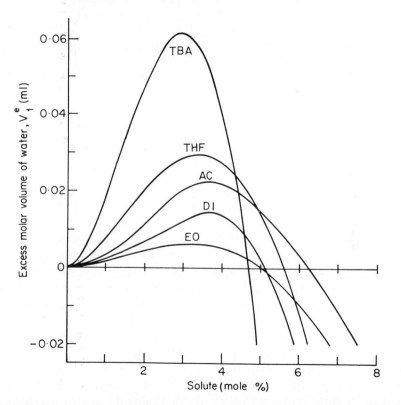

**Figure 14.** Excess molar volumes of water in dilute aqueous solutions of ethylene oxide (EO) at 10°, dioxane (DI) at 25°, acetone (AC) at 0°, tetrahydrofuran (THF) at 10° and tertiary butyl alcohol (TBA) at 15°. (From D. N. Glew and coworkers, 1968)

E

solutes (Glew, 1962). For methane, for example, when one mole of gas is dissolved in liquid water $\Delta H°$ is $-4621$ calories; correspondingly the $\Delta H°$ when one mole of gas is incorporated into solid ice to form the hydrate, is $-4553$ calories. It is apparent, therefore, that the energy of interaction of methane molecules in aqueous solution with the water in the hydration shell is essentially the same as that of methane with the water molecules in the solid polyhedra of the hydrate clathrate. This equivalence implies that the environment in both cases, in solution and in the solid hydrate, is similar with respect to coordination and spatial orientation of the water molecules in the hydration shells. Likewise the correspondence in the values of the two columns of $\Delta H°$ listed in table 6 for the other molecules implies that for these more polarizable solutes also the structure of the hydration shell in the liquid is similar to that in the solid hydrate. Thus these solutes stabilize and orient water molecules in their hydration shells in aqueous solution in the same way that water molecules do in the hydrogen-bonded solid hydrate lattices.

More recently, other physical–chemical properties of aqueous solutions of one clathrate former, ethylene oxide, have been studied in great detail and with high precision (Glew and coworkers, 1967). These studies have revealed a variety of anomalous properties of water in these solutions, with maximum deviations occurring at about 4 mole per cent concentration of solute. One particularly striking example is shown in figure 14 which presents the excess partial molar volumes of water in these solutions of non-polar molecules. The positive values of these molar volumes indicate that water occupies a larger volume in these solutions than in the pure liquid. The expansion of the water arises from the orientation of the molecules in the hydration shell in a manner similar to that found in solid hydrates. Solute molecules occupy interstitial spaces in this hydration shell.

Thermodynamic properties such as the activity coefficients of water in these solutions also pass through extrema (Glew and coworkers, 1967), again indicating maximum thermodynamic stabilization of water at a concentration of ethylene oxide at about 4 mole per cent (figure 15). Nuclear magnetic resonance of the water protons shows shifts toward lower magnetic fields (figure 15) as solutes such as ethylene oxide, acetone, dioxane or tetrahydrofuran are added to the solvent (Glew and coworkers, 1967). Such a shift indicates a strengthening of the water structure; again the maximum effect is observed near 4 mole per cent of solute concentration.

Interestingly enough, the concentration at which maxima or minima in physical–chemical properties occur, about 4 mole per cent, corresponds to approximately 24–28 water molecules per molecule of solute, a coordination number which is very near that found in the solid clathrate hydrate. It thus seems that these dilute aqueous solutions of a non-polar molecule consist of

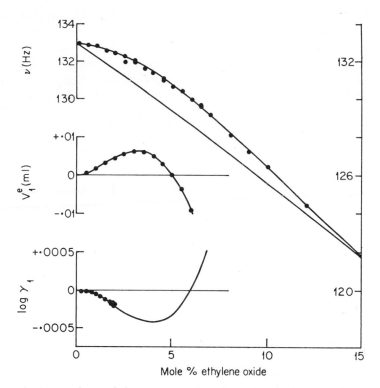

**Figure 15.** Comparison of the concentration dependence of n.m.r. chemical shift of water protons (top), excess partial molar volume of water (middle), and water activity coefficients (bottom) in aqueous solutions of ethylene oxide. (From D. N. Glew and coworkers, 1967).

solute molecules surrounded by a solvation shell consisting of water in ordered structures similar to those found in the solid hydrates.

Heat capacity measurements in solutions of apolar solutes also point to ordered water solvation shells. For example, the apparent heat capacity of alkyl ammonium salts in dilute aqueous solution increases by 15–20 cal/mole deg for each $CH_2$ group introduced (Ackermann and Schreiner, 1958). The heat capacity increment per $CH_2$ to be expected from well-established additivity rules is 5–10 cal/mole deg. Similar increments have been found for fatty acids in aqueous solution. Thus a residual of 10 cal remains to be accounted for. If we assume that apolar groups, such as $C_4H_9$, attached to a solute molecule stabilize ice-like solvent clusters, then in effect the extent and lifetime of these ordered regions will be increased. Consequently an extra apparent molal heat capacity must appear in solutions of such apolar solutes since the additional 'ice-likeness' must be melted when the temperature of the solution is raised.

An interesting confirmation of this structural view is available from dielectric relation studies of aqueous solutions. The relaxation times of ice-like clusters may be identified with those observed in dielectric constant measurements at high frequencies (wavelengths of 1–10 cm). Water molecules in the pure liquid have a relaxation time (Hasted and Roderick, 1958) of $0.8 \times 10^{-11}$ sec at 25°. Solutes which stabilize the ice-like clusters of water ought to lengthen the half-life and thus the relaxation time. Such indeed is the case for apolar solutes in water (table 7); in these solutions the relaxation wavelength (which is proportional to relaxation time) is increased, by an amount $\delta\lambda_S$, over that of pure solvent (Haggis and coworkers, 1952). The structure of water molecules in the hydration region is thus more rigid than in the pure liquid. Contrariwise, negative values are observed for $\delta\lambda_S$ for aqueous solutions of salts (table 7), which is consistent with the structure-breaking effects of ionic solutes on water. In these salt solutions water molecules can reorient themselves more readily than in the pure liquid.

**Table 7.** Change of relaxation wave length in aqueous solutions

| Solute | $\delta\lambda_S \times 10^2$(cm) | Solute | $\delta\lambda_S \times 10^2$(cm) |
|--------|--------|--------|--------|
| LiCl | —10 | $(CH_3)_4NBr$ | + 5 |
| NaF | — 5 | $(CH_3)_4NI$ | +10 |
| NaCl | —20 | $(C_2H_5)_4NCl$ | +30 |
| NaBr | —20 | $n\text{-}C_3H_7OH$ | +35 |
| NaI | —25 | $t\text{-}C_4H_9OH$ | +40 |
| KF | —10 | $C_2H_5NH_2$ | +20 |
| KCl | —20 | $(C_2H_5)_3NHCl$ | +20 |
|  |  | $C_2H_5COOH$ | +20 |

From G. H. Haggis and coworkers, *J. Chem. Phys.*, **20**, 1452 (1952)

### 3. *Macromolecules*

It is thus apparent that water is a remarkably versatile substance in regard to hydration, being capable of forming a large variety of cage-like structures to accommodate itself to a whole gamut of apolar groups. We have taken the view (Klotz, 1960), therefore, that similar structures should form around apolar groups projecting from a macromolecule, particularly since the local concentration of such groups in the polymer region is high and cooperative interactions should be possible. Particularly in the case of proteins, there is a remarkable parallelism between its apolar side chains and corresponding small molecules known to form clathrate-like hydrates (table 8). If small molecules such as $CH_3—CH(CH_3)_2$ and $CH_3—SH$ form stable water cages, it seems reasonable that leucyl [$—CH_2—CH(CH_3)_2$] and cysteinyl ($—CH_2—SH$) side-chains of proteins should be able to do so also. In the protein, in addition, the local surface concentration of side chains may be high and one would expect a cooperative effect of adjacent apolar ones to induce a stabilized arrangement of water. Furthermore, with increasing indications that amine hydrates are also cage-like structures (McMullan and coworkers, 1967; Jeffrey and McMullan, 1967) it seems possible that corresponding side chains in proteins would have water shells that could interdigitate with those of neighboring apolar groups.

On this basis one can understand the normal effects of temperature in denaturing proteins. The effect of temperature could be attributed to the disorganization or melting of the 'hydrotactoids'. Likewise the denaturing effect of urea could be due to its ability to disorganize the hydration lattice and thereby decrease the stabilization of the protein originating from the cooperative interactions in the hydration shell.

Evidence of ordered water around macromolecules is being provided by several physical and chemical probes. Particularly interesting are recent

**Table 8.** Comparison of hydrate formers with amino acid residues of proteins

| Some molecules forming crystal hydrates | Some amino acid side chains |
|---|---|
| $CH_4$ | $—CH_3$ (Ala) |
| $CH_2\begin{smallmatrix}\diagup CH_3 \\ \diagdown CH_3\end{smallmatrix}$ | $—CH\begin{smallmatrix}\diagup CH_3 \\ \diagdown CH_3\end{smallmatrix}$ (Val) |
| $CH_3—CH\begin{smallmatrix}\diagup CH_3 \\ \diagdown CH_3\end{smallmatrix}$ | $—CH_2—CH\begin{smallmatrix}\diagup CH_3 \\ \diagdown CH_3\end{smallmatrix}$ (Leu) |
| $CH_3—SH$ | $—CH_3—SH$ (Cys) |
| $CH_3—S—CH_3$ | $—CH_2—CH_2—S—CH_3$ (Met) |
| $C_6H_6$ (benzene) | $—CH_2—C_6H_5$ (Phe) |

measurements of ultrasonic attenuation in aqueous solutions of polymers (Hammes and Schimmel, 1967). From the frequency-dependence of the sound absorption one can calculate a relaxation time (of the order of $10^{-9}$ sec) which reflects the structure of the solvent and responds to perturbations of it. In the presence of a polymer, the (single) relaxation time, $\tau$, is increased, the increment depending on the molecular weight of the polymer until a plateau value is reached (figure 16). This increase in $\tau$ probably reflects the

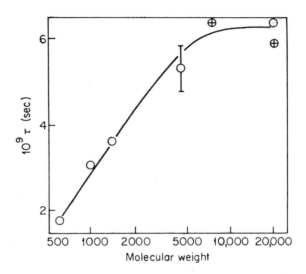

**Figure 16.** Relaxation time, $\tau$, against polymer molecular weight in aqueous solutions of polyethylene glycol. The error bracket corresponds to a 10% error in the relaxation time. (From Hammes and Schimmel, 1967)

cooperative formation and breakdown of the structured hydration layer around the polymer. Interestingly enough, upon addition of increasing quantities of urea, $\tau$ shows no change until a concentration near 2 M is reached, whereupon $\tau$ drops steeply as the molarity of urea rises and then levels off once again (figure 17). Such behavior clearly suggests that urea is perturbing the solvent structure.

Many investigations on bound water in biological macromolecules have been carried out with nuclear magnetic resonance. In principle this is a very attractive method since the proton resonance should definitely reflect any change in the state of the water. The expected broadenings in signal have been observed repeatedly. (For references to some of these studies see Klotz, 1965). Nevertheless, it is still not certain whether or not the increased

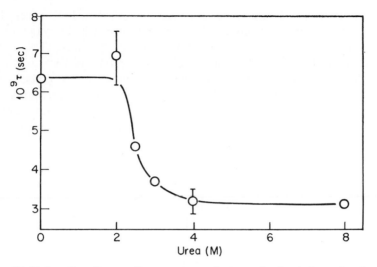

**Figure 17.** Relaxation time, $\tau$, for aqueous solutions of polyethylene glycol against molality of urea. The error bracket corresponds to a 10% error in the relaxation time. (From Hammes and Schimmel, 1967)

line widths reflect only the immobilization of water molecules. Far more detailed studies have been made for collagen (Berendsen, 1962) in which both the magnitude and angular dependence of the splitting of proton resonance have been measured. These indicate that a large part of the water associated with the collagen is reoriented, restrictively, so that chains of water molecules are formed parallel to the fibre axis with a repeat distance of 28·6 Å. Such studies, however, cannot be carried over to aqueous solution.

Chemical probes also indicate that water at the surface of a polymer is different from bulk water. For example, the exchange of labile H atoms by a D atom from $D_2O$ in CONH groups is very much slower when these groups are attached to a polymeric matrix as compared to corresponding small molecules (Klotz, 1968). The transfer of H (or D) to, and from, $D_2O$ is modified in the neighborhood of the polymer, and it seems very likely that the self-dissociation constant, $K_w$, of the polymer hydration water is lowered substantially.

At somewhat higher levels of molecular size than polymers there have been repeated suggestions that water at the surface of particles, from clays to viruses (Forslind, 1954; Bernal, 1959; Johnson and coworkers, 1966) is modified, in comparison to bulk water. Special properties of these systems can be rationalized if it is assumed that the hydration water is more rigid or ice-like (although not necessarily as hexagonal ice) than bulk water. These ideas have received much support from the description by Deryaguin (1966) of experiments with water condensed in capillary tubes. The unusual behavior of such water (indications of a density of 1·4 and a viscosity fifteen times that of ordinary water) has been ascribed to a new phase, named 'ortho water'; the formation of this from ordinary water is facilitated at an interface. Should these conclusions be verified by X-ray and other physico-chemical studies, it would then be reasonable to assume that similar phases of water appear at biological interfaces also.

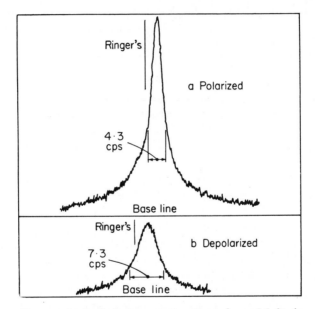

**Figure 18.** a: Proton magnetic resonance spectrum for a $Mn^{2+}$-doped polarized nerve trunk (17°C). b: Proton magnetic resonance spectrum for a $Mn^{2+}$-doped depolarized nerve trunk (13·5°C).

### 4. *Tissues and Membranes*

It is very difficult to determine the state of water in tissues or membranes because there are no simple probes. Nevertheless, some attempts have been made using nuclear magnetic resonance methods. Nuclear magnetic reson-ance measurements in biological matrices, indicate immobilization of water molecules. Muscle, for example, shows a well-defined, broadened proton resonance whose line width is narrowed during tetanic isometric contraction (Bratton and coworkers, 1965). The line widths and relaxation times indicate that some of the intracellular water molecules are relatively immobile and that the restrictions on rotational freedom decrease during contraction.

A recent study has also examined the state of water in frog nerve fibres by an alternate nuclear resonance technique (Fritz and Swift, 1967). It has proved possible to discriminate between extracellular and intracellular water by adding a paramagnetic ion, $Mn^{2+}$, to the Ringer's solution and it was found that the line-width of the resonance signal (see figure 18) changed on depolarization of the nerve fibre. This is a definite indication that very marked changes have occurred in the state of the intracellular water as a result of depolarization.

## IV. CONCLUSION

The results obtained with a variety of experimental probes have shown that macromolecular conformation is determined not only by its *intrinsic* primary chemical structure (*autoplastic effects*) but also by its *extrinsic* interactions with water molecules of the aqueous solvent (*alloplastic effects*) (Klotz, 1966). Conformational changes of biological macromolecules are at the heart of regulatory control in a host of biosynthetic processes. It seems very likely that similar phenomena play a governing role in membrane behaviour. Experimentation in this field is only beginning, and if there is to be any adequate interpretation of both new and old observations, then a better understanding of the structure of liquid water and of macromolecular hydration must first be achieved.

### REFERENCES

Ackermann, T. and F. Schreiner (1958) *Z. Electrochem. Ber. Bunsenges. Physik. Chem.*, **62**, 1143
Berendsen, H. J. C. (1962) *J. Chem. Phys.*, **36**, 3297
Bernal, J. D. (1959) In D. Hadzi (Ed.), *Hydrogen Bonding*, Pergamon Press, New York pp. 7–22
Bernal, J. D. and R. H. Fowler (1933) *J. Chem. Phys.*, **1**, 515
Beurskens, P. T. and G. A. Jeffrey (1964) *J. Chem. Phys.*, **40**, 2800
Brady, G. W. and J. T. Krause (1957) *J. Chem. Phys.*, **27**, 304
Bratton, C. B., A. L. Hopkins and J. W. Weinberg (1965) *Science*, **147**, 738
Claussen, W. F. (1951) *J. Chem. Phys.*, **19**, 1425
Collie, J. N. and T. Tickle (1899) *J. Chem. Soc.*, **75**, 710
Danford, M. D. and H. A. Levy (1962) *J. Amer. Chem. Soc.*, **84**, 3965
Darling, B. T. and D. M. Dennison (1940) *Phys. Rev.*, **57**, 128
Davy, H. (1811) *Ann. Chim. (Paris)*, **78**, 298; **79**, 5 (see page 26).
de Forcrand, R. (1903) *Ann. Chim. (Paris)*, **28**, 384
de Forcrand, R. (1925) *Compt. Rend.*, **181**, 15
Deryaguin, B. V. (1966) *Discussions Faraday Soc.*, **42**, 109
Eigen, M. (1964) *Angew. Chem. Intern. Ed. Engl.*, **3**, 1

Eucken, A. (1948) *Z. Elektrochem.*, **52**, 255
Eucken, A. (1949) *Z. Elektrochem.*, **53**, 102
Faraday, M. (1823) *Quart. J. Sci.*, **15**, 71
Feil, D. and G. A. Jeffrey (1961) *J. Chem. Phys.*, **35**, 1863
Fiat, D. and R. E. Connick (1968) *J. Amer. Chem. Soc.*, **90**, 608
Forslind, E. (1954) In V. G. N. Harrison (Ed.), *Proceedings Second International Congress on Rheology*, Academic Press, New York, p. 50
Frank, H. S. and M. W. Evans (1945) *J. Chem. Phys.*, **13**, 507
Frank, H. S. and W. Y. Wen (1957) *Discussions Faraday Soc.*, **24**, 133
Fritz, O. G., Jr. and T. J. Swift (1967) *Biophys. J.*, **7**, 675
Glew, D. N. (1962) *J. Phys. Chem.*, **66**, 605
Glew, D. N., H. D. Mak and N. S. Rath (1967) *Can. J. Chem.*, **45**, 3059
Glew, D. N., H. D. Mak and N. S. Rath (1968) In A. K. Covington and P. Jones (Eds.), *Hydrogen-Bonded Solvent Systems*, Taylor and Francis, London p. 195
Hagan, M. (1962) *Clathrate Inclusion Compounds*, Reinhold Publishing Co., New York
Haggis, G. H., J. B. Hasted and T. J. Buchanan (1952) *J. Chem. Phys.*, **20**, 1452
Hammes, G. G. and P. R. Schimmel (1967) *J. Amer. Chem. Soc.*, **89**, 442
Hanlon, S. (1966) *Biochem. Biophys. Res. Commun.*, **23**, 861
Hasted, J. B. and G. W. Roderick (1958) *J. Chem. Phys.*, **29**, 17
Henderson, L. J. (1913) *The Fitness of the Environment*, Macmillan, New York
Hückel, E. (1928) *Z. Elektrochem.*, **34**, 546
Hunt, J. P. (1963) *Metal Ions in Aqueous Solution*, Benjamin, New York, chaps. 2 and 3
Jeffrey, G. A. and R. K. McMullan (1967) *Progr. Inorg. Chem.*, **8**, 43
Johnson, G. A., S. M. A. Lecchini, E. G. Smith, J. Clifford and B. A. Pethica (1966) *Discussions Faraday Soc.*, **42**, 120
Kamb, B. (1965) *Science*, **150**, 205
Klotz, I. M. (1960) *Brookhaven Symp. Biol.*, **13**, 25
Klotz, I. M. (1965) *Federation Proc.*, **24**, No. 2, Part 3, Supplement 15, 524
Klotz, I. M. (1966) *Arch. Biochem. Biophys.*, **116**, 92
Klotz, I. M. (1968) *J. Colloid Interface Sci.*, **27**, 804
Lennard-Jones, J. and J. A. Pople (1951) *Proc. Roy. Soc.*, **A205**, 155
Lonsdale, K. (1958) *Proc. Roy. Soc. (London)*, Ser. *A*, **247**, 424
Luck, W. (1964) *Fortschr. Chem. Forsch.*, **4**, 653
McMullan, R. K. and G. A. Jeffrey (1959) *J. Chem. Phys.*, **31**, 1231
McMullan, R. K., T. H. Jordan and G. A. Jeffrey (1967) *J. Chem. Phys.*, **47**, 1218
Morgan, J. and B. E. Warren (1938) *J. Chem. Phys.*, **6**, 666
Pauling, L. and R. E. Marsh (1952) *Proc. Natl. Acad. Sci. U.S.*, **38**, 112
Peterson, S. W. and H. A. Levy (1957) *Acta Cryst.*, **10**, 70
Pickering, S. U. (1893) *Trans. Chem. Soc.*, **63**, I, 141
Pople, J. A. (1953) *J. Chem. Phys.*, **21**, 2234
Samoilov, O. Y, (1965) *Hydration of Ions*, Translated by D. J. G. Ives, Consultants Bureau, New York
Stackelberg, M. V. and H. R. Müller (1954) *Z. Elektrochem.*, **58** 25
Stokes, R. H. (1964) *J. Amer. Chem. Soc.*, **86**, 979
van Panthaleon van Eck, C. L. H. Mendel and J. Fahrenfort (1958) *Proc. Roy. Soc. (London) Ser. A*, **247**, 472
Villard, P. (1896) *Compt. Rend.*, **123**, 377
von Grotthus, C. J. T. (1806) *Ann. chim. (Paris)*, **58**, 54
Wicke, E. (1966) *Angew. Chem. Intern. Ed. Engl.*, **5**, No. 1, 106

# II
# Theoretical Aspects of Transport Phenomena

# CHAPTER 5

# Irreversible thermodynamics as applied to biological systems

## Russell Paterson

*Department of Chemistry, The University, Glasgow, Scotland*

## I. INTRODUCTION

Thermodynamic arguments were first applied to an irreversible process by Lord Kelvin (1854, 1857) in a study of thermoelectric phenomena. The phenomenon of coupling, so characteristic of irreversible thermodynamics, was first established in his formulation of the relationship between the thermoelectric potential of a thermocouple and its Peltier heat. This relationship was finally justified by Onsager (1931a,b), who showed that it was a general consequence of the principle of microscopic reversibility, which is based upon the invariance of the microscopic equations of motion under time reversal.

The development of a coordinated theory of irreversible processes requires a general statement of the second law of thermodynamics and as early as 1850, Clausius introduced the concept of uncompensated heat as a measure of irreversibility in closed systems. De Donder (1927) was able to relate the irreversibility in a chemical reaction to the chemical affinity but the subject began to take its present form when Onsager (1931a,b) established his celebrated reciprocal relations. Meixner (1941, 1942, 1943), Casimir (1945) and Prigogine (1947) expanded and generalized upon this work and set up the present phenomenological theory of irreversible processes as we now know it. The general formulation of the subject is most adequately treated in a number of excellent treatises; De Groot (1951), De Groot and Mazur (1962), Fitts (1962), and Prigogine (1967), while the specific requirements of the biophysicist are covered in a subsequent text by Katchalsky and Curran (1965).

Application of thermodynamic principles to biology has, in the past, presented a number of major conceptual difficulties, which have been mainly concerned with the problem of entropy or disorder. The most general conclusion of thermodynamics is that the energy of the world is constant, while the entropy of the world tends to a maximum and life, which is characterized by ordered structures, is in continuing opposition to the general drive towards increasing disorder. A living organism is, furthermore, an open system continuously exchanging both energy and matter with its surroundings and so it is to irreversible thermodynamics that we must look in our effort to understand the energy balances necessary for life. The broad concepts, including the stability of the steady state and the concept of minimum entropy production may be applied to the entire organism in its mature or adult state but a more definitive explanation of function must be sought at the cellular level, since it is the selectivity and function of cellular and other biological membranes, which ultimately control life.

Application of irreversible thermodynamics to the flows and forces across biological and other membranes gives purely phenomenological results which are far removed from microscopic or molecular models, which represent the ultimate goal of the biologist. This is not quite the disadvantage that, at first sight it would appear, since the infrastructure of biological membranes is far from being resolved to a degree where such an analysis would be possible. The biological membrane is the fundamental unit in any living system, regulating and directing the flow of nutrients, water and other solutes from the blood to the extracellular fluids and tissue cells. In the reverse direction these same membranes must remove metabolites and poisons, which would otherwise accumulate. On quite another plane, electrical nerve communication is a question of ionic transport processes across nerve membranes and the rhythmical frequence modulated 'spikes' represent one of the most complex of all membrane processes. The work of Teorell (1959, 1962, 1966) has shown that this type of phenomenon may be demonstrated in the laboratory with a simple model system and from a theoretical point of view, it may be shown to be a property of systems far from equilibrium and without the range of the simple theory of irreversible thermodynamics. Prigogine (1967) and Bak (1963) have made considerable progress in the extension of theory into this, the 'non-linear' region of irreversible thermodynamics.

In certain cases permeant species are required to be concentrated and are transported against their concentration gradients. This process is entirely contrary to our simple concepts of diffusion and cannot be explained without the application of irreversible thermodynamics and the concept of coupling. Any thermodynamic approach is concerned primarily with energetics and never with the individual kinetic pathways and, in the former example, it becomes obvious that such 'active' transport must involve an input of energy from another source—usually from a chemical or metabolic reaction. The formalism of irreversible thermodynamics immediately provides us with the concept of coupling between the process of diffusion and chemical reaction and at the same time lays down that such an effect will require an asymmetrical membrane. These facts create a simple picture of the basic requirements of the process and in doing so lay down a guide which must condition the thinking of the experimental or molecular biologist.

## II. THERMODYNAMIC PRINCIPLES

### A. General Considerations

Thermodynamics may be applied to any well-defined geometrical volume of macroscopic dimensions. A thermodynamic system is that part of the observable world which has been isolated for a thermodynamic study and may be classified according to the permitted exchanges of energy and matter through their boundaries. Three such systems are possible. The first is an isolated or adiabatic system, which may exchange neither energy nor matter. Such a system may, however, perform mechanical work and so interact with the external world. The second is the closed system which may exchange energy but not matter and the third, the open system, which constantly exchanges both energy and matter with its invironment. Open systems are therefore of most interest in biological applications, since a living organism or indeed any section of that organism must exchange constantly both energy and matter with its environment.

Classical thermodynamics is concerned mainly with closed systems and it is the most striking feature of non-equilibrium thermodynamics that we may now generalize the methods of classical thermodynamics to include open systems with the specific inclusion of time as a variable.

Thermodynamic quantities may be divided into two classes, extensive and intensive properties. Extensive variables of a system include mass, volume, enthalpy, free energy and entropy, and it is clear that such variables are additive. Intensive variables, characterized by pressure or temperature and concentration, take well-defined values at each point of the system. The state of a thermodynamic system may be specified by a number of such variables and an extensive variable, $Z$, may be defined as a function of state if it may be expressed as a function of pressure, volume and the number of moles present.

$$dZ = \left(\frac{\partial Z}{\partial P}\right)_{V,n_i} dP + \left(\frac{\partial Z}{\partial V}\right)_{P,n_i} dV + \sum_{i=1}^{k} \left(\frac{dZ}{\partial n_i}\right)_{P,V,n_j} dn_i \qquad (1)$$

An important property of the state function is that in a transition it will change by an amount which is independent of the method or pathway.

The second law of thermodynamics postulates the existence of entropy, $S$. Entropy is an extensive property and is a function of state. In the general case the change in entropy $dS$ of a system may be split into two parts such that

$$dS = d_e S + d_i S \qquad (2)$$

where $d_e S$ is the change in entropy due to a reversible interaction with the exterior and $d_i S$ due to production of entropy within the system itself. $d_i S$ is zero when the system undergoes reversible changes but is positive if such changes occur irreversibly. Entropy is not conserved in an irreversible process. It is the aim of irreversible thermodynamics to relate this entropy production to exchanges of matter and energy in a system close to equilibrium. In an isolated system (since neither mass nor energy may be exchanged) $dS = d_i S \geqslant 0$.

If two such systems I and II are enclosed to form an isolated system and in each some irreversible process may take place, then

$$dS = dS^I + dS^{II}$$

$$= d_e S^I + d_i S^I + d_e S^{II} + d_i S^{II}$$

Since the reversibly exchanged entropies are conserved

$$d_eS^I = -d_eS^{II}$$

so that

$$dS = d_iS^I + d_iS^{II} \geqslant 0$$

We must further postulate that $d_iS^I \geqslant 0$ and $d_iS^{II} \geqslant 0$

a physical situation, in which $d_iS^I < 0$ and $d_iS^{II} > 0$

is excluded. In every macroscopic region of the system the entropy production is positive.

It is the non-conservation of entropy which produces the inequalities of the classical thermodynamic treatment when applied to irreversible systems and it is the primary achievement of irreversible thermodynamics that we may replace these inequalities by quantitative relationships between the flows and thermodynamic forces involved.

The concept of thermodynamic coupling will be discussed at greater length in subsequent sections. It is, however, useful to introduce the concept of chemical affinity and the coupling of chemical reactions at this point.

Consider a closed system containing $n$ components in which there occurs a chemical reaction. For this purpose, the degree of advancement of a chemical reaction is defined as $d\xi$, such that $d\xi/dt = J_{Ch}$ the velocity of the chemical reaction, (De Donder, 1936).

For a general chemical reaction

$$v_A A + v_B B + \ldots \rightarrow v_C C + v_D D + \ldots$$

$v_i$ is the stoichiometric coefficient in the chemical equation and is taken as negative for reactants and positive for products, while $dn_i$ is the increment in moles of $i$ due to reaction. Although the system is closed, the work term will include chemical work, $-\mu_i dn_i$ in addition to the normal pressure-volume term, $PdV$. The expression for entropy change may then be written

$$dS = \left(\frac{dU + PdV}{T}\right) - \sum_{i=1}^{n} \frac{\mu_i dn_i}{T} \tag{3}$$

Since the reversible entropy change with the exterior is given by the second law

the second law $d_eS = \left(\dfrac{dU + PdV}{T}\right) = \dfrac{dQ_{rev}}{T}$

so that

$$d_iS = -\sum_{i=1}^{n} \frac{\mu_i dn_i}{T} = -\sum_{i=1}^{n} \frac{v_i \mu_i d\xi}{T} \geqslant 0$$

and so $\qquad$ $d_i S = \dfrac{A d\xi}{T}$ $\qquad$ (4)

The affinity $A$ of the chemical reaction is defined as $A = -\sum\limits_{i=1}^{n} v_i d\mu_i$ and is the driving force of a chemical reaction process. Since the degree of advancement and $T$ are both positive, it is therefore necessary that $A > O$ for a spontaneous chemical reaction. This is a condition which is satisfied if the chemical potentials of the reactants are greater than the products.

The rate of production of entropy per unit time is given by

$$\frac{d_i S}{dt} = \frac{A}{T} \frac{d\xi}{dt} = \frac{A}{T} J_{Ch} \geqslant O \qquad (5)$$

If, however, two simultaneous reactions, 1 and 2 occur

$$\frac{d_i S}{dt} = \frac{1}{T} (A_1 J_{Ch1} + A_2 J_{Ch2}) \geqslant O \qquad (6)$$

It is then possible that $A_1 J_{Ch1}$ is positive and $A_2 J_{Ch2}$ is negative provided their sum is positive. The reactions are then coupled, reaction (1) proceeding spontaneously and driving the second against its natural or spontaneous tendency.

Such thermodynamic coupling is an essential feature of living systems and one such example was recognized many years ago by Van Rysselberghe (1936, 1937). In the normal liver, the synthesis of urea is coupled with the oxidation of glucose and other substances and may be represented schematically by

(1) Coupled reaction $2NH_3 + CO_2 \rightarrow (NH_2)_2 CO + H_2O$; affinity, $A_2$
(2) Driving reaction $\frac{1}{6} C_6 H_{12} O_6 + O_2 \rightarrow CO_2 + H_2O$; affinity, $A_1$

Under standard conditions the affinities are

$$A_2 = -11 \text{ kcal}$$
$$A_1 = +115 \text{ kcal}$$

By itself, reaction (2) would not proceed spontaneously in the direction shown.

The rate of the coupled reaction is constrained by the rate of the driving reaction according to the inequality $J_{Ch2} \leqslant J_{Ch1} A_1 / A_2$

In any thermodynamic system undergoing irreversible change, for example, diffusion, there will invariably be gradients of temperature, concentration, pressure and energy. The question must arise as to the validity of using the formalism of classical thermodynamics to describe the interrelation of thermodynamic variables. It is therefore an important postulate that, for a system in which irreversible processes occur, all thermodynamic functions of state exist for each element of the system. Local temperatures, chemical potentials and so on may therefore be defined by the same formalism as is given by equilibrium systems. Sometimes known as Postulate I, this amounts to the assumption of local equilibrium in limited but macroscopic elements of a system. Experimentally it is found to be sound, provided the gradients of the thermodynamic functions are small in the experimental system. For large gradients the assumption of local equilibrium fails. The validity of the postulate has been investigated by a statistical thermodynamic treatment (Prigogine, 1967).

## B. Special Considerations

### 1. *The Rate of Entropy Production and the Dissipation Function*

A particularly important quantity in this treatment is the local rate of production of entropy, $\sigma$, per elemental volume, $dV$. It is such that the overall rate of entropy production is

$$d_iS/dt = \int_v \sigma dV \tag{7}$$

The rate of production of entropy may be identified as a sum of terms each being a product of a flux and a conjugate thermodynamic force, related to the non-uniformity of the system. For example, in a system in mechanical equilibrium, the following relation holds:

$$\sigma = J_q \cdot \text{grad}\frac{1}{T} + \sum_{i=1}^{n} J_i \cdot \text{grad}\left(\frac{-\mu_i}{T}\right) + J_{\text{Ch}}\frac{A}{T} \tag{8}$$

where $J_q$, $J_i$ and $J_{\text{Ch}}$ are the fluxes of heat, matter and chemical reaction each multiplied by its conjugate driving force. The choice of flows and forces is to some extent arbitrary provided the products have the dimensions of entropy production and that for a given system, the sum of the products of fluxes and flows remains unaltered by the transformation.

One such transformation gives,

$$\sigma = \frac{J_S}{T} \cdot \text{grad}\left(-T\right) + \sum_{i=1}^{n} \frac{J_i}{T} \cdot \text{grad}\left(-\mu_i\right) + J_{\text{Ch}}\frac{A}{T}$$

(where $J_S$ is the entropy flow) (9)

It is often convenient to use $T\sigma = \Phi$, the dissipation function. This has the dimension of energy/unit time and is a measure of the rate of dissipation of free energy so that

$$\Phi = T\sigma = J_S \, \cdot \, \mathrm{grad}(-T) + \sum_{i=1}^{n} J_i \, \cdot \, \mathrm{grad}\,(-\mu_i) + J_{\mathrm{Ch}}A \qquad (10)$$

In the presence of ions the function may be written in terms of electrochemical potentials, $\tilde{\mu}_i$ and affinities, $\tilde{A}$, which include the energy contributions $z_i F\psi$, where $\psi$ is the local internal electrical potential of the phase.

$$\tilde{\mu}_i = \mu_i + z_i F\psi$$

$$\tilde{A} = -\sum_{i=1}^{n} v_i \tilde{\mu}_i = A - \sum_{i=1}^{n} v_i z_i F\psi \qquad (11)$$

In any practical application, it is necessary to define frames of reference for the flows, commonly taken relative to the local centre of mass. It may be shown (Prigogine, 1947) that provided the system is in a state of mechanical equilibrium, the formalism above, presented in terms of absolute flows, will also apply to all other reference frames.

In general therefore

$$\Phi = \sum_{i=1}^{n} J_i X_i \geqslant 0 \qquad (12)$$

It is to be noted that the summation does not require that each term individually be positive but only that their sum be positive. Should any individual term $J_i X_i < 0$, then the $i$th flow will proceed in a direction opposite to its thermodynamic force reducing the entropy production of the system. Such a term may be used within a cell system for organizational purposes. This is an important concept stemming the tide of entropic evolution yet in no way contradicting the second law. The synthesis of, for example, ribonucleic acids, involving the organization of constituent amino acids, is an entropy-reducing process and depends upon a degree of coupling with metabolic (chemical) reactions which provide sufficient entropy production for the overall dissipation function to be positive, i.e.

$$\Phi = \underset{\text{(negative)}}{J_{\text{synth}} \, \cdot \, A_{\text{synth}}} + \underset{\text{(positive)}}{J_{\text{metabolism}} X_{\text{metabolism}}} \geqslant 0 \qquad (13)$$

A similar situation must occur in the process of active transport where a metabolic reaction may be coupled to a diffusion flow against its chemical potential (or less accurately concentration) gradient. Biological systems consist of regions of matter enclosed within membranous walls. The structural and chemical nature of the membrane selectively controlling the exchange of energy and matter with the surroundings.

It is therefore to these membranes that we must look for a deeper understanding of biological function and, in particular, for the mechanism, organization, and coupling phenomena such as active transport.

## 2. *Phenomenological Equations and the Onsager Reciprocal Relations*

Flows and forces are related by the phenomenological equations which may be written.

$$J_i = \sum_{K=1}^{n} L_{iK} X_K \quad i = 1, 2, 3, \ldots n \tag{14}$$

These equations convey the possibility of cross-effects between various irreversible phenomena since each flux may, in principle, be a linear function of all the thermodynamic forces which characterize the entropy source. The direct coefficients $L_{ii}$ appear on the diagonal of the matrix of forces and in the absence of other interactions, $(L_{iK} = O)$, would provide the linear proportionality between flow and conjugate force described by the simple laws of Fick and Ohm for diffusion and electric current. The flow, $J_i$, will be affected by the other forces on the system provided the cross coefficients $L_{iK}$ are non-zero. The linear relations hold only in systems which are in a steady state and close to thermodynamic equilibrium. More rapid processes involve non-linear relationships.

An alternative representation of phenomenological coefficients has the form

$$X_i = \sum_{i=1}^{K} R_{iK} J_K \tag{15}$$

This expression is obtained by matrix inversion of equation (14) and expresses the forces as linear functions of the flows. The $R_{iK}$ coefficients have the dimension of force per unit flow and are generalized frictional coefficients.

To characterize a system of $n$ forces and flows, $n^2$ such coefficients would be required if it were not for certain symmetry requirements first shown by Onsager (1931a,b). Onsager's reciprocal relations state that

$$L_{iK} = L_{Ki} \quad (i \neq K)$$
$$R_{iK} = R_{Ki} \quad (i \neq K) \tag{16}$$

They hold provided the phenomenological equations are in accord with the dissipation function. As a result, the coefficients required to characterize the system are reduced from $n^2$ to $\frac{1}{2}n(n+1)$. Although originally proved by statistical mechanical models, these relations have been shown to hold experimentally well beyond the range of theoretical justification (Miller, 1960).

In all cases the cross coefficients are constrained by the inequalities,

$$L_{ii} L_{jj} \geqslant (L_{ij})^2 \text{ and } R_{ii} R_{jj} \geqslant (R_{ij})^2 \tag{17}$$

Direct coefficients must be positive but cross coefficients may be either positive or negative. Having established the degree of coupling between flows and non-conjugate forces, certain important limitations are sum-marized in the Curie (or Curie–Prigogine) principle, which states that flows and forces whose tensorial character differ by an odd integer cannot couple in an isotropic system. More specifically, it states that there can be no coupling between vectorial and scalar systems, since the flows and conjugate forces of heat and matter are vectors (tensorial order, 1) and chemical affinity is scalar (tensorial order, 0), there can be no coupling between these in an isotropic system. This limitation simplifies the phenomenological equations but leads to the postulation of anisotropy in systems displaying active transport; a process which, by definition, involves the use of chemical reaction to maintain diffusive flows against concentration gradients.

It should be emphasized that the phenomenological coefficients are functions of the parameters of state but not of their gradients. They have no direct dependence upon the flows or forces involved.

### 3. *The Steady State and Minimum Entropy Production*

The steady state is characterized by the requirement that all state variables be constant with time. Since entropy is produced from the irreversibility of the process, the rate of entropy production per unit volume must equal the flow of entropy from that volume and so

$$\sigma = \text{div } J_S \tag{18}$$

where $J_S$ is the flow of entropy.

Since the remainder of the state variables are conservative, their flows must be constant in the steady state, or more specifically

$$\text{div } J_i = \text{O} \tag{19}$$

The stationary state has a position in irreversible thermodynamics quite analogous to that held by equilibrium in the classical theory. If a chemical system is in a state of non-equilibrium and is left to age without external restriction, all flows and forces will eventually vanish and a state of equilibrium will be attained in time. If, however, certain forces are maintained constant and others allowed to adjust freely the system will, in time, obtain a steady state, in which the flows of species conjugate to the free forces will disappear. These forces will adjust to a steady, but not necessarily zero value, corresponding to zero conjugate flow and a time invariant system will result. Those forces held constant are sometimes stated to be at static head—a term derived from the analogy with the conditions for electroosmotic pressure in a plug or capillary system. Since

$$\sigma = \sum_i^n J_i X_i \text{ and } J_i = \sum_{K=1}^n L_{iK} X_K \, i = 1, 2, \ldots n$$

it may be shown quite generally that

$$\frac{\partial \sigma}{\partial X_i} = 2J_i \tag{20}$$

If $X_i$ is not constrained and so is allowed to adjust freely, then $J_i$ is zero in the steady state. The differential is zero and so corresponds to an extremum, which, since entropy production is positive definite, must correspond to a minimum. This principle of minimum entropy production is valid only when the phenomenological coefficients are constants and independent of the forces, the Onsager relations hold and the phenomenological equations are linear. It may be further demonstrated that the stationary state has the property of stability, since fluctuations from it generate forces, which bring the system back to its original state. Any reductions in the number of restraints will cause a reduction in the entropy production of the stationary state so that the ultimate stationary state of equilibrium is achieved in an aged system under zero restraints and in which the entropy production is zero.

Prigogine and Wiame (1946) have applied this principle to the biological host. Maturity corresponds to a steady state which is compatible with the external restraints such as availability of food, temperature and other variables of the environment. The stationary state is also the most economical since it corresponds to a minimal loss of free energy from the host.

## 4. *Application to Membrane Systems*

In biological systems, transport processes involve exchange of matter or energy through membranes. These membranes are complex in structure and selective in their function. As yet structural investigations lack sufficient detail to allow an intimate analysis of the transport mechanism in terms of specific diffusion paths or interactions within the membrane phase. Experimental investigations are restricted to an examination of the phenomena manifest in the boundary phases.

Flows and chemical potential in these phases are usually the only observable data.

The membrane is therefore very much a 'black box' and we may only deduce its mechanism and function from observations of input and output data. This is a position which is rather analogous to that of an industrial spy who, being excluded from a factory, must deduce its organization and function from an examination of the raw materials entering and the quality and output of the finished products. His conclusions would obviously be of a rather general nature. With these restrictions, it is inevitable that a phenomenological rather than a molecular or microscopic analysis is adopted. The formalism of irreversible thermodynamics must therefore serve to organize thought in membrane science, in much the way that classical thermo-dynamics served chemistry before the advent of sophisticated atomic and molecular theory.

In the stationary state the flows are constant through the membrane path and the state variables are constant with time at any point. We may therefore apply the dissipation function, for example, to any volume element of unit area and thickness $dx$; where the $x$ coordinate is normal to its surface. Since flows are constant in a stationary state it is possible to integrate from surface $o$ to surface $x$ and calculate the total dissipation function of unit area of membrane.

$$\Phi_{\text{total}} = \int_0^x \Phi dx = \sum_{i=1}^n J_i \int_0^x \text{grad}\,(-\tilde{\mu}_i) = \sum_{i=1}^n J_i(\tilde{\mu}_i^o - \tilde{\mu}_i^x) \tag{21}$$

The unknown internal gradients are replaced by the differences in chemical potentials at the membrane surfaces. Assuming local equilibrium at the membrane/solution interfaces (Kirkwood, 1954), these chemical potentials may be replaced by those of the species in the bathing solutions at the interfaces. In this way the dissipation function of the membrane may be deduced from a knowledge of the external phases. It remains to formulate the phenomenological equations, and in particular, to establish the validity of the Onsager reciprocal relations which from the symmetry will reduce (from $n^2$ to $\frac{1}{2}n(n+1)$ the number of phenomenological coefficients required for characterization).

Since the biological membrane is in itself a complex structure, it is important to establish the general conditions for the validity of the reciprocal relations. Many authors have either assumed or proved that such a symmetry is obtained for their particular model (Staverman, 1952; Kirkwood, 1954; Spiegler, 1958; Kedem and Katchalsky, 1961, 1963; Dainty and Ginzberg, 1963; and Snell, 1965).

A general treatment has been applied to systems in mechanical equilibrium (Snell and Steen, 1966). The general requirements may be summarized as follows:

i) Flows are normal to the membrane surface.

ii) Forces are everywhere the same over the surface.

iii) All forces bear a constant relationship to their potential difference across the membrane. This may be taken as a requirement for equal path lengths across the membrane. It also eliminates the possibility of sources and sinks of individual components and so chemical reaction within the membrane.

These conditions are obviously met by a homogeneous planar membrane of isotropic structure but also include those membranes which are hetero-geneous in a direction only normal or only parallel to their surfaces. Any greater degree of anisotropy is excluded. In the absence of internal chemical reaction of transported species, the Onsager relations will hold for composite membranes composed of regions in series or in parallel array. It will therefore accommodate most of the classical and current models of the biological membrane based upon a layered structure of protein and phospholipid (Danielli, 1935).

A parallel treatment, which includes stationary states where mechanical equilibrium does not exist, has been given (Mickulecky and Caplan, 1966). In this it is shown that the dissipative contributions of viscous, mechanical, kinetic and intrinsic rotational energies would be zero or vanishingly small in a membrane which was homogeneous or which does not have density variations in the direction, normal to its surface plane. Once again the dissipation function may be represented by equation (10) where all flows are measured relative to a stationary membrane. In the model systems to be discussed subsequently, the Onsager relations will be assumed. It is, how-ever, most important that they be justified in an experimental system.

## III. ACTIVE TRANSPORT

### A. Definitions

It is obvious from the general discussion that one might not expect a multiflow system to strictly obey simple diffusion laws. Striking deviations occur in many biological systems where a species may be transported against the gradient of its conjugate force, or alternatively, a force (or concentration gradient) may be shown to exist without net conjugate flow. These processes fall under the general heading of active transport. This has been defined as a flow 'dependent on the activity or energy change in another system' (Conway, 1954). A dominance of coupling is implied. Membrane systems which exhibit this phenomenon specifically for one or more permeant species are generally termed 'pumps'. Sodium, potassium and other types of specific ion pumps are well known. The classic work of Ussing and Zerahn on frog skin has served to establish the experimental criteria of such 'pumps' showing them to involve a specific coupling between transport and metabolic (chemical) reactions taking place within the membrane (Ussing and Zerahn, 1951; Zerahn, 1956; Ussing, 1958).

If the entire pump mechanism is contained within the membrane, Curie's principle requires that this mechanism of direct coupling will involve a considerable anisotropy within the membrane structure. This feature has been further stressed by Mitchell (1957, 1961) in a discussion of direct coupling of ions and a metabolic reaction catalysed by an orientated enzyme.

The first phenomenological treatment involving direct coupling of this type is due to Kedem (1960). The phenomenological equations take the general form,

$$X_i = \sum_j R_{ij}J_j + R_{ir}J_{Ch}$$

$$A = \sum_i R_{ri}J_i + R_{rr}J_{Ch} \tag{22}$$

The flows of solutes and water are represented by $J_j$ and of reaction, $J_{Ch} = d\xi/dt$, where $d\xi$ is the degree of advancement of the chemical reaction. The metabolically induced transport of a species $i$ is therefore dependent upon the rate of chemical reaction and the cross coefficient, $R_{ir}$. Since $A$ and $J_{Ch}$ are scalars and $X_j$ and $J_j$ are vectors, $R_{ij}$ and $R_{rr}$ are scalars. All $R_{ir}$ are vectors, however, and must be zero in an isotropic medium. The essential anisotropy of the membrane phase is stressed once again.

In the simple case analysed by Kedem (1960), the membrane components were considered as the ions 1 and 2 of a 1:1 electrolyte, and water flow was considered negligible. For further simplicity it was assumed that there was no coupling between anion, 2, and the chemical reaction or the flow of cation, 1, so that $R_{2r} = R_{12} = O$. This former condition limits the model to active transport of cation. Indeed, in its formulation the experiments of Ussing and Zerahn were borne in mind. Under these rather severe restrictions the phenomenological equations become

$$\Delta\tilde{\mu}_1 = R_{11}J_1 + R_{1r}J_{Ch} \tag{23}$$

$$\Delta\tilde{\mu}_2 = R_{22}J_2$$

$$A = R_{r1}J_1 + R_{rr}J_{Ch}$$

In the frog skin experiments the bathing solutions were identical chloride-Ringer's solutions. Reversible electrodes were placed on either side of the membrane and short circuited. Under these conditions a short circuit current, I, equal to $(z_1J_1 + z_2J_2)$ F, will flow. Since the two bathing solutions are identical, $\Delta\tilde{\mu}_1 = \Delta\tilde{\mu}_2 = 0$, and so from equation (23), $J_2 = 0$. The short circuit current is due solely to cation flow so that,

$$I = z_1J_1F = -\frac{R_{1r}}{R_{11}}J_{Ch}F \tag{24}$$

Since $R_{11}$, $J_{Ch}$ and F are positive, the current will be positive for a negative value of $R_{1r}$. The existence of coupling in active transport reduces the force required to drive $J_1$, since $R_{1r}J_{Ch}$ is negative. The short circuit current is therefore equal to the ionic current through the membrane and proportional to the rate of metabolic reaction.

These are the essential conclusions of Ussing and Zerahn (1951) in their measurements of net sodium flux under short circuit conditions (see table 1). The further proportionality of I to the rate of oxygen consumption bears out the general form of equation (24) (table 1).

The existence of direct coupling does not exclude the possibility that indirect coupling may occur between a metabolic reaction.

**Table 1.** Short circuit current and sodium flux values for a number of short circuited frog skins (*Rana temporaria*) with Ringer's solution on both sides.

| Experiment | Sodium flux ($\mu A/cm^2$) | | | Current density |
|:---:|:---:|:---:|:---:|:---:|
| | in | out | net | |
| 1 | 20·1 | 2·4 | 17·7 | 17·8 |
| 2 | 11·1 | 1·5 | 9·6 | 9·9 |
| 3 | 40·1 | 0·89 | 39·2 | 38·6 |
| 4 | 62·5 | 2·2 | 60·3 | 56·8 |
| 5 | 47·9 | 2·5 | 45·4 | 44·3 |

(From Ussing, 1957)

## B. Diffusion in the Presence of Chemical Reactions—Systems without Direct Coupling

In a general treatment (Baranowski and Popielawski, 1964) consider a simple biological cell which is taken to be in contact with its cell membrane. It is supposed that transport takes place at a much slower rate in the membrane than in the adjacent solution phases. No interfacial gradients of chemical potential will result under such conditions and the total thermodynamic forces will be defined as the differences in bulk chemical potentials of the bathing solutions. It is further assumed that chemical reactions take place within the cell and that the composition of the external environment is constant due to its greater expanse. In the general case there may be $N$ components of which a number, $L$, may be involved in chemical reactions within the cell, such that

$$\sum_{i=1}^{L} v_i m_i = O \tag{25}$$

This equation represents the condition for conservation of mass in the reaction, where $v_i$ is the stoichiometric coefficient and $m_i$, the molecular mass of the species $i$. This reaction formally expresses the metabolism of the cell.

The dissipation function is therefore

$$\Phi = \sum_{i=1}^{N} J_i X_i + J_{Ch} A \geqslant 0 \tag{26}$$

$J_i$ and $J_{Ch}$ are the diffusional and chemical flows respectively.

$X_i = \Delta \tilde{\mu}_i$ is the difference in electrochemical potential of $i$ across the membrane and $A$, the chemical affinity of the metabolism given by

$$A = -\sum_{i=1}^{L} v_i \tilde{\mu}_i \tag{27}$$

Since no direct coupling of chemical and diffusional flows has been introduced the phenomenological equations are

$$J_i = \sum_{K=1}^{N} L_{iK} X_K \tag{28}$$

and

$$J_{Ch} = L_{Ch} A$$

or in inverse form

$$X_i = \sum_{K=1}^{N} R_{iK} J_K \tag{29}$$

$$A = R_{Ch} J_{Ch}$$

In its original form, the paper of Baranowski and Popielawski uses the direct form of these equations. The equally valid representation in terms of generalized frictional coefficients will be used in this presentation, since it gives a rather simpler form of the final equations.

Since the environment is large all electrical and chemical potentials outwith the cell itself are assumed constant so that

$$A - \sum_{i=1}^{L} v_i X_i = \text{constant} \tag{30}$$

This equation stresses the relationship between diffusional and chemical driving forces stressed by Jardetsky and Snell (1960). Applying the principle of minimum entropy production to the system (Prigogine, 1947), the following stationary state conditions result.

$$J_i = -v_i J_{Ch} \quad 1 \leqslant i \leqslant L \tag{31}$$

$$J_i = 0 \qquad L+1 \leqslant i \leqslant N \tag{32}$$

(Note: The relationship between $J_i$ and $J_{Ch}$ should be $J_i = (a/v)v_i J_{Ch}$ where $a/v$ is the ratio of the membrane area to the volume of the cell. For a spherical cell this ratio is $3/r$ where $r$ is the radius of the cell and will be a very large number for a single biological cell. We shall however assume it to be unity for the purpose of a general discussion.)

It is obvious from equations (31) and (32) that the flows in the stationary state are either zero, if the species is not involved in the metabolism, or non-zero and proportional to the rate of the metabolic reaction.

For a two component system, in which species 1 and 2 may freely diffuse through the membrane and are also involved in a reaction,

$$v_1 m_1 \rightarrow v_2 m_2 \tag{33}$$

Within the cell the phenomenological equations are

$$X_1 = -J_{Ch}(v_1 R_{11} + v_2 R_{12}) \text{ or } \left(R_{11} + \frac{v_2}{v_1} R_{12}\right) J_1$$

$$\tag{34}$$

$$X_2 = -J_{Ch}(v_1 R_{21} + v_2 R_{22}) \text{ or } \left(\frac{v_2}{v_1} R_{21} + R_{22}\right) J_2$$

The gradients $X_1$ and $X_2$ developed in the steady state are therefore proportional to the rate of chemical reaction inside the cell.

Extending this treatment to include a third species, 3, with the metabolic reaction as before, it is possible to establish the general conditions for the development of a concentration gradient of 3 in the steady state. The steady state conditions are now

$$J_1 = -v_1 J_{Ch}$$

$$J_2 = -v_2 J_{Ch} \text{ and } J_3 = O \tag{35}$$

The phenomenological equations are

$$X_1 = R_{11} J_1 + R_{12} J_2$$

$$X_2 = R_{21} J_1 + R_{22} J_2 \tag{36}$$

$$X_3 = R_{31} J_1 + R_{32} J_2$$

$$A = R_{Ch} J_{Ch}$$

The relationship of $X_1$ and $X_2$ to $J_{Ch}$ is formally identical to the first analysis but on substitution

$$X_3 = -(R_{31} v_1 + R_{32} v_2) J_{Ch} \tag{37}$$

but

$$X_3 = -\Delta \mu_3 = -(\mu_3^{Soln} - \mu_3^{cell})$$

$$= -RT \ln \frac{a_3^{Soln}}{a_3^{cell}}$$

so that

$$RT \ln \frac{a_3^{Soln}}{a_3^{cell}} = (R_{31} v_1 + R_{32} v_2) J_{Ch} \tag{38}$$

If there is coupling between either of the chemical reactants and the species 3, there will be a concentration (activity) gradient of 3 across the membrane which is directly proportional to the rate of metabolic reaction. For a symmetrical reaction where $|v_1| = |v_2|$ coupling will result if $(R_{13} - R_{32}) \neq 0$. A second restriction may be placed upon the system by imposing the condition that $X_3 = 0$ by external means. The system is then short circuited with respect to $J_3$. Under this limitation, the steady state conditions are

$$J_1 = -v_1 J_{Ch} \text{ and } J_2 = -v_2 J_{Ch} \tag{39}$$

And so, from the phenomenological equations

$$X_3 = R_{31}(-v_1 J_{Ch}) + R_{32}(-v_2 J_{Ch}) + R_{33} J_3 = 0 \tag{40}$$

so that

$$J_3 = +\frac{(v_1 R_{31} + v_2 R_{32}) J_{Ch}}{R_{33}} \tag{41}$$

The flow $J_3$ is not zero when conjugate force, $X_3$, is zero provided that coupling occurs between 1 and 3 or 2 and 3. With the specific condition that $v_1 R_{31} = -v_2 R_{32}$ zero flow would result. The direction of flow will depend upon the sign and magnitude of the resistance coefficients.

To examine the condition in which a short circuit current is obtained, we may consider a system of four species, 1, 2, 3, and 4, in which species 1 and 2 are electrically neutral and take part in a chemical reaction within the cell. Species 3 and 4 are ions of the salt 3, 4 with valencies $Z_3$ and $Z_4$. For electroneutrality

$$\alpha_3 Z_3 + \alpha_4 Z_4 = 0 \tag{42}$$

where $\alpha_3$ and $\alpha_4$ are the stoichiometric coefficients for the dissociation of the salt. The dissipation function will be

$$\Phi = J_1 X_1 + J_2 X_2 + J_3 X_3 + J_4 X_4 + J_{Ch} A \tag{43}$$

If electrodes reversible to ion 4 are placed in the cell and environment there will develop a potential $E$, such that

$$E = \frac{-X_4}{Z_4 F} = \frac{-\Delta \tilde{\mu}_4}{Z_4 F} \tag{44}$$

The flows and forces may be transposed into a more convenient form using the relationships (Katchalsky and Curran, 1965)

$$\alpha_3 J_y = J_3 \text{ and } \alpha_4 J_y = J_4 \tag{45}$$

where $J_y$ is the flow of salt

$$I = (Z_3 J_3 + Z_4 J_4) F \tag{46}$$

where $I$ is the current density in the system,

and $\qquad X_y = \alpha_3 X_3 + \alpha_4 X_4$ where $X_y$ is the force on the salt $\qquad$ (47)

The resultant form of the dissipation function is therefore

$$\Phi = J_1 X_1 + J_2 X_2 + J_y X_y + IE + J_{Ch} A \qquad (48)$$

The corresponding phenomenological equations become

$$X_i = R_{i1} J_1 + R_{i2} J_2 + R_{iy} J_y + R_{ie} I \quad (i = 1,\ 2,\ \text{and}\ y) \qquad (49)$$
$$E = R_{e1} J_1 + R_{e2} J_2 + R_{ey} J_y + R_{ee} I$$
$$A = R_{Ch} J_{Ch}$$

Under steady state conditions of electrical short circuit, $E = 0$ and

$$J_1 = v_1 J_{Ch}, \ J_2 = -v_2 J_{Ch}$$

so that $\qquad I = \left( \dfrac{R_{e1} v_1 + R_{e2} v_2}{R_{ee}} \right) J_{Ch} \qquad (50)$

There will be a short circuit current proportional to the rate of chemical reaction provided there is a non-zero frictional interaction (and hence coupling) between the ions of the salt and the reactants and products of the chemical reaction.

## C. The Carrier Concept and Facilitated Transport

The concepts of the previous section are applicable to an isolated cell but there remains the problem of an isolated membrane which will transport actively without chemical reaction in the solutions on either side.

The concept of a carrier was introduced into membrane work by Danielli (1954) in order to explain the abnormally high permeability of constituents with low solubilities in the membrane. Such systems were also found to have a highly specific mode of operation, which was limited to a number of chemically similar species, suggesting a specific chemical reaction within the membrane phase. Their large deviations from Fick's law and the phenomenon of saturation flows at high concentrations of permeant, all suggested a specific interaction of permeant with a selective carrier with a finite number of reaction sites. The permeant in combination with the carrier would have a reduced resistance and enhanced flow in the membrane. This mechanism constitutes a model for facilitated transport.

Blumental and Katchalsky (1968) have re-examined the kinetic model of Rosenberg and Wilbrandt (1951) using the simple model system shown in figure 1. The membrane system is composed of two compartments separated by a membrane permeable to the carrier, C, the permeant, S and to the

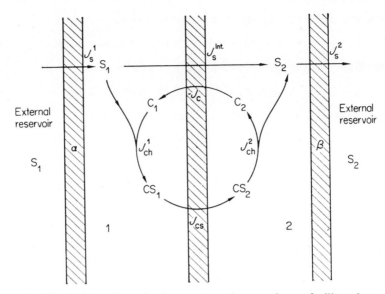

**Figure 1.** Circulation of carrier in a composite membrane-facilitated transport system. (From Katchalsky and Spangler, 1968).

permeant–carrier complex, CS. Equilibrium is assumed between permeant in compartment (1) and the external reservoir concentration, $S_1$, and similarly for compartment (2). The outer membranes, $\alpha$ and $\beta$ are considered impermeable to the carrier and the permeant–carrier complex. Assuming simple rate laws the reaction rates $J_{Ch}{}^1$ and $J_{Ch}{}^2$ may be expressed

$$J_{Ch}^1 = K_1 C_1 S_1 - K_{-1} CS_1 \tag{51}$$

$$\text{and} \quad J_{Ch}^2 = K_1 C_2 S_2 - K_{-1} CS_2$$

Where $K_1$ and $K_{-1}$ are the rate constants for the reactions

$$C_i + S_i \underset{K_{-1}}{\overset{K_1}{\rightleftharpoons}} CS_i \quad i = 1 \text{ or } 2 \tag{52}$$

The flows of C and CS are assumed to be simply proportional to their concentration differences across the membrane so that

$$J_C = P(C_1 - C_2) \tag{53}$$

$$J_{CS} = P(CS_1 - CS_2)$$

While the flow of free permeants, $J_s^{int}$ is given by

$$J_s^{int} = P_S (S_1 - S_2) \tag{54}$$

(The permeability coefficient $P$ is assumed to be the same for C and CS.) In the steady state all concentrations are constant, and so,

$$-J_{Ch}^1 - J_C = 0 \tag{55}$$
$$J_C - J_{Ch}^2 = 0$$
$$J_{Ch}^1 - J_{CS} = 0$$
$$J_{CS} - J_{Ch}^2 = 0$$

and for the circulation process,

$$J_{S_1} - J_S^{int} - J_{Ch}^1 = 0 \tag{56}$$
$$\text{and} - J_{S_2} + J_S^{int} - J_{Ch}^2 = 0$$

Applying the stationary state conditions to concentrations and flows it is obvious that

$$J_C = J_{Ch}^2 = -J_{CS} = J_{Ch}^1$$
$$\text{and } J_{S_1} = J_S^{int} + J_{CS} = J_{S_2} = J_S^{ext} \tag{57}$$

This system is one of five diffusional flows and two reaction flows. In a macroscopic examination one would simply observe an external driving force $\Delta\mu_S^{ext}$ and one flow $J_S^{ext}$, so that the dissipation function would be

$$\Phi = J_S^{ext}\Delta\mu_S^{ext} \tag{58}$$

while the more detailed mechanism would contain all elements of the model, so that

$$\Phi = J_{S_1}\Delta\mu_{S_1} + J_S^{int}\Delta\mu_S + J_{S_2}\Delta\mu_{S_2} + J_C\Delta\mu_C + J_{CS}\Delta\mu_{CS} + J_{Ch}^1 A_1 + J_{Ch}^2 A_2 \tag{59}$$

from equation (57)

$$\Phi = J_S^{ext}(\Delta\mu_{S_1} + \Delta\mu_S + \Delta\mu_{S_2}) + J_{CS}(\Delta\mu_{CS} - \Delta\mu_C - \Delta\mu_S + A_1 - A_2) \tag{60}$$

Expanding the changes of chemical potentials one obtains once more the expression given in equation (58).

The two approaches are therefore quite identical and the corresponding phenomenological equation may be represented as

$$J_S^{ext} = L\Delta\mu_S^{ext} \tag{61}$$

and expansion of $L$ in terms of the model shows that

$$L = (L_p + L_f) \tag{62}$$

where $L_p$ is the term for passive transport independent of carrier mechanism and proportional to the permeability of S, $P_S$. $L_f$ is the contribution due to circulation increasing the overall permeability of the membrane to S. (The phenomenon of saturation may also be described by this model.)

F

### D. A Molecular Model for the Sodium Pump

Katchalsky and Spangler (1968) have evaluated a simple model system for active transport which incorporates the essential thinking of Glynn (1957), Skou (1957, 1960), Post and coworkers (1960), Dunham and Glynn (1961), Whittam (1964) and Opit and Charnock (1965).

In this work a considerable weight of experimental evidence is provided for the presence of an enzyme system known as $Na^+$ and $K^+$-dependent adenosine triphosphatase which is an essential constituent in most or all biological membranes which exhibit active transport of sodium and potassium ions across cellular membranes. The enzyme system is membrane-bound and ubiquitous in animal cells (Skou, 1962). It requires the presence of both sodium and potassium ions. In the erythrocyte, enzyme activation occurs only with sodium inside and potassium outside the cell, and may be inhibited

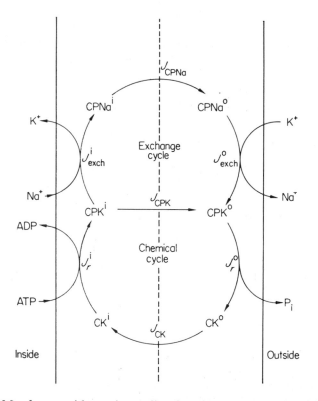

**Figure 2.** Membrane with carrier-mediated active transport showing coupling between chemical reaction cycle and exchange cycle. (From Katchalsky and Spangler, 1968).

by the cardiac glycoside ouabain. Opit and Charnock consider a unit membrane consisting of a double protein layer with lipid interposed, the inner protein structure containing ATPase enzyme. This structure is considered as a highly ordered, polarizable chain with many surface-orientated cationic and anionic groups. Among such groups are situated a much smaller number of amino acid groups which constitute the active centre of the enzyme. The sodium ions of the substrate are considered to activate the enzyme into forming a phosphorylated intermediate. On phosphorylation the protein folds or deforms thereby rotating the anionic sites outwards and propelling the sodium counter-ions to the outside of the membrane. In this deformed state the anionic sites of the phosphorylated enzyme strongly prefer potassium. A process of ion exchange of sodium for potassium occurs and the resultant redistribution of electron density renders the phosphate groups more vulnerable to hydrolytic attack. On releasing phosphate (inside) the protein returns to its original conformation and in doing so, transfers the potassium ions to the internal phase. The concept is to some degree hypothetical but is still basically a model involving a carrier, although the carrier is an integral part of the membrane structure.

Since the enzyme is considered to be an integral part of the inside layer of protein, the essential asymmetry required by the Curie principle is incorporated.

Whether the carrier is a mobile molecule or a flexing protein structure is quite immaterial to the formal analysis, although the model of Katchalsky and Spangler differs in the side at which phosphate is released. The hypothetical carrier mechanism is shown in figure 2. It consists of two cycles. The first is a chemical cycle involving the phosphorylation of the carrier and its subsequent hydrolysis.

$$C+ATP \rightarrow CP+ADP \quad \text{inside} \tag{63}$$
$$CP+H_2O \rightarrow C+P \quad \text{outside (K. and S.)}$$

With this chemical alteration the carrier CP is presumed selective towards sodium ion, while C is selective towards potassium. The second is the exchange cycle based upon the ion exchange reaction.

$$CPK+Na^+ \rightleftharpoons CPNa+K^+ \tag{64}$$

Through chemical coupling, the chemical cycle drives the exchange cycle, which may be entropy reducing. The model also incorporates a direct flow of CPK. In the steady state the concentrations of all species are constant with time at any point in the system. Applying this condition, one may equate a number of flows within the system.

$$J^i_{Ch} = J^o_{Ch} = J_{CK} \text{ since } \frac{dCK}{dt} = \frac{dCK^o}{dt} = 0 \tag{65}$$

$$J^i_{exch} = J^o_{exch} = J_{CPNa} \text{ since } \frac{dCPNa^i}{dt} = \frac{dCPNa^o}{dt} = 0$$

and $J_{CPK} = J^i_{Ch} - J^i_{exch} = J^o_{Ch} - J^o_{exch}$

from the time constancy of concentration, CPK.

The detailed dissipation function becomes

$$\Phi = J_{CK}\Delta\mu_{CK} + J^i_{Ch}A^i + J^o_{Ch}A^o + J_{CPK}\Delta\mu_{CPK}$$
$$+ J_{CPNa}\Delta\mu_{CPNa} + J^i_{exch}A^i_{exch} + J^o_{exch}A^o_{exch} \tag{66}$$

We may expand the forces of equation (66) as follows:

$$\Delta\mu_{CK} = \mu^o_{CK} - \mu^i_{CK}, \ \Delta\mu_{CPK} = \mu^o_{CPK} - \mu^i_{CPK}$$
$$\Delta\mu_{CPNa} = \mu^i_{CPNa} - \mu^o_{CPNa}, \ A^i = \mu_{ATP} - \mu^i_{CK} - \mu_{ADP} - \mu^i_{CPK} \tag{67}$$
$$A^o = \mu^o_{CPK} + \mu_{H2O} - \mu^o_{CK} - \mu^i_P$$
$$A^i_{exch} = \mu^i_{CPK} + \tilde{\mu}^i_{Na} - \mu^i_{CPNa} - \tilde{\mu}^i_K$$
$$\text{and } A^o_{exch} = \mu^o_{CPNa} - \tilde{\mu}^o_K + \tilde{\mu}^o_{Na} + \mu^o_{CPK}$$

Inserting these forces and the corresponding steady state flows into equation (66), the dissipation function becomes

$$\Phi = J_{Ch}(\mu_{ATP} + \mu_{H2O} - \mu_{ADP} - \mu^i_P) + J_{exch}[(\tilde{\mu}^i_{Na} - \tilde{\mu}^o_{Na}) - (\tilde{\mu}^i_K - \tilde{\mu}^o_K)]$$
$$= J_{Ch}A_{Ch} + J_{exch}X_{exch} \tag{68}$$

where $A_{Ch}$ is the chemical affinity of the hydrolysis of ATP. The second term contains a new force, which is the force of the ion exchange reaction, $X_{exch}$ where

$$X_{exch} = \Delta\tilde{\mu}_{Na} - \Delta\tilde{\mu}_K \tag{69}$$

The macroscopic dissipation function is given by equation (68) solely in terms of external forces, although based upon an intimate model of membrane function.

Once more, as in the model for facilitated transport, the microscopic model yields an equivalent statement of the dissipation function in terms of the observable forces in the external world. The exchange force $X_{exch}$ is negative when the pump removes the minority ion from its bathing solution. For this reason the observed exchange flow of the pumped ions is opposed to its conjugate driving force and the term $J_{exch}X_{exch}$ is negative and so entropy consuming.

The phenomenological equations corresponding to the macroscopic dissipation function are

$$J_{exch} = L_{11}X_{exch} + L_{12}A_{Ch} \qquad (70)$$
$$J_{Ch} = L_{21}X_{exch} + L_{22}A_{Ch}$$

$L_{11}$ and $L_{22}$ are the direct coefficients for exchange and chemical reaction. The cross coefficient, $L_{12}$, is the sole source of active transport. If it is zero $J_{exch} X_{exch}$ will be positive since $L_{11}$ is positive, as can be seen from substituting in equation (70).

If $J_{exch}$ is zero, the chemical reaction is just sufficient to maintain a steady state as may be the case in the 'resting' cell so that

$$X_{exch} = -\frac{L_{12}}{L_{11}}A_{Ch} = -RT \ln \Gamma \; (J_{exch} = 0) \qquad (71)$$

and $\Gamma$ is defined as the activity quotient

$$\Gamma = \frac{a_K{}^i \cdot a_{Na}{}^o}{a_K{}^o \cdot a_{Na}{}^i} = \exp\left(\frac{L_{12}}{L_{11}}\frac{A_{Ch}}{RT}\right) \approx \frac{[K^i][Na^o]}{[K^o][Na^i]} \qquad (72)$$

Here concentrations may replace activities to a first approximation. For red blood cells the ratio, $\Gamma$, is around 220.

Katchalsky and Spangler continued this analysis with a statistical thermodynamic evaluation of the $L$ coefficients and with a simple analysis of the frequency of carrier transport $v$ as

$$v = \frac{\text{sodium exchange flow}}{\text{surface density of charge carrier}} \approx \frac{6 \times 10^{10}}{6 \times 10^9} \approx 10\text{--}100 \text{ cps} \qquad (73)$$

Each carrier must oscillate at around ten cycles per second, a figure which is reasonable for the rate of configurational changes in biopolymers.

## E. Active Transport and the Efficiency of Energy Conversion

In a simple binary system in which there is active transport, the dissipation function may be represented in terms as

$$\sigma = J_1 X_1 + J_2 X_2 \geqslant 0 \qquad (74)$$

and the phenomenological equations as

$$J_1 = L_{11}X_1 + L_{12}X_2 \qquad (75)$$
$$J_2 = L_{21}X_1 + L_{22}X_2$$

The term $J_1 X_1$ is negative for the actively transported component, which therefore consumes entropy at the expense of the driving process, process 2.

A proportion of the entropy, and so the free energy dissipated by process 2, is converted into a form characteristic of process 1. The energy so withdrawn may be stored up within the system in the form of a static head of force, $X_1$, or it may be dissipated externally by a continuing flow, $J_1$.

The more general concepts of efficiency and coupling in such processes is clearly developed by Kedem and Caplan (1965) and Caplan (1966), where the efficiency of energy conversion is defined as $\eta$, where

$$\eta = -\frac{J_1 X_1}{J_2 X_2} = 1 - \frac{\sigma}{J_2 X_2} \tag{76}$$

This function represents the fraction of the entropy production of process 2, which is consumed by process 1. When process 1 is driven by 2 in a direction opposed to its own driving force, the efficiency is in the range $0 \leqslant \eta \leqslant 1$. The efficiency is zero both at level flow, when the driving force, $X_2$, is held constant and $X_1$ is zero and at static head when, as before, $X_2$ is constant and $J_1$ is zero. Between these two limits the efficiency passes through a maximum defined by

$$\eta_{max} = \frac{q^2}{(1 + \sqrt{1 - q^2})^2} \tag{77}$$

where $q$ is the degree of coupling, defined as

$$q = \frac{L_{12}}{\sqrt{L_{11} L_{22}}} = \frac{-R_{12}}{\sqrt{R_{11} R_{22}}} \tag{78}$$

The direct coefficients, are, by definition positive, while the cross coefficients may be either positive or negative so that the value of $q$ must fall within the limits, $1 \geqslant q \geqslant -1$ and be zero when there is no coupling between the processes since under these conditions $L_{12} = R_{12} = 0$.

Figure 3 due to Kedem and Caplan (1965) shows the dependence of the efficiency, $\eta$, on the force ratio, $Zx$, at different degrees of coupling.

The force ratio is defined as $Zx$ where

$$Zx = \sqrt{\frac{L_{11}}{L_{22}}} \cdot \frac{X_1}{X_2} \tag{79}$$

The forces are weighted by their corresponding direct mobility terms. The right hand origin in figure 3 corresponds to the level flow condition when $X_1$ is zero, while the intercepts on the abscissa correspond to the various static head situations, since $J_1$ is zero, when $q$ equals $-Zx$.

As $q$ decreases there is a very rapid reduction in the maximum efficiency possible and in the area of the driving region. These general observations imply that biological systems, showing active transport, must require a large degree of coupling.

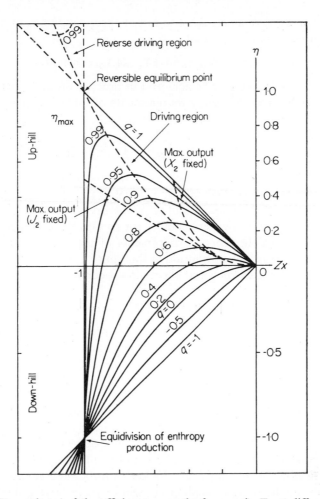

**Figure 3.** Dependence of the efficiency $\eta$ on the force ratio $Zx$ at different degrees of coupling $q$. The driving region, which refers to positive coupling only, lies between the origin and $Zx = -1$, $\eta = 1$. The mirror-image of this diagram about the $\eta$ axis contains the driving region for negative coupling. The loci of maximum efficiency and maximum output are shown by broken lines. (From Kedem and Caplan, 1965).

The value of $q$ is therefore of fundamental importance and fortunately may be obtained without a complete analysis of the system. It is necessary only to measure the flow $J_2$ of the driving process (metabolic reaction) under conditions where the active transport is at level flow and at static head, while

$X_2$ is held constant. A simple substitution of these conditions in the pheno-menological equations shows that

$$1-q^2 = (J_2)_{J_1=0}/(J_2)_{X_1=0} \quad (X_2, \text{constant}) \qquad (80)$$

Since such restrictions may be achieved in practice on many real systems, it is useful to examine the energy requirements of these stationary states. It may be shown that

$$J_2X_2 = X_1^2 L_{11} \left( \frac{1}{q^2} - 1 \right) \quad (J_1 = 0) \qquad (81)$$

under static head. The energy required for maintenance of a concentration (or chemical potential) gradient is proportional to the square of that gradient and is a function of the direct mobility of the driven process, $L_{11}$, and the degree of coupling, $q$. A low permeability for species 1 and a large degree of coupling will reduce the energy requirements of the driving force—a con-clusion which is in accord with intuitive thinking.

Under level flow, also known as short circuit conditions, a similar substitu-tion yields

$$J_2X_2 = J_1^2/L_{11}q^2 \quad (X_1 = 0) \qquad (82)$$

The energy required to cause a level flow, $J_1$, is proportional to $J_1{}^2$ and may be reduced by a large degree of coupling and a large direct mobility, $L_{11}$. If the flow $J_1$ is caused by its conjugate force alone, then from equation (75)

$$\Phi = J_1X_1 = J_1^2/L_{11} \qquad (83)$$

and is a measure of the direct energy requirement to produce $J_1$. From a comparison of equations (82) and (83), the direct energy requirement is reduced by a factor of $q^2$ if the flow is caused by its own conjugate force.

This treatment has been extended to multiple-flow systems by Caplan (1966). The degree of coupling is redefined as $q$ on a more general basis and is a measure of the overall coupling as 'seen' by the driving flow and the same formalism applies.

## IV.   ISOTOPIC FLOWS AND FLUX RATIO EXPRESSIONS

### A.  General Considerations

Once radioactive isotopes became commonly available it became possible to follow the flows of individual membrane permeants without serious perturbation of the system. The most common experiments involve a measure of tracer influx or efflux across a test membrane. The first theoretical interpretation of such experiments was put forward by Ussing (1952), and since that time a number of papers have appeared which have expanded, refined and tested these theories. In particular, the work of Meares and Ussing (1959a,b) has shown the applicability of a *quasi*thermodynamic approach to ion exchange membranes under both concentration and electrical gradients. They first applied irreversible thermodynamics, using the Spiegler model and obtained very satisfactory agreement between observed and calculated counterion flux ratios. More recently, Nims (1961), Hoshiko and Lindley (1964), Kedem and Essig (1965) and Coster and George (1968) have expanded and refined the earlier concepts and applied them to more complicated model systems. The treatment of Kedem and Essig is rigorous and includes systems in which isotope–isotope interactions may be important. For this reason it serves as the basis of the remainder of this section.

In a membrane system consisting of a homogeneous array of parallel elements, the test substance is designated 0, and considered to consist of an abundant isotope 1, and tracer isotopes, 2 and 3. For any local region within the membrane the thermodynamic force is the local gradient of electro-chemical potential, $\text{grad}(-\tilde{\mu}_i)$, and may be related to the flows of all membrane species by phenomenological equations of the type

$$\text{grad}\,(-\tilde{\mu}_i) = \sum_{K=0}^{K=3} r_{iK}J_K + \sum_{j=4}^{n} r_{ij}J_j \quad i = 1,2,3. \tag{84}$$

The local frictional coefficients between isotopes of the test substance are designated $r_{iK}$ and all may be shown to be equal, provided the isotopic tracer components are kinetically and thermodynamically identical to that of the abundant isotope of the test species, so that

$$r_{iK} = r_{10} = r_{20} = r_{12} = r_{13} = r_{23} \tag{85}$$

The second summation represents the contribution of interactions between the test and all other mobile species in the membrane, and will include such terms as $r_{i\text{Ch}}J_{\text{Ch}}$ which represent the contributions of active transport. Since the total flow of test species is the sum of its components

$$J_0 = J_1 + J_2 + J_3 \tag{86}$$

so that from equation (84)

$$\text{grad}\,(-\tilde{\mu}_i) = (r_{ii} - r_{iK})J_i + r_{iK}J_0 + \sum_{j=4}^{n} r_{ij}J_j \tag{87}$$

and for total test species

$$\text{grad}\,(-\tilde{\mu}_0) = r_{00}J_0 + \sum_{j=4}^{n} r_{0j}J_j \tag{88}$$

If we may make the common assumption that there will be no isotope separation and that the specific activity, $\rho_i$, of the isotope is constant across the membrane (no isotope exchange), then

$$J_i = \rho_i J_o \text{ where } \rho_i = C_i/C_o \tag{89}$$

$$\text{and grad} (-\tilde{\mu}_o) = \text{grad} (-\tilde{\mu}_i) \quad i = 1,2,3 \tag{90}$$

From equations (87), (88) and (90)

$$\frac{J_i}{J_0} = \frac{(r_{00} - r_{iK})}{(r_{ii} - r_{iK})} + \sum_{j=4}^{n} \frac{(r_{0j} - r_{ij})}{(r_{ii} - r_{iK})J_0} = \rho_i \tag{91}$$

but since all $r_{ij}$ and $r_{oj}$ are equal since the isotopes are chemically identical

$$(r_{00} - r_{iK})/(r_{ii} - r_{iK}) = \rho_i \tag{92}$$

(In the absence of isotope–isotope coupling, $r_{iK}$ equals zero). For identical isotopes all thermodynamic parameters except concentration are identical, so that

$$\text{grad} (-\tilde{\mu}_i) - \text{grad} (-\tilde{\mu}_0) = RT \left( \frac{d \ln \rho_i}{dx} \right) \tag{93}$$

This equation applies in the general case in which the specific activity is no longer constant across the membrane, as in an isotope diffusion experiment. From equations (87), (88) and (93)

$$RT \left( \frac{d \ln \rho_i}{dx} \right) = (r_{00} - r_{iK})J_0 + (r_{ii} - r_{iK})J_i \tag{94}$$

Dividing by $r_{ii} - r_{iK}$ and introducing the specific activity from equation (92)*

$$J_i - \rho_i J_0 = \frac{-RT}{(r_{00} - r_{iK})} \left( \frac{d\rho_i}{dx} \right) \tag{95}$$

## B. Special Considerations

### 1. *The Flux Ratio Expression*

Consider a test membrane of thickness $\Delta x$ measured from outside to inside surface. Assuming local equilibrium at the interfaces, the specific activity at the membrane surfaces will equal those of the adjacent solutions. If the specific activity of isotope $i$ is $\rho_{io}$ at $X = 0$ (outside) and $\rho_{il}$ at $x = \Delta x$ (inside), integration of equation (95) in the steady state gives

$$J_0 \int_0^{\Delta x} \frac{(r_{00} - r_{iK})dx}{RT} = \frac{J_0 R^x}{RT} = \ln \frac{J_i - \rho_{il}J_0}{J_i - \rho_{io}J_0} \tag{96}$$

---

*Equation (92) is derived by Kedem and Essig for very limiting conditions under which the specific activity is maintained constant across the membrane. This is not generally the case in isotopic experiments. It may be shown, however, that equation (92) is quite general by application of equation (105) to bulk test species, 0, and to isotope, $i$ (1, 2 or 3).

where $R^x = \int\limits^{\Delta x}(r_{00} - r_{iK})dx$ is the exchange resistance and is a measure of the total frictional interaction of test species, and includes interactions with component isotopes.

If two isotopic tracers 2 and 3, are present in the outside and inside compartments, respectively, then

$$\frac{J_0 R^x}{RT} = \ln\frac{J_2 - \rho_{21} J_0}{J_2 - \rho_{20} J_0} = \ln\frac{J_3 - \rho_{31} J_0}{J_3 - \rho_{30} J_0} \tag{97}$$

In most experiments of this sort, the system is arranged so that the inside compartment is effectively an infinite sink for 2 and similarly the outside compartment for isotope 3. Under these conditions $\rho_{30} = \rho_{21} = 0$, so that equation (97) gives

$$\frac{J_0 R^x}{RT} = \ln\frac{(J_2/\rho_{20})}{(J_2/\rho_{20} - J_0)} \tag{98}$$

and $\dfrac{J_2}{\rho_{20}} + \dfrac{J_3}{\rho_{31}} = J_0$ (influx $-$ outflux $=$ net flux) $\tag{99}$

Substituting equation (99) into (98) one obtains an expression for the flux ratio, $f$.

$$\ln f = \ln\frac{J_2/\rho_{20}}{-J_3/\rho_{31}} = \frac{J_0 R^x}{RT} \tag{100}$$

The exchange resistance may therefore be determined from a knowledge of the flux ratio or one isotopic flow and net flow of test species. It is not necessary to have any knowledge of the structure, function or other intimate details of the membrane system. The factors influencing the magnitude of the flux ratio are more clearly seen if a value of $J_0$ is substituted from the integral form of equation (88) so that

$$f = \exp R^x / R \frac{(X - \int\limits_0^{\Delta x} \sum\limits_{j=4}^{n} r_{0j} J_j dx)}{RT} \tag{101}$$

where $R = \int\limits_0^{\Delta x}(r_{00})dx$ and $X = \int\limits_0^{\Delta x} \mathrm{grad}\,(-\mu_0)dx$

It is obvious that the flux ratio will be determined not only by the applied force $X$, but also by coupling of flows, active transport and isotope interaction. In the absence of coupling the flux ratio would be unity when the applied force was zero.

Since little is known of biological membrane systems, it is useful to consider the relative importance of the terms of equation (101) for some simpler systems where more data is available. Miller (1966) has provided a complete analysis of Onsager frictional and mobility coefficients for a number of simple chloride salts in aqueous solution at $25\,°C$. It is instructive to consider the effect of the coupling term in a solution 0·1 molar in sodium chloride under an electrical force, $z_i F\left(-\dfrac{\partial \psi}{\partial x}\right)$. The contribution of coupling may be represented by the fraction

$$\frac{-\int_0^{\Delta x} \sum_4^n r_{0j} J_j \mathrm{d}x}{X} = \frac{-r_{12} J_2}{z_1 F(-\partial \psi/\partial x)} = \frac{-r_{12} t_2 \kappa}{z_1 z_2 F^2} = -1\cdot04 \times 10^{-1}$$

(where 1 and 2 refer to sodium and chloride ions, respectively). $r_{12}$ is the frictional coefficient between sodium and chloride ions and $t_2$, $\kappa$, $z_1$ and F are the transport number of chloride, the specific conductivity, valency (including sign) and the Faraday constant, respectively. The effect of coupling amounts to only some ten percent of the applied force, but it is in the opposite direction to that force. Since $r_{12}$ is negative, the flow of anion tends to drag the cation with it to some degree.

From this argument, it may well be permissible to ignore the effect of coupling between a test ion and its environment, provided interaction between the ion and the membrane pore wall is negligibly small.

The ratio $R^x/R$ may also be obtained for several simple aqueous solutions, provided both frictional coefficients and isotopic diffusion coefficients are available. Isotopic or self-diffusion experiments are conducted in an experimental system in which all bulk flows and chemical potential gradients are zero. The thermodynamic force on the isotopic tracer is then

$$\mathrm{grad}\,(-\tilde{\mu}_i) = -RT\frac{\partial \ln C_i}{\partial x}$$

and so the sole driving force for the tracer is the gradient of its concentration. Under these restrictions equation (87) becomes

$$J_i = \frac{RT}{C_i(r_{ii} - r_{iK})} \cdot \left(-\frac{\partial C_i}{\partial x}\right) \tag{102}$$

and from equations (89) and (92)

$$J_i = \frac{RT}{C_0(r_{00} - r_{iK})} \cdot \left(-\frac{\partial C_i}{\partial x}\right) \tag{103}$$

$$\left[\text{or in integral form } J_i = \left(\frac{RT}{C_0 R^x}\right) \cdot (-\Delta C_i)\right]$$

The isotopic diffusion coefficient, $D_{ii}$, is therefore

$$D_{ii} = \frac{RT}{C_0(r_{00} - r_{iK})} \quad (104)$$

Combining the isotopic diffusion coefficients of Wang (1952) with the Onsager frictional coefficients of Miller (1966), a comparison of $R^x$ and $R$ may be made for sodium chloride solutions at 25 °C. The results are shown in table 2. It is obvious that the ratio $R^x/R$ is of the order of unity for dilute solutions so that the flux ratio is again only slightly affected by isotope–isotope interactions. Again it would appear that the assumption of zero coupling in a simple aqueous pore or leak would be good to a first approximation (Kedem and Essig, 1965). It is also interesting to note that $r_{iK}$ is positive in this system where there is a repulsive interaction between like-charged ions. The positive sign indicates that the coupling is negative so that the flow of isotope 1 diminishes the flow of isotope 2. In the case of water, one would expect $r_{iK}$ to be negative since there is a mutual attraction between water molecules, causing a positive coupling of flows. The fact that a particular isotope is present in trace amounts has no effect upon the value of $r_{iK}$, since this parameter measures the frictional coefficient between one mole of $i$ and one mole of $K$ in the immediate surroundings—a point to which we shall return in the next section.

**Table 2.** A comparison of the exchange resistance and direct frictional coefficients for sodium and chloride ions in aqueous sodium chloride solutions

| Molarity of NaCl | $r_{11}$ $\times 10^{-11}$ | $r_{22}$ $\times 10^{-11}$ | $D_{11}$ $\times 10^5$ | $D_{22}$ $\times 10^5$ (cm²/sec) | $(r_{11} - r_{ik})$ $\times 10^{-11}$ | $(r_{22} - r_{ik})$ $\times 10^{-11}$ | $\dfrac{(r_{11} - r_{ik})^*}{r_{11}}$ | $\dfrac{(r_{22} - r_{ik})^*}{r_{22}}$ |
|---|---|---|---|---|---|---|---|---|
| 0·01 | 191·8 | 125·6 | 1·31 | — | 189·2 | — | 0·99 | — |
| 0·05 | 39·69 | 25·96 | 1·30 | 1·96 | 38·14 | 25·3 | 0·96 | 0·97 |
| 0·10 | 20·28 | 13·26 | 1·31 | 1·94 | 18·90 | 12·8 | 0·93 | 0·96 |
| 0·50 | 4·43 | 2·87 | 1·32 | 1·85 | 3·76 | 2·7 | 0·84 | 0·93 |

Species 1 and 2 are sodium and chloride ions, respectively, and the dimensions of the resistance coefficients, $r$, are joule . cm . sec/mole².
*The ratios in the last two columns may be considered as equal to the integral ratio $R/R^x$.
From the original experimental data of Wang (1952), Wang and Miller (1952), and Miller (1966).

## 2. Isotopic Effects in Water Diffusion Studies

From the arguments given above it would appear that the contribution of isotope–isotope friction might well be secondary in the analysis of flux ratios for simple inorganic ions. This does not appear to be the case for water permeation. Hevesy and coworkers (1935) found that the osmotic permeability of frog skin was some five times greater than the permeability obtained from tracer studies with deuterium oxide. This observation has been supported by a number of similar investigations on a variety of membranes. Mauro (1957) reported a ratio of permeability coefficients as large as seven hundred. Nims and Blaudeau (1968) have applied these techniques to Millepore filters and cellulose membranes. In this work the permeability by isotopic diffusion of HDO ($3\%$) was compared with that obtained by simple water flow under a pressure gradient. In this simple system in which there is only water and the membrane and no other components, $R$ may be obtained from the integrated form of equation (88) and $R^x$ from the self diffusion expression, equation (103). (For the Millepore filters the integration is taken across the filter thickness, which is quoted as 130 $\mu$ in the Millepore catalogue.) The results of this work are shown in table 3. In this case the ratio $R^x/R$ is large, $5 \times 10^6$ for the Millepore filters, while for cellophane the ratio is only 11·2. The implication is that the water isotope–isotope friction is indeed negative, as suggested above, and is much larger than $r_{oo}$.

**Table 3.** Magnitudes of the integral, $R$, and exchange resistance coefficients, $R^x$, for water

| Membrane | $R^x$ | $R$ | $R^x/R$ |
|---|---|---|---|
| Pure water | $2\cdot47 \times 10^7$ | 0 | $\infty$ |
| Millipore 0·3 | $5\cdot48 \times 10^7$ | $1\cdot14 \times 10^1$ | $4\cdot82 \times 10^6$ |
| Millipore 0·45 | $6\cdot06 \times 10^7$ | $1\cdot09 \times 10^1$ | $5\cdot57 \times 10^6$ |
| Cellophane 450 | $7\cdot99 \times 10^7$ | $7\cdot11 \times 10^6$ | 11·2 |
| Phenol sulphonic cation exchange membrane | $7\cdot33 \times 10^7$ | $3\cdot32 \times 10^6$ | 22·1 |

(The units of $R$ are joules . $cm^2$ . $sec/moles^2$.)

Data for Millipore and cellophane membranes are taken from Nims and Blaudeau (1968) and those for pure water calculated from the isotopic diffusion data of Wang and coworkers (1953). Those for the cation exchange membrane are taken from experimental diffusion coefficients of Mackay (1960) and the resistance coefficients of Mackay and Meares (1959). In all cases the thickness, $\delta$, is taken as 130$\mu$ so that a direct comparison may be made with the Millepore results.

To understand the enormous difference between $R^x$ and $R$ it is again useful to consider data which is available for simple aqueous solutions. The direct coefficient is seen to be zero in all aqueous salt solutions at infinite dilution (Miller, 1966). This apparently remarkable result may be rationalized in terms of the expansion $r_{oo}$, which is applicable to all vectorial systems,

$$C_0 r_{00} = - \sum_{j=1}^{n} C_j r_{0j} \qquad (105)$$

The terms $C_j r_{oj}$ correspond to the frictional interaction coefficients between one mole of $O$ and those species, $j$, per unit volume of the adjacent solution. This term is the negative of the Spiegler frictional coefficient $X_{oj}$ (Spiegler, 1958). The summation therefore measures the total interaction of the test species, with all other components in the system but not with itself. For this reason $r_{oo}$ is zero for water in pure water. The non-zero but small value obtained for the Millepore experiments represents the frictional interaction, $C_m r_{om}$, between water and the porous membrane, $m$.

Isotopic diffusion coefficients for deuterated, tritiated and oxygen-18 enriched water are available with values of $2\cdot34, 2\cdot44,$ and $2\cdot66 \times 10^{-5} cm^2/sec$ at 25 °C, (Wang and coworkers, 1953). The particular isotope used has a secondary effect upon the observed diffusion coefficient and so it might be supposed that the kinetic and thermodynamic identity of isotopic and bulk water may be assumed to a first degree. From equation 103 the value of $(r_{oo}-r_{iK})$ may be calculated for DHO. The tabulated value, $R^x$, is taken as $(r_{oo}-r_{iK})$ $\delta$ where $\delta$, for comparison, is taken as the reported thickness of the Millepore filter, 130 $\mu$. As can be seen from table 3, this computed value, $2\cdot5 \times 10^7$, is of the same magnitude as that obtained in the Millepore experiments and indicates the dominance of water isotope–isotope contributions. From equation (105)

$$\frac{R^x}{R} = \frac{\int_0^{\Delta x}(C_0 r_{iK} + C_m r_{0m} + \sum_j C_j r_{0j})dx}{\int_0^{\Delta x}(C_m r_{0m} + \sum_j C_j r_{0j})dx} \tag{106}$$

For the cellophane membrane in pure water, alone, the general summation terms are zero. $C_0$ is reduced and $r_{0m}$ may well be increased due to a smaller pore size and a more intimate interaction between the water and cellulose. Both effects must tend to reduce the ratio in comparison to the Millepore results.

It is interesting to consider one further model system. Mackay and Meares (1959) have obtained a complete analysis of the frictional coefficients of the sodium-form of a Zeokarb 315 P.S.A. membrane in contact with 0·05 molar sodium chloride solution. (The analysis follows that of Spiegler (1958) and assumes negligible isotopic effects for the sodium counterion and the chloride coion). Mackay (1960) has measured the isotopic diffusion coefficient for HDO in the leached sodium-form of the same membrane. Calculation again gives at least an approximate value for $R^x$ and $R$ (see table 3). In this case the ratio of 22·1 is comparable to the cellulose results although the water content of the membrane, 40·7 moles/l of membrane, is similar to that of the Millepore filter, 42·7 moles/l, when calculated on the basis of a 77% volume porosity. The contributions of water interactions with all other components are much larger in this case and since these contribute to both $R^x$ and $R$ their ratio will tend more closely to unity. A ratio of unity or near unity which has been obtained in at least one biological membrane (Gutkneckt, 1967) may be due to a very small water content within the membrane or to a very large degree of coupling between water and membrane components or both effects in combination.

### 3. *Composite Membranes*

Composite membranes formally introduce the possibility of application of the above arguments to heterogeneous membrane systems. For simplicity Kedem and Essig (1966) consider the heterogeneity to be limited to a system of two discrete parallel arrays designated $\alpha$ and $\beta$. Each array is considered to consist of identical elements. In a system of two tracers 2 and 3, as before, the net flux in $\alpha$ (per unit area) is $J_\alpha$, with influx $J_{2\alpha}/\rho_{20}$ and outflux $J_{2\alpha}/\rho_{20} - J_\alpha$ and similarly for the $\beta$ array. For this system the observed flux ratio, $f$, is

$$f = (\text{influx } \alpha + \text{ influx } \beta)/(\text{outflux } \alpha + \text{ outflux}\beta) \tag{107}$$

$$f = \left(\frac{J_{2\alpha}}{\rho_{20}} + \frac{J_{2\beta}}{\rho_{20}}\right) \Bigg/ \left[\left(\frac{J_{2\alpha}}{\rho_{20}} - J_\alpha\right) + \left(\frac{J_{2\beta}}{\rho_{20}} - J_\beta\right)\right]$$

Expressing $f$ in terms of the individual flux ratios across the separate arrays $f_\alpha$ and $f_\beta$ equation (107) becomes

$$f = \frac{(f_\alpha/f_\alpha - 1)\ln f_\alpha/R_\alpha^x + (f_\beta/f_\beta - 1)\ln f_\beta/R_\beta^x}{(1/f_\alpha - 1)\ln f_\alpha/R_\alpha^x + (1/f_\beta - 1)\ln f_\beta/R_\beta^x} \tag{108}$$

If the $\alpha$ array is considered to be a passive leak then $\ln f_\alpha = \ln f_p = X/RT$. As $X$, the force across the membrane tends to zero, $f_p \to 1$ and $\ln f_p \to f_p - 1$. Under these conditions

$$f_{X=0} = \frac{n + (f_\beta/f_\beta - 1)\ln f_\beta}{n + (1/f_\beta - 1)\ln f_\beta} \tag{109}$$

where $n$ is defined as the ratio $R_\beta^x/R_\alpha^x (= R_\beta^x/R_p)$

This equation is applicable to short circuit conditions across a membrane in which one of the parallel arrays ($\alpha$) represents a passive leak with zero coupling. In particular, when the $\beta$ array constitutes an active transport path, equation (109) shows the net flux ratio, $f$, as a function of that in the active path, $f_\beta$. A plot of $f$ versus $f_\beta$, shown in figure 4, indicates that these are equal only if $n$ is zero and so $R_\beta$ is infinite. Physically this means that there is no passive leak. It is also obvious that flux ratios of twenty or greater are not to be expected if $n$ is greater than 0·5, even though the flux ratio in the active

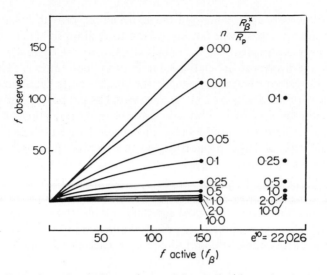

**Figure 4.** Effect of passive leak on observed flux ratio (short-circuited membrane). (From Kedem and Essig, 1966).

array is $\approx$ 20,000. The implications for biological systems, where observed flux ratios may be high under short circuit conditions, are obvious. In such systems the passive leak must be relatively insignificant.

Essig (1968) has applied the same theory to a discussion of the pump-leak model for symmetrical membranes where there is no net flow of species in the steady state. The presence of active transport is demonstrated by the development of electrical and chemical potentials across the membrane.

The validity of the pump-leak model for sodium transport across muscle has been evaluated from energetic considerations. This analysis requires an estimate of the electrochemical potential difference, $-X$, across the membrane and the rate of active transport of the ion, $J^a$. In the steady state their product is a measure of the minimal requirements for energy expenditure and may be compared with estimates of supply of energy to the system from metabolic sources. Experimentally it is rather difficult to measure these quantities precisely. The rate of active transport is usually determined by tracer experiments in which either the influx from an isotopically traced external solution or the efflux from a preloaded tissue is measured. Taking the commonly accepted value of $-X/F$ as 120 mV in the muscle tissue the passive influx of sodium across a simple leak pathway will be greater than the passive efflux, since the electrical potential is negative within the tissue. Since the net flux of sodium is zero in the steady state, the passive influx of sodium will closely approximate to the active efflux, provided that the pump is unidirectional and so will not contribute to sodium influx. In the second method the measurement of tracer efflux will also approximate to the net rate of active transport under the same limiting assumptions. The rates of active transport so obtained predict rather high energy requirements when combined with cited values of $X$. Ling (1962) has proposed that this rather high estimate may be due to an over-estimation of the value of $X$. Ussing and coworkers have proposed a mechanism of exchange diffusion in which it is assumed that a large proportion of the isotope flow occurs without expenditure of metabolic energy. Levi and Ussing (1948) suggested a thermal mechanism by which this might occur, at least in principle.

Essig (1968) has reassessed the experimental assumptions, in particular, allowing that the active transport mechanism will allow passage of tracer (i.e. sodium) in either direction, provided the gradient of electrochemical potential is of a suitable size and direction to successfully oppose metabolic forces.

Using the treatment of Kedem and Essig (1965), outlined above, we may consider the influx as the sum of the influxes by active and passive paths, so that

$$-\text{influx} = \frac{J^x}{\rho} = \frac{J^{ax}}{\rho} + \frac{J^{px}}{\rho} \qquad (110)$$

where $J_i^{ax}/\rho$ and $J_i^{px}/\rho$ are the influx by active and passive paths, respectively. Defining the flux ratio in the active path by $f^a$, then

$$f^a = -\frac{\text{efflux}}{\text{influx}} = \frac{J^{ax}/\rho - J^a}{J_{ax}/\rho} \quad \text{where } J^a \text{ is the net active flow.} \qquad (111)$$

$$\text{so that } J^{ax}/\rho = J^a/1 - f^a \qquad (112)$$

and a similar expression relates passive influx, $J/^{px}\rho$, to net passive flow, $J^p$.

Substituting equation (112) into equation (110)

$$J^x/\rho = J^a/1 - f^a + J^p/1 - f^p \qquad (113)$$

If it is assumed that there is zero coupling in the passive leak,

$$f^p = \exp X/RT, \ J^p = X/R^p \qquad (114)$$

and from equation (100)

$$f^a = \exp(J^a R^{ax}/RT) \qquad (115)$$

Since there is no net flow in the steady state

$$J^a + J^p = 0 \qquad (116)$$

From equations (113) and (116)

$$\frac{J^x}{\rho} = J^a \left( \frac{1}{1-f^a} - \frac{1}{1-f^p} \right) \tag{117}$$

and so, from equations (114) and (115),

$$\frac{J^x}{\rho J^a} = \frac{1}{1+\exp(-R^{ax}/R^p \, X/RT} - \frac{1}{1-\exp(X/RT)} \tag{118}$$

This expression relates the observed influx, $J^x/\rho$, to the rate of active transport, $J^a$.

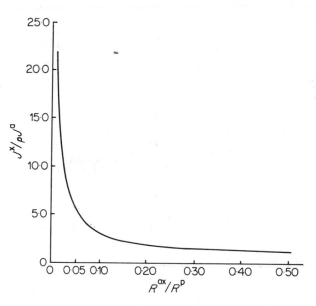

**Figure 5.** Ratio of rate of influx to rate of active transport $(-X/F = 120 \text{ mV})$. (From Essig, 1968).

Since no experimental values of the ratio $R^{ax}/R^p$ are available, it is of interest to plot $J^x/\rho J^a$ as a function of variable $R^{ax}/R^p$, taking $-X/F$ as 120 mV. The results obtained are shown in figure 5. It is obvious that the influx may be many times the rate of active flow when $R^{ax}/R^p$ is small (or if $X$ is small). When $R^{ax}/R^p$ is small the passive leak becomes negligible and a large proportion of the influx occurs by the active pathway. If, on the other hand, the value of $X$ becomes small, the flux ratio in the passive leak will approach unity and the observed influx (now mainly by the leak pathway) will be greater than the net outflux.

This treatment clearly shows the ability of the pump-leak model to describe active transport across symmetrical membranes and indicates the fundamental error in the usual interpretation of influx or efflux as a measure of the net rate of active transport. It cannot make further contribution to the arguments concerning the energy requirements of the pump without a knowledge of the ratio $R^{ax}/R^p$. Essig tentatively estimates this ratio as 0·2 for a typical frog skin. Should this be the case, the influx would be a little more than twice the true rate of active transport.

The value of $X$ will usually be determined by the electrical potential set up by the active pump, which by removing sodium ions will create a negative electrical potential inside the tissue. In the presence of a leak pathway the value of $X$ will be large only if $R^p$ is also large, since the leak is quite analogous to an electrical short-circuit. There is therefore some limitation to the minimal value of the ratio $\dfrac{X}{RT} \cdot \dfrac{R^{ax}}{R^p}$ of equation (118). Essig has shown that any measure which will reduce $J^a$ (without altering $R^{ax}$) will increase the influx via the active transport path while at the same time decreasing the influx in the leak through its effect upon $X$. If such is the effect of certain poisons upon the membrane system the resultant effect upon the observed influx might well be small due to this compensation effect.

## REFERENCES

Bak, T. A. (1963) *Contributions to the Theory of Chemical Kinetics*, W. A. Benjamin, New York
Baranowski, B. and J. Popielawski (1964) *Roczniki Chem.*, **38**, 483
Blumental, R. A. and A. Katchalsky (1968) In preparation
Caplan, S. R. (1966) *J. Theoret. Biol.*, **10**, 209
Casimir, H. B. G. (1945) *Rev. Mod. Phys.*, **17**, 343
Conway, E. J. (1954) *Symp. Soc. Exptl. Biol.*, **8**, 297
Coster, H. G. L. and E. P. George (1968) *Biophys. J.*, **8**, 457
Dainty, J. and B. Z. Ginzburg (1963) *J. Theoret. Biol.*, **5**, 256
Danielli, J. F. (1935) *J. Gen. Physiol.*, **19**, 19
Danielli, J. F. (1954) *Symp. Soc. Exptl. Biol.*, **8**, 502
De Donder, Th. (1927) *L'Affinite*, Gauthier Villers, Paris
De Donder, Th. and P. Van Rysselberghe (1936) *Thermodynamic Theory of Affinity*, Stanford University Press, Stanford
Dunham, E. T. and I. M. Glynn (1961) *J. Physiol. (London)*, **156**, 274
Essig, A. (1968) *Biophys. J.*, **8**, 53
Fitts, D. D. (1962) *Non-equilibrium Thermodynamics*, McGraw-Hill, New York
Glynn, I. (1957) *Prog. Biophys. Biophys. Chem.*, **8**, 241
De Groot, S. R. (1951) *Thermodynamics of Irreversible Processes*, North-Holland, Amsterdam

De Groot, S. R. and P. Mazur (1962) *Non-equilibrium Thermodynamics*, North-Holland, Amsterdam
Gutknecht, J. (1967) *Science*, **158**, 787
Hevesy, G. V., E. Hofer and A. Krogh (1935) *Scand. Arch. Physiol.*, **72**, 199
Hoshiko, T. and B. D. Lindley (1964) *Biochem. Biophys. Acta*, **79**, 301
Jardetsky, O. and F. M. Snell (1960) *Proc. Natl. Acad. Sci. U.S.*, **46**, 616
Katchalsky, A. and P. Curran (1965) *Non-Equilibrium Thermodynamics in Biophysics*, Harvard University Press, Cambridge
Katchalsky, A. and R. Spangler (1968) *Quart. Rev. Biophys.*, **1**, 127
Kedem, O. (1960) In A. Kleinzeller and A. Kotyk (Eds.), *Membrane Transport and Metabolism*, Academic Press, New York
Kedem, O. and S. R. Caplan (1965) *Trans. Faraday Soc.*, **61**, 1897
Kedem, O. and A. Katchalsky (1961) *J. Gen. Physiol.*, **45**, 143
Kedem, O. and A. Essig (1965) *J. Gen. Physiol.*, **48**, 1047
Kedem, O. and I. Michaeli (1961) *Trans. Faraday Soc.*, **57**, 1185
Kedem, O. and A. Katchalsky (1963) *Trans. Faraday Soc.*, **59**, 1918, 1931, 1941
Kirkwood, J. G. (1954) In T. T. Clarke and D. Nachmanson (Eds.), *Ion Transport Across Membranes*, Academic Press, New York, p. 119
Levi, H. and H. H. Ussing (1948) *Acta Physiol. Scand.*, **16**, 232
Ling, G. N. (1962) *A Physical Theory of the Living State: The Association–Induction Hypothesis*, Blaisdell, New York
Mackay, D. (1960) *J. Phys. Chem.*, **64**, 1718
Mackay, D. and P. Meares (1959) *Trans. Faraday Soc.*, **55**, 1221
Mauro, A. (1957) *Science*, **126**, 252
Meares, P. and H. H. Ussing (1959a) *Trans. Faraday Soc.*, **55**, 142
Meares, P. and H. H. Ussing (1959b) *Trans. Faraday Soc.*, **55**, 244
Meixner, J. (1941) *Ann. Physik*, **39**, 333
Meixner, J. (1942) *Ann. Physik*, **41**, 409
Meixner, J. (1943) *Ann Physik*, **43**, 244
Mickulecky, D. C. and S. R. Caplan (1966) *J. Phys. Chem.*, **70**, 3049
Miller, D. G. (1960) *Chem. Rev.*, **60**, 15
Miller, D. G. (1966) *J. Phys. Chem.*, **70**, 2639
Mitchell, P. (1962) *Biochem. Soc. Symp. (Cambridge, Engl.)*, **22**, 142
Nims, L. F. (1962) *Science*, **137**, 130
Nims, L. F. and G. Blaudeau (1968) *J. Appl. Physiol.*, in press
Onsager, L. (1931a) *Phys. Rev.*, **37**, 405
Onsager, L. (1931b) *Phys. Rev.*, **37**, 2265
Opit, L. J. and J. S. Charnock (1965) *Nature*, **208**, 471
Post, R. L., C. R. Merrett, C. R. Kinsolving and C. D. Albright (1960) *J. Biol. Chem.*, **235**, 1796
Prigogine, I. (1947 )*Étude Thermodynamique des Phénomènes Irreversibles*, Dunod, Paris and Desoer, Liège
Prigogine, I. (1967) *Introduction to Irreversible Thermodynamics*, 3rd Edn., Interscience, New York
Prigogine, I. and J. M. Wiame (1946) *Experientia*, **2**, 451
Rosenberg, T. and W. Wilbrandt (1951) *Helv. Physiol. Pharmacol. Acta.*, **9**, C86
Skou, J. C. (1957) *Biochim. Biophys. Acta.*, **23**, 394
Skou, J. C. (1960) *Biochim. Biophys. Acta.*, **42**, 6
Skou, J. C. (1962) *Biochim. Biophys. Acta.*, **58**, 314
Snell, F. M. (1965) *J. Phys. Chem.*, **69**, 2479
Snell, F. M. and B. Stein (1966) *J. Theoret. Biol.*, **10**, 177
Spiegler, K. S. (1958) *Trans. Faraday Soc.*, **54**, 1409
Staverman, A. J. (1952) *Trans. Faraday Soc.*, **48**, 176
Teorell, T. (1959) *J. Gen. Physiol.*, **42**, 831
Teorell, T. (1962) *Biophys. J.*, **2**, 27
Teorell, T. (1966) *Ann. N.Y. Acad. Sci.*, **137**, 950

Thomson, W. (Lord Kelvin) (1854) *Proc. Roy. Soc. Edinburgh*, **3**, 225
Thomson, W. (Lord Kelvin) (1857) *Trans. Roy. Soc. Edinburgh*, **21**, 123
Ussing, H. H. (1949) *Acta Physiol. Scand.*, **19**, 43
Ussing, H. H. (1957) In Q. R. Murphy (Ed.), *Metabolic Aspects of Transport Across Cell Membranes*, The University of Wisconsin Press, Madison
Ussing, H. H. and K. Zerahn (1951) *Acta Physiol. Scand.*, **23**, 110
Van Rysselberghe, P. and Th. De Donder (1936) *Thermodynamic Theory of Affinity*, Stanford University Press, Stanford
Van Rysselberghe (1936) *Bull. Classe Sci. Acad. Roy. Belg.*, **22**, 1330
Van Rysselberghe, P. (1937) *Bull. Classe Sci. Acad. Roy. Belg.*, **23**, 416
Whittam, R. (1964) *Transport and Diffusion in Red Blood Cells*, Edward Arnold, London
Wang, J. H. (1952) *J. Amer. Chem. Soc.*, **74**, 1612
Wang, J. H. and S. Miller (1952) *J. Amer. Chem. Soc.*, **74**, 1611
Wang, J. H., C. V. Robinson and I. S. Edelman (1953) *J. Amer. Chem. Soc.*, **75**, 466
Zerahn, K. (1956) *Acta Physiol. Scand.*, **23**, 347

CHAPTER 6

# A thermodynamic appraisal of oxidative phosphorylation with special reference to ion transport by mitochondria

H. Rottenberg    S. R. Caplan

*Biophysical Laboratory,*
*Harvard Medical School,*
*Boston, Massachusetts, U.S.A.*

A. Essig

*Renal Service,*
*New England Medical Center Hospitals*
*and Tufts University School of Medicine,*
*Boston, Massachusetts, U.S.A.*

# I.  INTRODUCTION

It is generally recognized that mitochondrial oxidative phosphorylation and ion transport are linked inseparably. The substrates and products of both oxidation and phosphorylation are ions, and their transport must affect the rates of these processes. Ions which are not metabolized are also transported and thereby affect the coupling between oxidation and phosphorylation. In addition, the necessity for a precise arrangement of the enzymes in a functional mitochondrial membrane suggests strongly the importance of transport phenomena.

Many models have been proposed to account for the link between ion transport and oxidative phosphorylation. A detailed molecular description of so complicated a system is hard to test experimentally, and standard biochemical techniques have failed to distinguish among the alternatives. Little attention has been paid to the thermodynamic aspects of the models, despite the fact that physical intuition is frequently insufficient to overcome conceptual difficulties. Thermodynamics can lead to greater clarity, since it imposes self-consistent logical requirements. The usefulness of a macroscopic thermodynamic approach in both testing and constructing molecular models is the subject of this chapter.

We shall begin by reviewing briefly considerations relating to both oxidative phosphorylation and the associated ion transport, and shall then demonstrate how a thermodynamic approach can help to distinguish between different theories.

# II.  OXIDATIVE PHOSPHORYLATION

Oxidative phosphorylation is the process in which oxidation of Krebs cycle intermediates (or other reducing agents), catalysed by the respiratory system, results in the synthesis of ATP (figure 1). Most thermodynamic treatments have been based on the determination of a 'stoichiometric ratio', i.e. the ratio between the rates of phosphorylation and oxidation (P/O ratio). Normally, the *maximal* experimental ratio, rounded off to an integer, is considered to represent the stoichiometry which would obtain ideally; taken together with other evidence it is assumed to indicate the number of 'coupling sites' (Slater, 1966). The free energies of the reactions are compared and the efficiency of the process computed. The second law of thermodynamics then provides a criterion for the self-consistency of the data. Thus if $n$ is the presumed stoichiometric ratio, it is then necessary that

$$-\Delta G_{\text{oxidation}} \geq n\Delta G_{\text{phosphorylation}} \tag{1}$$

or in other words the efficiency obeys the relation

$$-n\Delta G_{\text{phosphorylation}}/\Delta G_{\text{oxidation}} \leq 1 \tag{2}$$

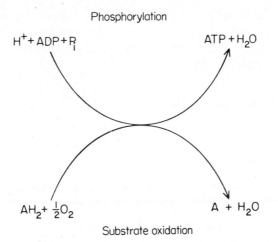

**Figure 1.** General scheme of oxidative phosphorylation

This consideration is also used to examine the possibility of coupling in different regions of the respiratory chain. Thus, assuming $n = 1$ at a given site, the oxido-reduction potentials are compared with the phosphate potential* (Lehninger, 1964; Slater, 1966).

As we have pointed out (Rottenberg and coworkers, 1967), the above considerations are unambiguous only if there is complete coupling.

Since this is obviously not the case† stoichiometric ratios and efficiencies cannot be discussed without reference to the state of the system—otherwise such calculations only accidentally reflect its physical parameters. It is noteworthy that the determinations of $n$, $\Delta G_{\text{oxidation}}$, and $\Delta G_{\text{phosphorylation}}$ are seldom performed simultaneously, and calculations are based on extrapolated and estimated values not necessarily referring to the same state.

Similarly, there are inadequacies in the usual energetic analyses of data obtained in state 4 (oxidation in the absence of phosphorylation, owing to a limiting amount of ADP—Chance and Williams, 1956) and state 3 (steady state phosphorylation after the addition of ADP). The classical treatment leads to the 'respiratory ratio' for these states (Chance and Williams, 1956), but does not make use of all the relevant information. It is quite clear that a system in state 4, in which the P/O ratio is zero, reflects the energy conversion mechanism just as well as a system in state 3. We shall attempt to utilize all

---

*In principle, $n$ need not necessarily equal 1 at a given site. It is conceivable that oxidative phosphorylation might occur at sites not ordinarily considered, with values of $n$ less than 1.

†Under many conditions respiration can take place at reasonable rates without net phosphorylation. The uncoupling may be the consequence of side reactions or leaks.

the available data in our treatment below, employing nonequilibrium thermodynamics. However, first we ought to examine some of the methods used for the determination of the P/O ratio and the free energies of reactions (see for example Slater, 1966).

## A. The Hexokinase Trap Method

In this method mitochondria are incubated in the presence of an excess of hexokinase catalysing the phosphorylation of glucose by ATP. This reaction keeps the phosphate potential small and thus allows continuous phosphorylation. (In the absence of the hexokinase reaction phosphorylation would rapidly come to a halt.) Apparently the free energy of ADP phosphorylation has never been determined in this system; the estimate of 9 kcal/mole (Lehninger, 1964), considered adequate for most conditions, is not appropriate here.

The appropriate value can be computed according to the following scheme:

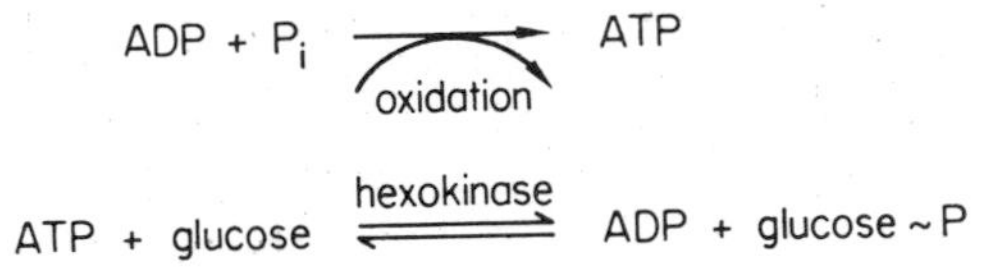

$$ADP + P_i \xrightarrow{\text{oxidation}} ATP \tag{3}$$

$$ATP + glucose \underset{\text{hexokinase}}{\rightleftharpoons} ADP + glucose \sim P \tag{4}$$

The hexokinase reaction is assumed to be in equilibrium. The initial concentrations of $P_i([P_i]_0)$, $ATP([ATP]_0)$, and glucose ($[glucose]_0$) are known and the concentration of $P_i$ is determined during the phosphorylation. The standard free energies of glucose phosphorylation and ADP phosphorylation at pH 7·0, room temperature, and in the presence of $Mg^{2+}$ are $-4·6$ kcal/mole and 7·0 kcal/mole, respectively (Robbins and Boyer, 1957; Benzinger and coworkers, 1959). Since the initial ADP concentration is essentially zero,

$$[ADP] = [ATP]_0 - [ATP] \tag{5}$$

Also

$$[glucose \sim P] = ([ATP]_0 - [ATP]) + ([P_i]_0 - [P_i]) = ([ATP]_0 - [ATP]) + \Delta P_i \tag{6}$$

and

$$[glucose] = [glucose]_0 - [glucose \sim P] \tag{7}$$

For the equilibrium of the hexokinase reaction,

$$2·3 \, RT \log ([glucose \sim P][ADP]/[glucose][ATP]) = -\Delta G^0_{\text{glucose phosphorylation}}$$
$$= 4·6 \text{ kcal/mole} \tag{8}$$

Finally,

$$\Delta G_{\text{ADP phosphorylation}} = \Delta G^0_{\text{ADP phosphorylation}} + 2\cdot3\, RT \log([\text{ATP}]/[\text{ADP}][\text{P}_i])\,(9)$$
(where $\Delta G^0_{\text{ADP phosphorylation}} = 7\cdot0$ kcal/mole).

This set of five equations in five unknowns ([ADP], [ATP], [glucose], [glucose$\sim$P] and $\Delta G_{\text{ADP phosphorylation}}$) can be solved easily. Substituting equations (5), (6), and (7) into (8) gives

$$2\cdot3\, RT \log \frac{(\Delta \text{P}_i + [\text{ATP}]_0 - [\text{ATP}])\,([\text{ATP}]_0 - [\text{ATP}])}{([\text{glucose}]_0 - \Delta \text{P}_i - [\text{ATP}]_0 + [\text{ATP}])\,[\text{ATP}]} = 4\cdot6 \text{ kcal/mole} \quad (10)$$

and under most conditions in experiments of this kind one may use the approximation

$$[\text{ATP}] = \frac{(\Delta \text{P}_i + [\text{ATP}]_0)\,[\text{ATP}]_0}{([\text{glucose}]_0 - \Delta \text{P}_i - [\text{ATP}]_0)\,(2\cdot3 \times 10^3)} \quad (11)$$

It is then found by solving equation (9) that under normal conditions, for instance when $[\text{P}_i]_0 = 5$ mM, $[\text{ATP}]_0 = 1$ mM, and $[\text{glucose}]_0 = 10$ mM, after a 1 mM decrease in $[\text{P}_i]$ the free energy of ADP phosphorylation is only $4\cdot8$ kcal/mole. This is about half of the value normally assumed.

## B. The Respiratory Control Method

In this method mitochondria are first maintained in state 4 (oxidation in the absence of net phosphorylation—this is a case of 'static head', as will be discussed in Section IV). A known amount of ADP is added, causing an increased rate of respiration (and presumably phosphorylation). When the respiration rate declines to its state 4 value, it is assumed that all the added ADP has been phosphorylated. Even if we consider state 3 to be a steady state, there are numerous difficulties in treating these data. A complete analysis requires knowledge of the rates and free energies of oxidation and phosphorylation throughout the course of the reaction. In fact no direct measurement of the rate of phosphorylation is available. In practice it is assumed that the P/O ratio obtaining throughout the reaction is given by the ratio of added ADP to the amount of oxygen consumed during state 3 respiration. However, according to the formulation to be presented in Sections IV and V we would expect the P/O ratio to vary as the free energies change. For example, we would expect that the P/O ratios found by the hexokinase method would be larger than those found by the respiratory control method, and it is our impression that this is so (Slater, 1966). Similarly, it is important to use the appropriate value for the free energy of phosphorylation. There are only a few studies in which the free energy of phosphorylation has been measured or estimated under the conditions of the respiratory control experiment (Slater, 1966; Cockrell and coworkers, 1966). The value seems significantly higher than that which we have calculated for the hexokinase trap method. In Section VI we shall show how to extract enough information from a single respiratory control experiment to approximately evaluate the energetics.

We have not as yet commented on the mechanism of coupling in oxidative phosphorylation. Although thermodynamic considerations *per se* are independent of mechanisms, a particular molecular model often carries thermodynamic implications which can be tested experimentally. The many models and theories that have been proposed for oxidative phosphorylation fall into two categories: the 'chemical' hypothesis and the 'chemiosmotic' hypothesis.

The chemical hypothesis postulates direct chemical coupling: oxidation results in the formation of a high-energy intermediate which drives phosphorylation (Slater, 1966). Since it is now known that mitochondria can maintain concentration gradients of ions (including protons) (Lehninger and coworkers, 1967), the hypothesis is usually extended to include coupling of ion pumps to the intermediate.

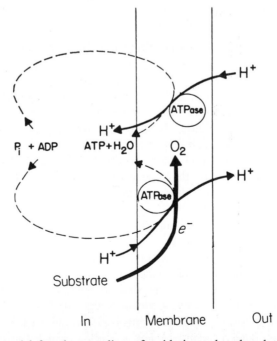

**Figure 2.** A model for the coupling of oxidation, phosphorylation and proton translocation. (From Rottenberg and coworkers, 1967)

In his chemiosmotic hypothesis, Mitchell (1961, 1966) suggests that oxidation is coupled to ejection of protons from the mitochondria. The resultant electrochemical potential gradient of hydrogen drives phosphorylation in the membrane by a reversal of an ATPase–hydrogen pump. Cation concentration gradients are a direct result of the electrical membrane potential produced by proton transport.

Both theories link ion transport to oxidative phosphorylation, but whereas in the chemical theory coupling results from the sharing of a high-energy intermediate, in the chemiosmotic theory ion flows are intrinsic to the very mechanism of oxidative phosphorylation. The thermodynamic implications of the two alternatives are quite different. However, as we discussed elsewhere (Rottenberg and coworkers, 1967), the mechanisms are not mutually exclusive and it is possible that both are operative, as indicated in figure 2.

In order to examine these matters further we must consider the models for ion transport in greater detail.

## III. THE LINK BETWEEN ION TRANSPORT AND OXIDATIVE PHOSPHORYLATION

Ion transport in mitochondria has been reviewed by Lehninger and co-workers (1967) and is also discussed in Volume 2 of this work. We shall therefore limit ourselves to considerations relevant to oxidative phosphorylation. Mitochondria can accumulate large quantities of salts and maintain concentration gradients of both cations and anions. This uptake can be supported either by hydrolysis of ATP or by substrate oxidation. Proponents of the chemical hypothesis assume that the high-energy intermediate of oxidative phosphorylation drives a pump which transports certain cations inward. Such a model has been proposed for potassium (Cockrell and coworkers, 1966) and for calcium (Rasmussen and coworkers, 1965). On the other hand, it has been suggested that the cation pump is non-specific, and that specificity results from a permeability barrier (Chance, 1967). Other models assume that the high-energy intermediate drives a hydrogen pump which transports protons out of the mitochondria (Chappell and Crofts, 1965). In Mitchell's model (Mitchell, 1966) the hydrogen pump is driven directly by oxidation. The resultant electrochemical potential difference of hydrogen ion is the driving force for phosphorylation (by reversal of an ATPase–hydrogen pump) and the electrical potential influences cation uptake. Most of the models mentioned postulate exchange diffusion systems which exchange cations for protons and/or anions for hydroxyl ions. Another hydrogen pump model involving no exchange diffusion has been suggested by Rottenberg (1968). This model which is described in figure 3, assumes that a pump ejects protons from the mitochondria, resulting in a membrane

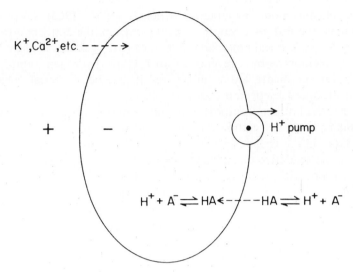

**Figure 3.** A model for ion transport in mitochondria. (After Rottenberg, 1968) In the steady state, after all net ion flows have vanished:

$$\Delta\psi < 0 \qquad \frac{RT}{F} \ln \frac{[\text{cation}]_{in}}{[\text{cation}]_{ex}} \simeq \Delta\psi$$

$$\Delta pH > 0 \qquad \frac{[A^-]_{in}}{[A^-]_{ex}} = \frac{[H^+]_{ex}}{[H^+]_{in}}$$

$$\frac{[A^{-2}]_{in}}{[A^{-2}]_{ex}} = \left(\frac{[H^+]_{ex}}{[H^+]_{in}}\right)^2$$

potential and a pH difference. The potential can drive cations upward and the pH difference can promote the accumulation of weak acids. Ion transport rates are determined by membrane permeability and affect both pH and electrical potential gradients.

The fact that external cations can cause uncoupling of phosphorylation is explained on the chemical theory by a competition for the high-energy intermediate; increasing the external cation concentration is assumed to increase the rate of active transport. In the chemiosmotic theory this uncoupling is assumed to be the result of an increase in membrane conductance which reduces the membrane potential. The resultant decrease in the electrochemical potential gradient of the hydrogen ion slows phosphorylation.* The

---

*In Mitchell's model the influx of cations stimulates the exchange of internal cations for external hydrogen ions; this exchange diffusion reduces $\Delta pH$.

difference between the chemical and the chemiosmotic theories regarding the mechanism of uncoupling by cations suggests a possible means of distinguishing between the two models.

It is known that valinomycin can completely uncouple phosphorylation and that the crucial factor in this uncoupling is the external concentration of potassium (Höfer and Pressman, 1966). Only in the presence of relatively high concentrations of potassium does uncoupling occur, while at low concentrations (up to 10 mM) the P/O ratio is not affected (the rate of phosphorylation might even increase with the addition of potassium). Höfer and Pressman attribute these findings to an increased rate of transport at high potassium concentrations (and a consequent depletion of the intermediate). However, Rottenberg's (1969a) investigation of uncoupling by valinomycin in the presence of potassium does not verify this explanation. It was found that the rate of net potassium transport does not increase with external potassium concentration. Since it seems unlikely that increasing the external potassium concentration would increase potassium leakage from the mitochondria, these findings argue against the view that uncoupling is due to increased potassium pumping. Moreover, since the concentration gradient decreases with increase of external potassium, the work performed by the potassium pump, according to the assumptions of Cockrell and coworkers (1966), would decrease as well. Table 1 shows the work of potassium transport calculated on this basis for various potassium concentrations. The potassium transport work is the product of the number of moles of potassium transported and the free energy expenditure per mole, $-\Delta G_{potassium}$. The calculation assumes that $-\Delta G_{potassium} = RT \ln ([K^+]_{in}/[K^+]_{ex})$, i.e. that the electrical potential difference across the membrane is zero. It can be seen that as external potassium is increased there is a decrease both in potassium transport work and in phosphorylation work. Moreover, the phosphorylation work is much larger than the potassium transport work even when the P/O ratio becomes very low. If one considers the possibility of an electrical potential difference as well, with the outside positive as proposed by Mitchell

**Table 1.** Comparison of potassium transport work and phosphorylation work at various degrees of uncoupling by valinomycin and potassium

| External potassium mM/l | P/O ratio | Potassium transport work cal. $\times 10^{-6}$/mg dry wt | Phosphorylation work cal. $\times 10^{-6}$/mg dry wt |
|---|---|---|---|
| 10·3 | 1·8 | 246 | 1090 |
| 20·3 | 0·7 | 41 | 366 |
| 29·8 | 0·4 | 38 | 222 |
| 103 | 0·0 | 1 | 0 |

(From Rottenberg, 1969a)

(1966), the calculated potassium transport work becomes even smaller. A further consideration is that additions of potassium in concentrations which result in uncoupling (10 to 30 mM, see table 1) fail to alter the rate of respiration. This again indicates that the uncoupling is not due to increased energy consumption by a potassium pump.

Cockrell and coworkers (1967) have reported synthesis of ATP by mitochondria exposed to potassium-free media in the presence of valinomycin. Their interpretation is that valinomycin facilitates the access of internal potassium to an energy-linked cation transport system, transport being tightly coupled to ATP synthesis. However, these data, like those above, may also be interpreted as a consequence of effects on the membrane potential (Glynn, 1967).

It will be seen in the following sections that the determination of the electrochemical potential difference of hydrogen ion ($\Delta\tilde{\mu}_H$) plays a central role in the evaluation of oxidative phosphorylation models. This quantity is given by

$$\Delta\tilde{\mu}_H = \tilde{\mu}_H^{in} - \tilde{\mu}_H^{ex} = RT\ln([H^+]_{in}/[H^+]_{ex}) + F\Delta\psi = -2\cdot3\,RT\Delta pH + F\Delta\psi \quad (12)$$

Methods for the determination of $\Delta pH$ have been developed by Chance (1967) and Addanki and coworkers (1968). The electrical potential difference can be calculated from the distribution of potassium in the presence of an excess of valinomycin; the assumption here, which is at variance with the assumptions of Cockrell and coworkers, is that under these conditions the mitochondria are sufficiently permeable to potassium so that $\Delta\tilde{\mu}_K \simeq 0$. This assumption is based on recent experimental observations (Rottenberg, 1969b). Thus,

$$\Delta\tilde{\mu}_K = RT\ln([K^+]_{in}/[K^+]_{ex}) + F\Delta\psi \simeq 0 \quad (13)$$

and

$$\Delta\psi \simeq -(RT/F)\ln([K^+]_{in}/[K^+]_{ex}) \quad (14)$$

Figure 4 shows the relation between $\Delta\tilde{\mu}_H$ and the P/O ratio at various potassium concentrations. It is seen, as expected from the chemiosmotic theory, that the P/O ratio decreases with a decrease in $-\Delta\tilde{\mu}_H$. In particular, when $\Delta\tilde{\mu}_H \simeq 0$, P/O $= 0$, which is the most obvious prediction of the chemiosmotic theory.

## IV. GENERAL THERMODYNAMIC CONSIDERATIONS

Much of the difficulty in deciding between alternative models for oxidative phosphorylation results from the analysis of energetics exclusively in terms of equilibrium thermodynamics. As Lehninger and coworkers (1967) have pointed out, 'It is misleading . . . . . to think of the (chemiosmotic) hypothesis in terms of equilibrium thermodynamics, and there appears to be considerable

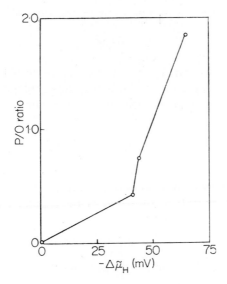

**Figure 4.** The effect of $\Delta\tilde{\mu}_H$ on the P/O ratio. (From Rottenberg, 1969a)

misunderstanding . . . . . because of this fact. The processes implicit in the chemiosmotic hypothesis are more properly analysed by non-equilibrium thermodynamic principles'. Thus an appropriate treatment requires the simultaneous determination of flows and forces. Much experimental work centres on the determination of flows alone. However, the pertinent considerations are the rates of input and output of energy, which necessitate a knowledge of the forces as well.

Although nonequilibrium thermodynamics has been discussed in the preceding chapter we should like to present here a formulation convenient for the consideration of oxidative phosphorylation. Nonequilibrium thermodynamics deals with relationships between generalized flows and forces. These flows and forces may or may not have a directional character, i.e. they may be vectorial or scalar (Katchalsky and Curran, 1965). In mitochondria the flows comprise vectorial flows of ions and scalar flows of metabolic reactions. The corresponding forces are the vectorial electrochemical potential differences and the scalar reaction affinities* (ordinarily the negative free energies of reaction are identical to the reaction affinities, i.e. $-\Delta G = A$).

The starting point of a nonequilibrium thermodynamic treatment at

*The affinity of a reaction is defined as $A = -\sum_i \nu_i \mu_i$, where $\nu_i$ stands for stoichiometric coefficients, positive for products and negative for reactants, and $\mu_i$ stands for chemical potentials (Prigogine, 1955).

G

constant temperature is the specification of the rate of dissipation of free energy $\Phi$, given by

$$\Phi = \sum_i J_i X_i \tag{15}$$

where $J_i$ is the $i$th flow and $X_i$ its conjugate force. $\Phi$ is important for two reasons. First, it defines a set of forces and flows adequate to completely characterize the system. For example, if it can be shown that the complete rate of dissipation of free energy is given by

$$\Phi = J_1 X_1 + J_2 X_2 + J_3 X_3 \tag{16}$$

any combination of three flows and/or forces from this set will determine the values of all the remaining flows and forces. In this way the dissipation function provides a secure basis for the design of experiments (see Section V). Secondly, according to the second law,

$$\Phi \geq 0 \tag{17}$$

Thus, for example, for the coupling of phosphorylation and oxidation,

$$J_P A_P + J_O A_O \geq 0 \tag{18}$$

Although this equation requires that the sum of the two terms be greater than or equal to zero, it places no restriction on the sign or magnitude of each term considered separately. In the case of oxidative phosphorylation, oxidation occurs spontaneously, so that $J_O A_O$ will be positive. Since the phosphorylation is coupled to oxidation, this energy may be used to drive $J_P$ against its conjugate force $A_P$, i.e. in a direction opposite to that which would be considered spontaneous. If so, $J_P A_P$ is negative. Equation (18) shows that if an oxidative reaction of affinity $A_O$ takes place at a rate $J_O$, the product $J_O A_O$ represents a limit to the rate of performance of phosphorylation work, $-J_P A_P$. In practice, since no process of interest can be thermodynamically reversible (there will always be a dissipation of free energy), $-J_P A_P$ must be less than $J_O A_O$, i.e. input must exceed output.

Having specified the dissipation function, we are able to describe each of the flows as a function of all the forces, or *vice versa*. Sufficiently close to equilibrium the relations will be linear. For example, for any two-flow system, expressing the relation between flows and forces in terms of conductance coefficients we write

$$J_1 = L_{11} X_1 + L_{12} X_2 \tag{19}$$

$$J_2 = L_{21} X_1 + L_{22} X_2 \tag{20}$$

Alternatively, in terms of resistance coefficients,

$$X_1 = R_{11} J_1 + R_{12} J_2 \tag{21}$$

$$X_2 = R_{21}J_1 + R_{22}J_2 \tag{22}$$

The phenomenological coefficients, like the flows and forces, may be either scalar or vectorial, as appropriate. According to Onsager's hypothesis, the matrix of phenomenological coefficients is symmetrical, i.e. $L_{12} = L_{21}$ and $R_{12} = R_{21}$. The Onsager relationship reduces the number of independent measurements necessary for a complete characterization of the system, but is not necessary for the analysis to be presented. Although the Onsager relations are widely applicable, their validity for metabolic coupling is as yet unknown. The problem has been studied, however, in a model membrane system in which the enzymatic hydrolysis of an amide to form a salt results in a current flow simulating active transport. In this system the relationship was found to be valid (Blumenthal and coworkers, 1967).

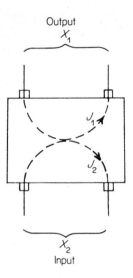

**Figure 5.** A diagrammatic representation of an energy converter.

It is often assumed that strict stoichiometry holds, i.e. that the output flow is an integral and constant multiple of the input flow whatever the forces. Although specific mechanisms may suggest stoichiometric ratios (see, for example, Slater, 1966; Mitchell, 1966) they are not required by thermodynamic considerations. In Section II we have pointed out that the coupling between oxidation and phosphorylation is in fact incomplete. For incompletely coupled systems it is useful to define a generalized parameter which determines the extent to which input drives output under all conditions of operation. This parameter has been called the 'degree of coupling' (Kedem and Caplan, 1965). For a two-flow system such as that shown in figure 5, the degree of coupling is given by

$$q = L_{12}/\sqrt{L_{11}L_{22}} = -R_{12}/\sqrt{R_{11}R_{22}} \tag{23}$$

The quantity $q$ is dimensionless. A corollary of equation (17) is that the matrix of phenomenological coefficients is 'positive-definite'. From this it is easily shown that

$$-1 \leqslant q \leqslant 1 \tag{24}$$

For completely uncoupled processes $q = 0$, for completely coupled processes (stoichiometry) $q = \pm 1$. At intermediate values of $q$ the ratio of rates

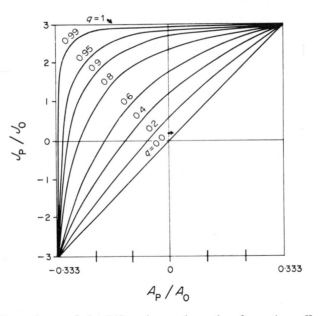

**Figure 6.** Dependence of the P/O ratio on the ratio of reaction affinities. The pertinent reaction affinities are those in the external solutions.
In the left upper quadrant oxidation 'drives' phosphorylation. (For complete coupling the P/O ratio has been taken to be 3.)

i) $J_H = 0$: $q = q_{PO}$

ii) $\Delta\tilde\mu_H = 0$: $q = \dfrac{(q_{PO} + q_{PH}q_{OH})}{\sqrt{(1 - q_{PH}^2)(1 - q_{OH}^2)}}$

(After Caplan and Essig, 1969)

of the two processes will be a function of the ratio of forces. All this is shown in figure 6, in which these concepts have been applied to the case of oxidative phosphorylation. Here $J_P/J_O$, the P/O ratio, is plotted as a function of $A_P/A_O$, the ratio of the affinities for phosphorylation and oxidation. The region of primary interest is the left upper quadrant, where oxidation 'drives' phosphorylation. It is seen that if coupling is incomplete the P/O ratio is not constant. However, if the coupling is not too incomplete, studies confined to a limited range of the ratio of forces might appear to indicate stoichiometry. Nevertheless, even for high degrees of coupling, studies over a wider range will reveal marked deviations from stoichiometry.

In the application of nonequilibrium thermodynamics considered here we confine ourselves to stationary states, i.e. states in which parameters such as temperature, pressure, concentrations, and electrical potentials do not

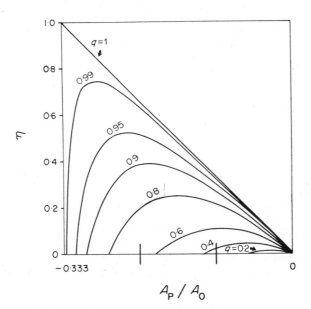

**Figure 7.** Dependence of the efficiency on the ratio of reaction affinities. The pertinent reaction affinities are those in the external solutions.
(For complete coupling the P/O ratio has been taken to be 3)

vary with time. For such states the flows also are independent of time. If $X_1$ and $X_2$ are both fixed experimentally, $J_1$ and $J_2$ will ultimately reach the stationary values given by equations (19) and (20). If, on the other hand, only one force is fixed, the flow conjugate to the other will vanish in the stationary state. For example, if only $X_2$ is fixed, $J_1$ will eventually become zero. Stationary states of particular interest are those of level flow and static head (Kedem and Caplan, 1965). Level flow is a state in which the output force $X_1$ is zero. Conversely, static head is a state in which the output flow $J_1$ is zero. Both level flow and static head occur commonly, for example $Na^+$ transport in the proximal tubule of mammalian kidney takes place at near level flow, as does an unloaded muscle contraction, while net ion transport in symmetrical cells such as the red cell is at static head.

A fundamental consideration is the adequacy of utilization of metabolic energy. No criterion is uniquely suitable; the conditions of operation determine the criterion to be employed (Essig and Caplan, 1968). The standard means of evaluating energy utilization is the efficiency, i.e. output/input. In thermodynamic terms this is given by

$$\eta = -\frac{J_1 X_1}{J_2 X_2} \tag{25}$$

The efficiency of a system is of course dependent on the degree of coupling as well as the ratio of forces, as shown in figure 7 for the case of oxidative phosphorylation. The maximum efficiency is solely dependent on the degree of coupling, being given by (Kedem and Caplan, 1965)

$$\eta_{\max} = q^2/(1+\sqrt{1-q^2})^2 \tag{26}$$

The efficiency is of interest if a system converts one form of energy into another. Two obvious examples are an electrochemical cell operating into a finite load, and a muscle raising a weight. In certain states oxidative phosphorylation also may result in appreciable conversion of one form of energy into another, and hence the efficiency may be important. On the other hand, at level flow or static head energy conversion and $\eta$ become zero. This does not mean, however, that the system is not utilizing energy adequately. In these circumstances, the function of the system is not energy conversion but the maintenance of large flows or forces, and hence the efficiency is of no significance.

The above formalism can be extended to multiple-flow systems; in the case of three flows, the forces are given by

$$X_1 = R_{11}J_1 + R_{12}J_2 + R_{13}J_3 \tag{27}$$

$$X_2 = R_{21}J_1 + R_{22}J_2 + R_{23}J_3 \tag{28}$$

$$X_3 = R_{31}J_1 + R_{32}J_2 + R_{33}J_3 \tag{29}$$

For a multiple-flow system the degree of coupling $q_{ij}$ is defined by

$$q_{ij} = -R_{ij}/\sqrt{R_{ii}R_{jj}} \tag{30}$$

and measures the extent to which the $i$th flow 'drags' the $j$th flow when the $j$th force and other flows are zero (Caplan, 1966). In contrast to the case of a two-flow system, $q_{ij}$ is not given by $L_{ij}/\sqrt{L_{ii}L_{jj}}$. Again for any pair of flows equation (24) continues to apply. As mentioned earlier, if any forces are left uncontrolled, the system approaches a stationary state in which the conjugate flows vanish (Katchalsky and Curran, 1965). This reduces the number of terms in the dissipation function and hence the number of independent variables characteristic of the system. Such simplification may be achieved by an appropriate choice of experimental conditions.

## V. NONEQUILIBRIUM THERMODYNAMIC ANALYSIS OF OXIDATIVE PHOSPHORYLATION AND ION FLOWS

A good deal of effort aimed at distinguishing between models of oxidative phosphorylation has involved attempts to isolate high-energy intermediates. The above general nonequilibrium thermodynamic considerations lead to criteria which do not depend on the isolation of an intermediate; neither is it necessary to observe a component of $H^+$ flow used for phosphorylation (Caplan and Essig, 1969). Although it is necessary to determine $\Delta\tilde{\mu}_H$ and possibly $\Delta\tilde{\mu}_K$ (see Section III), the other parameters are evaluated externally. In principle, it is possible to envisage conditions in which $J_H$ may be maintained non-zero and constant for brief periods by appropriately setting $\Delta\tilde{\mu}_H$. For this system, under isothermal conditions, it is reasonable to write

$$\Phi = J_P A_P + J_H \Delta\tilde{\mu}_H + J_O A_O \tag{31}$$

Here the subscripts P, H, and O refer respectively to phosphorylation, net $H^+$ flow, and substrate oxidation. Despite the complexity of the system, equation (31) is adequate; we consider only steady states in which flows such as those of $Ca^{2+}$, $Mg^{2+}$, and $H_2O$ will have come to a halt because their conjugate forces, i.e. their electrochemical potential gradients, are not experimentally controlled.*

---

*It is clear that electroneutrality necessitates that a net flow of $H^+$ be compensated by an equivalent flow of anions and/or cations. If these flows occur through leak pathways they are not coupled to the metabolic processes and so need not be included in this dissipation function, i.e. each flow of equation (31) is a function of only three forces. If some compensating flow is coupled to metabolism, its contribution may be completely eliminated from the dissipation function by making its conjugate force zero. It seems that the important compensatory ion is potassium. We have discussed in Section III our reasons for believing that potassium flow is not coupled to metabolism. However, even if it were, the means may possibly exist for making $\Delta\tilde{\mu}_K$ extremely small (e.g. exposure to valinomycin).

A related consideration is the possibility of an $H^+/K^+$ exchange. This would tend to keep $\Delta pH$ small, but we do not feel that the evidence cited for such exchange diffusion is compelling. If in fact it exists equation (31) will continue to apply providing $J_K \Delta\tilde{\mu}_K$ is zero.

The above dissipation function is ambiguous as to whether the affinities are to be measured internally or externally. Furthermore, since the flows may be non-conservative, i.e. a substance may be metabolized within the composite membrane, the rate at which substances enter at one surface may not equal the rate at which they leave at the other. Hence there is also ambiguity as to the definition of the flows. Clearly, it would be of great convenience to evaluate the affinities and the reaction rates externally, since internal activities are difficult to determine. This is found to be appropriate, as shown below, providing $J_H$ is taken as the rate of decrease of $H^+$ content inside the mitochondrion, and $\Delta\tilde{\mu}_H$ is defined as $\tilde{\mu}_H^{in}$-$\tilde{\mu}_H^{ex}$. (In the absence of a reaction involving $H^+$ in the interior of the mitochondrion, $J_H$ is simply the net outward flux across the inner surface of the membrane.)

---

Consider mitochondrial oxidation of a substrate to a product. The rate of dissipation of free energy isothermally when the membrane is in a stationary state is

$$\Phi = -\sum_j(\dot{n}_j^{in}\tilde{\mu}_j^{in}+\dot{n}_j^{ex}\tilde{\mu}_j^{ex}) \tag{32}$$

in which $\tilde{\mu}_j$ and $n_j$ represent electrochemical potential and number of moles of species $j(\dot{n}_j = dn_j/dt)$. The species $j$ include substrate, product, $O_2$, $CO_2$, $H_2O$, ATP, ADP, $P_i$, $H^+$, and possibly additional components such as $K^+$, $Ca^{2+}$, and $Mg^{2+}$. Designating the $r$th metabolic reaction by $J_r$ and the stoichiometric coefficients by $\nu_{jr}$,

$$\sum_r\nu_{jr}J_r = \sum_r(\dot{n}_{jr}^{in}+\dot{n}_{jr}^{ex}) = (\dot{n}_j^{in}+\dot{n}_j^{ex}) \tag{33}$$

It is immaterial whether the reactions take place entirely within the mitochondrial membrane or to some extent within the interior of the mitochondrion as well. Substituting equation (33) in equation (32),

$$\Phi = -\sum_j(\dot{n}_j^{in}\Delta\tilde{\mu}_j+\sum_r\nu_{jr}J_r\tilde{\mu}_j^{ex}) \tag{34}$$

The affinity $A_r$ in any region is defined as

$$A_r = -\sum_j\nu_{jr}\tilde{\mu}_j \tag{35}$$

Combining equations (34) and (35),

$$\Phi = -\sum_j\dot{n}_j^{in}\Delta\tilde{\mu}_j+\sum_r J_r A_r^{ex} \tag{36}$$

When the interior of the mitochondrion is in a stationary state for all components, $\dot{n}_j^{in} = 0$ for all $j$, and consequently for our example,

$$\Phi = \sum_r J_r A_r^{ex} = J_P A_P^{ex}+J_O A_O^{ex} \tag{37}$$

Thus, in this stationary state it suffices to measure changes in the external solution only, although in general $J_P$ and $J_O$ are given by the sum of the rates of change on both sides of the membrane. Equation (37) is an example of the general principle that if any forces are not controlled the system will reach a stationary state in which their conjugate flows are zero. The flows $J_P$ and $J_O$ do not vanish since $A_P^{ex}$ and $A_O^{ex}$ are fixed experimentally.

If we consider the stationary state in which both $\Delta \tilde{\mu}_H$ and $J_H$ are non-zero, equation (36) gives

$$\Phi = J_P A_P^{ex} + J_H \Delta \tilde{\mu}_H + J_O A_O^{ex} \tag{38}$$

In the approach to the above stationary states equation (36) applies, i.e. each flow is a function of several forces. However, in the stationary state of equation (38), each flow is a function of only three forces. If $\Delta \tilde{\mu}_H$ is not directly controlled, $J_H$ becomes zero, giving the stationary state of equation (37), in which each flow is a function of only two forces. Although neither flow is now an explicit function of $\Delta \tilde{\mu}_H$, the existence of the $H^+$ transport mechanism influences the values of the phenomenological coefficients (Katchalsky and Spangler, 1968).

As indicated earlier, the application of the dissipation function of equation (31) requires a decision as to whether flows and affinities should be evaluated internally or externally. Equation (37) indicates that the reaction flows and affinities should be evaluated externally. Since the whole mitochondrion is here considered in a stationary state, this might seem self-evident. However, the situation is complicated if there is non-conservative flow, e.g. that of $H^+$, in that it is not evident intuitively whether the flow should be evaluated from the change in internal or external hydrogen content. This question is resolved by equation (36): if the affinities are defined externally, $H^+$ flow must be defined in terms of internal changes. Although this quantity would be difficult to measure, for most purposes such measurements are unnecessary.

Symmetry considerations show that we could have expressed the dissipation function in terms of external flows and internal affinities. Although this formulation is equally valid, it would be impracticable for several reasons. There are obvious difficulties in evaluating internal affinities. The mitochondrion is highly compartmentalized and knowledge of internal activities is imprecise.* Further, even when internal concentrations are constant, it would be necessary to consider many flows and forces, since the external flows, unlike the internal flows, need not be zero when the mitochondrion is in a stationary state.

---

*Note that, except at equilibrium, internal and external reaction affinities will be unequal. Consider an enclosed region into which a reactant $\alpha$ diffuses and forms a product $\beta$ which diffuses outward. Then $\mu_\alpha^{ex} > \mu_\alpha^{in}$ and $\mu_\beta^{in} > \mu_\beta^{ex}$. Adding these, $\mu_\alpha^{ex} - \mu_\beta^{ex} > \mu_\alpha^{in} - \mu_\beta^{in}$, i.e. $A^{ex} > A^{in}$. A similar consideration applies to oxidative phosphorylation.

Having interpreted our dissipation function in this way, we can now write phenomenological equations:

$$J_P = L_P A_P^{ex} + L_{PH} \Delta \tilde{\mu}_H + L_{PO} A_O^{ex} \qquad (39)$$

$$J_H = L_{PH} A_P^{ex} + L_H \Delta \tilde{\mu}_H + L_{OH} A_O^{ex} \qquad (40)$$

$$J_O = L_{PO} A_P^{ex} + L_{OH} \Delta \tilde{\mu}_H + L_O A_O^{ex} \qquad (41)$$

This general formalism should be applicable near equilibrium whichever hypothesis is correct. For simplicity, we have made the usual assumptions of Onsager symmetry and linearity.

### A. Chemical Hypothesis

In this view, since substrate oxidation produces a high-energy intermediate, which can drive either phosphorylation or outward $H^+$ transport, $J_O$ is positively coupled to both $J_P$ and $J_H$, as shown in figure 8. However, since outward $H^+$ transport may also result from the hydrolysis of ATP, $J_P$ and $J_H$ are negatively coupled. It is clear that for this system each flow may be influenced by every force; in phenomenological terms, all coefficients of equations (39–41) are non-zero in the most general case.

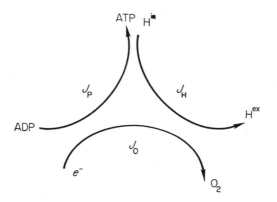

**Figure 8.** Chemical hypothesis. (From Caplan and Essig, 1969)

### B. Chemiosmotic Hypothesis

Site I in figure 9 is formally equivalent to Mitchell's 'proton-translocating oxido-reduction chain' while site II corresponds to his 'proton-translocating ATPase system' (Mitchell, 1961, 1966). $J_P$ and $J_O$ are coupled only through

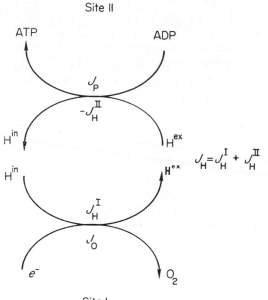

Site II

**Figure 9.** Chemiosmotic hypothesis. (From Caplan and Essig, 1969)

circulation of $H^+$, rather than through a high-energy intermediate (although, of course, the formalism does not exclude possible participation of a high-energy intermediate at either site).

Equations (39–41) are again applicable. However, in this case we can characterize the over-all coefficients in terms of those of the sites.

Site I:

$$J_H^I = L_H^I \Delta \tilde{\mu}_H + L_{OH}^I A_O^{ex} \tag{42}$$

$$J_O = L_{OH}^I \Delta \tilde{\mu}_H + L_O^I A_O^{ex} \tag{43}$$

Site II:

$$J_P = L_P^{II} A_P^{ex} + L_{PH}^{II} \Delta \tilde{\mu}_H \tag{44}$$

$$J_H^{II} = L_{PH}^{II} A_P^{ex} + L_H^{II} \Delta \tilde{\mu}_H \tag{45}$$

At site I oxidation results in the transport of $H^+$ against $\Delta \tilde{\mu}_H$, while at site II the spontaneous flow of $H^+$ results in phosphorylation.

The net flow of $H^+$ is given by

$$J_H = J_H^I + J_H^{II} \tag{46}$$

so that adding equations (42) and (45) gives

$$J_H = L_{PH}^{II} A_P^{ex} + (L_H^I + L_H^{II}) \Delta \tilde{\mu}_H + L_{OH}^I A_O^{ex} \tag{47}$$

Equations (44), (47) and (43) correspond respectively to equations (39), (40) and (41). Thus, the straight coefficients $L_O$ and $L_P$ and the cross coefficients $L_{OH}$ and $L_{PH}$ are simply those of the appropriate elemental sites, while $L_H$ is the sum $L_H^I + L_H^{II}$. The important observation is that

$$L_{PO} = 0 \text{ (chemiosmotic hypothesis)} \tag{48}$$

i.e. if $\Delta \tilde{\mu}_H$ were maintained constant experimentally, phosphorylation would be independent of oxidation, and *vice versa*. In particular, if $\Delta \tilde{\mu}_H = 0$ uncoupling is complete, and $J_P = 0$ if $A_P^{ex} = 0$, irrespective of $J_O$, while $J_O = 0$ if $A_O^{ex} = 0$, irrespective of $J_P$.*

Obviously there are substantial experimental difficulties in these applications of the phenomenological equations. However, it should be possible to vary and measure the affinities $A_O^{ex}$ and $A_P^{ex}$ over a considerable range. Furthermore, both the pH difference and the electrical potential difference may be evaluated and regulated in functioning mitochondria (Rottenberg, 1968, 1969a). Although it may be difficult to fix $\Delta \tilde{\mu}_H$ exactly at zero, it seems possible to make it very small as seen in Section III.

## C. Degree of Coupling and Effectiveness of Energy Utilization

For oxidative phosphorylation we are concerned with three degrees of coupling, $q_{PH}$, $q_{OH}$ and $q_{PO}$. These are readily derived from the inverse phenomenological equations:

$$A_P^{ex} = R_P J_P + R_{PH} J_H + R_{PO} J_O \tag{49}$$

$$\Delta \tilde{\mu}_H = R_{PH} J_P + R_H J_H + R_{OH} J_O \tag{50}$$

$$A_O^{ex} = R_{PO} J_P + R_{OH} J_H + R_O J_O \tag{51}$$

If the chemiosmotic hypothesis holds, a special relationship applies†, viz.:

$$q_{PO} = -q_{PH} q_{OH} \tag{52}$$

(Since $J_P$ and $J_H$ are negatively coupled, $q_{PH}$ is negative.) Any factors which decrease $q_{OH}$ or $-q_{PH}$ will decrease $q_{PO}$ and hence the effectiveness with which oxidative energy is used for phosphorylation.

---

*For simplicity the argument is developed with linear equations. However, since there is no direct interaction between oxidation and phosphorylation, higher order terms such as $L_{POO} A_P^{ex} A_O^{ex}$ must also be zero.

†$L_{ij} = R_{ij}'/D$, where $R_{ij}'$ is the cofactor of $R_{ij}$, and $D$ the determinant of the resistance matrix. Since $L_{PO} = 0$ for the chemiosmotic hypothesis, $R_{PO}' = R_{PH} R_{OH} - R_{PO} R_H = 0$. Dividing through by $R_H \sqrt{R_P R_O}$ leads to equation (52).

## D. Applications of the Above Considerations

The above treatment should be applicable to the steady state analysis of oxidative phosphorylation, irrespective of the detailed mechanisms. In particular, such questions as the possible existence of high-energy intermediates are irrelevant. A similar treatment may be applicable to photophosphorylation, despite the differences in specific features of the systems (Jagendorf and Uribe, 1967).

The system can be analysed in a variety of experimental circumstances, all describable in terms of this formulation. It is simplest to consider situations in which the dissipation function is reduced to two terms, so that each flow depends on only two forces. The degree of coupling $q$ which characterizes the resulting two-flow system is determined by the specific conditions considered (Caplan, 1966). Since in general we are primarily interested in the conversion of free energy of oxidation into free energy of phosphorylation, we shall consider only those cases in which the term $J_H \Delta \tilde{\mu}_H$ is zero, i.e. static head and level flow of $H^+$.

---

i) $J_H = 0$ (static head). Equations (49) and (51) then become

$$A_P^{ex} = R_P J_P + R_{PO} J_O \tag{53}$$

$$A_O^{ex} = R_{PO} J_P + R_O J_O \tag{54}$$

Consequently, the effective degree of coupling is given by

$$q = -R_{PO}/\sqrt{R_P R_O} = q_{PO} \quad (J_H = 0) \tag{55}$$

ii) $\Delta \tilde{\mu}_H = 0$ (level flow). Equations (49), (50) and (51) then give

$$A_P^{ex} = R_P(1 - q_{PH}^2)J_P - \sqrt{R_P R_O}(q_{PO} + q_{PH}q_{OH})J_O \tag{56}$$

$$A_O^{ex} = -\sqrt{R_P R_O}(q_{PO} + q_{PH}q_{OH})J_P + R_O(1 - q_{OH}^2)J_O \tag{57}$$

Consequently,

$$q = \frac{q_{PO} + q_{PH}q_{OH}}{\sqrt{(1 - q_{PH}^2)(1 - q_{OH}^2)}} \quad (\Delta \tilde{\mu}_H = 0) \tag{58}$$

---

If $A_O^{ex}$ is maintained constant, either by experimental manipulation or by the *in vivo* characteristics of the system, and $\Delta \tilde{\mu}_H$ is not controlled, $J_H$ will become zero in the stationary state. For this case $q = q_{PO}$. On the other hand, it would be useful to carry out experiments in which $\Delta \tilde{\mu}_H$ is maintained at zero.* In this case $q = (q_{PO} + q_{PH} q_{OH})/\sqrt{(1 - q_{PH}^2)(1 - q_{OHH}^2)}$. Figures 6 and 7 apply to both these states. Introducing equation (52) shows that for the chemiosmotic hypothesis, setting $\Delta\tilde{\mu}_H = 0$ effectively uncouples oxidation from phosphorylation, so that in the range where the phenomenological equations are linear $J_P/J_O$ is directly proportional to $A_P^{ex}/A_O^{ex}$, as seen in figure 6. With non-zero but constant $\Delta\tilde{\mu}_H/A_O^{ex}$, proportionality would be lost, but linearity retained. To the extent that the relation between $J_P/J_O$ and $A_P^{ex}/A_O^{ex}$ becomes non-linear (i.e. $q$ differs from zero), the system deviates from the chemiosmotic theory, which is therefore seen to be a limiting case. If the system were not chemiosmotic, oxidation could drive phosphorylation even when $\Delta\tilde{\mu}_H = 0$. In this case, either the chemical mechanism or a combination of mechanisms might be operative.

## VI. CONCLUSIONS

The thermodynamic considerations discussed above may be applied to the testing and construction of models. In practice, however, this is not a simple matter. Even though we reduce the number of independent variables to a minimum, the determination of free energies of reactions and electro-chemical potential differences across the membrane involves various difficulties. We have presented evidence that in the steady state it is legitimate to measure forces (other than $\Delta\tilde{\mu}_H$) externally. Thus it seems that these determinations are possible in principle, at least under some experimental conditions. The main point of this analysis is the importance of the determination of the forces (free energies, potentials, etc.) as well as the flows. For example, if the chemiosmotic hypothesis is correct, then the ratio of the calculated $\Delta\tilde{\mu}_H$ to the phosphorylation potential in the experiment shown in figure 4 suggests that the H/~P ratio must necessarily be greater than 4·5 (Mitchell claims a value of 2). Alternatively, the H/~P ratio may be evaluated from data obtained in state 4. We estimate that in state 4 the phosphorylation potential is 18·4 kcal/mole (see below). Although no precise data on $\Delta\tilde{\mu}_H$ are available under these conditions, it has been estimated (Rottenberg, 1969a) that $-\Delta\psi$ is no greater than 130 mV and $\Delta$pH is no greater than 0·5. This would give a $-\Delta\tilde{\mu}_H$ of about 4 kcal/mole, again pointing to a minimal value of H/~P of about 4·5.

The usefulness of the degree of coupling in characterizing the energetics of mitochondria unambiguously was pointed out in Section V. The over-all

*Here we refer to techniques affecting the force but not the elemental phenomenological coefficients. Although such an experiment may not be practicable at present, it remains possible in principle to devise one, particularly if high-speed techniques become available. In the valinomycin experiments at high potassium concentrations described in Section III, $\Delta\tilde{\mu}_H$ approximates zero as a consequence of an alteration in membrane permeability factors which tend to uncouple site I (see equations (42) and (43)).

degree of coupling of oxidative phosphorylation can be evaluated easily from results of respiratory control experiments. In addition to the respiratory control ratio and the P/O ratio, it is necessary to determine the phosphate potential and the substrate oxidation potential in both states 3 and 4. If such information were available we could calculate all the phenomenological coefficients without the assumption of the Onsager relation. In the absence of such information we shall make a calculation on the basis of available data, assuming Onsager symmetry. The data are taken from a respiratory control experiment with rat liver mitochondria (Estabrook, 1967) described in figure 10. We assume the standard affinity of succinate oxidation to be

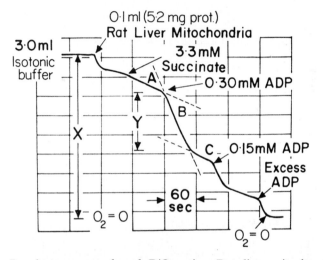

**Figure 10.** Respiratory control and P/O ratio. Rat liver mitochondria were diluted with an isotonic buffer containing 0·225M sucrose, 10mM potassium phosphate, pH 7·4, 5mM MgCl₂, 20mM KCl and 20mM triethanolamine buffer, pH 7·4. Succinate and ADP were added as indicated. The recorder deflection $X$ expresses the total oxygen content of the reaction medium. (From Estabrook, 1967)

35·7 kcal/mole and that of phosphorylation to be −7·0 kcal/mole. The respiratory control ratio (rate B/rate C) is 5·9 and the P/O ratio is 1·7. From these figures, expressing all rates as multiples of the oxidation rate in state 4, the respiration rate in state 3 is 5·9 and the phosphorylation rate is 10·0. The succinate oxidation potential is calculated to have an average value of 37·9 kcal/mole in state 3 and 37·5 kcal/mole in state 4. Using equations (53) and (54):

State 4:

$$(A_P^{ex})_4 = R_{PO}(J_O)_4 = R_{PO} \tag{59}$$

$$(A_O^{ex})_4 = R_O(J_O)_4 = R_O = 37 \cdot 5 \tag{60}$$

State 3:

$$(\overline{A}_P^{ex})_3 = R_P(J_P)_3 + R_{PO}(J_O)_3 = 10 \cdot 0 R_P + 5 \cdot 9 R_{PO} \tag{61}$$

$$(\overline{A}_O^{ex})_3 = R_{PO}(J_P)_3 + R_O(J_O)_3 = 10 \cdot 0 R_{PO} + 221 \cdot 5 = 37 \cdot 9 \tag{62}$$

These give

$$R_{PO} = -18 \cdot 36 = (A_P^{ex})_4 \tag{63}$$

The average value for $(A_P^{ex})_3$ when half the added ADP has been phosphorylated is

$$(\overline{A}_P^{ex})_3 = -9 \cdot 72 \text{ kcal/mole} \tag{64}$$

Introducing equations (63) and (64) into (61) gives

$$R_P = 9 \cdot 84 \tag{65}$$

Equation (23) then allows the calculation of $q$:

$$q = \frac{18 \cdot 36}{\sqrt{37 \cdot 5 \times 9 \cdot 84}} = 0 \cdot 96 \tag{66}$$

(Both $R_P$ and $q$ are insensitive to errors in the estimation of $(\overline{A}_P^{ex})_3$.) The maximum efficiency is obtained from equation (26):

$$\eta_{max} = 0 \cdot 56 \tag{67}$$

During state 3, the average efficiency is given by

$$(\overline{\eta})_3 = -\frac{(J_P)_3(\overline{A}^{ex}_P)_3}{(J_O)_3(\overline{A}^{ex}_O)_3} = 0.44 \tag{68}$$

It should be stressed that these figures are estimates and are given as an example only. Although it appears that the degree of coupling is high, this still represents a significant deviation from absolute stoichiometry, as demonstrated by a maximum efficiency of only 56%. It is interesting to note that the average efficiency in state 3 is close to the maximum possible value. On the other hand, the efficiency of ATP synthesis during a typical hexokinase trap experiment is only about 0·24.

Recently there has been considerable discussion as to the value of the phosphorylation potential during state 4 (Cockrell and coworkers, 1966; Slater, 1967). It should be pointed out that our estimate (equation (63)) is based on the measured P/O ratio, the respiratory control ratio and the estimated free energy of succinate oxidation, without using values for the standard free energy of phosphorylation or the ATP/ADP ratio. The calculated value of 18·4 kcal/mole is somewhat higher than the highest previous estimate, viz. 15·6 kcal/mole (Cockrell and coworkers, 1966).

## Acknowledgements

This study was supported by U.S.P.H.S. Career Development Awards to Dr. Caplan (1-K03-GM-35,292,02) and to Dr. Essig (5-K3-HE-24,481-03,04) and grants from the U.S. Atomic Energy Commission (Contract No. AT (30-1)-2453), the Office of Saline Water, U.S. Department of the Interior (No. 14-01-0001-2148), and the U.S.P.H.S. (GM 12852 and HE 00759-19).

## REFERENCES

Addanki, S., F. D. Cahill and J. F. Sotos (1968) *J. Biol. Chem.*, **243**, 2337
Benzinger, T., C. Kitzinger, R. Hems and K. Burton (1959) *Biochem. J.*, **71**, 400
Blumenthal, R., S. R. Caplan and O. Kedem (1967) *Biophys. J.*, **7**, 735
Caplan, S. R. (1966) *J. Theoret. Biol.*, **10**, 209; *Errata* **11**, 346
Caplan, S. R. and A. Essig (1969) *Proc. Natl. Acad. Sci. U.S.*, **64**, 211
Chance, B. and G. R. Williams (1956) In F. F. Nord (Ed.), *Advances in Enzymology*, Vol. XVII, Interscience, New York p. 65
Chance, B. (1967) In E. C. Slater, Z. Kaniuga and L. Wojtczak (Eds.), *Biochemistry of Mitochondria*, Academic Press, New York, p. 93
Chappell, J. B. and A. R. Crofts (1965) *Biochem. J.*, **95**, 393
Cockrell, R. S., E. J. Harris and B. C. Pressman (1966) *Biochemistry*, **5**, 2326
Cockrell, R. S., E. J. Harris and B. C. Pressman (1967) *Nature*, **215**, 1487
Essig, A. and S. R. Caplan (1968) *Biophys. J.*, **8**, 1434
Estabrook, R. W. (1967) In R. W. Estabrook and M. E. Pullman (Eds.), *Methods in Enzymology*, Vol. X, Academic Press, New York, p. 41
Glynn, I. M. (1967) *Nature*, **216**, 1318
Höfer, M. and B. C. Pressman (1966) *Biochemistry*, **5**, 3919
Jagendorf, A. T. and E. Uribe (1967) *Brookhaven Symp. Biol.*, **19**, 215
Katchalsky, A. and P. F. Curran (1965) *Nonequilibrium Thermodynamics in Biophysics*, Harvard University Press, Cambridge, Mass.
Katchalsky, A. and R. Spangler (1968) *Quart. Rev. Biophys.*, **1**, 127
Kedem, O. and S. R. Caplan (1965) *Trans. Faraday Soc.*, **61**, 1897
Lehninger, A. L. (1964) *The Mitochondrion: Molecular Basis of Structure and Function*, W. A. Benjamin, New York
Lehninger, A. L., E. Carafoli and C. S. Rossi (1967) In F. F. Nord (Ed.), *Advances in Enzymology*, Vol. XXIX, Interscience, New York, p. 259
Mitchell, P. (1961) *Nature*, **191**, 144
Mitchell, P. (1966) *Biol. Rev.*, **41**, 445
Prigogine, I. (1955) *Introduction to Thermodynamics of Irreversible Processes*, Interscience, John Wiley, New York
Rasmussen, H., B. Chance and E. Ogata (1965) *Proc. Natl. Acad. Sci. U.S.*, **53**, 1069
Robbins, E. A. and P. D. Boyer (1957) *J. Biol. Chem.*, **224**, 121
Rottenberg, H., S. R. Caplan and A. Essig (1967) *Nature*, **216**, 610
Rottenberg, H. (1968) Ph.D. Thesis, Harvard University, Cambridge, Mass.
Rottenberg, H. (1969a) (In preparation)
Rottenberg, H. (1969b) (In preparation)
Slater, E. C. (1966) In M. Florkin and E. H. Stotz (Eds.), *Comprehensive Biochemistry*, Vol. XIV, Elsevier, New York, p. 327
Slater, E. C. (1967) *Eur. J. Biochem.*, **1**, 317

# CHAPTER 7

# Reversible coupling between transport and chemical reactions

## Peter Mitchell

*Glynn Research Laboratories*
*Bodmin, Cornwall, England*

# I. INTRODUCTION

## A. Catalysis of Transformation and Translocation

The present day biochemical knowledge of the mechanisms of chemical synthesis and maintenance in living cells has stemmed largely from the study of the equilibration of chemical groups between group donors and acceptors in multi-enzyme systems, as discussed by Lipmann in his remarkable review on group potential and the thermodynamic principles of group transfer in 1941. In the same year, Glasstone and coworkers reviewed the 'theory of absolute reaction rates' which is applicable, not only to chemical reactions, but to 'any process involving a rearrangement of matter'. The fact that the theory of absolute reaction rates (Evans and Polyani, 1925; Rice and Gershinowitz, 1934; Eyring, 1935; Glasstone and coworkers, 1941) treated chemical transformation and physical diffusion as proceeding by fundamentally similar thermally activated translocation reactions was of great biochemical significance; but it is only comparatively recently that the implications have reached the stage of practical utility and have begun to be generally appreciated by biochemists.

The elementary thermodynamic basis of most of the description of coupled transport reactions and chemical reactions given in this chapter follows logically from the general treatment of chemical cells and circuits by Guggenheim (1933) and from the application of this type of equilibrium thermodynamic treatment to biological transport reactions described by Rosenberg (1948), which I have discussed elsewhere (Mitchell, 1954a, 1961a, 1968). I have also attempted here to take account of the general way in which the levels of so-called energy barriers, determined by topological and chemical features of the system, define or control the thermodynamic process as a whole, in accordance with or stemming from the principles reviewed by Lipmann (1941), by Glasstone and coworkers (1941), by Teorell (1949) and by Pauling (1950, 1956).

One of the main objects in pursuing this rather elementary general theoretical treatment of biological transformation and translocation processes is to emphasize that as soon as the general principle of enzyme action was understood in terms of the catalysis of equilibrium of certain chemical reactions, great advances were possible in the subject of metabolic enzymology even though there was little or no knowledge of the detailed molecular mechanisms by which the specific chemical reactions were facilitated by enzymes. Likewise, the understanding of the general principles of the catalysis of equilibrium of certain chemical particles in specific translocation reactions provides the basis for substantial advances in the knowledge of the metabolic and physiological role of transport reactions even though there is little detailed knowledge of the molecular mechanisms of the catalysis of translocation (Mitchell, 1963a).

As Teorell (1949) has emphasized, the rate of flow of a component A through a transport reaction is given by the product of the concentration (denoted by [ ]) of A and the net velocity of movement of A. The net velocity of movement can generally be described as the ratio of the total chemical potential gradient $(d\bar{\mu}_A/dx)$, which is the driving force, and a frictional coefficient $(f_A)$, which defines the retarding force per unit velocity. Thus,

$$\frac{dn_A}{dt} = \frac{[A]}{f_A} \times \frac{d\bar{\mu}_A}{dx} \qquad (1)$$

Group transfer (i.e. chemical reaction) can be described similarly. When the transfer or translocation of A occurs through a series of such reactions, in which the component A interacts in various ways with the environment, the value of $\bar{\mu}_A$ at equilibrium will be constant throughout the system, but the value of $[A]/f_A$ will depend on the interactions of A with the environment. The efficient transmission of A through the system when the equilibrium is disturbed will obviously depend on factors that maximize [A] and minimize $f_A$, and such factors can be described as facilitating (or catalysing) the transfer or translocation of A (see Wilbrandt and Rosenberg, 1961; Mitchell, 1967a).

When we say that a reversible reaction involving transfer of A is impeded by a free-energy barrier in a system close to or at equilibrium, it is necessary to bear in mind that we do not mean that the component A, exchanging across or undergoing slow net transfer across the energy barrier in the free state or in intermediary complexes or compounds, actually has a higher time-average energy when in a transition state on the barrier. For, as mentioned above, the total chemical potential of A must be the same at all points along the reaction pathway at equilibrium and must be nearly the same at all points when near to equilibrium. The idea of the energy barrier is related to the Boltzmann distribution relationship (see Moelwyn-Hughes, 1947) according to which particles of the component A will distribute at equilibrium between states L and R at probabilities $\{A\}_L$ and $\{A\}_R$ when the average energy of a particle of A in state L is greater than that in state R by an amount $\varepsilon$, or

$$\{A\}_L = \{A\}_R \times \exp\left(-\varepsilon/kT\right) \tag{2}$$

The energy attributable to the concentration of A is omitted from the value of $\varepsilon$ because equation (2) is concerned with the time-distribution of energy of a single particle, or, according to a principle known as the ergodic hypothesis (Lindsay and Margenau, 1936), with the average energy of a set of such individually considered particles at any one time. The so-called energy of activation represented by $\varepsilon$, or the height of an energy barrier (or depth of an energy channel) as usually defined, therefore does not include the concentration term that is an essential part of the total chemical potential (see equation (14)).

It follows from these considerations that, according to equation (1), the transmission of the component A through a series of reactions will occur with equal facility along the pathway when the factor $[A]/f_A$ and the driving force $d\bar{\mu}_A/dx$ do not vary. In practice the factor $f_A$ generally varies little compared with [A], and thus it is roughly true that favourable conditions for the reversible transmission of A along the reaction coordinate can be represented as a deep energy channel with a uniformly inclined floor not containing humps that are large compared with $kT$. This is only another way of saying that the time-average concentration of A all along the reaction pathway should be sufficiently large to facilitate the rapid passage of A along the whole pathway under the influence of a small total chemical potential gradient of A.

## B. Enzymes and Porters

In the chemical transformations catalysed by enzymes, the course of the chemical reaction is determined by the pathway of group transfer, specified by the catalytic property of the active centres of the enzymes, because a given group G is covalently bonded alternatively to a donor group D or to an acceptor group A, and work far in excess of the thermal energy $kT$ is required to enable G to diffuse between D and A at a significant rate except in the special environment of the active centre of the appropriate enzyme, as shown, for example, by Ogston (1955) in a discussion of activation and inhibition of enzymes. When the reactants (DG+A) and resultants (D+AG) are not intrinsically stable with respect to the migration of G, (or, in other words, when the energy barrier opposing the migration of G between D and A is not large compared with $kT$), the reaction goes spontaneously; and enzymic control of reaction pathways in systems of such unstable reactants is relatively limited. In practice, most metabolites are very stable covalent compounds, and thus the pathways of group transfer in metabolic sequences and networks depend almost entirely on the facilitation of diffusion of the appropriate chemical groups in the active centre regions of the enzymes.

In reactions classified as biological transport, we are concerned, not with the scalar aspect of chemical transformation resulting from the diffusion of chemical groups from donor to acceptor species, but with the vectorial physical translocation of chemical particles from one location to another. In physical translocation, as in chemical transformation, the reaction pathway may be effectively determined when (and only when) the initial and final states of the particles transferred in the thermally activated reaction are intrinsically relatively stable. Thus, the course of biological transport processes may be specified by the catalysts of translocation equilibrium, conveniently called porters (Mitchell, 1967a), because a given type of passenger S may alternatively be located in one region L or another region R, separated by a diffusion barrier M; and work far in excess of the thermal energy $kT$ is required to enable S to diffuse at a significant rate between L and R except via regions in the barrier M where the appropriate porter provides a special environment for diffusion (see Danielli, 1954; Wilbrandt and Rosenberg, 1961; Mitchell, 1961a).

There is a fundamental similarity between the chemical transformation catalysed by the enzyme E in which the chemical group G migrates from donor component D to acceptor component A, and the solute translocation reaction catalysed by the porter P in which the solute S migrates from the left hand phase L to the right hand phase R, across the barrier M. There is a fundamental difference in that whereas the initial and final states of the group G are determined by covalent bonding in the stoichiometric *compounds* DG and AG present at certain concentrations (activities) in the one phase system, the initial and final states of the solute S are determined by the concentrations of S in the two *phases* L and R separated by the osmotic barrier phase M. The donor and acceptor components D and A that may be alternatively occupied by G in the chemical reaction are equivalent to the phases L and R that may alternatively be occupied by S in the translocation reaction. The transfer of the group G from D to A involves the opening and closing of covalent bonds as well as of secondary bonds in the active-centre region of the enzyme, and it is the process of covalent bond exchange that generally presents the main energy barrier in chemical transformation reactions. The main function of enzymes is thus to lower the free energy of activation of covalent bond exchange required for the diffusion of certain chemical groups between certain substrates. On the other hand, the translocation of the solute S between phases L and R need not generally involve the exchange of covalent bonds but depends on the opening and closing of the secondary and ionic bonds that limit the diffusion of solutes across the barrier M. The main function of porters is thus to lower the free energy of activation of secondary and ionic bond exchange required for the translocation of certain solutes between pairs of phases separated by osmotic barriers. In general, therefore, while the basic property of enzymes is to catalyse the covalent bond exchange required for chemical transformation, the basic property of porters is to catalyse the secondary and ionic bond exchanges required for physical translocation in condensed media.

The substrate specificity of enzyme-catalysed group transfer is dependent on the formation of complex three-dimensional bonding relationships (transition state complexes) between enzyme and substrate, the special property of which is to permit degrees of freedom, corresponding to the diffusion of the transferable group (along the reaction coordinate) between donor and acceptor groups, not involving free energy changes much in excess of $kT$. There is little doubt that the passenger specificity of porter-catalysed translocation reactions in lipid membranes is similarly dependent on complex bonding relationships between porter and passenger that permit degrees of freedom, corresponding to the diffusion of the passenger between the donor and acceptor media, not involving free energy changes much in excess of $kT$. The narrowness of substrate and passenger selectivity of enzymes and porters is thus attributable to the narrowness of the tolerances permitted in the requirement for packing and bonding relationships of minimum free energy (or of maximum probability or concentration) in the transition state complexes.

The general mechanism by which the chemical or translocation reaction is facilitated by the formation of the transition state complex with the catalytic system presumably depends on a mutual balancing of free energy changes in different parts of the complex such that, although the diffusion of the group or passenger along the reaction coordinate of the transition state complex may require an amount of work that is large in comparison to $kT$ to open or stretch *certain* bonds, the condensed structure of the complex and the mutual dependence of articulations between its atomic constituents involves a simultaneous adjustment of *other* bonds which thereby provide a compensating amount of work. This type of mechanism—described explicitly by Hammes (1964) for enzyme-catalysed reactions, and implicit in an earlier argument by Pauling (1950) which showed that enzymes must have a higher affinity for transitional configurations of their substrates than for the normal reactant and resultant species—is consistent with the requirement of a close-packed structure of transition state complexes with particular degrees of freedom in substrate-specific enzyme systems and passenger-specific porter systems. The essential point is that the specific catalytic activity of enzymes and porters depends on complex thermally activated movements within the catalytic system by which the passenger species move through transition states not involving large free energy changes compared with $kT$ *in the system as a whole*. The composition of the transitional intermediates is specified within certain tolerances by bonding and close-packing relationships that may be localized not only in the active-centre region of the enzyme or in the carrier-centre region of the porter (Mitchell, 1967a), but also in other more widely distributed regions of the catalytic system. Incidentally, it should be remarked that the transition state complexes described in this context represent a more extensive set of states along the reaction coordinate than the unique transition state or activated complex (Eyring, 1935) confined to the top of the energy barrier.

Koshland (1960) and others (see Jencks, 1966; and Koshland and Neet, 1968) have emphasized the extent to which the catalytic activity of enzymes may involve movements within the enzyme protein as a whole, described as conformational changes. The conformational changes of haemoglobin molecules on oxygenation (Muirhead and Perutz, 1963) are the prototype example of the extent to which small passenger molecules may influence close-packing and bonding relationships in relatively large complexes. Wyman (1948, 1964) has introduced a simple linked-function concept to describe quantitatively how the bonding of one ligand in haemoglobin molecules, or in other similar close-packed complexes, reciprocally affects the bonding of other ligands, irrespective of the distance separating the sites of bonding of the ligands in the complex. This concept depends on the fact, inherent in the definition of chemical potential (Guggenheim, 1949), that for any two components $i$ and $j$ in a thermodynamic system at equilibrium at constant temperature and pressure,

$$\frac{\partial \mu_i}{\partial n_j} = \frac{\partial \mu_j}{\partial n_i} \tag{3}$$

where $\mu$ and $n$ denote the chemical potentials and quantities of the components respectively in the system, and the partial differentials refer to constant composition with respect to all components other than that shown in the denominator of the differential. When $i$ and $j$ represent two ligands that participate in a transitional complex, it follows from equation (3), that the binding of the ligand may be mutually inclusive or mutually exclusive, and the effective attraction or repulsion between the ligands must be strictly reciprocal, as though the ligands interacted directly, even though the interaction may be communicated between distant binding sites in the complex.

## II.  COUPLING RELATIONSHIPS BETWEEN CHEMICAL REACTIONS AND OSMOTIC REACTIONS

### A. General Principles and Evolution

One reaction is said to be coupled to another in a biochemical system when the flow of chemical particles through either reaction influences the flow of chemical particles through the other. Any such dependence between the flows in an equilibrium system must be strictly reciprocal because the transitional intermediates involved in either process will interact according to the principle described by equation (3). When the interaction between the transitional intermediates is so strong that the progress of either complex along the reaction coordinate would involve large free energy changes compared with k$T$ when not accompanied by corresponding progress of the other complex along the reaction coordinate, the two complexes behave effectively as one and the reactions are tightly coupled. The uniqueness of the transitional pathways and the mutual coupling of the flows of passenger species through such complex transitional intermediates in condensed catalytic systems depends on the bonding and close-packing relationships in the complexes which have been evolved by living organisms through the process of natural selection. The types of bonding involved in the specific conduction of the passenger species through the catalytic system is, of course, immaterial provided only that the escape of the passenger from the prescribed intermediate pathway would involve an increase in free energy which is large compared with k$T$. Owing to the way in which the catalytic systems are likely to have evolved through stages of increasing complexity, there was presumably a tendency for the ordinary chemical reactivities of the small molecular weight components involved as metabolites in the biologically catalysed transformation and translocation reactions to be utilized initially. These *ordinary* reactivities, by which we mean the primary reactivities exhibited between pairs of components of relatively low molecular weight, have presumably become modulated by the secondary bonding and secondary and tertiary packing relationships exhibited by the more complex catalytic components of relatively high molecular or particle weight that have been evolved. The extent of this modulation has come to play a more or less dominant part in the channelling of the over-all reactions at the present stage of evolution.

As we are concerned with *reversible* coupling between translocation and transformation reactions we shall focus attention on idealized enzyme and porter systems in which side reactions are absent and the flows through pairs of reactions are either coupled or independent.

## B. Basic Symbol System

The conceptual framework outlined above enables us to describe chemical transformation and physical translocation reactions and the coupling relationships between them by means of a simple type of flow diagram that has come into general use in this laboratory, but for which we make no proprietary claim. We represent reversible reactions between and within aqueous phases L and R on the left and right of a non-aqueous barrier phase M by arrows that represent the transfer or translocation of given chemical particles. Arrows with ends in the same phase represent macroscopic chemical transformation (i.e. group transfer), and arrows with ends in different phases represent macroscopic physical translocation. The latter may also represent chemical transformation when the species translocated is an electron or chemical group. The tangential meeting of two arrows represents tight coupling between the reactions (i.e. coupling between the flows of chemical particles). Arrows that are circular represent transitional intermediates that may be well-defined chemical species or may be transition state complexes. The barbs on the arrows represent the conventional forward direction of the reversible reactions. The position of the symbols on the page may often indicate the appropriate reactions between the left hand, middle and right-hand phases without the symbols L, M and R being required to denote the phases; but it is sometimes helpful to indicate the position of the M phase with a vertical line, especially when representing transformations localized only or mainly in one or other aqueous phase.

## C. Chemical Coupling

Two chemical reactions are said to be chemically coupled when the flow of chemical groups through each reaction is dependent on the flow through the other because a chemical compound or complex produced in the one reaction is consumed in the other. The coupling is said to be primary when the intermediate is a chemical compound (involving covalent bonds) and secondary when the intermediate is a chemical complex (not involving covalent bonds).

### 1. *Non-coupled Reactions*

Under certain conditions, the flow through a chemical reaction can proceed with little coupling to other chemical flows. For example, in the hydrolytic reactions of digestion, represented by

$$AB + H_2O \rightarrow AH + BOH \tag{4}$$

the availability of $H_2O$ at almost constant potential in the usual aqueous media, and the capacity of the system for accumulating AH and BOH without appreciable back pressure on the reaction, generally permits the flow through the hydrolytic process to be determined largely by the availability of AB.

## 2. *Primary and Secondary Chemical Coupling in Coenzyme-linked or Substrate-linked Systems*

When two chemical (group transfer) reactions share a group donor or acceptor, the reactions become chemically coupled, as first shown by Green and coworkers (1934) for two dehydrogenases sharing a coenzyme (see also Green, 1937). For example the coupling by NAD of $AH_2$ oxidation and B reduction may be represented as follows

$$(5)$$

where E/E—2H and E'/E'—2H represent transitional complexes of the two dehydrogenases. In this type of coupling, a given species of chemical group flows from initial donor to final acceptor species via transitional intermediates and via one or more intermediate carrier compounds. When we say that the successive reactions are coupled, we mean that the flow of the given species of chemical group through the first reaction influences that through the second, and in particular that when there are no side reactions the quantity of the group flowing per unit time into the first reaction pathway is equal to that flowing out of the second pathway in the steady state. Using the example of equation (3), the thermodynamic consequence of the coupling between the reactions is seen to be that the thermodynamic force on the first is $(\bar{\mu}_{AH_2} - \bar{\mu}_A) - (\bar{\mu}_{(NADH + H^+)} - \bar{\mu}_{NAD^+})$, while the thermodynamic force on the second is $(\bar{\mu}_{(NADH + H^+)} - \bar{\mu}_{NAD^+}) - (\bar{\mu}_{BH_2} - \bar{\mu}_B)$; and thus the pressure [represented by the hydrogen group potential $(\bar{\mu}_{(NADH + H^+)} - \bar{\mu}_{NAD^+})$] tending to drive the second reaction forward is equal to that tending to hold the first reaction back. In this example it can also be seen that the transfer of the thermodynamic force along the reaction pathway is due to two different types of bonding relationship. The transfer through the intermediate couple $NAD^+/(NADH+H^+)$ is due to the cyclic diffusion of the components of this couple through the aqueous medium between the A-specific and B-specific dehydrogenase molecules. The transfer of the hydrogen groups between the two dehydrogenases is thus dependent on primary bonding in the chemically defined carrier couple, and the coupling process achieved through this pathway may appropriately be called primary chemical coupling. On the other hand, the transfer of the hydrogen groups in the special environments of the active centre regions of the dehydrogenases involves complex secondary bonding interactions which play an important part in determining the chemical potentials of the transitional intermediates along these regions of the transfer pathway, as discussed above. Thus the coupling process achieved through such regions of the pathway may appropriately be called secondary chemical coupling. The secondary type of chemical coupling generally differs from the primary type because the secondary bonds between individual atoms or groups generally have low energies compared with primary bonds. In order that the transitional intermediate should be defined precisely enough to prevent spillage into side reactions, the secondary bonding relationships must concentrate the transitional complexes in energy valleys along the reaction coordinate, and the depth of these valleys must be large compared with $kT$. This generally requires cooperation between several secondary bonding groups, and the chemical species involved in secondary chemical coupling tend to be correspondingly complex, as in the well known case of enzymes.

The view that emerges from these elementary considerations is that the coupling phenomenon is concerned with the transmission of a thermodynamic force through the flow of specific chemical particles along deep energy channels (or through a continuous set of states of relatively high probability or concentration) defined by appropriate primary and secondary bonding relationships. When the energy channel is deep and the floor of the channel does not contain humps that are high compared with $kT$, or there are no transitional states of very low probability or concentration, conditions are appropriate for tight coupling and freedom from side reactions.

We have so far concentrated attention on a particular chemical group (e.g., the 2H group) flowing through two successive reactions, exemplified by equation (5). This flow, however, is coupled to cyclic flows of the carrier species, and the coupling between the flows of the passenger and the carrier species may appropriately be described as primary or secondary according to the type of bonding between passenger and carrier.

The primary chemical coupling between the flows of two different species of chemical particle is the classical basis of the coupling between pairs of metabolic reactions in the field of so-called 'energy metabolism'. The prototype example of substrate-level oxidative phosphorylation, the phosphorylation of ADP coupled to the NAD-linked oxidation of 3-phosphoglyceraldehyde (PG) to 3-phosphoglycerate (PGA) via 1,3-diphosphoglycerate (PGAP), may be represented as follows

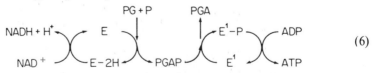

$$\tag{6}$$

where $E/E{-}2H$ and $E^1/E^1{-}P$ stand for 3-phosphoglyceraldehyde dehydrogenase and 3-phosphoglycerate kinase, respectively (Racker, 1961, 1965). The primary coupling of the flows through the oxidation and phosphorylation reactions by the flow of the common covalent intermediate PGAP depends on the same thermodynamic principles as we have already discussed in connection with equation (5). However, we should perhaps emphasize that efficient (reversible) coupling between the two types of reaction in the system of equation (6) depends on the requirement that the paths of the chemical particles transferred in the reaction system should follow deep energy channels with a fairly level floor, or in other words that the concentration of every component involved in transfer along the reaction channels must be sufficient to permit a rate of throughput along the channels that is fast compared with side reactions when the equilibrium is disturbed. This is the basis of the better-known rough requirement for the reversibility of pairs of coupled reactions exemplified by equation (6) that the standard free energy of the one reaction (the difference between oxidation of $NADH+H^+$ and oxidation of PG to PGA) should be about the same as the standard free energy of the other reaction (the hydrolysis of ATP), when we define the standard state of $H^+$ ions as that of water at pH 7, the standard state of water as that of the pure fluid, and the standard states of the other components as molal aqueous solutions.

## 3. *Primary Chemical Coupling at the Enzyme or Carrier Protein Level*

Coupling through a covalent enzyme intermediate is involved in the 3-phosphoglyceraldehyde dehydrogenase reaction (Racker, 1965; Park, 1966) but enzyme-level primary coupling has been demonstrated more obviously, for example, in the oxidative acylation reactions catalysed by the α-ketoacid dehydrogenase complexes (Gunsalus, 1954; Reed and Cox, 1966; Williams and coworkers, 1967; Hirashima and coworkers, 1967). The flow process involved in the decarboxylation of pyruvate (Py) or α-oxoglutarate (α-Og) coupled to NAD reduction and CoA acylation is illustrated in the following equation

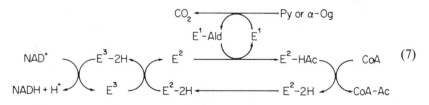

$$(7)$$

In this scheme $E^1$, $E^2$ and $E^3$ represent pyruvate or α-oxoglutarate decarboxylase containing thiamine pyrophosphate, dihydrolipoyl transacetylase or transsuccinylase containing covalently bound lipoate, and dihydrolipoyl dehydrogenase containing flavine adenine dinucleotide, respectively. Ald stands for the aldehyde corresponding to Py or α-Og and Ac stands for acyl (acetyl or succinyl).

The oxidative acylation systems represented by equation (7) consist of complexes of $E^1$, $E^2$ and $E^3$ and, as Reed and Cox (1966) have pointed out, the process of group transfer must involve the interaction of a lipoyl moiety, which is covalently bound to one enzyme, with the prosthetic groups of the other two enzymes, and 'these interactions between prosthetic groups of separate enzymes occur within a complex in which the movements of the individual enzymes is restricted and from which the intermediates (with the exception of acetyl or succinyl CoA) do not dissociate. Highly specific positioning of the three enzymatic components of the complex and, by inference, of the prosthetic groups of these components, is indicated by the resolution and reconstitution experiments'.

As in the cases already discussed, the efficient transmission of chemical particles through the reactions of equation (7) presumably depends on appropriate mobilities and articulations within the complex of enzymes that permit all intermediate states along the transfer pathways to occur with relatively high probability or at relatively high concentration. There is no fundamental thermodynamic difference between this type of primary coupling and the type due to free diffusion of coenzyme or substrate in an aqueous phase; but there is obviously a more stringent kinetic specification of the diffusion pathway of the chemical groups undergoing transfer when they are translocated within the special environment of an enzyme complex than when this translocation occurs through a homogeneous aqueous medium.

The catalytic transfer of hydrogen atoms and electrons through the complex known as the respiratory chain is another example of primary chemical coupling at the enzyme or carrier protein level, and some of the kinetic requirements have been well discussed by Chance and coworkers (1958).

## 4. *Secondary Chemical Coupling at the Enzyme or Carrier Protein Level*

In respiratory chain-linked oxidative phosphorylation (see Pullman and Schatz, 1967) and in the corresponding photosynthetic phosphorylation, the reaction

$$\text{SH}_2 \diagdown \diagup \text{A} \diagdown \diagup \text{BH}_2 \qquad\qquad (8)$$
$$\text{S} \diagup \diagdown \text{AH}_2 \diagup \diagdown \text{B}$$

or its electronic equivalent is apparently coupled to the reaction

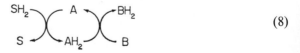

$$\qquad\qquad (9)$$

where A is a carrier in the respiratory or photosynthetic oxidoreduction chain, and $E^1$ is a reversible ATPase. According to the chemical hypothesis as formulated by Lipmann (1946) and Slater (1953) and reviewed by Pullman and Schatz (1967), reactions 8 and 9 are coupled by a primary mechanism involving a chemical intermediate $A \sim C$ formed by the oxidoreduction carrier A and an unidentified group C, the coupling intermediate $A \sim C$ being produced by the oxidoreduction reaction 8 and consumed by the phosphorylation (or hydrodehydration) reaction 9, very much as in substrate-level phosphorylation. However, owing to the lack of evidence for the existence of any covalent compound corresponding to $A \sim C$, Boyer (1965) suggested that $A \sim C$ might represent an energized form or state, and that the energized state could involve the conformation change of a protein. This is equivalent to the suggestion that the $\sim$ bond in $A \sim C$ is not a covalent bond but represents a secondary bonding relationship of equivalent free energy of formation. Thus, there is no formal difference between the flow pattern for primary chemical coupling of oxidoreduction to phosphorylation via the hypothetical covalent $X \sim I$ intermediate and the flow pattern for secondary chemical coupling via the hypothetical secondarily bonded $X \sim I$ intermediate complex suggested by Boyer (1965). The following equation represents the type of flow system required

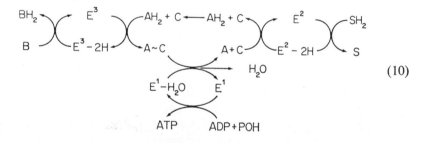

$$(10)$$

where $E^1$ stands for $A \sim C$ synthetase, and $E^2$ and $E^3$ stand for oxidoreduction enzymes. We might alternatively have represented the substrate of the $A \sim C$ synthetase and of $E^2$ as $(A+C)H_2O$, rather than $A+C$, in which case the $H_2O$ would be liberated by oxidation of $(AH_2+C)H_2O$ by $E^3$ in accordance with the scheme of Lipmann (1946).

As in the well-documented primary chemical coupling at the enzyme or carrier protein level described by equation (7), the hypothetical secondary chemical coupling system represented by equation (10) requires the intermediate coupling complex $A \sim C$ to make close molecular contact with the enzymes (or catalytic carriers) catalysing the flows of hydrogen atoms (or electrons) and water (or anhydro bonds) between which the coupling must occur. If the opening of the $A \sim C$ bonding relationship involved a large

configurational change of the A–C complex, the resulting cyclic translocations within the enzyme complex as a whole could couple the flows of
electrons or hydrogen atoms through the respiratory chain to those of water
through the reversible ATPase over a distance comparable to the dimensions
of the A–C complex by the type of linked-function mechanism defined by
Wyman (1948, 1964). But at all events, the secondary chemical coupling
phenomenon would require the direct transmission of tightly articulated
bond exchanges involving both oxidoreduction and hydrodehydration
reactions in an integral enzyme complex.

## D. Chemical Flow and Energy Transduction

### 1. *Optimum Poise of Intermediate Couples*

   The basic principle that the efficient transmission of chemical flows and
the efficient (reversible) coupling between the flows of chemical particles in a
thermodynamic system depends on the maximization of the concentrations
of *all* the intermediates in the flow pathways requires some qualification
because, as discussed by Chance and coworkers (1958) for the case of the
respiratory chain, in the normal two-channel coupled flow of a circulating
carrier (Wilbrandt and Rosenberg, 1961; Mitchell, 1967a), the concentration
of loaded carrier in one pathway is inversely related to the concentration of
unloaded carrier in the other pathway when the total carrier concentration
is constant. We can best illustrate the basic properties of coupled two-
channel flow by reference to the simple coenzyme-coupled dehydrogenase
system of equation (5).
   At first sight, it might be thought that if the hydrogen potentials of the
$AH_2/A$ and $BH_2/B$ couples of the system of equation (5) were increased, the
rate of transmission of 2H groups through the carrier pathway under a given
potential gradient would necessarily be increased because the concentration
of the intermediate (NADH) would be increased. This is not, however, the
case because the carrier function of the coenzyme in the steady state requires
that the rate of flow of NADH one way shall be equal to the rate of flow of
$NAD^+$ the other; and thus, since the frictional factors (equation (1)) for
NADH and $NAD^+$ diffusion in aqueous media are about the same, the
fastest (and most efficient) passage of 2H groups through the coenzyme
channel under a given potential gradient would occur when the concentrations of NADH and $NAD^+$ were about equal, or in other words, when the
standard hydrogen (oxidoreduction) potentials of the couples $AH_2/A$ and
$BH_2/B$ were about the same as that of the linking coenzyme couple. The fact
that transmission is most efficient when the carrier couple is centrally poised
is shown quantitatively by the following equation (Mitchell, 1967a)

$$\frac{dn_A}{dt} = -\frac{d\bar{\mu}_A}{dx} \times \frac{[\bar{X}]}{f_X} \times \frac{K_A[A]}{(K_A+[A])^2} \tag{11}$$

where $dn_A/dt$ is the rate of transfer or translocation of A via the monovalent carrier X, and $[\bar{X}]$, $f_X$ and $K_A$ are the total concentration, the frictional coefficient and the dissociation constant of the carrier, respectively. The frictional coefficient of the loaded carrier is assumed to be the same as that of the unloaded carrier. It can readily be shown that $dn_A/dt$ in equation (11) has a maximum value, given by

$$\frac{dn_A}{dt} = -\frac{d\bar{\mu}_A}{dx} \times \frac{[\bar{X}]}{4f_X} \tag{12}$$

when $K_A=[A]$, or when $RT \ln K_A = RT \ln [A]$. Thus the fastest steady rate of net transfer of A is one quarter the gross rate of X and $X_A$ translocation, and requires that the chemical potential at or near which A is transferred shall correspond to the mid-point potential or central poise of the $X_A/X$ couple.

It follows that the statement that flow rates and efficiency or reversibility of coupling are maximized by maximizing the concentrations of all the intermediates must be taken to refer to the maximization of each intermediate concentration separately. The above discussion shows that reversible coupling will generally be improved by maximizing the total concentrations of loaded and unloaded carrier species (corresponding to $[\bar{X}]$), and by adjusting the mid-point potentials of the carrier couples to coincide approximately with the potentials of the chemical particles undergoing transfer when the frictional coefficients $f_X$ and $f_{XA}$ are about equal.

These considerations lead to the important conclusion that the requirement that an intermediate, such as the hypothetical $A \sim C$ complex in equation (10) should be 'energized' or should contain 'energy-rich bonds' should not be taken to imply that 'energy coupling' is due to the carriage of packets of bond energy of fixed or standard value from one reaction to the other. If this implication were correct, it would follow that the higher the standard free energy of formation of the bond or bonding relationship in the intermediate, the greater would be the quantity of energy transferred per mole of the intermediate or per unit time or both. In reality, when the intermediate comes into equilibrium with donor and acceptor systems between which it acts as the coupling link, the *actual* free energy of formation of the $\sim$ bonds will be determined by the potentials of the donor and acceptor systems and will be independent of the *standard* free energy of formation of the compound or complex. Further, for the kinetic (and not thermodynamic) reasons given above, the rate of transfer of energy via $X \sim I$ will be maximal when the standard free energy of $X \sim I$ formation has an intermediate value

H

such that the poises $[A \sim C]/([A]x[C])$, $[AH_2]/[A]$, $[BH_2]/[B]$ and $[ATP]/[ADP]$ are all fairly central when equilibrated according to equation (10).

## 2. So-called Energy Coupling

The process commonly called 'energy coupling' (see Pullman and Schatz, 1967) in which one reaction is linked to another through a so-called energy-rich intermediate has often been discussed as though it could involve the transfer of *energy* between the one reaction and the other without a corresponding participation of *material* in the coupling of the reactions. For example, according to Dixon (1951), 'when two enzyme reactions are coupled together it means that the second is dependent on the first . . . . This dependence is due to two quite distinct causes. The second reaction may be dependent on the first for its chemical material . . . Or it may depend on the first reaction not for its material but for its *energy*. In this case the substances produced by the first reaction may not enter into the second at all, but the latter must have energy in order to take place'. Energy transfer can, of course, take place by electromagnetic radiation from one molecule to another after electronic excitation of one of them (see Guéron and Schulman, 1968). But 'energy coupling' in metabolism is not generally intended to refer to this type of process. Vernon (1960) has stated in a criticism of abuses of the $\sim$-bond concept that chemically coupled metabolic systems involving 'energy rich' intermediates are coupled only because of the involvement of the product of the one reaction in the other. The correctness of Vernon's statement in the present context can be illustrated as follows.

Equation (10) may be simplified to the following essential scheme for coupling an oxidation reaction to a dehydration reaction through the intermediate X which has an 'energized' form $\sim$ X

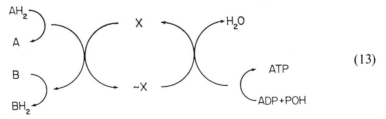

(13)

One might possibly describe this scheme by saying that the substances produced by the first reaction $(A + BH_2)$ do not enter into the second at all, and that energy coupling is effected by transferring the energy in $\sim$ X to the second reaction in order to enable the latter to proceed. This view, however, overlooks the fact that whatever X and $\sim$ X may be, $\sim$ X is a product of the first reaction and is a reactant in the second.

In the system of equation (13), the forward pressure on the oxidoreduction reaction is the total chemical potential difference of the hydrogen group $\Delta\bar{\mu}_{2H}$ between the couples $AH_2/A$ and $BH_2/B$, and the backward pressure on this reaction is the total chemical potential difference $\Delta\bar{\mu}_X$ of X in the couple $\sim X/X$. Similarly, the forward pressure on the dehydration reaction is $\Delta\bar{\mu}_X$ and the backward pressure on this reaction is the total chemical potential difference $\Delta\bar{\mu}_{H_2O}$ between water in the physiological aqueous media and water in equilibrium with the $(ATP+H_2O)/(ADP+POH)$ couple. The potential differences $\Delta\bar{\mu}_{2H}$, $\Delta\bar{\mu}_X$ and $\Delta\bar{\mu}_{H_2O}$ will be the same at equilibrium; when steady flow through the system of equation (13) occurs near equilibrium, the rate of energy transduction is the rate of flow of 2H, X or $H_2O$ multiplied by the potential difference $\Delta\bar{\mu}_{2H}$, $\Delta\bar{\mu}_X$ or $\Delta\bar{\mu}_{H_2O}$.

In general, total chemical potential differences $\Delta\bar{\mu}_i$ can be divided into several terms, the more important of which can be represented as follows in condensed systems at constant temperature and pressure

$$\Delta\bar{\mu}_i = z_i F\Delta\psi_i + RT\Delta \ln [i] + \Delta\mu_i' \tag{14}$$

where $\Delta$ represents the difference of a potential of $i$ between pairs of chemical components or phases contributing to the total potential difference $\Delta\bar{\mu}_i$, where $z_i$ and $[i]$ represent the ionic valency and concentration (or probability) of $i$, and $\mu_i'$ and $\psi_i$ represent the purely chemical potential of $i$ and the electric potential at the alternative sites occupied by $i$, respectively. The total driving potential $\Delta\bar{\mu}_i$ is thus made up of a purely electrical part $z_i F\Delta\psi$ (or $\Delta\mu_i^{\pm}$), a concentration part $RT\Delta \ln [i]$ (or $\Delta\mu_i^*$), and a purely chemical part $\Delta\mu_i'$. We have included in the chemical and electrical potential terms the effects of all the chemical and electrical forces, respectively, that reversibly influence the distribution of $i$ at equilibrium, and the usual activity coefficient or partition coefficient factors are thus contained in $z_i F\Delta\psi_i$ and $\Delta\mu_i'$ (see Section IIE2, p. 211).

Returning to the system of equation (13), it is evident that the difference between X and $\sim X$ which defines $\Delta\bar{\mu}_X$ may be due to a primary or secondary chemical transformation in X that permits a distinction between the total chemical potentials of X and $\sim X$ in the same phase, or it may be due to the translocation of X across an osmotic barrier that permits a distinction between the total chemical potentials of X in the phases on either side. The latter type of coupling has been called chemiosmotic because it requires, not a chemical distinction between X and $\sim X$ as in primary or secondary chemical coupling, but an osmotic distinction between X in one phase (L) and X in another phase (R). In any case, it is evident that reversible coupling of chemical reactions through an intermediate by the general type of flow system exemplified by equation (13) requires an obligatory coupling of the flows of chemical particles of both the input (2H) and output ($H_2O$) systems with the flow of the particles of the intermediate X system. The mechanism of this obligatory coupling depends, as discussed above, on the participation of catalytic complexes in which the appropriate transfer and translocation reactions are favoured by transitional intermediates of low free energy (high concentration) and high mobility along the prescribed reaction co-ordinates. The flow of matter is thus specifically channelled in space between donor and acceptor sites, and the net direction of flow along the reaction coordinates is determined by the electrical, concentration and chemical forces which define the total chemical potentials. Energy transduction in this type of equilibrium thermodynamic system is due to the transmission of forces through a continuous series of interactions between the chemical particles deployed along the reaction pathways.

It appears that there has been a tendency to describe chemical coupling as energy coupling when there was doubt about (or indifference to) the identity of the chemical intermediates actually involved. A similar tendency with regard to the use of the phrase active transport has been discussed previously (Mitchell, 1961a, 1967b).

## 3. *Backlash and Tightness of Coupling*

We generally speak of the tightness of coupling in the same sense as the reversibility in the steady state. There is, however, another usage, and care must be taken to avoid confusion. When two reactions are coupled by an intermediate, although the efficiency of transmission through the intermediate in the steady state may be virtually independent of the concentration of the intermediate couple when this couple is present at high concentration, there may be a looseness of coupling, especially at higher concentrations of the intermediate couple, in the sense that a change of the potential (and rate) of the input reaction is followed by a lag before the corresponding new steady output potential (and rate) is achieved. The lag is due to the fact that the change of the total chemical potential of the intermediate per unit quantity of input is inversely proportional to the thermodynamic capacity (or buffering power) of the intermediate. This kind of effect is well known in transmission systems in engineering, and is called backlash (see Mitchell, 1966, 1968).

It is particularly important to note that the existence of backlash during transient activity of coupled systems does not generally militate against high efficiency of chemical and energetic transmission in the steady state.

Some apparent disagreement about this point (Tager and coworkers, 1966; Slater, 1967) appears to have arisen from different interpretations of the meaning of the word backlash (Mitchell and Moyle, 1967d).

### E. Osmotic Coupling

The osmotic processes or reactions discussed here concern the translocation equilibrium of chemical components between phases L and R, separated by an osmotic barrier phase M. One translocation reaction is said to be coupled to another when the flow of a chemical component through each reaction is dependent on a flow through the other. The coupling of the flows may occur at the porter level because a chemical compound or complex (or state) produced in the porter system in the M phase by the one reaction is consumed in the other. Alternatively, coupling may occur because the translocation of a given component from phase R to phase L results in a shift of the thermodynamic poise of one or more other components that are free to equilibrate across the M phase. The latter case differs from the former in that the components between which coupling occurs need not be translocated through the M phase by the same route.

### 1. *Water Translocation*

Although the bulk of the M phase of biological lipid membranes generally has a very low permeability to most metabolites (Davis, 1958; Mitchell, 1967a), it is comparatively permeable to water (Dick, 1966). For this reason, the net translocation of a solute across the M phase is generally accompanied by the net translocation of an osmotically equivalent amount of water; and there thus tends to be an obligatory coupling between net solute translocation and net water translocation. This coupling does not occur at the porter level but is a direct consequence of the equilibration of the total chemical potential of water between phases L and R in response to a disequilibration of this potential by the redistribution of other components across the M phase. The present treatment of osmotic reactions is primarily concerned with coupling via the passenger-specific porters and we shall therefore consider the osmotic equilibration of water only inasmuch as it is relevant to the translocation equilibria of solutes.

### 2. *Translocation of Individual Ionic Species and the Phase-specific Membrane Potential*

When an osmotic process or reaction involves the translocation of individual ionic species, the net translocation of electric charge across the M phase results in an electric potential difference between the bulk of phases L and R and causes a corresponding electric force across the thickness of the M phase and in the surface layers of phases L and R on either side of the M phase.

The transfer of individual ionic species between one region of a system and another has long been recognized in the development of potentials across interfaces, for example, electrode potentials, contact potentials, Nernst potentials and Donnan potentials (Adam, 1941), and electric potentials in excitable membrane systems have been studied very successfully (see Cole, 1968). Nevertheless, the relationships between thermodynamic chemical potentials and electric potentials in inhomogeneous chemical systems have continued to be somewhat obscure. Since an explicit statement of these relationships is a prerequisite for the formulation of any theory of energy transduction that involves the transmission of a force by a train of ions, some explicit justification for the use of equation (14) and for the general viewpoint outlined previously (Mitchell, 1961c, 1966, 1968) and adopted in this article appears to be required.

Making use of the relationship

$$\mu = RT \ln a, \tag{15}$$

between the chemical potential $\mu$ and the corresponding absolute activity $a$, equation (14) is derived from the equation

$$\Delta \bar{\mu}_i = z_i F \Delta \psi_i + \Delta \mu_i \tag{16}$$

in which the total potential difference of $i$ is described as being divisible into an electrical part $z_i F \Delta \psi_i$ and another part $\Delta \mu_i$. The potential difference $\Delta \mu_i$ can be given in terms of the corresponding absolute activity $a_i$, and $a_i$ may be represented as the product of the concentration of $i$ (in all forms in equilibrium with pure $i$) and the absolute activity coefficient $\gamma_i$. Thus

$$\Delta \bar{\mu}_i = z_i F \Delta \psi_i + RT \Delta \ln [i] + RT \Delta \ln \gamma_i \tag{17}$$

It will be noted that in equation (17) the purely chemical term $\Delta \mu_i$ of equation (14) is replaced by the term containing $\gamma_i$ which therefore corresponds to the absolute chemical activity coefficient of $i$. In a system at equilibrium with respect to $i$, if it is possible to determine the value of $[i]$ in different regions, since $\Delta \bar{\mu}_i = O$, we may determine the sum $z_i F \Delta \psi_i + \Delta \mu_i'$. To obtain the individual values of $z_i F \Delta \psi_i$ and $\Delta \mu_i'$ (and incidentally to define the purely electric potential difference of an ionic component between two phases or regions of different chemical composition) it is necessary to distinguish between the electrostatic forces determining $z_i F \Delta \psi_i$ and the other chemical forces determining $\Delta \mu_i'$. In practice this can be done approximately because the electrostatic forces are of much longer range than the other chemical forces (Adam, 1941). The attempt to do so is worth while because a phase or region can be characterized by a statistically defined electric potential $\Delta \psi$ that has an effect on all the ionic components present, in the same way as a phase or region can be characterized by a hydrostatic pressure. In either case, a complication arises in condensed systems because of the occurrence of local foci of intrinsic potential or pressure. In an electrolyte phase, the electrical environmental potential is not statistically the same for ions of different charge or radius because positive ions tend to condense upon and be shielded by negative ions and *vice versa*. Thus, the electric potential in a phase or region has a different value for each species of ion when defined with respect to the total electrostatic energy change involved in transferring the pure ion between the given region and another region of standard (homogeneous) electric potential (Debye and Hückel, 1923; MacInnes, 1939). However, we can attribute an average electric potential to an electrolyte phase or region, and base the individual electric potentials for different pure ionic species on this average. This is generally achieved (implicitly) by including the microscopic electrostatic ion-pairing effects with the non-electrostatic chemical forces which are represented by the chemical activity coefficient $\gamma_i$.

The purely electric potential difference of electrons or of ions between pairs of phases or regions of virtually identical composition may be measured by classical electrochemical methods—for example, by using reversible ion-specific electrodes—because the purely chemical potential differences are zero (Guggenheim, 1933, 1949). In many contexts in cell physiology and biochemistry, however, it has become the general practice to use salt bridges to measure electric potential differences between aqueous phases that differ in composition. The theoretical basis of the use of a bridge of concentrated salt solution is that the ions of this concentrated salt are present in large excess at the junction and consequently carry almost the whole of the current across the boundary. The conditions are thus made somewhat similar to those existing when the electrolyte is the same on both sides of the junction (Nernst, 1897). The potential difference between two phases or regions measured in this way represents an empirical average difference that has proved to be very useful in practice (Davies and Rideal, 1963), especially in electrophysiology (see Cole, 1968).

A discussion of the question of electrical neutrality by Guggenheim (1949), which has been quoted by Robinson and Stokes (1959), appears to imply that the transfer of individual ionic species is impossible in practice and virtually impossible in theory because it would result in an enormous increase in the energy of the solution due to the self-energy of the electric charge involved. However, Guggenheim's calculation is based on the fact that the energy change that would result from the transfer of charge to a given phase depends on the electric capacity of the phase. As I have shown previously (Mitchell, 1968), it follows that in a disperse phase of relatively high electric capacity, the change of electric potential and energy that would result from the transfer of a chemically measurable quantity of a single ionic species to or from the phase need not be impossibly large (see also Hodgkin,

1951; Patlak, 1956). This can readily be demonstrated from the following differential form of equation (16)

$$\frac{d(\Delta\bar{\mu}_i)}{d\overleftarrow{n}_i} = z_i F \frac{d(\Delta\psi_i)}{d\overleftarrow{n}_i} + \frac{d(\Delta\mu_i)}{d\overleftarrow{n}_i} \qquad (18)$$

where $\overleftarrow{n}_i$ denotes the translocation of a quantity of the single ionic species $i$ of valency $+z_i$ from phase R to phase L. An electric capacity $M$ of the osmotic system is given by

$$\frac{z_i}{M_i} = \frac{d(\Delta\psi_i)}{d\overleftarrow{n}_i} \qquad (19)$$

and equation (18) can therefore be written as

$$\frac{d(\Delta\bar{\mu}_i)}{d\overleftarrow{n}_i} = z_i^2 \frac{F}{M_i} + \frac{d(\Delta\mu_i)}{d\overleftarrow{n}_i} \qquad (20)$$

Thus, the change of the total chemical potential difference of the ionic species $i$ across phase M during the passage of $n_i$ moles from phase R to phase L would be given by

$$\Delta\bar{\mu}_i = z_i^2 F \int_0^{\overleftarrow{n}_i} \frac{d\overleftarrow{n}_i}{M_i} + \Delta\mu_i \qquad (21)$$

If phase R were disperse and [as in organelles like mitochondrial cristae vesicles, chloroplast grana and bacterial protoplasts (see Mitchell, 1966, 1968)], if the area of the M phase were some 100 m² per ml of phase R and if the electric capacity were some 1 microfarad per cm² of phase M (i.e. about 1 farad per ml of phase R), the translocation across phase M of 1 m equiv. of ionic charge per litre of phase R would cause the electrical component $(F/M_i)$ of $\bar{\mu}_i$ to change by an amount corresponding to only about 100 mV. The precision with which it is possible to distinguish between the electrical component of $\bar{\mu}_i$ and the sum of the osmotic and chemical components represented by $\Delta\mu_i$ in equation (21) is dependent on our ability to define and measure the electrical capacity $M_i$, and this must inevitably involve an agreed distinction between the comparatively long-range electrical forces and the other chemical forces affecting the total ionic chemical potentials. At all events, however, the total chemical potential difference of the single ionic species $i$ across the osmotic system is well defined and corresponds to the total driving force on the ion, no matter whether phases L and R are chemically similar or different. In a perfectly reversible electro-chemical cell, for example, the total electrochemical potential difference $\Delta\bar{\mu}_i$

of electrons across the electrodes on open circuit will be $\Delta G'/n$, where $\Delta G'$ is the free energy per mole of the electrochemical reaction and $n$ is the number of electrons transferred across the system per mole of the reaction. In the special case when the electrodes are of the same composition, the purely electrical potential across the electrodes will correspond to $\Delta \bar{\mu}_i$ (i.e. $\Delta \bar{\mu}_i = F\Delta \psi_i$), but otherwise part of the well-defined total chemical potential difference of the ion will be electrical and part will be osmotic and chemical (i.e. as $\Delta \mu_i$). The same will hold true for a cell that is motive for any kind of ion, as shown by Guggenheim's general treatment of chemicomotive cells (Guggenheim, 1933).

The distinction between $\Delta \psi_i$ and $\Delta \mu_i'$ depends on the distinction between the comparatively long-range electrical forces and the other chemical forces. In practice, for pairs of phases or regions of different composition, the value of $\Delta \psi_i$ is different for each ionic species. We can, however, attribute an average potential difference $\Delta \psi$ to a pair of phases or regions, and specify the sum $z_i F\Delta \psi_i + \Delta \mu_i'$ as a sum $z_i F\Delta \psi + \Delta \mu_i''$ in which part of the electrostatic term is added to $\Delta \mu_i'$ to give $\Delta \mu_i''$, provided that it is possible to describe a relationship between the average value $\Delta \psi$ and the individual $\Delta \psi_i$ values, or that it is possible to describe a method of arriving at an empirical average value for $\Delta \psi$ that can be used consistently with corresponding values of the individual chemical activities (or activity coefficients $\gamma_i'$) obtained under similar conditions such that the total ionic chemical potential difference is the sum of the osmotic and the (empirical) electrical and chemical terms. This is the justification for using salt bridges to estimate the (empirical) electric potential differences described as membrane potentials in electrophysiology or in Nernst or Donnan equilibria, and for utilizing, consistently with such studies, estimates of (empirical) individual ionic chemical activities, using reversible ion-specific electrodes and salt bridges, as in techniques for pH measurement widely employed in biochemical and physiological laboratories. Accordingly, equation (21) can be written as

$$\Delta \bar{\mu}_i = z_i^2 F \int_0^{\overleftarrow{n_i}} \frac{d\overleftarrow{n_i}}{M} + RT\Delta \ln [i] + RT\Delta \ln \gamma_i' \tag{22}$$

where the value of the ratio $M/z_i$ (which corresponds to the average potential $\Delta \psi$) is independent of the ionic species translocated, but is dependent on the composition of phases L, M and R. The electric capacity $M$ represents the average value that would ideally be observed by alternating current bridge measurements in which the alternating electric field acts on all ionic species present.

We can alternatively write equation (22) as

$$\frac{\Delta\bar{\mu}_i}{F} = z_i\Delta\psi - Z\Delta p([i]\gamma'_i) \tag{23}$$

where $Z = 2.303\ RT/F$ and p stands for $-\log_{10}$. In this equation, $\Delta\psi$ can be described as

$$\Delta\psi = \Sigma z_{i,\ j}\int_o^{\overset{\leftarrow}{n}_{i,j}} \frac{d\overset{\leftarrow}{n}_{i,j}}{M} \tag{24}$$

where the contribution to $\Delta\psi$ of the translocation of all mobile ionic species $i$, $j$ across phase M is taken into consideration in the summation denoted by $\Sigma$. If $M$ is constant, or can be given as an appropriate average value,

$$\Delta\psi = \frac{1}{M}\left(z_i\overset{\leftarrow}{n}_i + \Sigma z_j\overset{\leftarrow}{n}_j\right) \tag{25}$$

In this equation the quantity in the bracket represents the total displacement of mobile ionic charge (due to $i$ and to the other species $j$, respectively) from phase R to phase L. In equation (23) the quantity $([i]\gamma_i')$ represents the total ionic activity difference of $i$ between phases L and R that would be measured with a reversible $i$-sensitive electrode and with the same type of salt bridge as is used to measure $\Delta\psi$. The potential difference $-Z\Delta p([i]\gamma_i')$ is mainly determined by the value of $\Delta p[i]$ in the case of most strong electrolyte ion species; but in the case of certain ions—particularly $H^+$ and $OH^-$ ions in aqueous media—the potential difference $-Z\Delta p([i]\gamma'_i)$ is mainly determined by $\Delta p(\gamma_i')$. Taking account of the translocation of species $i$ alone, for the concentration term $\Delta p[i]$,

$$d\Delta p[i] = -\frac{d\overset{\leftarrow}{n}_i}{N_i} \tag{26}$$

where

$$\frac{dp[i]}{dn_i} = -\frac{2\cdot303}{n_i} \tag{27}$$

for phases L and R individually, and where

$$\frac{0\cdot435}{N_i} = \left(\frac{1}{n_i}\right)_L + \left(\frac{1}{n_i}\right)_R \tag{28}$$

Likewise, for the activity term $\Delta p(a_i')$,

$$d\Delta p(\ '\gamma = -\frac{d\overset{\leftarrow}{n}_i}{B_i} \tag{29}$$

where the individual 'buffering powers' (Van Slyke, 1922) of phases L and R are given by

$$\frac{dp(\gamma_i')}{dn_i} = -\frac{1}{b_i} \tag{30}$$

and

$$\frac{1}{B_i} = \left(\frac{1}{b_i}\right)_L + \left(\frac{1}{b_i}\right)_R \tag{31}$$

Making the appropriate substitutions, equation (18) can therefore be given in a useful form (see Mitchell and Moyle, 1969a), as follows,

$$d\frac{\Delta\mu_i}{F} = \left(\frac{z_i^2}{M} + \frac{Z}{N_i} + \frac{Z}{B_i}\right) d\bar{n}_i \tag{32}$$

This equation describes changes in the total chemical potential difference of $i$ between phases L and R of the osmotic system in terms of the displacement of $i$ across phase M and the experimentally available factors $M$, $N_i$ and $B_i$. In the special case that $\Delta\bar{\mu}_i$ is zero for a particular ion, any one of the three terms on the right of equation (32) can be evaluated from measurements of the other two. Thus, the value of $\Delta\psi$ can be obtained from an integrated form of equation (32), provided that we know $N_i$ and $B_i$ and have a boundary condition corresponding to a known value of $\Delta\psi$, determined by the displacement of ions other than $i$ across phase M.

It should be remarked that phases L and R are homogeneous with respect to the total chemical potentials of the ionic species $i$, $j$, etc., but that the surface regions of phases L and R, neighbouring the surfaces of phase M, will be subject to the electric field associated with the electric potential difference $\Delta\psi$ between the bulk of phases L and R: the bulk of the phases means the regions that are relatively distant from the surfaces. Thus, as discussed previously (Mitchell, 1968), it is necessary, for some purposes, to distinguish between surface phases L and R, which are of the same order of thickness as the double electrical layers (Verwey and Overbeek, 1948; see also Eisenman and coworkers, 1968), and bulk phases L and R. In the bulk phases the values of $\psi$ (and $\mu^* + \mu''$) are virtually constant in three dimensions, but in the surface phases the values of $\psi$ are constant in two dimensions but variable in the third. When the surface phases represent only a small proportion of the bulk phase volumes—as is usually the case with the type of system that we are considering in this article (Mitchell, 1968)—we may neglect the correspondingly small ambiguity in the definition of $\Delta\psi$ (and of $\Delta\mu^* + \Delta\mu''$). For the highly disperse systems in which the surface phases predominate, a range of values of $\Delta\psi$ would be required to define the electrical state of the equipotential surfaces at different distances from the surfaces of the M phase.

It is noteworthy, incidentally, that when an individual ionic species is transferred from phase R to phase L, the corresponding change of non-electroneutrality is mainly confined to the surface phases L and R; but the species of ion transferred across the M phase will nevertheless be displaced mainly from bulk phase R to bulk phase L if it is a minority ion (e.g. 1 $\mu$M $H_3O^+$) in phases L and R containing electrolyte (e.g. 1 mM KCl). The word 'mainly' is required here because the double electrical layer which distinguishes the surface from the bulk phase in this context has a diffuse boundary.

Summarizing the main point of this section: we may express $\Delta\bar{\mu}_i$ for a single ionic species either in terms of the component-specific electric potential $\psi_i$ and the purely chemical activity coefficient $\gamma_i$, as described by equation (17),

$$\Delta\bar{\mu}_i = z_i F \Delta\psi_i + RT\Delta \ln [i] + RT\Delta \ln \gamma_i \tag{17}$$

or we may express $\Delta\bar{\mu}_i$ in terms of a phase-specific electric potential $\psi$ and an equivalently adjusted chemical activity $\gamma_i'$ in which the effects of electrical ion-pairing and shielding have been included,

$$\Delta\bar{\mu}_i = z_i F \Delta\psi + RT\Delta \ln [i] + RT\Delta \ln \gamma_i' \tag{33}$$

The three terms on the right hand side of equation (33) are alternatively written as $\Delta\mu_i^{\pm}$, $\Delta\mu_i^*$ and $\Delta\mu_i''$, respectively.

## 3. *Translocation not Coupled at the Porter Level: Uniport*

The class of translocation catalyst defined previously as uniporters (Mitchell, 1967a) catalyses the non-coupled translocation of individual chemical components or species. That is to say, the translocation reaction catalysed across the M phase by the uniporter involves net displacement across the M phase only of that single component for which the porter is specific. As indicated in the previous two sections, the flow of any one component across the M phase will tend to influence the thermodynamic poise of the activity of water and of other solutes across the M phase both by concentration and by electrical effects. Corresponding flows will occur only if permitted by the properties of the M-phase system. In practice, natural lipid membranes have a comparatively high permeability to water (Dick, 1966) and thus a flow of water (passing elsewhere through the membrane) generally accompanies the translocation of a given solute via a uniporter.

As discussed previously, the best characterized examples of the membrane-located uniporters of physiologically normal membranes are the systems for D-glucose and for L-leucine translocation in the mature red blood cells of mammals (Mitchell, 1967a). Further work on these and other related systems has been reviewed by Rothstein (1968).

Certain physiologically abnormal uniporters have recently been identified, which promise to give new impetus to the knowledge of the mechanisms of specific translocation reactions in lipid membranes; these substances are particularly relevant to our discussion because they appear to illustrate important general principles of translocation catalysis (Mitchell, 1968; Pressman, 1968).

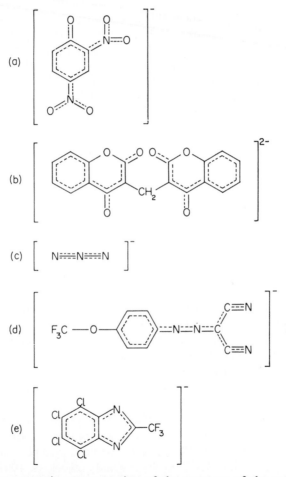

**Figure 1.** Diagrammatic representation of the structure of the anions: (a) 2,4-dinitrophenate; (b) dicoumarate; (c) azide; (d) carbonyl cyanide *p*-trifluoro-methoxyphenylhydrazone; (e) tetrachloro-2-trifluoromethylbensimidazole. The broken lines represent the $\pi$-orbitals over which the electrons carrying the anionic charge are distributed.

*Proton conductors or proton uniporters.* The classical uncoupling agents, 2,4-dinitrophenol (DNP), dicoumarol and azide were the first pharmacologically active ion conductors to be recognized (Mitchell, 1961d). Their identification as proton conductors depended on the knowledge that these catalytic agents are mixtures of a weak acid form and a correspondingly deprotonated anionic form, characterized by a $\pi$-bonded unsaturated structure, and on the observation that they catalysed acid–base titration of the inner phase buffers of bacterial and mitochondrial suspensions in KCl media at acid pH when the permeability to the chloride ion was expected to be comparatively high (Mitchell, 1961d, 1963a,b; Carafoli and Rossi, 1967). The structure of the anions, 2,4-dinitrophenate, dicoumarate, and azide can be diagrammatically represented as shown in figure 1. Two of the more recently discovered proton-conducting agents of very high activity (Heytler and Prichard, 1962; Jones and Watson, 1965), carbonylcyanide p-trifluoromethoxy phenylhydrazone (FCCP) and tetrachloro-2-trifluoromethyl benzimidazole (TTFB) are also shown for comparison. It was thought that the $\pi$-orbital system that is characteristic of the classical uncoupling agents would minimize the electric field at the surface of the anion (because of the delocalization of the electron and the effective spreading of the charge over the ion) and would thus favour the penetration of the anion into the nonpolar M phase without a charge-neutralizing partner (Mitchell, 1966, 1968). The translocation of $H^+$ across the M phase was attributed to a circulating carrier type of mechanism, as illustrated by the following equation

$$H^+ \diagdown \quad A^- \quad \diagup H^. \tag{34}$$
$$AH$$

where AH represents the protonated form of the proton-conducting agent and $A^-$ represents the corresponding lipid-soluble anion. This interpretation has been amply confirmed by studies of the proton-conducting effects of DNP and other classical uncoupling agents, such as FCCP, on mitochondrial cristae membranes (Carafoli and Rossi, 1967; Mitchell and Moyle, 1967a, b,c), on chloroplast grana membranes (Jagendorf and Neumann, 1965; Rumberg and coworkers, 1968), on chromatophore membranes (Jackson and coworkers, 1968), on bacterial membranes (Harold and Baarda, 1968; Pavlasova and Harold, 1969), on red cell membranes (Chappell and Crofts, 1966; Harris and Pressman, 1967) and on artificial phospholipid membranes (Chappell and Haarhoff, 1967; Skulachev and coworkers, 1967; Hopfer and coworkers, 1968; Liberman and Topaly, 1968; Liberman and coworkers, 1968). In particular, Hopfer and coworkers (1968) and Liberman and Topaly (1968) have shown that the agents of the DNP class exhibit a high degree of specificity for $H^+$ translocation. The latter workers have observed that the proton conductance is optimal when AH is about half dissociated to A$^-$, thus adding support to the circulating carrier type of mechanism described by equation (34). It is interesting to note that the conduction of H$^+$ ions across the M phase by the DNP class of agent involves the formation and breakdown of a covalent chemical intermediate AH; but it is relevant that protonation and deprotonation in aqueous media are unusually fast chemical reactions (Bell, 1959).

*Alkali metal cation conductors or cation uniporters.* Following the discovery by Pressman and coworkers that valinomycin and the gramicidins have a profound effect on ion translocation in mitochondria (Pressman, 1963; Moore and Pressman, 1964), Chappell and Crofts (1965, 1966) observed that these antibiotics affected mitochondrial swelling as though they cata-lysed specific cation uniport reactions. This interpretation was confirmed by observations on the cation conductance of artificial lipid membranes (Chappell and Haarhoff, 1967; Lev and Buzhinsky, 1967; Mueller and Rudin, 1967; Andreoli and coworkers, 1967). The literature on valinomycin, the macrotetralide actins, the enniatins, the gramicidins and certain synthetic analogues which act as cation conductors has recently been well reviewed by Eisenman and coworkers (1968), Lardy (1968), Pedersen (1968), Pressman (1968), Liberman and Topaly (1968) and Henderson and coworkers (1969).

It is now generally agreed that the specific conduction of cations across lipid membranes by the macrocyclic antibiotics and chemically simpler polyethers occurs as a result of the abilities expected of these substances to solubilize cations of appropriate size in low dielectric media and thus act as carriers for such ions through the non-polar medium of the M phase (Eisenman and coworkers, 1968; Liberman and Topaly, 1968; Mitchell, 1968; Tosteson, 1968; Henderson and coworkers, 1969; and see Ilani and Tzivoni, 1968). Since these agents are themselves uncharged and their cation complexes are cations of large radius, there is a fairly close analogy (see Mitchell, 1968) between the mechanism by which the proton conductors of the DNP class catalyse proton conduction and the mechanism by which the alkali metal ion conductors of the valinomycin class catalyse the conduction of other cations across the M phase. As indicated by the following equation, representing the circulating carrier mechanism for catalysis of M$^+$ ion translocation by the valinomycin class of agent (V),

$$\tag{35}$$

the mechanisms of equations (34) and (35) differ in the sign of the charged species crossing the M phase. There are, of course, other distinguishing features. The specificity of bonding of $H^+$ by the DNP type of proton-conducting agent depends largely on the chemical property required for the formation of the covalent acid—a property which is lacking in the valino-mycin type of reagent. The specificity of binding of $M^+$ with the valinomycin type of agent depends on ion–dipole interactions responsible for hydration of $M^+$ (and of V) on the one hand, and for association between $M^+$ and V on the other. As Pressman (1968) has emphasized, the polar–nonpolar inter-action energies associated with the conformations of V and $VM^+$ also contribute substantially to the total free energy balance determining the binding between $M^+$ and V, especially in the case of the more conforma-tionally flexible agents. Certain of the cation-conducting agents of low specificity, notably the gramicidins, bind (and conduct) $H_3O^+$ ions as well as alkali metal ions (Henderson and coworkers, 1969).

*Effects of ion or charged-site concentrations and of space-charge in the M phase.* According to the circulating carrier type of proton and cation conduc-tion mechanisms represented by equations (34) and (35), the general principles of catalysis discussed in the earlier sections of this article indicate that the velocity of conduction will depend on the concentrations and mobilities of the components of the $A/A^-$ and $V/VM^+$ couples in the M phase (equation (11))—in much the same way as the velocity of an NAD-linked dehydrogenase reaction depends on the concentrations and mobilities of the components of the $NAD^+/(NADH + H^+)$ couple in the aqueous phase. There are, however, special circumstances that tend to be dominant in determining the concen-trations of $A^-$ and $VM^+$ in the M phase of the systems of equations (34) and (35). The total quantity of $A^-$ or $VM^+$ that can reside in the M phase depends on: i) the quantity of charge-neutralizing sites or ions present in the M phase with $A^-$ or $VM^+$; and ii) the quantity of space-charge (i.e. the extent of non-electroneutrality) in phase M at a given electric potential. If, as assumed in a quantitative theory developed by Eisenman and coworkers (1968), there are no charge-neutralizing ions, the space-charge effect will determine the concentration of $A^-$ or of $VM^+$ in the M phase, and this will play a dominant part in determining the maximum rate of $H^+$ or $M^+$ conduction by the specific agents. The effect of the lipophilic laurate (Pressman and coworkers, 1967) or thiocyanate (Pressman, 1968) ion on solubilization of complexes of the $VM^+$ class in a bulk organic phase, and the enhancement of the conductance of bimolecular membranes to picrate and $H^+$ ions by the lipophilic base decylamine (Liberman and Topaly, 1968) illustrate that the presence of charge-neutralizing ions (or fixed sites) distributed across the M phase will, in general, raise the concentration of the charged form of the catalytic carrier couple in the M phase and improve the conductivity. Conversely, the presence of a space-charge of the same sign as that of the charged species of the carrier couple in the M phase would generally be expected to have an inhibitory effect.

It follows from these considerations of the space-charge effects, that the potency of proton-conducting agents such as DNP or FCCP (which contribute to a space-charge of negative sign), and the potency of cation-conducting agents of the valinomycin class (which, in the form of the cation complex, contribute to a space-charge of positive sign) will depend on the intrinsic polarity of the regions of the M phase through which translocation of the charge conductor (A⁻ or VM⁺) may occur. This may be an important factor contributing to the relative susceptibility of different natural membranes to these agents. For example, the relatively high concentrations of DNP or FCCP required to collapse the electrochemical potential difference of H⁺ ions and uncouple photosynthetic phosphorylation in chloroplast grana might be accounted for by this type of space-charge effect. If this were the case, a potentiation of the effect of DNP or FCCP by positively charged lipid-soluble agents, such as valinomycin in presence of a potassium salt, would be expected. Just such a potentiation has, in fact, been observed by Karlish and Avron (1968), although they have not sought to explain it in this way.

The space-charge effects that limit the activity of the non-physiological ion-conducting couples of the DNP/DNP⁻ and valinomycin/K⁺-valinomycin classes by resisting their entry into the membrane draw attention to the importance of the factors that are responsible for locating or anchoring catalytic components in normal physiological systems (Green, 1957; Mitchell, 1957, 1959a,b, 1963a).

It will be seen in Section IIE5 (p. 225) that the uniport of a composite substrate is equivalent to the symport of its components; and that, in particular, the uniport of a given acid may alternatively be described as the symport of H⁺ ions with the corresponding deprotonated or base form.

## 4. *Translocation Coupled at the Phase Level*

When two reactions are each coupled at the enzyme or porter level (Section IIIA and B, pp. 240–248 and Section IIE5, p. 225) to the translocation of a common component, the reactions exhibit a mutual coupling that is due to the closed cyclic translocation of the common or coupling component between phases L and R. The transmission of power per mole of the coupling component cyclically translocated will be equal to the total chemical potential difference $\Delta\bar{\mu}_i$ of the coupling component across the M phase, and, as shown above (p. 209), the total chemical potential difference can be attributed to the sum of electrical, concentration and chemical parts as follows

$$\Delta\bar{\mu}_i = z_i F \Delta\psi + RT\Delta \ln [i] + RT\Delta \ln \gamma_i' \tag{33}$$

The relative contributions of the electrical, concentration and chemical forces to the total force represented by $\Delta\bar{\mu}_i$ is, of course, determined by the properties of the component $i$ and by certain properties of the osmotic system in which it acts as the coupling link.

*Coupling by an uncharged component.* If the coupling component were uncharged and formed ideal solutions in phases L and R, $\Delta\bar{\mu}_i$ would be given only by the concentration term, and for an energy transduction of 5000 cal/mole of cyclic translocation of $i$, the concentration ratio of $i$ in phases L and R would be about $10^4$. If the concentration of $i$ in the phase at lower concentration were $0\cdot1$ mM, the concentration in the other phase would have to be 1 M. Owing to water movement to the phase of higher osmotic pressure, this concentration difference could not generally be achieved without some compensating solute translocation (see below), but certain bacteria might be exceptional in this respect (Mitchell and Moyle, 1956a). Thus, certain restrictions would generally be imposed on energy transduction by this type of osmotic coupling mechanism in the usual biological systems where large hydrostatic pressure differences cannot be maintained.

*Coupling by a strong electrolyte ionic component.* In the case of an ionic coupling component, such as $Na^+$, that forms semi-ideal solutions, most of the coupling force across the M phase could be electric or due to concentration or both, depending on the influence of other ionized components in the system. When an electric potential difference was established between phases L and R by the displacement of a given ionic component across the M phase, there would be a force on all ionic components in the system, and a *transient* coupling effect would occur with respect to any ionic component that could be displaced across the M phase. If electrical neutrality were restored in this way, $\Delta\psi$ would subside and only the concentration component of $\Delta\bar{\mu}_i$ would remain (see Mitchell, 1968). If, for example, the ion mainly responsible for electrical neutralization were $K^+$ when the coupling component was $Na^+$, there need be no imbalance of osmotic pressure (of water), even at very high $[Na^+]_L/[Na^+]_R$ ratios—a circumstance that is relevant to the state of affairs in many types of cell, where the $Na^+/K^+$ antiporter ATPase is responsible for maintaining approximately reciprocal concentration differences of $Na^+$ and $K^+$ across the cell membranes (Glynn, 1968; Potts, 1968). It is interesting to note, incidentally, that if both $Na^+$ and $K^+$ were involved as coupling components circulating in opposite directions between a pair of reactions, a considerable improvement of power transmission would be achieved. In such push–pull coupling, the concentration ratios of $Na^+$ and $K^+$ would have to be only 1/100 and 100/1 to given an energy transduction of about 5000 cal/mole of $Na^+$ and $K^+$ circulating. The practical reversibility of the reaction catalysed by the $Na^+/K^+$ antiporter ATPase (Garrahan and Glynn, 1967) appears to be favoured by the push–pull properties of this enzyme-linked translocation reaction.

*Coupling by protons.* The proton has unique properties as a coupling component between aqueous media in much the same way as the electron has unique properties as a coupling component between metallic media. The total proton concentration in aqueous media is virtually constant, and proton transfer reactions proceed extremely fast in aqueous media, even though the 'free' proton concentration is extremely small (Bell, 1959). Thus, the concentration term in equation (35) is negligible in the case of coupling by a circulating proton current. This fact may seem surprising at first sight (Baum, 1967), but can be justified readily in terms of the known high proton conductance of aqueous media (Mitchell, 1967d). It follows that the coupling force across the M phase can be electrical or chemical or both. The chemical force can be given in terms of $\Delta pH'$, defined in conformity with the requirements discussed in Section IIE2 (p. 211). As shown elsewhere (Mitchell, 1961c, 1966, 1968)

$$\frac{\Delta \bar{\mu}_{H^+}}{F} = \Delta \psi' - Z \Delta pH' \tag{36}$$

where $Z = 2 \cdot 303 \, RT/F$. The numerical value of $Z$ is about 59 at 25° when $\Delta \psi'$ is given in $mV$ and pH is given in the usual units. If the membrane potential $\Delta \psi'$ established by the net displacement of protons across the M phase were collapsed by the migration of other ions between phases L and R, only the chemical force represented by $-\Delta pH'$ would remain. The M phase cannot, in practice, be completely impermeable to any ion, and the potential $\Delta \psi'$ must therefore fall towards a value dependent on the Donnan equilibrium distribution of the permeating ions as coupling by the proton current approaches the steady state (Mitchell, 1968), unless subsidiary ion translocating systems in the M phase compensate for the ion leakage down the electric gradient created by the proton displacement. This compensation may readily be catalysed by electrically neutral proton/cation antiporters and proton/anion symporters (or the equivalent hydroxyl/anion antiporters) which (see Section IIE5 p. 225), owing to their electrical neutrality would appropriately equilibrate the concentration+chemical potentials of the protons and other ions, but not their electrical potentials (Mitchell, 1961b, 1966, 1968). Owing to the comparatively low concentration of free $H^+$ or $H_3O^+$ ions, relatively little imbalance of osmotic pressure (of water) across the M phase need be directly involved in coupling by a proton current.

## 5. *Translocation Coupled at the Porter Level: Antiport and Symport*

The concept of exchange-diffusion, introduced by Ussing (1947, 1949) and the complementary concept of codiffusion (see Crane and coworkers, 1961) are particular cases of the general principle that the escaping tendencies and the corresponding flows of pairs of chemical particles will be sym-coupled if the particles reside together in a mobile complex in a membrane and will be anti-coupled if they compete with one another. As I have pointed out previously (Mitchell, 1963a), this general principle of coupling, which is related to the linked-function concept of Wyman (1948), does not necessarily require any structural similarity between the ions or molecules undergoing antiport or symport. The following equations illustrate antiport and symport of $n$ moles of A and $m$ moles of B, respectively, via the translocator X.

$$nA \diagdown \diagup^{XB_m} \diagdown \diagup^{nA} \qquad (37)$$
$$mB \diagup \diagdown_{XA_n} \diagup \diagdown_{mB}$$

$$nA \diagdown \diagup^{X} \diagdown \diagup^{nA} \qquad (38)$$
$$mB \diagup \diagdown_{XA_nB_m} \diagup \diagdown_{mB}$$

When a component A is equilibrated between phases L and R by a uniporter,

$$\Delta \bar{\mu}_A = 0$$

When components A and B are equilibrated between phases L and R by an antiporter or by a symporter catalysing the reactions described by equations (37) and (38), respectively, it follows (Mitchell, 1968) that for $nA$–$mB$ symport

$$n\Delta \bar{\mu}_A + m\Delta \bar{\mu}_B = 0 \qquad (39)$$

and for $nA/mB$ antiport

$$n\Delta \bar{\mu}_A - m\Delta \bar{\mu}_B = 0 \qquad (40)$$

Thus, the driving force on the translocation of A is balanced against that on the translocation of B by means of the coupling porter—just as the driving force on one chemical reaction is balanced against that on another by means of a coupling enzyme, as discussed in sections IIC3 and 4 (pp. 203–204).

A number of antiporters and symporters have been described and discussed in the literature (see Albers, 1967; Heinz, 1967; Lehninger and coworkers, 1967; Chappell and Robinson, 1968; Klingenberg and Pfaff, 1968; Rothstein, 1968). Our special interest stems from the desire to identify the functional significance of these translocation catalysts in the coupling between one translocation reaction and another, and in the coupling between metabolism and transport (Mitchell, 1959a, 1963a, 1967a). At the present time two main groups of translocation reactions coupled by porters are being investigated: those associated with sugar and amino acid absorption in mammalian tissues, which are thought to be $Na^+$-sugar and $Na^+$-amino acid symport reactions (Crane and coworkers, 1961; Crane, 1965; Heinz, 1967; Kohn and coworkers, 1968; Munck, 1968; Koopman and Schultz, 1969; and see Rothstein, 1968) or possibly $Na^+$ symport/$K^+$ antiport push–pull (see p. 224) type reactions (Kuchler, 1967; Eddy, 1968a,b; Eddy and coworkers, 1969); and those associated with substrate and ion translocation in bacteria and mitochondria, which are thought to be proton-coupled symport and antiport reactions (Mitchell, 1961b, 1963a, 1967a, 1968; Chappell and Crofts, 1966; Chappell and Haarhoff, 1967; Lehninger, Carafoli and Rossi, 1967; Chappell and coworkers, 1968; Harold and Baarda, 1968; Jackson and coworkers, 1968, 1969; Mitchell and Moyle, 1969b; Tyler, 1969). As Lehninger (1966) has pointed out, 'the growing abundance of information on ion movements in isolated mitochondria, together with the large amount of information available on electron transport and oxidative phosphorylation, makes it possible to study active ion transport in this organelle in the most fundamental manner'. We shall accordingly concentrate attention on the proton-coupled symporters and antiporters of mitochondria, and on certain analogous systems in microorganisms.

## 6. *Proton-linked Symport and Antiport*

It has long been recognized (Jacobs, 1927, 1940; Davson and Danielli, 1943) that the ammonium salts of weak acids, such as acetic acid (HAc), permeate natural lipid membranes by the following type of mechanism

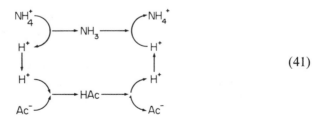

$$(41)$$

in which the upper reaction is effectively $NH_4^+/H^+$ antiport and the lower reaction is $H^+$–$Ac^-$ symport. This mechanism arises from the fact that the non-polar $NH_3$ and HAc molecules dissolve in and permeate the lipid phase of the membrane much more readily than the polar $NH_4^+$ and $Ac^-$ ions (Davson and Danielli, 1943). In mammalian red blood cells, Jacobs (1922, 1924, 1940) observed that the ammonium salts of relatively strong acids, such as hydrochloric acid, also permeate. He suggested the following type of mechanism

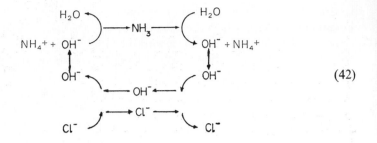

$$\text{(42)}$$

in which there is $OH^-/Cl^-$ antiport (not coupled at the porter level). Owing to the relatively high permeability of natural membranes to water, hydroxyl ion-coupled anion antiport (such as $OH^-/Cl^-$ antiport) has the same net effect as proton-coupled anion symport (such as $H^+-Cl^-$ symport). The latter, it was suggested, might occur indirectly via $CO_2$ and bicarbonate in red cell suspensions (see Keilin and Mann, 1941; Jacobs and Stewart, 1942) and Keilin privately suggested to the author in 1947 that the $HCO_3^-/Cl^-$ antiport might be tightly coupled at the level of the porter in the following type of scheme,

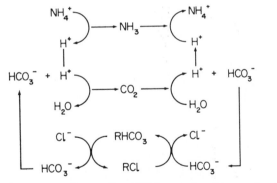

$$\text{(43)}$$

The reaction in the centre of equation (43) is catalysed by carbonic anhydrase (Keilin and Mann, 1941; Jacobs and Stewart, 1942) where R represents the translocater centre of the $HCO_3^-/Cl^-$ antiporter. Recent work on red blood cells (Chappell and Haarhoff, 1967; Harris and Pressman, 1967; Hunter, 1967; Henderson and coworkers, 1969) appears to be consistent with this type of mechanism.

Using light scattering ('absorbance') to observe the swelling that accompanies salt entry in non-metabolizing mitochondria, Chappell and Crofts (1966) showed that not only does ammonium acetate enter rapidly, presumably by the well-known mechanism of equation (41), but ammonium phosphate also enters quite rapidly. They proposed a mechanism corresponding to that of equation (42) for ammonium phosphate translocation, in which $Cl^-$ was replaced by the monovalent $H_2PO_4^-$ ion, and the $OH^-/H_2PO_4^-$ antiport was supposed to be tightly coupled at the porter level. As indicated above (and see Mitchell, 1967a, 1968), the same net result would be obtained by proton–phosphate symport (corresponding to phosphoric acid uniport) represented as follows

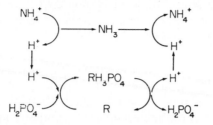

$$(44)$$

The objection of Chappell and Crofts (1966) that 'it seems unlikely that $H_3PO_4$ could be present at sufficiently high concentration for it to be the penetrant' is irrelevant when it is recognized that the reaction occurs via the translocater centre R of a specific porter, and that a normal function of catalysts is to change the local concentrations of particular species of reactant, as discussed in the earlier sections of this article. When, as in this case, the experimental information is not yet available to permit a distinction to be made between hydroxyl antiport and proton symport it is conceptually simpler to refer coupling to the proton.

It can readily be shown that, at pH values near neutrality, where the total inorganic phosphate concentration $[PO_4]$ is nearly equal to the sum of the individual ionic concentrations $[H_2PO_4^-]+[HPO_4^{2-}]$, and assuming that the activity coefficients are the same in phases L and R, when phosphoric acid is equilibrated between the two aqueous phases, the total phosphate concentration ratio is given by

$$\frac{[PO_4]_L}{[PO_4]_R} = \frac{\{H^+\}_R^2}{\{H^+\}_L^2} \frac{K+\{H^+\}_L}{K+\{H^+\}_R} \tag{45}$$

where $\{\}$ denotes activity and $K$ is the second dissociation constant of phosphoric acid. This equation may be derived from relationships corresponding to equation (40) which can be written as

$$\Delta p(H_2PO_4^-) = -\Delta pH \tag{46}$$

$$\Delta p(HPO_4^{2-}) = -2\Delta pH \tag{47}$$

where p stands for $-\log_{10}$ of the chemical activity (Mitchell, 1968). Thus, it is evident that a catalytic phosphoric acid porter would couple changes of $\Delta pH$ to changes of $\Delta p$(phosphate) across the membrane. In other words, a step or pulse change imposed on $\Delta pH$ would exhibit a decay, dependent on the translocation of $H^+$ with $H_2PO_4^-$ (or $2H^+$ with $HPO_4^{2-}$) across the membrane. In suspensions of rat liver mitochondria, a phosphate-dependent $\Delta pH$ decay phenomenon has been observed (Mitchell and Moyle, 1967b, 1969b) which is attributable to a phosphoric acid porter of high temperature coefficient. Measurements with a $K^+$ ion-sensitive electrode have shown that the phosphate-dependent catalysis of $H^+$ ion translocation does not directly involve $K^+$ ion translocation, and confirm that the porter catalyses net translocation of $H_3PO_4$ and not of $H_2KPO_4$.

In suspensions of rat liver mitochondria, the $Na^+$-dependent and $K^+$-dependent catalysis of $H^+$ ion translocation (Mitchell and Moyle, 1967b) has likewise been attributed to one or more proton/cation antiporters, the existence of which was postulated previously (Mitchell, 1961c). The proton/cation antiport, like the proton-phosphate symport, appears to exhibit a high temperature coefficient (Mitchell and Moyle, 1967b). This attribute would be consistent with the requirement for rather complex conformational changes in the porter system during translocation.

Further evidence for tightly coupled $H^+/Na^+$ antiport and probably also for $H^+/K^+$ antiport across the cristae membrane of non-metabolizing rat liver mitochondria has recently been obtained (Mitchell and Moyle, 1969b), using osmotic techniques similar to those of Chappell and Crofts (1966) and Chappell and Haarhoff (1967). We confirmed that the ammonium salts of acetic or phosphoric acids equilibrate rapidly across the cristae membrane and that the corresponding choline salts do not. It was found, however, that the sodium salts equilibrate at a rate comparable to that of the ammonium salts. A rapid equilibration of sodium acetate was likewise observed in beef heart mitochondria (Brierley and coworkers, 1968). This, of course, would be expected, given the presence of the $H^+/Na^+$ antiporter, and the mechanism is as follows for sodium acetate equilibration

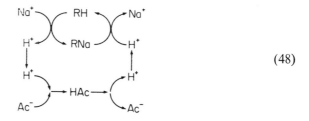

(48)

where R stands for the translocater centre of the $H^+/Na^+$ antiporter. At a given pH, a similar, but slower equilibration of potassium acetate (Mitchell and Moyle, 1969b) confirmed the earlier conclusion (Mitchell and Moyle, 1967b) that the $H^+/Na^+$ antiporter or a separate $K^+$-specific porter catalyses $H^+/K^+$ antiport. Both $H^+/Na^+$ and $H^+/K^+$ antiport exhibit acid pH optima at high $Na^+$ and $K^+$ concentration (150 mM), respectively.

Recent work on antibiotics of the nigericin class (Lardy and coworkers, 1967; and see Pressman, 1968) has added substantially to our knowledge of the type of mechanism that may give rise to tightly coupled $H^+$/cation antiport. Although the mechanism of action of nigericin and similar antibiotics was originally suggested by analogy with the behaviour of the physiologically natural $H^+$/cation antiporters discussed above (see Mitchell, 1968), these antibiotics may well turn out to provide a valid model for the physiologically normal porters.

The members of the nigericin class of antibiotics are distinguished in that they contain a dissociable carboxyl group, and that whereas tight cation binding (especially of $K^+$), occurs when the carboxyl group is deprotonated, it does not occur when the carboxyl group is in the protonated form (Pressman and coworkers, 1967; Pressman, 1968; Steinrauf and coworkers, 1968). Thus, nigericin induces tightly coupled $H^+$/$K^+$ antiport of high activity in mitochondria (Pressman and coworkers, 1967), erythrocytes (Harris and Pressman, 1967), chloroplasts (Packer, 1967; Shavit and coworkers, 1968) and chromatophores (Jackson and coworkers, 1968). The agents of this class do not increase the net ion conductance of artificial membranes (Mueller and Rudin, quoted by Pressman and coworkers, 1967), confirming the electrical neutrality of the $H^+$/cation antiport reaction. The electrical neutrality of the phosphoric acid porter of rat liver mitochondria has been confirmed by the observation that whereas potassium phosphate normally equilibrates very slowly across the cristae membrane and the addition of valinomycin has little effect, the addition of nigericin catalyses fast equilibration (Mitchell and Moyle, 1969b), presumably by the following mechanism

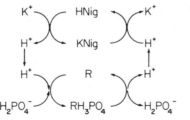 (49)

where Nig stands for nigericin and R stands for the translocater centre of the phosphoric acid porter. Also, as expected, nigericin (but not valinomycin) catalyses fast equilibration of potassium acetate in rat liver mitochondria at neutral pH, presumably by a mechanism analogous to that of equation (48). In this case, the nigericin simply supplements the natural $H^+$/$K^+$ antiporter, which has a relatively low activity at pH 7.

In addition to the properties described above, the mitochondrial phosphoric acid porter exhibits an alkaline pH optimum at high phosphate concentration (Mitchell and Moyle, 1969b), and reacts with arsenate as well as with phosphate (Chappell and Crofts, 1966; Chappell and Haarhoff, 1967; Tyler, 1968, 1969), and it is specifically inhibited by mercurials and by certain other —SH reactors (Tyler, 1968, 1969; Fonyo, 1968). In all these respects the mitochondrial phosphoric acid porter resembles the system that catalyses phosphate exchange and uptake across the plasma membrane of *Staphylococcus aureus* (Mitchell, 1953, 1954a,b). The work on the $PO_4$ porter of *Staph. aureus* led to the conclusion that $H_2PO_4^-$ was translocated as a tightly bonded complex or covalent intermediate (Mitchell, 1954a,b; Mitchell and Moyle, 1956b), and the data are consistent with the proposition that $H_2PO_4^-$ is translocated as the covalent intermediate $H_3PO_4$, via a specific phosphoric acid porter. The work on the bacterial transport systems also prompted the suggestion (Mitchell, 1959a) that phosphate may act as an exchanger, passing out on one translocater and, after exchange with a reactant on the outer surface, being reabsorbed through a second translocater while the reactant passes in on the first. Thus, the linkage would be similar in principle to that shown in equation (49), but would involve the cyclic translocation of phosphate according to the following type of scheme

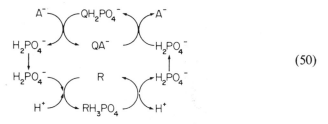

$$(50)$$

where Q stands for the translocater centre of the $H_2PO_4^-/A^-$ antiporter.

Chappell and Haarhoff (1967) have identified several proton-linked (or hydroxyl-linked) anion porter systems specific for dicarboxylic and tricarboxylic acids in mitochondria (see also, Azzi and Azzone, 1967; Haslam and Griffiths, 1968; Azzone and coworkers, 1969). The properties of these systems broadly resemble those of the phosphoric acid porter (Chappell and Robinson, 1968; Chappell, 1968), but the reactions catalysed by the individual porters have not yet been worked out satisfactorily. The fact that net dicarboxylate (e.g. succinate) translocation requires 'activation' by phosphate, and that tricarboxylate (e.g. citrate) and oxoglutarate translocation requires 'activation' by phosphate + L-malate (Chappell and Haarhoff, 1967; Chappell and Robinson, 1968; Chappell and coworkers, 1968; De Haan and Tager, 1968; Harris, 1968; Meijer and Tager, 1969) suggests that the substrate-specific porters may not be carboxylic acid uniporters. The dicarboxylate translocations requiring phosphate may be attributable to electrically neutral phosphate/dicarboxylate antiport. The tricarboxylate and oxoglutarate translocations requiring L-malate may be attributable to electrically neutral L-malate/tricarboxylate antiport (Chappell and Robinson, 1968). There is also some evidence for a glutamate/aspartate antiporter (Chappell and coworkers, 1968). The phosphate-linked and L-malate-linked anion antiporters discussed by Chappell and his collaborators depend on the same principle as the phosphate-linked system (Mitchell, 1959a) mentioned above and illustrated by equation (50). It is important to bear in mind, however, that while there is some evidence for electrical neutrality of the proposed antiport reactions it is not yet known in practice what phosphate/ anion stoichiometries such systems may have, or in what state of ionization the phosphate and anions might be translocated.

The translocation of sulphate across the cristae membrane of rat liver mitochondria, like that of phosphate, is catalysed by a system that appears to be specific for net sulphuric acid translocation (Mitchell and Moyle, 1969b). It is not known whether this system is phosphate 'activated'.

A strictly coupled mitochondrial ATP/ADP antiporter of high specificity has been extensively studied by several research groups (see Dueé and Vignais, 1968; Klingenberg and Pfaff, 1968; Pfaff and Klingenberg, 1968; Winkler and Lehninger, 1968; Winkler and coworkers, 1968), but the question of the net stoichiometry of the translocation reaction—in particular, whether ATP/ADP antiport is electrically neutral—has not yet been satisfactorily resolved. Also, it is not clear to what extent the system responsible for the rapid, strictly reciprocal, ATP/ADP exchange across the membrane may be involved in the slower net translocation of nucleotide into mitochondria (Pfaff and Schwalbach, 1967; Meisner and Klingenberg, 1968; Vignais and coworkers, 1968).

Further circumstantial evidence has been produced in favour of the suggestion (Mitchell, 1963a) that β-galactoside translocation in *Escherichia coli* may be catalysed by a proton–β–galactoside symporter (Winkler and Wilson, 1966; Pavlasova and Harold, 1969; Robbie and Wilson, 1969).

In this and in the previous section, our main object has been to define certain specific types of translocation reaction and to show how the translocation of one type of chemical particle may be coupled to that of another through the physical and chemical specificities of the catalytic porters in the membrane. These observations and other work beyond the scope of this review, add substantial support to the general thesis that the physiological organization and control of ion and substrate translocation depends, in part, on the coupling between pairs or groups of specific porter-catalysed translocation reactions involving common ions or substrates. The mechanism is strictly analogous to that of the coupling between pairs of enzyme-catalysed chemical reactions via common reactants. It must be carefully borne in mind, however, that the porters are distinct from enzymes in that the net reaction catalysed by a porter involves a rearrangement of spatial and secondary and ionic bonding relationships whereas the net reaction catalysed by an enzyme involves a rearrangement of covalent bonding, In conformity with the recommendations of the Commission on Enzyme Nomenclature (1965) we have avoided the use of the termination 'ase' in describing the catalysts of the exchange of secondary and ionic bonds.

## F. Chemiosmotic Coupling

When a chemical reaction and a translocation reaction are mutually dependent, the integral process has been described as a chemiosmotic process (Mitchell, 1959a), and the coupling relationship has been called chemiosmotic coupling (Mitchell, 1961b,c). This nomenclature was introduced to help to overcome the difficulties inherent in the concept of active transport (Ussing, 1949; Rosenberg, 1954; Christensen, 1962; Mitchell, 1967b) and to develop a self-consistent rationale for describing the integral physiological process of vectorial metabolism (Mitchell, 1962a, 1963a).

*Primary chemiosmotic coupling by group translocation.* When—as originally suggested by Lundegardh (1945) for electron transfer (see also Davies and Ogston, 1950; Conway, 1955; Robertson, 1960)—the group donor D and acceptor A participating in a chemical reaction are situated on opposite sides of a membrane, the transfer of the group G involves its translocation across the membrane (Mitchell, 1957, 1959a; Mitchell and Moyle, 1958a), or

$$
\begin{array}{c}
\text{DG} \\
\text{D}
\end{array}
\Big) \longrightarrow \text{G} \longrightarrow \Big(
\begin{array}{c}
\text{AG} \\
\text{A}
\end{array}
\qquad (51)
$$

Thus, the force on the chemical group undergoing translocation is directly attributable to the primary electrochemical field across the membrane, and the chemical and osmotic driving forces can come into equilibrium, as in a fuel cell (Mitchell, 1967c; Liebhafsky and Cairns, 1968).

Group translocation reactions involving the elements of water are especially interesting in this context because they can be arranged in pairs or loops (Mitchell, 1963a, 1966, 1968) so that the net result is the translocation of protons. Oxidoreduction reactions and hydrodehydration reactions can be arranged in this way, as shown by the following equations,

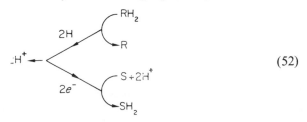

$$(52)$$

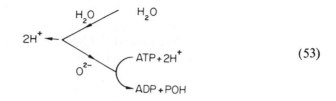

$$(53)$$

I have shown in detail elsewhere (Mitchell, 1968) how the electrochemical field across the membrane system would come directly into equilibrium with the particles undergoing translocation. It has been suggested that the respiratory chain systems, located in the membranes of mitochondria and bacteria, and the photo-oxidoreduction chain systems of the membranes of chloroplasts and chromatophores are made up of several proton-translocating oxidoreduction loops, like that of equation (52), arranged in series (Mitchell, 1961c, 1962b, 1963a,b, 1966). Likewise it has been suggested that the reversible ATPases, located in the same membranes, correspond to the proton-translocating hydrodehydration loop of equation (53). It is beyond the scope of the present article to give a detailed appraisal of the experimental evidence for and against these suggestions, and the reader is referred to reviews by Mitchell (1966, 1967c, 1968, 1969a,b), Slater (1967), Jagendorf and Uribe (1967), Robertson (1968), Witt and coworkers, (1968) and Greville (1969). It is, however, appropriate to mention that some recent observations on the cytochrome oxidase region of the respiratory chain of rat liver mitochondria (Mitchell and Moyle, 1970; Hinkle and Mitchell, 1970) leave little doubt that the arrangement in the cristae membrane of intact mitochondria can be represented as follows

$$
\begin{array}{c}
2H \\
2H^+ \longleftarrow {\Large<} \quad \frac{1}{2}O_2 + 2H^+ \\
2e^- \\
\qquad H_2O
\end{array}
\qquad (54)
$$

where the electrons are carried inwards across the membrane through the lower arm of the loop via cytochromes $c_1$, $c$, $a$ and $a_3$, the reduction of oxygen absorbs $H^+$ ions at the inner side of the membrane, and the oxidation of a hydrogenated carrier (probably coenzyme $QH_2$) by cytochrome $c_1$ or $c$ liberates $H^+$ ions at the outer side of the membrane (and see Mitchell, 1963b). Thus, respiration and proton translocation occur by one and the same process in the oxygen-terminal part of the respiratory chain system.

The establishment of this fact eliminates the possibility that respiration-driven proton translocation generally occurs by means of a proton pump actuated by a chemical intermediate of the $X \sim I$ type (Slater, 1967). The oxidoreduction loop mechanism therefore appears to afford the only chemically defined explanation of the observed proton translocation stoichiometries (Mitchell and Moyle, 1967b,c; Izawa and Hind, 1967; Dilley and Shavit, 1968; Schwartz, 1968; Witt and coworkers, 1968) that is at present available.

Little is known of the details of the reversible reaction (Reid and coworkers, 1966) catalysed by the proton-translocating ATPase system represented by equation (53), but the proton-translocation stoichiometry (Mitchell and Moyle, 1968; Schwartz, 1968), the resolution of the system into two major components and the susceptibility to inhibitors and other characteristics of these components (Racker, 1965, 1967; Selwyn, 1967; Bulos and Racker, 1968; Kaplan and Jagendorf, 1968; McCarty and Racker, 1968; MacLennan and Tzagoloff, 1968; Tzagoloff and coworkers, 1968) suggest that the ATPase may consist essentially of an $X—I$ hydrolase and an $X—I$ synthetase connected by an $X—I/(X^-+IO^-)$ antiport system (Mitchell, 1966) which may be represented as follows

$$2H^+ \quad \diagup\!\!\!\diagdown \; X^- + IO^- \; \diagup\!\!\!\diagdown \; ATP + 2H^+$$
$$H_2O \quad \diagdown\!\!\!\diagup \; X-I \; \diagdown\!\!\!\diagup \; ADP+POH \tag{55}$$

In this diagram the $X—I$ synthetase reaction is shown on the right and the $X—I$ hydrolase on the left. The right side would be the inside of the inner membrane of intact mitochondria or bacteria, or the outside of the grana membrane of chloroplasts. There do not appear to have been any direct attempts to test this suggested mechanism but it is compatible with the experimental facts as far as they are known at present.

*Secondary chemiosmotic coupling.* The intensive study of the $Na^+/K^+$ antiporter-ATPase of plasma membranes (Skou, 1965, 1969; Glynn, 1968; Post, 1968) has emphasized the fact that the translocation of passengers that do not directly participate in the covalent bond exchanges of a chemical reaction can be coupled reversibly (Garrahan and Glynn, 1967) to the process of group transfer. This type of coupling is more complicated than primary chemiosmotic coupling because it requires a system of secondary and electrovalent bonding and packing relationships that are so articulated as to translocate the passenger species in a particular direction relative to the flow of the chemical reactants that participate in the primary bond exchanges (Mitchell, 1963a, 1967a). Although there is evidence that a phosphorylated intermediate is involved in the $Na^+/K^+$ antiporter–ATPase reaction (Bader and coworkers, 1968; Hegvary and Post, 1969) and that $Na^+$ ions promote the synthesis of the intermediate, while $K^+$ ions promote its hydrolysis, the details of the mechanisms by which the translocations of $Na^+$ and $K^+$ are linked to the overall ATPase reaction are far from clear. As I have pointed out previously (Mitchell, 1963a), during the chemical reaction that is secondarily coupled to a translocation, the movements of different parts of the catalytic system (including the reactants) will be expected to have different directions in space, depending upon the way the components are packed together as they pass through the transition states. Consequently the direction of translocation of an ion or molecule involved (secondarily) in the transitional processes can as readily be at right angles to the direction of movement of a given group undergoing primary bond exchange as parallel to it. This circumstance adds to the difficulty of relating the translocation of $Na^+$ and $K^+$ to particular chemical events in the case of the $Na^+/K^+$ antiporter-ATPase.

Following the scheme originally suggested by Shaw (quoted by Glynn, 1957), the chemical transformations involved in the ATPase reaction—including the transformation of a component from a $K^+$-carrying form to a $Na^+$-carrying form and back again—were assumed to be separate in space and time from the physical translocation of the $Na^+$ and $K^+$ ions across the membrane. I have advocated the alternative view that the $Na^+$ and $K^+$ ions may be involved in the transitional intermediates of the ATPase reaction and that the translocation of $Na^+$ and $K^+$ may therefore occur in concert with the processes of group transfer catalysed by the ATPase system (Mitchell, 1961e, 1963e). Experimental observations in favour of this view were described by Baker and Connelly (1966) and by Garrahan and Glynn (1967), and the alternative interpretations have been discussed further by Mitchell (1967a), Stone (1968) and Glynn (1968). The following alternative basic reaction mechanisms appear to be in accord with the experimental facts as far as they are known at present (Fahn and coworkers, 1968; Glynn, 1968; Lindenmayer and coworkers, 1968; Priestland and Whittam, 1968; Hegvary and Post, 1969; Yoshida and coworkers, 1969), where XH stands

$$3Na^+ + XP \quad \rightleftarrows \quad \begin{array}{c} XH + ATP + 3Na^+ \\ ADP \end{array}$$

$$2K^+ + XP \quad \rightleftarrows \quad \begin{array}{c} H_2O \\ XH + POH + 2K^+ \end{array}$$

(56a)

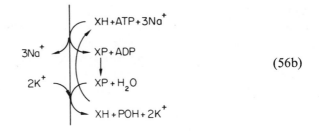

$$\text{(56b)}$$

for a group that can be phosphorylated and P stands for phosphoryl. In each equation the upper reaction represents the $Na^+$-activated XH phosphokinase activity and the lower reaction represents the $K^+$-activated XP phosphatase activity of the ATPase system. The scheme of equation (56a) shows the translocation of phosphoryl across the membrane outwards in the phosphokinase reaction and inwards in the phosphatase reaction. This scheme was suggested previously (Mitchell, 1967a) and corresponds to the basis of that elaborated by Stone (1968). It has the distinction (shared by the scheme of equation (56b)) of accounting for the observed ATP-requiring $Na^+$ exchange across the membrane in absence of external $K^+$ and at low internal $Na^+$ concentration, when the XP/XH couple would be expected to be fairly centrally poised in equilibrium with the $Na^+$-translocating phosphokinase reaction [see Mitchell (1967a); Stone (1968)]. The scheme of equation (56b) illustrates that there need not be any net macroscopic vector component of phosphoryl group transfer *across the membrane*; and it emphasizes at the same time that the translocater centre (of the phosphokinase) involved in $Na^+$ translocation may be distinct from the translocater centre (of the phosphatase) involved in $K^+$ translocation in the sense that the one translocater centre does not have to be the chemical or physical derivative of the other.

At the level of dimensions of the individual enzyme complexes in the scheme of equation (56b), the electrochemical field responsible for phosphoryl transfer (and translocation) must be orientated relative to the direction of $Na^+$ and $K^+$ translocation—presumably by virtue of appropriate articulations between the parts of each enzyme complex—because a chemical reaction that is scalar (i.e. not spatially orientated with respect to some frame of reference) cannot, according to the Curie symmetry principle (see Mitchell, 1967b), drive a spatially orientated translocation. Thus, in each enzyme complex there must be a polarity of phosphoryl translocation and this may be orientated tangential to the membrane (but may be orientated at random in the plane of the membrane) while the polarity of $Na^+$ or $K^+$ translocation is normal to the membrane. Thus, it is possible that the alkali metal ion translocation may not be adequately explained without an understanding of the complex conformational changes undergone by the XH phosphokinase and XP phosphatase complexes during the catalysis of phosphoryl group transfer (see for example, Matsui and Schwartz, 1968).

*Intrinsic anisotropy of vectorial enzyme systems catalysing chemiosmotic reactions.* In the case of chemiosmotic coupling by group translocation, the groups translocated across the membrane during the chemical reaction move in the electrochemical field between the group donors and group acceptors which are on opposite sides of the M phase. As I have pointed out previously (Mitchell, 1961a), the anisotropy of the reaction may be due either to the anisotropic distribution of the donor and acceptor species across a catalytically isotropic membrane, or may be due to the anisotropy of the group transfer enzymes in the membrane with respect to reaction with the specific donor and acceptor species. There appears to be little doubt that the proton-translocating respiratory chain, photo-oxidoreduction and ATPase are macroscopically anisotropic enzyme systems because they exhibit an intrinsic sidedness with respect to the direction of proton translocation and with respect to the accessibility of the other reactants (Low and Vallin, 1963; Lee and Ernster, 1966; Mitchell, 1966, 1968, 1969b; Malviya and coworkers, 1968; Scholes and coworkers, 1969). For example, in intact rat liver mitochondria (Mitchell and Moyle, 1967c) and in certain bacteria (Mitchell, 1962b, 1963b), the reduction of ferricyanade on the outside of the membrane is coupled to the oxidation of succinate or NADH on the inside of the membrane and to outward proton translocation.

In the case of secondary chemiosmotic coupling, however, as we have indicated in the previous section, the chemical field associated with group transfer must be orientated relative to the direction of the secondarily coupled solute translocation in *each individual* catalytic complex. However, in an assembly of complexes in the membrane there need be no macroscopic orientation of the chemical field associated with group transfer—although there must, of course, be a macroscopic orientation of the secondary field associated with the solute translocation across the membrane. Thus, the catalytic complex required for secondary chemiosmotic coupling—or enzyme-linked solute translocation, as it was previously called (Mitchell, 1967a)—would be equivalent to an enzyme–porter complex and an assembly of such complexes in a membrane would have to be macroscopically anisotropic with regard to the polarity of the porter in the membrane, but not with regard to the *primary* chemical polarity of the enzyme.

The porters that catalyse the symport and antiport reactions discussed in Section IIE5 (p. 225) are not intrinsically anisotropic, but the osmotic force on either passenger species arises extrinsically as a result of the anisotropic distribution of the other passenger across the membrane. In the case of the enzyme–porter complex mentioned above, part of the enzyme, associated

J

with the porter, would act as the second passenger and would be the cause of the polarity of the porter in the membrane.

## III.  PHYSIOLOGICAL FUNCTION OF MULTIENZYME-MULTIPORTER SYSTEMS

The rationale of the classification of the catalysts of chemical, osmotic and chemiosmotic reactions in the earlier sections of this article was based on the premise that coupling phenomena in chemical and osmotic reactions depend on the catalysis of the equilibration of chemical groups and solutes between donor and acceptor groups and between donor and acceptor phases by appropriate assemblies of enzymes and porters in osmotic systems. In spite of the comparatively rudimentary state of knowledge of porters and vectorial enzyme systems compared with that of many metabolic enzymes and enzyme systems, our object in this section is to attempt to obtain a meaningful, if rather speculative, synthesis of some of the facts; and to attempt to attribute some physiological significance to multienzyme-multiporter systems. The general treatment of the problem follows from earlier discussions of vectorial metabolism (see Mitchell, 1957, 1959a,b, 1963a, 1967a,b; Siekevitz, 1959; Robertson, 1960, 1968), the essential point being that a prerequisite of the coupling of a chemical reaction (covalent bond exchange) with a transloca-tion reaction is that the electrochemical field associated with the chemical reaction be directed in space relative to the direction of translocation. In order to permit the author to dispense with repetitious protestations of doubt, the reader is asked to recognize the fragmentary nature of much of the evidence that forms the basis of the argument that follows in this section, and to treat the conclusions more as a working hypothesis than as an established theory.

### A.  Power Transmission by Means of a Proton Current

The proton-translocating respiratory chain system of mitochondria and certain microorganisms, and the proton-translocating photo-oxidoreduction chain system of certain photosynthetic microorganisms catalyse an out-wardly directed translocation of protons across the coupling membrane (the cristae membrane or plasma membrane) during respiration or illumination (Mitchell, 1962b, 1963b; Mitchell and Moyle, 1965, 1967b; Rossi and Azzone, 1965; Carafoli and coworkers, 1967; Cummins and coworkers, 1969; Edwards and Bovell, 1969; Scholes and coworkers, 1969; and see Greville, 1969). Likewise the proton-translocating ATPase systems of mitochondria and microorganisms catalyse an outwardly directed trans-location of protons during ATP hydrolysis (Mitchell and Moyle, 1965, 1969a; Rossi and coworkers, 1967; Scholes and coworkers, 1969). In this

Proton-translocating respiratory chain system

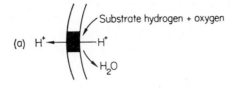

(a)

Proton-translocating ATPase system

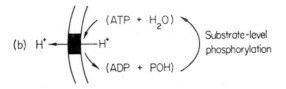

(b)

Proton-translocating anaerobic o/r system

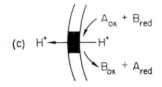

(c)

**Figure 2.** Proton translocation coupled to oxidoreduction and hydrodehydration reactions: (a) in respiratory chain system; (b) in ATPase system, which may be supplied with ATP by substrate-level phosphorylation; (c) in anaerobic oxidoreduction loops system. The stoichiometry of proton translocation is not indicated.

case, oxidoreduction coupled through substrate-level phosphorylation could result in outward proton translocation through the mediation of ATP. Thus, we deduce that under normal physiological conditions, oxidoreductive metabolism—either directly by means of proton-translocating oxidoreduction loops, or indirectly through substrate-level phosphorylation and the proton-translocating ATPase—maintains an electrochemical potential difference of protons (protonmotive force) across the coupling membrane, as illustrated in figures 2a and b. Fermentative metabolism (Stadtman, 1966) and the corresponding anaerobic inorganic oxidoreductive reactions of chemolithotrophic metabolism (Peck, 1968) may, I suggest, also be coupled through proton-translocating oxidoreduction loops in certain cases, and may thus directly maintain the protonmotive force across the coupling membrane (figure 2c).

The relatively low ion permeability of the coupling membrane enables the proton-translocating oxidoreduction and ATPase reactions to approach equilibrium with the protonmotive force across the membrane. When proton-translocating oxidoreduction and ATPase reactions are catalysed in the same coupling membrane they approach equilibrium with each other and oxidoreductive metabolism may thus poise the $[ATP]/[ADP] \times [POH]$ level via the poise of the couples feeding across the oxidoreduction loops, so that the utilization of ATP results in its resynthesis by the circulation of a proton current between the oxidoreduction loops and the ATPase system (Mitchell, 1966, 1968).

In respiring rat liver mitochondria, the protonmotive force is estimated to be about 230 mV when ATP is not rapidly utilized and when the slowing of respiration (respiratory control) indicates that the oxidoreduction reactions of the respiratory chain approach equilibrium with the protonmotive force across the membrane (Mitchell and Moyle, 1969a; and see Rossi and Azzone, 1969). The protonmotive force $\Delta p$ (conveniently given in mV) is made up of the membrane potential $\Delta \psi$ (inner phase negative) and the chemical activity difference $-Z\Delta pH$ ($\Delta pH$ denoting outer pH — inner pH) as described by equation (23). In mitochondria and certain microorganisms $\Delta \psi$ is thought to be the major component of $\Delta p$ under normal physiological conditions (Mitchell, 1961c, 1966, 1968; Jackson and coworkers, 1968), but in micro-organisms that grow in very acid environments (e.g. at pH3 or 5), $-Z\Delta pH$ may be the main component of $\Delta p$. For the oxidoreduction loop and ATPase systems described by equations (52) and (53) the value of $\Delta p$ in equilibrium with oxidoreduction is equal to the oxidoreduction potential span across the loop; and, taking the $\Delta G'$ of hydrolysis of ATP to be 9,400 cal/mole, when [POH] was 10 mM, the value of $[ATP]/[ADP]$ would be about unity when $\Delta p$ was 210 mV, and would increase by a factor of 10 for each 30 mV increase of $\Delta p$ (Mitchell, 1966, 1968).

Since the environment generally contains facilities for maintaining oxido-reductive metabolism, either in the form of photons or in the form of a supply of oxidants and reductants, chemiosmotic coupling between proton-translocating oxidoreduction loops and the reversible proton-translocating ATPase system provides a remarkably versatile method for coupling oxido-reductive metabolism to ATP synthesis, and may be of almost universal occurrence amongst animals, plants and microorganisms. As I have pointed out previously (Mitchell, 1968), it is conceivable that the proton-translocating oxidoreduction loop system and the reversible proton-translocating ATPase evolved separately as alternatives for generating the protonmotive force required for nutrient uptake and ionic regulation in primitive cells (figure 2a and b) and the occurrence of both systems in the same cell may have accidentally provided the means of storing the free energy of oxidoreduction in ATP synthesized by the reversal of the ATPase or possibly in some other anhydride produced by a similar mechanism.

It should be remarked that the polarity of the coupling membrane of the grana of the chloroplasts of higher plants is the opposite of that of the plasma membrane of certain microorganisms or of the cristae membrane of mitochondria (Jagendorf and Uribe, 1967). The chromatophores of certain photosynthetic bacteria resemble chloroplast grana with respect to the polarity of proton translocation (Jackson and coworkers, 1968; Scholes and coworkers, 1969); and this may be explained by the mechanism of formation of the chromatophores which appear to be pinched off from invaginations of the plasma membrane (Hickman and Frenkel, 1965). The polarity of the grana of higher plants may have originated by a similar mechanism.

The electrogenic outward translocation of protons by the vectorial metabolic systems of the coupling membrane of mitochondria and microorganisms in a salt medium must result in the leakage of cations inwards through the membrane and must cause a corresponding rise in the inner pH unless these cations are continuously translocated out or are metabolized (Mitchell, 1961c, 1968). In practice alkali metal ions appear to be translocated out of mitochondria, and may likewise be translocated out of certain microorganisms (Rothstein, 1968), by $H^+$/cation antiporters such as the $H^+/Na^+$ antiporter and $H^+/K^+$ antiporter of rat liver mitochondria (Mitchell and Moyle, 1967b, 1969b). As shown by equation (39) and discussed in detail elsewhere (Mitchell, 1968), this type of electrically neutral system has the effect of collapsing the $\Delta$pH component of the protonmotive force and enabling the $\Delta\psi$ component to increase correspondingly in the partial equilibrium state. The level of concentration of a given cation under partial equilibrium (or steady state) conditions is, in practice, regulated by the balance between the rate of entry (in the cationic form) down the electric gradient and the rate of exit by proton-linked antiport. As illustrated by the effects of valinomycin, entry of cations down the electric gradient may be specifically catalysed by appropriate uniporters.

Anions will tend to leak out of the inner phase of mitochondria and microorganisms when a large membrane potential is maintained. In the case of a membrane potential of 180 mV, the concentration of a divalent anion at equilibrium in the inner phase would be about $10^{-6}$ times the concentration in the outer medium. The uptake of the required inorganic and organic anions appears to be catalysed by proton–anion symport (or acid uniport systems which rapidly equilibrate the effectively protonated form of the anions across the coupling membrane by electrically neutral reactions (see Chappell, 1968; Chappell and Robinson, 1968). Thus, as discussed in Section IIE6 (p. 227) phosphate uptake in rat liver mitochondria and probably in certain bacteria, is catalysed by a phosphoric acid uniporter. Supposing that about 60 mV of $\Delta p$ were in the form of $-Z\Delta pH$ in a medium near pH7, so that the pH of the inner phase would be about 8, the total inner phosphate concentration would be some 50 times the outer concentration at equilibrium (see equation 45), provided that the permeability of the membrane to phosphate ions was relatively low. Sulphate uptake in rat liver mitochondria resembles phosphate uptake and may be attributable to a sulphuric acid uniporter (Mitchell and Moyle, 1969b). Exit of metabolically-produced $CO_2$ through the membrane, facilitated by the presence of carbonic anhydrase (see p. 228), is equivalent to proton–bicarbonate symport. In mitochondria from mammalian tissues specific dicarboxylate and tricarboxylate porter systems catalyse the uptake of Krebs cycle anions and there is good evidence (Palmieri and Quagliariello, 1969) that the degree of accumulation of the anions in the mitochondria depends on the pH difference across the membrane in qualitative agreement with equation (40) or (59). The 'activation' characteristics of these systems suggest that the dicarboxylate anions may be translocated via a phosphate/dicarboxylate antiporter and the tricarboxylate anions via a malate/tricarboxylate antiporter, the phosphate and malate being utilized as circulating intermediates (Mitchell, 1959a; Chappell and Haarhoff, 1967); but at all events the translocation of the anions is coupled to proton translocation. The Krebs cycle acid porters do not appear to be present in insect mitochondria (Chappell and Haarhoff, 1967; Chappell and Robinson, 1968), but there is an $\alpha$-glycerophosphate porter that is 'activated' by $Ca^{2+}$ (Hansford and Chappell, 1967; and see Mitchell, 1968). In certain microorganisms, translocation systems for the uptake of Krebs cycle acids (Kogut and Podoski, 1953; Barrett and Kallio, 1953; Mitchell and Moyle, 1958b; Clarke and Meadow, 1959), for pyruvate (Kornberg and Smith, 1967) and for a $\alpha$-glycerophosphate (Hayashi, and coworkers 1964; Koch and coworkers, 1964) exhibit similar properties to the corresponding mitochondrial systems. It appears that the proton-coupled porter type of mechanism may be of widespread occurrence.

The basic thesis is that the protonmotive force created across the membrane by the vectorial oxidoreduction or ATPase reactions and by the exit of $CO_2$ may be utilized to cause an electrochemical potential difference of a variety of passengers across the membrane by an appropriate proton-coupled porter (or porter system) in the membrane. This duplex type of system has obvious facilities for adaptability because, although the vectorial enzyme system required to generate the protonmotive force between the phases must have rather complex properties, including intrinsic structural and functional anisotropy, the porters need not be intrinsically anisotropic and are simply required to couple the flow of $H^+$ ions to the flows of specific passengers by secondary mechanisms such as those described in Section IIE6.

As suggested earlier (Mitchell, 1963a), the uptake of lactose and other β-galactosides by constitutive or inducible systems in *Escherichia coli* may be catalysed by a proton-β-galactoside symporter. Kennedy and his collaborators (Jones and Kennedy, 1968; Scarborough and coworkers, 1968) isolated and purified a protein (the M protein) which appears to be implicated in β-galactoside translocation but which is devoid of chemical catalytic activity. Kennedy (1966) showed in a lucid discussion of the idea of permease (Cohen and Monod, 1957; Kepes and Cohen, 1962) in relation to experimental knowledge concerning β-galactoside translocation that the role of the M protein cannot readily be fitted into the kind of formulation favoured by the Parisian school (Kepes, 1964; Koch, 1964). The M protein may, however, be fitted into the duplex system (Mitchell, 1963a) discussed here; and the same conclusion was almost reached by Kennedy (1966). Having pointed out that the transport of oxygen by haemoglobin is influenced by pH because the reaction of haemoglobin with $H^+$ ions lowers its affinity for oxygen, Kennedy (1966) suggested that the M protein 'is roughly analogous to hemoglobin in its presumptive function, rather than to hexokinase', but he concluded with the reservation that the reaction responsible for coupling the translocation of the β-glactoside–M protein complex to metabolism

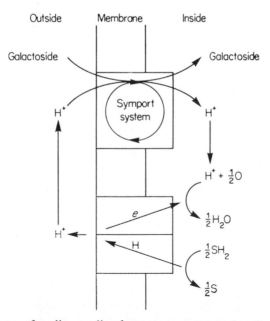

**Figure 3.** Diagram of cyclic coupling between proton translocation and proton-β-galactoside symport. (From Mitchell, 1963a).

'presumably . . . . is not simply a reaction with hydrogen ion'. Without this unexplained reservation, Kennedy (1966) would have reached the conclusion that the M protein corresponds to the proton–β–galactoside symporter (Mitchell, 1963a) or to an essential part of it. Other recent work (Winkler and Wilson, 1966, 1967; Pavlasova and Harold, 1969; Robbie and Wilson, 1969) is consistent with my formulation of β-galactoside translocation and accumulation in *E. coli*, illustrated in figure 3.

The sulphate translocation system of *Salmonella typhimurium* described by Dreyfuss (1964) has yielded a sulphate-binding protein (Pardee, 1966) which appears to be situated in the surface of the plasma membrane of the cells (Pardee and Wanatabe, 1968). Pardee (1966) observed that the uptake of sulphate by the sulphate-binding protein was not much affected by pH over the range 5·3 to 8·3 in his assay system. Nevertheless, I suggest that the sulphate-binding protein may effectively bind sulphuric acid *in situ* in the membrane and may correspond to the sulphuric acid porter of rat liver mitochondria (Mitchell and Moyle, 1969b).

The reader may find it somewhat surprising that the metabolically generated anisotropy across the membrane, described by the components $\Delta\psi$ and $-Z\Delta pH$ of the total protonmotive force $\Delta p$ may give rise to a force on a given passenger component, the magnitude and direction of which is determined by the porter. For $nH^+–I^{z+}$ symport and $nH^+/I^{z+}$ antiport, respectively (see equations (39) and (40); and Mitchell, 1968) the equilibrium distribution of the passenger I of valency $+z$ is given by

$$(z+n)\Delta\psi - Z(\Delta pI + n\Delta pH) = 0 \qquad (57)$$

$$(z-n)\Delta\psi - Z(\Delta pI - n\Delta pH) = 0 \qquad (58)$$

or when the reactions are electrically neutral

$$\Delta pI + n\Delta pH = 0 \qquad (59)$$

$$\Delta pI - n\Delta pH = 0 \qquad (60)$$

Thus, for example, an ionic component that leaks across the membrane in one direction under the influence of $\Delta\psi$ may nevertheless pass across the membrane spontaneously in the opposite direction in a porter-catalysed electrically neutral reaction under the influence of $\Delta pH$—illustrating the important fact that the $\Delta\psi$ and $-Z\Delta pH$ components of $\Delta p$ represent partially independent phase-specific potentials, and that changes in the relative contributions of $\Delta\psi$ and $-Z\Delta pH$ to $\Delta p$ may give rise to subtle changes in the distribution of passengers across the coupling membrane (Harris and Manger, 1968; Palmieri and Quagliariello, 1969). The part played by the potential difference and flow of $H^+$ ions between the two aqueous phases in the osmotically coupled porter systems described here is analogous to the part played by the potential difference and flow of phosphoryl groups between high potential phosphoryl acceptors such as ADP and $H_2O$ in the chemically coupled enzyme systems of the type discussed by Lipmann (1941, 1946). Our treatment of power transmission in the osmotically and chemiosmotically coupled systems, like the usual treatment of metabolic chemically coupled systems, has been based on idealized cases in which certain reactions are very fast compared with others, so that partial equilibria can be well defined, and in which the protonmotive force (like the phosphoryl group potential) is not dissipated by cyclic reactions. This type of treatment can be adapted to describe the actual partial equilibria in

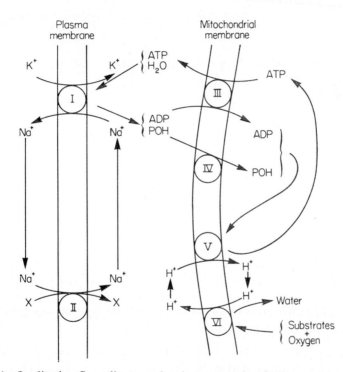

**Figure 4.** Qualitative flow diagram showing connection between proton-coupled mitochondrial translocation system and $Na^+/K^+$-coupled plasma-membrane system. The following porters or translocation systems are represented: I: the $Na^+/K^+$ antiporter ATPase; II: a $Na^+$-X symporter; III: the atractyloside-sensitive ATP/ADP antiporter; IV: the phosphoric acid uniporter; V: the oligomycin-sensitive $H^+$-translocating ATPase; VI: the $H^+$-translocating respiratory chain system. The entry of substrates (other than X) and exit of $CO_2$ are not shown and the stoichiometries of translocation are not represented in this simplified diagram.

the integral process of metabolism and transport by well-known kinetic techniques. A more radical non-equilibrium treatment of transport reactions is described in Chapter 5.

These examples of porters for the uptake of anionic, cationic and electrically neutral passengers must suffice to illustrate the versatility of the duplex proton-coupled physiological power transmission system.

It seems likely that in chemiosmotic and osmotic coupling, the proton is uniquely versatile as a means of power transmission. However, it is not the only component that can function in this way. Particularly in eukaryotic cells where the membrane system includes at least two osmotic barriers for solutes of small molecular weight (for example, the plasma membrane and the mitochondrial cristae membrane) conditions are suitable for the operation of several osmotic power transmission systems that may be mutually independent.

## B. Power Transmission by Means of Alkali Metal Ion Current

The $Na^+/K^+$ antiporter ATPase is present in the plasma membrane of a wide range of eukaryotic cells (see Glynn, 1968; Skou, 1969) and is responsible for maintaining electrochemical potential differences of $Na^+$ and $K^+$ between the outer medium and the cytoplasm. As is well known, the anisotropy between the outer and inner phases that is thus maintained by the vectorial ATPase reaction in the membrane is responsible for the 'excitability' and transmission of impulses in nerve, muscle and certain other cells (Cole, 1968). The anisotropic distribution of $Na^+$ and $K^+$ also appears to be responsible for the translocation of sugars and amino acids in certain eukaryotic cells as discussed in section IIE5 (p. 225). The general principles of coupling are the same as in the case of the proton-coupled reactions, and we shall not discuss further details here. Figure 4 illustrates the relationship between the $H^+$-coupled and $Na^+/K^+$ systems in some eukaryotic cells.

## C. Diversity of Chemiosmotic and Osmotic Coupling Systems and the Coupling of Metabolism to Transport

The uptake of a number of substrates may possibly be explained by a type of mechanism that is different from the duplex type on which we have chosen to focus attention here. For example, it has been proposed that certain sugars and sugar derivatives including methyl-α-glucoside are accumulated in *E. coli* and *Aerobacter aerogenes* by means of a phosphorylating enzyme system (Kundig and coworkers, 1966; Simoni and coworkers, 1967; Tanaka and Lin, 1967; Tanaka and coworkers, 1967; and see Fields and Luria, 1969a,b). The work of Laue and Macdonald (1968) on β-galactoside uptake in *Staphylococcus aureus* led them to conclude that 'the suggestion of Kennedy and Scarborough (1967) that the phosphorylated sugar is the substrate of the staphylococcal 'β-galactosidase', taken together with the experiments presented here, lead to the hypothesis that in *Staphylococcus*, galactoside permeation may depend upon the first reaction of galactose metabolism'. A similar conclusion has been reached in the case of deoxyglucose uptake by yeast cells (von Steveninck, 1968).

In this article the relationship between metabolism and transport has been described by treating the pathways of group transfer as extensions of the routes of escape of the chemical particles that are common to metabolism and transport, and we have attributed a spatially orientated (vectorial) group-translocation property to certain metabolic group-transfer enzymes or enzyme systems which may thereby act as prime movers of transport (Mitchell, 1963a). The coupling of transport to metabolism depends on the anisotropy of the chemical reaction system as a whole, and does not necessarily require group translocation at the level of the enzyme or enzyme complex (see Mitchell, 1961a; Kedem, 1961). Nevertheless, three basic types of catalytic unit appear to be involved in the physiological systems that organize the integral process of metabolism and transport:

i) Effectively isotropic enzymes or enzyme complexes catalysing chemical change at certain sites or in certain phases, providing a macroscopic source or sink for specific chemical groups;

ii) Enzymes or enzyme complexes catalysing the translocation equilibria of specific chemical groups across, or orientated relative to, an osmotic barrier;

iii) Porters or porter systems catalysing secondary and ionic bond exchanges required for uniport, symport and antiport reactions across an osmotic barrier. These basic catalytic units are the essential building modules which, when assembled in different permutations and combinations in physiological osmotic systems can display a variety of physiological characteristics such as 'ionic regulation', 'active transport' and 'energy transduction'.

## REFERENCES

Adam, N. K. (1941) *The Physics and Chemistry of Surfaces*, 3rd ed. University Press, Oxford
Ahmed, J. and I. Morris (1968) *Biochim. Biophys. Acta.*, **162**, 32
Albers, R. W. (1967) *Ann. Rev. Biochem.*, **36**, 727
Andreoli, T. E., M. Tieffenberg, and D. C. Tosteson (1967) *J. Gen. Physiol.*, **50**, 2527
Azzi, A. and G. F. Azzone (1967) *Biochim. Biophys. Acta.*, **131**, 468
Azzone, G. F., E. Rossi and A. Scarpa (1969) In J. Järnefelt (Ed.), *Regulatory Functions of Biological Membranes*, Elsevier, Amsterdam. p. 236
Bader, H., R. L. Post and G. H. Bond (1968) *Biochim. Biophys. Acta.*, **150**, 41
Baker, P. F. and C. M. Connelly (1966) *J. Physiol. (London)*, **185**, 270
Barrett, J. T. and R. E. Kallio (1963) *J. Bacteriol.*, **66**, 517
Baum, H. (1967) *Nature*, **214**, 1326

Bell, R. P. (1959) *The Proton in Chemistry*, Methuen, London

Boyer, P. D. (1965) In T. E. King and H. S. Mason (Eds.), *Oxidases and Related Redox Systems*, Vol. 2, Wiley, New York. p. 994

Brierley, G. T., C. T. Settlemire and V. A. Knight (1968) *Arch. Biochem. Biophys.*, **126**, 276

Bulos, B. and E. Racker (1968) *J. Biol. Chem.*, **243**, 3891

Carafoli, E. and C. S. Rossi (1967) *Biochem. Biophys. Res. Commun.*, **29**, 153

Carafoli, E., R. L. Gamble, C. S. Rossi and A. L. Lehninger (1967) *J. Biol. Chem.*, **242**, 1199

Chance, B., W. Holmes, J. Higgins and C. M. Connelly (1958) *Nature*, **182**, 1190

Chappell, J. B. (1968) *Brit. Med. Bull.*, **24**, 150

Chappell, J. B. and A. R. Crofts (1965) *Biochem. J.*, **95**, 393

Chappell, J. B. and A. R. Crofts (1966) In J. M. Tager, S. Papa, E. Quagliariello and E. C. Slater (Eds.), *Regulation of Metabolic Processes in Mitochondria*, Elsevier, Amsterdam. p. 293

Chappell, J. B. and K. N. Haarhoff (1967) In E. C. Slater, Z. Kaniuga and L. Wojtczak (Eds.), *Biochemistry of Mitochondria*, Academic Press, London. p. 75

Chappell, J. B., P. J. F. Henderson, J. D. McGivan and B. H. Robinson (1968) In P. N. Campbell (Ed.), *The Interaction of Drugs and Subcellular Components in Animal Cells*, Churchill, London, p. 71

Chappell, J. B. and B. H. Robinson (1968) *Biochem. Soc. Symp.*, **27**, 123

Christensen, H. N. (1962) *Biological Transport*, Benjamin, New York

Clarke, P. H. and P. M. Meadow (1959) *J. Gen. Microbiol.*, **20**, 144

Cohen, G. N. and J. Monod (1957) *Bacteriol Rev.*, **21**, 169

Cole, K. S. (1968) *Membranes, Ions and Impulses*, University of California Press, Berkeley

Conway, E. J. (1955) *Intern. Rev. Cytol.*, **4**, 377

Crane, R. K. (1965) *Federation Proc.*, **24**, 1000

Crane, R. K., D. Miller and I. Bihler (1961) In A. Kleinzeller and A. Kotyk (Eds.), *Membrane Transport and Metabolism*, Academic Press, New York, p. 439

Cummins, J. T., J. A. Strand and B. E. Vaughan (1969) *Biochim. Biophys. Acta.*, **173**, 198

Danielli, J. F. (1954) *Symp. Soc. Exptl. Biol.*, **8**, 502

Davies, R. E. and A. G. Ogston (1950) *Biochem. J.*, **46**, 324

Davies, J. T. and E. K. Rideal (1963) *Interfacial Phenomena*, Academic Press, New York

Davis, B. D. (1958) *Arch. Biochem. Biophys.*, **78**, 497

Davson, H. and J. F. Danielli (1943) *The Permeability of Natural Membranes*, University Press, Cambridge

Debye, P. and E. Hückel (1923) *Physik. Z.*, **24**, 185

De Haan, E. J. and J. M. Tager (1968) *Biochim. Biophys.*, *Acta.* **153**, 98

Dick, D. A. T. (1966) *Cell Water*, Butterworths, London

Dilley, R. A. and N. Shavit (1968) *Biochim. Biophys. Acta.*, **162**, 86

Dixon, M. (1951) *Multi-Enzyme Systems*, University Press, Cambridge, p. 10

Dreyfuss, J. (1964) *J. Biol. Chem.*, **239**, 2292

Duée, E. D. and P. V. Vignais (1968) *Biochem. Biophys. Res. Commun.*, **30**, 546

Eddy, A. A. (1968a) *Biochem. J.*, **108**, 195

Eddy, A. A. (1968b) *Biochem. J.*, **108**, 489

Eddy, A. A., C. Hogg and M. Reid (1969) *Biochem. J.*, **112**, 11P

Edwards, G. and C. R. Bovell (1969) *Biochim. Biophys. Acta.*, **172**, 126

Eisenman, G., S. M. Ciani and G. Szabo (1968) *Federation Proc.*, **27**, 1289

*Enzyme Nomenclature* (1965) Elsevier, Amsterdam

Evans, M. G. and M. Polanyi (1925) *Trans. Faraday Soc.*, **31**, 875

Eyring, H. (1935) *J. Chem. Phys.*, **3**, 107

Fahn, S., G. J. Koval and W. Albers (1968) *J. Biol. Chem.*, **243**, 1993

Fields, K. L. and S. E. Luria (1969a) *J. Bacteriol.*, **97**, 57

Fields, K. L. and S. E. Luria (1969b) *J. Bacteriol.*, **97**, 64

Fonyo, A. (1968) *Biochem. Biophys. Res. Commun.*, **32**, 624

Garrahan, P. J. and I. M. Glynn (1967) *J. Physiol.*, **192**, 237

Glasstone, S., K. J. Laidler and H. Eyring (1941) *The Theory of Rate Processes*, McGraw-Hill, New York

Glynn, I. M. (1957) *Progr. Biophys. Biophys. Chem.*, **8**, 241
Glynn, I. M. (1968) *Brit. Med. Bull.*, **24**, 165
Green, D. E. (1937) In J. Needham and D. E. Green (Eds.), *Perspectives in Biochemistry*, University Press, Cambridge, p. 175
Green, D. E. (1957) *Symp. Soc. Exptl. Biol.*, **10**, 30
Green, D. E., L. H. Stickland and H. L. Tarr (1934) *Biochem. J.*, **28**, 1812
Greville, G. D. (1969) In D. R. Sanadi (Ed.), *Current Topics in Bioenergetics*, Vol. 3, Academic Press, New York, p. 1
Guéron, M. and R. G. Schulman (1968) *Ann. Rev. Biochem.*, **37**, 571
Guggenheim, E. A. (1933) *Modern Thermodynamics by the Methods of Willard Gibbs*, Methuen, London
Guggenheim, E. A. (1949) *Thermodynamics: An Advanced Treatment for Chemists and Physicists*, North-Holland, Amsterdam
Gunsalus, I. C. (1954) In W. D. McElroy and B. Glass (Eds.), *The Mechanism of Enzyme Action*, Johns Hopkins, Baltimore, p. 545
Hammes, G. G. (1964) *Nature*, **204**, 342.
Hansford, R. G. and J. B. Chappell (1967) *Biochem. Biophys. Res. Commun.*, **27**, 686
Harold, F. M. and J. R. Baarda (1968) *J. Bacteriol.*, **96**, 2025
Harris, E. J. (1968) *Biochem. J.*, **109**, 247
Harris, E. J. and B. C. Pressman (1967) *Nature*, **216**, 918
Harris, E. J. and J. R. Manger (1968) *Biochem. J.*, **109**, 239
Haslam, J. M. and D. E. Griffiths (1968) *Biochem. J.*, **109**, 921
Hayashi, S., J. P. Koch and E. C. C. Lin (1964) *J. Biol. Chem.*, **239**, 3098
Hegvary, C. G. and R. L. Post (1969) In D. C. Tosteson (Ed.), *The Molecular Basis of Membrane Function*, Prentice-Hall, New Jersey, p. 519
Heinz, E. (1967) *Ann. Rev. Physiol.*, **29**, 21
Henderson, P. J. F., J. D. McGivan and J. B. Chappell (1969) *Biochem. J.*, **111**, 521
Heytler, P. G. and W. W. Prichard (1962) *Biochem. Biophys. Res. Commun.*, **7**, 272
Hickman, D. D. and A. W. Frenkel (1965) *J. Cell Biol.*, **25**, 279
Hinkle, P. and P. Mitchell (1970) *J. Bioenergetics*, **1**
Hirashima, M., T. Hoyakawa and M. Koike (1967) *J. Biol. Chem.*, **242**, 1356
Hodgkin, A. L. (1951) *Biol. Rev.*, **26**, 339
Hopfer, U., A. L. Lehninger and T. E. Thompson (1968) *Proc. Natl. Acad. Sci. U.S.*, **59**, 484
Hunter, F. R. (1967) *Biochim. Biophys. Acta.*, **135**, 787
Ilani, A. and D. Tzivoni (1968) *Biochim. Biophys. Acta.*, **163**, 429
Izawa, S. and G. Hind (1967) *Biochim. Biophys. Acta.*, **143**, 377
Jackson, J. B., A. R. Crofts and L. V. von Stedingk (1968) *Eur. J. Biochem.*, **6**, 41
Jacobs, M. H. (1922) *J. Gen. Physiol.*, **5**, 181
Jacobs, M. H. (1924) *Amer. J. Physiol.*, **68**, 134
Jacobs, M. H. (1927) *Harvey Lectures*, **22**, 146
Jacobs, M. H. (1940) *Cold Spring Harbor Symp. Quant. Biol.*, **8**, 30
Jacobs, M. H. and D. R. Stewart (1942) *J. Gen. Physiol.*, **25**, 539
Jagendorf, A. T. and J. Neumann (1965) *J. Biol. Chem.*, **240**, 3210
Jagendorf, A. T. and E. Uribe (1967) *Brookhaven Symp. Biol.*, **19**, 215
Jencks, W. P. (1966) In N. O. Kaplan and E. P. Kennedy (Eds.), *Current Aspects of Biochemical Energetics*, Academic Press, New York, p. 273
Jones, O. T. G. and W. A. Watson (1965) *Nature*, **208**, 1169
Jones, T. H. D. and E. P. Kennedy (1968) *Federation Proc.*, **27**, 644
Kaplan, J. H. and A. T. Jagendorf (1968) *J. Biol. Chem.*, **243**, 972
Karlish, S. J. D. and M. Avron (1968) *Fed. European Biochem. Soc. Letters*, **1**, 21
Kedem, O. (1961) In A. Kleinzeller and A. Kotyk (Eds.), *Membrane Transport and Metabolism*, Academic Press, New York, p. 87
Keilin, D. and T. Mann (1941) *Nature*, **148**, 493
Kennedy, E. P. (1966) In N. O. Kaplan and E. P. Kennedy (Eds.), *Current Aspects of Biochemical Energetics*, Academic Press, New York, p. 433
Kennedy, E. P. and G. A. Scarborough (1967) *Proc. Natl. Acad. Sci. U.S.*, **58**, 225

Kepes, A. (1964) In J. H. Hoffman (Ed.), *The Cellular Functions of Membrane Transport*, Prentice-Hall, New Jersey, p. 155
Kepes, A. and G. N. Cohen (1962) In I. C. Gunsalus and R. Stanier (Eds.), *The Bacteria*, Vol. 4, Academic Press, New York, p. 179
Klingenberg, M. and E. Pfaff (1968) *Biochem. Soc. Symp.*, **27**, 105
Koch, A. L. (1964) *Biochim. Biophys. Acta.*, **79**, 177
Koch, J. P., S. Hayashi and E. C. C. Lin (1964) *J. Biol. Chem.*, **239**, 3106
Kogut, M. and E. P. Podoski (1953) *Biochem. J.*, **55**, 800
Kohn, P. G., D. H. Smyth and E. M. Wright (1968) *J. Physiol. (London)*, **196**, 723
Koopman, W. and S. G. Schultz (1968) *Biochim. Biophys. Acta.*, **173**, 338
Kornberg, H. L. and J. Smith (1967) *Biochim. Biophys. Acta.*, **148**, 591
Koshland, D. E. (1960) *Advan. Enzymol.*, **22**, 45
Koshland, D. E. and K. E. Neet (1968) *Ann. Rev. Biochem.*, **37**, 359
Kuchler, R. J. (1967) *Biochim. Biophys. Acta.*, **136**, 473
Kundig, W., F. D. Kundig, B. Anderson and S. Roseman (1966) *J. Biol. Chem.*, **241**, 3243
Lardy, H. A. (1968) *Federation Proc.*, **27**, 1278
Lardy, H. A., S. N. Graven and S. Estrada-O (1967) *Federation Proc.*, **26**, 1355
Laue, P. and R. E. Macdonald (1968) *Biochim. Biophys. Acta.*, **165**, 410
Lee, C. P. and L. Ernster (1966) In J. M. Tager, S. Papa, E. Quagliariello and E. C. Slater (Eds.), *Regulation of Metabolic Processes in Mitochondria*, Elsevier, Amsterdam, p. 218
Lehninger, A. L. (1966) *Ann. N.Y. Acad. Sci.*, **137**, 700
Lehninger, A. L., E. Carafoli and C. S. Rossi (1967) *Advan. Enzymol.*, **29**, 259
Lev, A. A. and E. P. Buzhinsky (1967) *Cytology (U.S.S.R.)*, **9**, 102
Liberman, E. A., E. N. Mochova, V. P. Skulachev and V. P. Topaly (1968) *Biofizika (U.S.S.R.)*, **13**, 188
Liberman, E. A. and V. P. Topaly (1968) *Biochim. Biophys. Acta.*, **163**, 125
Liebhafsky, H. A. and E. J. Cairns (1968) *Fuel Cells and Fuel Batteries*, Wiley, New York
Lindenmayer, G. E., A. H. Laughter and A. Schwartz (1968) *Arch. Biochem. Biophys.*, **127**, 187
Lindsay, R. B. and H. Margenau (1936) *Foundations of Physics*, Wiley, New York, p. 178
Lipmann, F. (1941) *Advan. Enzymol.*, **1**, 99
Lipmann, F. (1946) In D. E. Green (Ed.), *Currents in Biochemical Research*, Interscience New York, p. 137
Low, H. and I. Vallin (1963) *Biochim. Biophys. Acta.*, **69**, 361
Lundegardh, H. (1945) *Ark. Bot.*, **32A**, **12**, 1
McCarty, R. E. and E. Racker (1968) *J. Biol. Chem.*, **243**, 129
MacInnes, D. A. (1939) *The Principles of Electrochemistry*, Reinhold, New York
MacLennan, D. H. and A. Tzagoloff (1968) *Biochemistry*, **7**, 1603
Malviya, A. N., B. Parsa, R. E. Yodaiken and W. B. Elliott (1968) *Biochim. Biophys. Acta.*, **162**, 195
Matsui, H. and A. Schwartz (1968) *Biochim. Biophys. Acta.*, **151**, 655
Meijer, A. J. and J. M. Tager (1969) *Biochim. Biophys. Acta*, **189**, 136
Meisner, H. and M. Klingenberg (1968) *J. Biol. Chem.*, **243**, 3631
Mitchell, P. (1953) *J. Gen. Microbiol.*, **9**, 237
Mitchell, P. (1954a) *Symp. Soc. Exptl. Biol.*, **8**, 254
Mitchell, P. (1954b) *J. Gen. Microbiol.*, **11**, 73
Mitchell, P. (1957) *Nature*, **180**, 134
Mitchell, P. (1959a) *Biochem. Soc. Symp.*, **16**, 73
Mitchell, P. (1959b) *Ann. Rev. Microbiol.*, **13**, 407
Mitchell, P. (1961a) In A. Kleinzeller and A. Kotyk (Eds.), *Membrane Transport and Metabolism*, Academic Press, New York, p. 22
Mitchell, P. (1961b) In T. W. Goodwin and O. Lindberg (Eds.), *Biological Structure and Function*, Vol. 2, Academic Press, London, p. 581
Mitchell, P. (1961c) *Nature*, **191**, 144
Mitchell, P. (1961d) *Biochem. J.*, **81**, 24P

Mitchell, P. (1961e) In A. Kleinzeller and A. Kotyk (Eds.), *Membrane Transport and Metabolism*, Academic Press, New York, p. 318

Mitchell, P. (1962a) *Biochem. J.*, **83**, 22P

Mitchell, P. (1962b) *J. Gen. Microbiol.*, **29**, 25

Mitchell, P. (1963a) *Biochem. Soc. Symp.*, **22**, 142

Mitchell, P. (1963b) In H. D. Brown (Ed), *Cell Interface Reactions*, Scholar's Library, New York, p. 33

Mitchell, P. (1966) *Chemiosmotic Coupling in Oxidative and Photosynthetic Phosphorylation*, Glynn Research, Bodmin

Mitchell, P. (1967a) *Advan. Enzymol.*, **29**, 33

Mitchell, P. (1967b) In M. Florkin and E. H. Stotz (Eds.), *Comprehensive Biochemistry*, Vol. 22, Elsevier, Amsterdam, p. 167

Mitchell, P. (1967c) *Federation Proc.*, **26**, 1370

Mitchell, P. (1967d) *Nature*, **214**, 1327

Mitchell, P. (1968) *Chemiosmotic Coupling and Energy Transduction*, Glynn Research, Bodmin

Mitchell, P. (1969a) *Fed. European Biochem. Soc. Symp.* **17**, 219

Mitchell, P. (1969b) In D. C. Tosteson (Ed.), *The Molecular Basis of Membrane Function*, Prentice-Hall, New Jersey, p. 483

Mitchell, P. and J. Moyle (1956a) *Symp. Soc. Gen. Microbiol.*, **6**, 150

Mitchell, P. and J. Moyle (1956b) *Discussions Faraday Soc.*, **21**, 258

Mitchell, P. and J. Moyle (1958a) *Nature*, **182**, 372

Mitchell, P. and J. Moyle (1958b) *Abstracts 7th Intern. Congr. Microbiol.*, Stockholm, 2g.

Mitchell, P. and J. Moyle (1965) *Nature*, **208**, 147

Mitchell, P. and J. Moyle (1967a) *Biochem. J.*, **104**, 588

Mitchell, P. and J. Moyle (1967b) *Biochem. J.*, **105**, 1147

Mitchell, P. and J. Moyle (1967c) In E. C. Slater, Z. Kaniuga and L. Wojtczak (Eds.), *Biochemistry of Mitochondria*, Academic Press, London, p. 53

Mitchell, P. and J. Moyle (1967d) *Nature*, **213**, 137

Mitchell, P. and J. Moyle (1968) *Eur. J. Biochem.*, **4**, 530

Mitchell, P. and J. Moyle (1969a) *Eur. J. Biochem.*, **7**, 471

Mitchell, P. and J. Moyle (1969b) *Eur. J. Biochem.*, **9**, 149

Mitchell, P. and J. Moyle (1970) *Round Table Discussion*, Bari, (In press)

Moelwyn-Hughes, E. A. (1947) *The Kinetics of Reactions in Solution*, 2nd ed. Clarendon Press, Oxford

Moore, C. and B. C. Pressman (1964) *Biochem. Biophys. Res. Commun.*, **15**, 562

Mueller, P and D. O. Rudin (1967) *Biochem. Biophys. Res. Commun.*, **26**, 398

Muirhead, H. and M. F. Perutz (1963) *Nature*, **199**, 633

Munck, B. G. (1968) *Biochim. Biophys. Acta.*, **156**, 192

Nernst, W. (1897) *Z. Elektrochem.*, **3**, 308

Ogston, A. G. (1955) *Discussions Faraday Soc.*, **20**, 161

Palmieri, F. and E. Quagliariello (1969) *Eur. J. Biochem* **8**, 473

Pardee, A. B. (1966) *J. Biol. Chem.*, **241**, 5886

Pardee, A. B. and K. Watanabe (1968) *J. Bacteriol.*, **96**, 1049

Park, J. H. (1966) In N. O. Kaplan and E. P Kennedy (Eds.), *Current Aspects of Biochemical Energetics*, Academic Press, New York. p. 299

Patlak, C. S. (1956) *Bull. Math. Biophys.*, **18**, 271

Pauling, L. (1950) *Ann. Rep. Smithsonian Institute*, 225

Pauling, L. (1956) In O. H. Gaebler (Ed.), *Enzymes: Units of Biological Structure and Function*, Academic Press, New York, p. 177

Pavlasova, E. and F. M. Harold (1969) *J. Bacteriol.*, **98**, 198

Peck, H. D. (1968) *Ann. Rev. Microbiol.*, **22**, 489

Pedersen, C. J. (1968) *Federation Proc.*, **27**, 1305

Pfaff, E. and K. Schwalbach (1967) In E. Quagliariello, S. Papa, E. C. Slater and J. M.Tager (Eds.), *Mitochondrial Structure and Compartmentation*, Adriatica Edrice, Bari, p. 346

Pfaff, E. and M. Klingenberg (1968) *Eur. J. Biochem.*, **6**, 66

Post, R. L. (1968) In J. Järnefelt (Ed.), *Regulatory Functions of Biological Membranes*, Elsevier, Amsterdam, p. 163
Potts, W. T. W. (1968) *Ann. Rev. Physiol.*, **30**, 73
Pressman, B. C. (1963) In B. Chance (Ed.), *Energy-Linked Functions of Mitochondria*, Academic Press, New York, p. 181
Pressman, B. C. (1968) *Federation Proc.*, **27**, 1283
Pressman, B. C., E. J. Harris, W. S. Jagger and J. H. Johnson (1967) *Proc. Natl. Acad. Sci. U.S.*, **58**, 1949
Priestland, R. N. and R. Whittam (1968) *Biochem. J.*, **109**, 369
Pullman, M. E. and G. Schatz (1967) *Ann. Rev. Biochem.*, **36**, 539
Racker, E. (1961) *Advan. Enzymol.*, **23**, 323
Racker, E. (1965) *Mechanisms in Bioenergetics*, Academic Press, New York
Racker, E. (1967) *Federation Proc.*, **26**, 1335
Reed, L. J. and D. J. Cox (1966) *Ann. Rev. Biochem.*, **35**, 57
Reid, R. A., J. Moyle and P. Mitchell (1966) *Nature*, 212, 257
Rice, O. K. and H. Gershinowitz (1934) *J. Chem. Phys.*, **2**, 853
Robbie, J. P. and T. H. Wilson (1969) *Biochim. Biophys. Acta.*, **173**, 234
Robertson, R. N. (1960) *Biol. Rev.*, **35**, 231
Robertson, R. N. (1968) *Protons, Electrons, Phosphorylation and Active Transport*, University Press, Cambridge
Robinson, R. A. and R. H. Stokes (1959) *Electrolyte Solutions*, 2nd ed. Butterworth, London
Rosenberg, T. (1948) *Acta Chem. Scand.*, **2**, 14
Rosenberg, T. (1954) *Symp. Soc. Exptl. Biol.*, **8**, 27
Rossi, C. and G. F. Azzone (1965) *Biochim. Biophys. Acta.*, **110**, 434
Rossi, E. and G. F. Azzone (1969) *Eur. J. Biochem.*, **7**, 418
Rossi, C. S., N. Siliprandi, E. Carafoli and A. L. Lehninger (1967) *J. Biochem.*, **2**, 332
Rothstein, A. (1968) *Ann. Rev. Physiol.*, **30**, 15
Rumberg, B., H. Schröder and U. Siggel (1968) *Naturwissenschaften*, **2**, 77
Scarborough, G. A., M. K. Rumley and E. P. Kennedy (1968) *Proc. Natl. Acad. Sci. U.S.*, **60**, 951
Scholes, P., P. Mitchell and J. Moyle (1969) *Eur. J. Biochem.*, **8**, 450
Schwartz, M. (1968) *Nature*, **219**, 915
Selwyn, M. J. (1967) *Biochem. J.*, **105**, 279
Siekevitz, P. (1959) In G. E. W. Wolstenholme and C. M. O'Connor (Eds.), *The Regulation of Cell Metabolism*, Churchill, London, p. 17
Simoni, R. D., M. Levinthal, F. D. Kundig, W. Kundig, B. Anderson, P. E. Hartman and S. Roseman (1967) *Proc. Natl. Acad. Sci. U.S.*, **58**, 1963
Skou, J. C. (1965) *Physiol. Rev.*, **45**, 596
Skou, J. C. (1969) In D. C. Tosteson (Ed.), *The Molecular Basis of Membrane Function*, Prentice-Hall, New Jersey, p. 455
Skulachev, V. P., A. A. Sharaf and E. A. Liberman (1967) *Nature*, **216**, 718
Slater, E. C. (1953) *Nature*, **172**, 975
Slater, E. C. (1967) *Eur. J. Biochem.*, **1**, 317
Stadtman, E. R. (1966) In N. O. Kaplan and E. P. Kennedy (Eds.), *Current Aspects of Biochemical Energetics*, Academic Press, New York, p. 39
Steinrauf, L. K., M. Pinkerton and J. W. Chamberlin (1968) *Biochem. Biophys. Res. Commun.*, **33**, 29
Stone, A. J. (1968) *Biochim. Biophys. Acta.*, **150**, 578
Tager, J. M., R. D. Veldsema-Currie and E. C. Slater (1966) *Nature*, 212, 376
Tanaka, S., D. G. Fraenkel and E. C. C. Lin (1967) *Biochem. Biophys. Res. Commun.*, **27**, 63
Tanaka, S. and E. C. C. Lin (1967) *Proc. Natl. Acad. Sci. U.S.*, **57**, 913
Teorell, T. (1949) *Ann. Rev. Physiol.*, **11**, 545
Tosteson, D. C. (1968) *Federation Proc.*, **27**, 1269
Tyler, D. D. (1968) *Biochem. J.*, **107**, 121

Tyler, D. Γ. (1969) *Biochem. J.*, **111**, 665

Tzagoloff, A., K. H. Byington and D. H. MacLennan (1968) *J. Biol. Chem.*, **243**, 2405

Ussing, H. H. (1947) *Nature*, **160**, 262

Ussing, H. H. (1949) *Physiol. Rev.*, **29**, 217

Van Slyke, D. D. (1922) *J. Biol. Chem.*, **52**, 525

van Steveninck, J. (1968) *Biochim. Biophys. Acta.*, **163**, 386

Vernon, C. A. (1960) In A Wasserman (Ed.), *Size and Shape Changes of Contractile Polymers*, Pergamon, Oxford, p. 109

Verwey, E. J. W. and J. T. G. Overbeek (1948) *Theory of the Stability of Lyophobic Colloids*, Elsevier, Amsterdam

Vignais, P. V., E. D. Duée and J. Huet (1968) *Life Sci.*, **7**, 641

Wilbrandt, W. and T. Rosenberg (1961) *Pharmacol. Rev.* **13**, 109

Williams, C. R., R. M. Oliver, H. R. Henney, B. B. Mukherjee and L. J. Reed (1967) *J. Biol. Chem.*, **242**, 889

Winkler, H. H., F. L. Bygrave and A. L. Lehninger (1968) *J. Biol. Chem.*, **243**, 20

Winkler, II. H. and A. L. Lehninger (1968) *J. Biol. Chem.*, **243**, 3000

Winkler, H. H. and T. H. Wilson (1966) *J. Biol. Chem.*, **241**, 2200

Winkler, H. H. and T. H. Wilson (1967) *Biochim. Biophys. Acta.*, **135**, 1030

Witt, H. T., B. Rumberg and W. Junge (1968) *Colloq. Ges. Physiol. Chem.*, **19**, 262

Wyman, J. (1948) *Advan. Protein Chem.*, **4**, 407

Wyman, J. (1964) *Advan. Protein Chem.*, **19**, 223

Yoshida, H., K. Nagi, T. Ohashi and Y. Nakagawa (1969) *Biochim. Biophys. Acta.*, **171**, 178

## GLOSSARY OF SYMBOLS

| | |
|---|---|
| $A_{OX}$ | oxidant form of component A |
| $A_{RED}$ | reductant form of component A |
| A | acceptor group of chemical compound |
| $a_i$ | chemical activity of component $i$ |
| $a_i'$ | chemical activity of component $i$ corrected for local electric potential relative to 0 |
| $b_i$ | buffering power with respect to component $i$ |
| $B_i$ | differential buffering power with respect to translocation of component $i$ between two phases |
| D | donor group of chemical compound |
| E | enzyme |
| $f_A$ | frictional coefficient for diffusion of component A |
| G | chemical group |
| $\Delta G'$ | Gibbs free energy per mole under non-standard conditions |
| $k$ | Boltzmann's constant |
| $K_A$ | dissociation constant for component A |
| L | left hand (aqueous) phase |
| M | middle or membrane (non-aqueous) phase |
| $M_i$ | electric capacity of system with respect to transfer of charged component $i$ |
| $N_i$ | coefficient defined by equation (28) |
| $\overleftarrow{n}_i$ | number of moles of component $i$ transferred from phase R to phase L |
| $\Delta p$ | protonmotive force |
| p | $-\log_{10}$ of quantity following |
| P | porter |
| R | right hand (aqueous) phase |

S            solute
X            carrier
[X]          concentration of (unoccupied) carrier
[$\overline{X}$]          total carrier concentration
$x$            distance
Z            coefficient equal to 2·303 $RT/F$
$\gamma_i$           chemical activity coefficient of component $i$
$\gamma_i'$           chemical activity coefficient of component $i$ corrected for local electric
             potential relative to 0
$\Delta$            amount by which a quantity is greater in phase L than in phase R
$\varepsilon$            amount by which average energy of a particle is greater in state L than
             in state R
$\mu_i$           purely chemical potential of component $i$
$\mu_i^*$          chemical potential corresponding to concentration of component $i$
$\mu_i'$          chemical potential corresponding to $\gamma_i$
$\mu_i^{\pm}$          chemical potential corresponding to $\psi_i$
$\mu_i''$          chemical potential corresponding to $\gamma_i'$
$\overline{\mu}_i$           total chemical potential of component $i$
$\psi$            internal electric potential of a phase (millivolt)
$\psi_i$           electric potential of component $i$
[#]          indicates concentration (distinction between concentration and activity
             if *not* neglected)
{#}          probability of a component being present per unit volume
$\sim$            bond of large free energy of hydrolysis or energized form of compound

# III

# Mechanisms of Active Transport Across Biological Membranes

CHAPTER 8

# Sodium–potassium activated adenosinetriphosphatase and cation transport

## Sjoerd L. Bonting

*Department of Biochemistry, University of Nijmegen,*
*Nijmegen, The Netherlands*

# I. INTRODUCTION

One of the great mysteries of nearly all living cells is how they maintain a relatively high potassium and low sodium content, while they are being surrounded by blood plasma, interstitial fluid or sea water with a reversed concentration ratio of the two cations (table 1). How are these cation gradients maintained by the cell? An early application of radioactive isotopes

**Table 1.** Intracellular cation concentrations in various tissues

| Tissue | Species | Na (meq/l cell water) | K (meq/l cell water) | Reference |
|---|---|---|---|---|
| Erythrocyte | man | 19 | 136 | Harris, 1956 |
| | cat | 142 | 8 | Harris, 1956 |
| Lens | toad | 17 | 112 | Duncan, 1968, 1969 |
| Nerve | cat | 28 | 150 | calculated from Krnjević, 1955; Dainty and Krnjević, 1955 |
| | squid | 48 | 406 | Harris, 1956 |
| Muscle | rat | 16 | 152 | Harris, 1956 |
| | frog | 10–16 | 128–152 | Harris, 1956 |
| Kidney | rabbit | 4–15 | 125–135 | Bojesen and Leyssac, 1965 |
| Liver | dog | 2·8 | 161 | Harris, 1956 |
| Adrenal | calf | 79 | 92 | Wellen and Benraad, 1969 |

gave the first clue. Cohn and Cohn (1939) injected $^{22}$Na into the bloodstream of a living dog and found a fairly rapid uptake into the red blood cells, indicating appreciable permeability of the cell membrane for Na⁺ ions. This ruled out the possibility that the gradients could be explained by selective permeability. More light was shed on the problem by the observation that maintaining red blood cells at low temperature tended to abolish the cation gradients, while subsequent incubation at physiological temperature re-established the gradients (Harris, 1941). Loss of potassium and gain of sodium by the cells was also seen when glucose was absent from the medium, or an inhibitor of energy metabolism like fluoride was present. These observations led to the so-called 'pump–leak concept': the cell membrane allows a passive (i.e. no energy required) leakage of both ions down their electrochemical gradients, while an active (i.e. energy required) pump system extrudes sodium ions and draws potassium ions into the cell against these gradients. Under equilibrium conditions there is a dynamic equilibrium between $Na^+$ ions leaking in and being pumped out and between $K^+$ ions leaking out and being accumulated.

The question which then arose was the identity of this pump system and its relation to the energy metabolism of the cell. The answer to the latter part of the question was solved first and led to an answer to the first part. In many cells, for example nerve, muscle and nucleated erythrocytes of birds and reptiles (Shanes, 1951; Maizels, 1954; Hodgkin and Keynes, 1955), oxidative metabolism is the source of energy for active cation transport, as shown by its inhibition under anaerobiosis or in the presence of cyanide or 2,4-dinitrophenol. On the other hand, in mammalian erythrocytes and lens, oxidative metabolism is absent or virtually absent. In these cases anaerobiosis does not inhibit cation transport, but removal of glucose or addition of glycolytic inhibitors, such as fluoride and iodoacetic acid, abolishes cation transport (Harris, 1941; Danowski, 1941; Maizels, 1951). This suggests that the pump is not linked directly to any of the individual stages in the glycolytic or oxidative phosphorylation chains, but that it is the adenosine-triphosphate (ATP) produced by both systems which provides the energy for ion transport. Direct evidence for this conclusion came from experiments in which ATP was placed inside cells lacking any other energy source. Injection of ATP into a squid giant axon poisoned with cyanide restored the sodium efflux (Caldwell and coworkers, 1960). Incorporation of ATP into erythrocyte ghosts that are in the process of being 'reconstituted' (Gardos, 1954) permits cation transport (Hoffman, Tosteson and Whittam, 1960). ATP has also been shown to supply the energy for active cation transport in Ehrlich ascites tumour cells (Weinstein and Hempling, 1964).

In directing himself to the first part of the question, Skou reasoned that an ATPase would be required to make the energy of ATP available to cation transport, and that this enzyme might require $Na^+$ and $K^+$ for activity. In a study of a crab nerve particulate fraction he was able to demonstrate a $Mg^{2+}$ activated ATPase activity, which was increased considerably upon addition of $Na^+$ and $K^+$ to the assay medium (Skou, 1957). He also showed that this additional ATPase activity was completely inhibited by the digitalis glycoside ouabain (Skou, 1960). Some years earlier, Schatzmann (1953) had observed that digitalis glycosides are powerful and specific inhibitors of active cation transport in erythrocytes. Therefore, Skou suggested that this ouabain-sensitive, $Na^+$–$K^+$ activated ATPase (Na–K ATPase) might be part of, or identical with, the cation transport system.

In the ensuing twelve years a great number of reports have been published confirming this conclusion for a variety of cells and tissues. It is the purpose of this chapter to review our knowledge of the relationship of the $Na^+$–$K^+$ activated ATPase system and active cation transport in the various cells and tissues. After a section on assay, properties and occurrence of the enzyme system, there will follow three sections on single cell systems, excitatory systems and secretory systems, and the chapter will be concluded by a section on the mechanism of the enzyme system.

## II.  ASSAY, PROPERTIES AND OCCURRENCE OF $Na^+$–$K^+$ ACTIVATED ADENOSINETRIPHOSPHATASE

### A.  Assay

#### 1. *Tissue Preparation*

This method of tissue preparation is of crucial importance. Whole cells should not be used, because ATP permeates the cell membrane very slowly and it is needed on the intracellular side of the membrane to be hydrolysed by the Na–K ATPase system. For example, in assaying intact frog retina (Scarpelli and Craig, 1963) less than 1 % of the activity in a lyophilised homogenate of the same tissue (Bonting, Caravaggio and Canady, 1964) was found. Highest activities were obtained (Yoshida and Fujisawa, 1962; Bonting and Caravaggio, 1963, pp. 43–44) when the tissue was homogenised in water, lyophilised at $-10$ to $-20\,^{\circ}C$ and reconstituted with water immediately before assay, as has been found for other enzymes as well (Bonting and Rosenthal, 1960). The lyophilised homogenate can be stored at $-25\,^{\circ}C$ for more than ten weeks without loss of activity (Bonting, Simon and Hawkins, 1961). Homogenisation in isotonic media gives lower values, presumably because folding up of membrane fragments reduces access of substrate to enzyme. This is indicated by the fact that detergents such as deoxycholate increase the measured activity in this case, but not after homogenisation in water, lyophilisation and reconstitution with water.

Since the enzyme system is localised in the cell membrane, isolation of the cell membrane rich fraction through centrifugal fractionation is used by many authors. This allows one to obtain a preparation of higher specific activity and to remove some non-$Na^+$–$K^+$ activated, $Mg^{2+}$ stimulated ATPase (Mg ATPase) activity, which is not attached to the cell membrane. However, a partial loss of Na–K ATPase activity is inevitable in this case, because the enzyme activity is never limited entirely to one fraction and its distribution over the centrifugal fractions varies in different tissues. In rat liver 93 % was found in the nuclear fraction, 3·7% each in the mitochondrial and microsomal fractions, while in rat brain these figures were 33, 10 and 57% respectively (Bonting, Caravaggio and Hawkins, 1962). Therefore, for determinations of the activity of Na–K ATPase in a tissue the use of whole homogenates is preferable. For studies of the properties fractionation is desirable when the percentage of Na–K ATPase activity in the whole homogenate is low. In whole rat liver this is 22%, while in the nuclear fraction it is 46% (Bonting, Caravaggio and Hawkins, 1962). Further fractionation to obtain a pure cell membrane fraction may, however, lead to a loss of Na–K ATPase activity: Emmelot and coworkers (1964) obtained a liver plasma membrane fraction in which only 23% of total $Mg^{2+}$ activated ATPase activity was activated by $Na^+ + K^+$.

Treatment with various agents may reduce the Mg ATPase activity without affecting the Na–K ATPase activity. Pretreatment of the homogenate with 1·5 M urea for 30–60 min at 0 °C may remove half of the Mg ATPase activity (Glynn and coworkers, 1965; Skou and Hilberg, 1965; Hafkenscheid and Bonting, 1968; Bakkeren and Bonting, 1968a). Pretreatment of the homogenate with 2 M NaI for 30 min at 0 °C had a similar effect in erythrocytes and brain (Nakao and coworkers, 1963, 1965) and in heart muscle (Matsui and Schwartz, 1966a). Pretreatment with deoxycholate for 5–7 days at 0 °C was used by Schoner and coworkers (1967) to prepare Na–K ATPase with less than 1% Mg ATPase activity from ox brain. This method should be applied with caution, because Shimizu, Horikawa and Yamazoep (1965) report a loss of inhibition by ouabain upon treatment of rabbit skeletal muscle with deoxycholate, probably because of phospholipid removal from the membrane complex (see Section IIB). Through a combination of treatments with NaI and the detergent Lubrol, followed by salt and isoelectric precipitation, ultracentrifugation and carboxymethylcellulose chromatography, Kahlenberg and associates (1969) have achieved a considerable degree of solubilisation and purification of brain Na–K ATPase.

## 2. Incubation Media

Since the Na–K ATPase activity is always accompanied by Mg ATPase activity, a differential assay is unavoidable. Many authors compare the activity in a medium containing only buffer, ATP and $Mg^{2+}$ with that after addition of $Na^+ + K^+$. However, this method may lead to considerable errors, because Mg ATPase in many tissues is activated to some extent by $Na^+$ ions (electric organ: Bonting and Caravaggio, 1963; rabbit lens: Bonting, Caravaggio and Hawkins, 1963; brain, ciliary body: Bonting, Hawkins and Canady, 1964; choroid plexus: Vates, Bonting and Oppelt, 1964; avian salt gland: Bonting and coworkers, 1964; retinal rods: Bonting, Caravaggio and Canady, 1964; elasmobranch rectal gland: Bonting, 1966; liver: Bak-keren and Bonting, 1968a; *E. coli*: Hafkenscheid and Bonting, 1969). It is, therefore, preferable to determine total $Mg^{2+}$ activated ATPase activity in a complete medium and compare this activity with media lacking $Na^+$ or $K^+$ or containing ouabain.

The compositions of the media used in our laboratory are listed in Table 2.

**Table 2.** Composition of incubation media for Na–K ATPase assay

All concentrations in mmole/l; $Cl^-$ is the only other ion present; final pH 7·5; ATP is present as the di-Na salt in media A, B, D and E, and as the tris-salt in medium C; average activity in media B, C, D and E is Mg ATPase activity; difference between activity in medium A and average activity in other media is Na–K ATPase activity

|              | A    | B    | C    | D    | E    |
|--------------|------|------|------|------|------|
| ATP          | 2    | 2    | 2    | 2    | 2    |
| $Mg^{2+}$    | 2    | 2    | 2    | 2    | 2    |
| $K^+$        | 5    | —    | 5    | 5    | —    |
| $Na^+$       | 60   | 60   | —    | 60   | 60   |
| EDTA         | 0·1  | 0·1  | 0·1  | 0·1  | 0·1  |
| Tris-buffer  | 92   | 92   | 151  | 92   | 92   |
| Ouabain      | —    | —    | —    | 0·1  | 0·1  |

Modified after Bonting and Caravaggio (1963)

A medium (E), lacking $K^+$ and containing ouabain, was added, because the $K_m$ value for $K^+$ activation is so low that substantial activation of the Na–K ATPase system may be caused by the tissue $K^+$ present in the homogenate (erythrocyte, frog muscle, squid axon, frog skin, toad bladder, electric organ: Bonting and Caravaggio, 1963; lens: Bonting, Caravaggio and Hawkins, 1963; brain, kidney, ciliary body and choroid plexus: Bonting, Hawkins and Canady, 1964; avian salt gland: Bonting and coworkers, 1964; leukocytes: Block and Bonting, 1964; elasmobranch rectal gland: Bonting, 1966; liver: Bakkeren and Bonting, 1968a). The cyanide, added to our earlier media (Bonting, Simon and Hawkins, 1961; Bonting, Caravaggio and Hawkins, 1963) in order to inhibit possible alkaline phosphatase activity, has been omitted because we have never detected significant activity of this enzyme at pH 7·5. The $Mg^{2+}$ concentration is made equal to that of ATP (2 mM), because a $Mg^{2+}$ : ATP ratio of 1 : 1 is generally optimal. The EDTA was added to bind traces of heavy metal ions, which are inhibitory to Na–K ATPase. The total osmolarity of the media is not critical.

When working with reconstituted lyophilised homogenates, we choose the homogenate concentration such that 1 volume of homogenate added to 15 volumes of medium gives not more than 40% substrate utilisation after 1 hr incubation at 37 °C. Under these conditions there is always a linear response with time for well over 1 hr. Below 40% substrate utilisation there is also a linear relation with enzyme concentration. The enzyme reaction is stopped by addition of a 4 to 5-fold volume of 10% (w/v) trichloroacetic acid to the incubation mixture. After centrifugation the supernatant is transferred to a tube containing an equal volume of a colour reagent, consisting of a 1% ammonium molybdate solution in 1·15 N $H_2SO_4$, in which 40 mg/ml $FeSO_4$ is dissolved immediately before use. This solution can be used for 2 hr after addition of $FeSO_4$. The resulting colour is read within 2 hr at 700 nm. Reagent blanks and inorganic phosphate standard are included in each experiment and serve to convert optical density into millimoles of inorganic phosphate released per gram of tissue (wet or dry weight) or protein per hr incubation at 37 °C, after correction with the reading obtained for non-incubated tubes containing tissue. The unincubated tubes permit correction for inorganic phosphate present in the tissue and for the slow non-enzymic hydrolysis of ATP catalysed by the molybdate reagent. For the latter correction it is necessary to treat the incubated and unincubated tubes for the same sample and medium in close succession. The method described here has proven to be

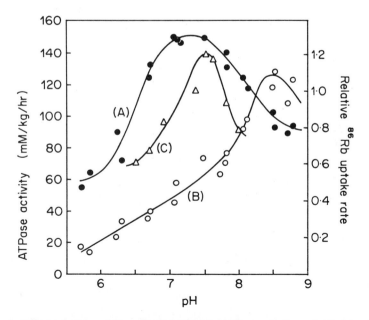

**Figure 1.** Effect of pH on (A) epithelial Na–K ATPase activity and (B) Mg ATPase activity compared (C) with the effect of pH on $^{86}$Rb uptake rate, all in rabbit lens. From Bonting (1965).

convenient and accurate and easily adaptable to ultramicro conditions and for use with microgram quantities of dissected fragments of lyophilised tissue (Bonting and Caravaggio, 1962, 1963; Kuijpers, van der Vleuten and Bonting, 1967; Kuijpers and Bonting, 1969). Automation of the Na–K ATPase assay has been achieved and applied by Stein, Glasky and Roderick (1965).

## B. Properties

The properties of the Na–K ATPase system from many tissues have been determined and found to be quite similar. We shall, therefore, illustrate each of these properties by a single example and summarise the quantitative aspects for various tissues in a table.

Figure 1 shows the pH-activity curves for Mg ATPase and Na–K ATPase in rabbit lens epithelium (Bonting, Caravaggio and Hawkins, 1963; Bonting, 1965). The optima are pH 8·4 and pH 7·3, respectively. The optimum for the active $^{86}$Rb uptake by the lens, also shown in figure 1, is 7·5, in agreement with the Na–K ATPase optimum. For comparison, table 3 lists pH optima for the Na–K ATPase and Mg ATPase systems of a number of tissues.

The effect of varying $Mg^{2+}$ concentrations on Mg ATPase and Na–K ATPase activities in rat liver after urea treatment and at a constant ATP level of 2 mM is displayed in figure 2 (Bakkeren and Bonting, 1968a). Both activities reached an optimal value at a Mg : ATP ratio of 1 : 1, which has been found in several other tissues. It suggests that the true substrate is

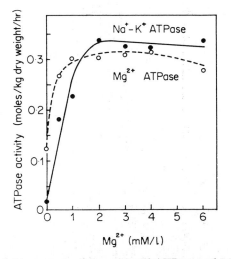

**Figure 2.** Effect of $Mg^{2+}$ concentration on Na–K ATPase and Mg ATPase activities in rat liver homogenate. (From Bakkeren and Bonting, 1968a).

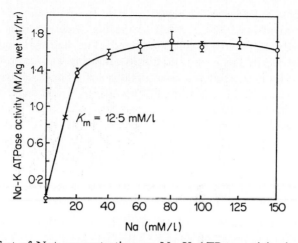

**Figure 3.** Effect of $Na^+$ concentration on Na–K ATPase activity in herring gull salt gland homogenate ($K^+$ constant at 5 mM). (From Bonting and coworkers, 1964).

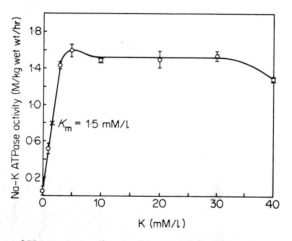

**Figure 4.** Effect of $K^+$ concentration on Na–K ATPase activity in herring gull salt gland homogenate ($Na^+$ constant at 60 mM). (From Bonting and coworkers, 1964).

Mg $ATP^{2-}$ or that the ATP is bound to the enzyme via a $Mg^{2+}$ ion. $Mn^{2+}$ ion can replace $Mg^{2+}$ in the activation of Na–K ATPase (Atkinson, Hunt and Lowe, 1968).

The Na+ activation curve for Na–K ATPase can be obtained by adding a graded amount of NaCl (0–150 mM) to the Na+-free medium C. Figure 3 shows the curve for herring gull salt gland. The half-maximal activation concentration $K_m$ in this case is 12·5 mM (Bonting and coworkers, 1964). Maximal activity is reached at a concentration of 60 mM; above 125 mM it begins to decrease slightly, presumably because K+ ions are then being replaced from their binding site by Na+ ions (Priestland and Whittam, 1968). Table 3 shows $K_m$ values for Na+ activation of Na–K ATPase in various tissues. Since the Mg ATPase is often affected by Na+ ions in the absence of K+ ions, it is necessary to carry out ATPase assays at various Na+ concentrations in the absence of K+ ions or in the presence of ouabain in order to correct the Na+ activation curve of Na–K ATPase for changes in the Mg ATPase activity.

The K+ activation curve for Na–K ATPase in herring gull salt gland, obtained by adding graded amounts of KCl (0–40 mM) to the K-free medium B, is shown in figure 4. The half-maximal activation concentration $K_m$ in this case is 1·5 mM. Maximal activity is reached at 5 mM and remains virtually constant until 30 mM. Beyond that concentration activity begins to decrease, presumably because K+ ions displace Na+ ions from their binding site. Table 3 shows $K_m$ values for K+ activation of Na–K ATPase in a number of tissues. It is noticeable that activation by K+ ions takes place at much lower concentrations than Na+ activation. Rb+ ions can replace K+ ions for activation (figure 5); the $K_m$ value is virtually the same and the maximal activity is

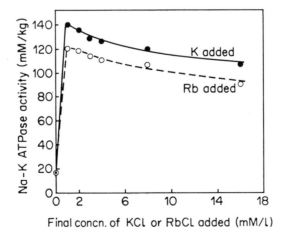

**Figure 5.** Effect of K+ and Rb+ on epithelial Na–K ATPase activity in the rabbit lens. (From Bonting, Caravaggio and Hawkins, 1963).

**Table. 3.** Properties of the Na-K ATPase system in various species and tissues

| Tissue | Species | Mg/ATP opt. | Na-K ATPase Na+ $K_m$ | K+ $K_m$ | pH opt. | ouabain $pI_{50}$ | activity MKH | Mg ATPase activity MKH | Mg ATPase pH opt. | Reference |
|---|---|---|---|---|---|---|---|---|---|---|
| Nerve | Crab | 2 | 6–8 | 1·8 | 7·2 | 3·9 | — | — | — | Skou (1957, 1960) |
| Intestine | Guinea pig | — | 1·0 | 0·5 | 7·5–8·0 | 5·4 | — | — | — | Taylor (1962) |
| Salt gland | Herring gull | 1·5 | 12·5 | 1·5 | 7·2 | 6·3 | 2·7 w | 1·4 w | 8·7 | Bonting and coworkers (1964) |
| Lens epithelium | Rabbit | — | — | 0·4 | 7·3 | 5·9 | 0·15 w | 0·06 w | 8·4 | Bonting (1965) |
| Rectal gland | Spiny dogfish | 1·5 | 11·7 | 1·0 | 7·0 | 6·8 | 5·7 d | 2·9 d | 8·9 | Bonting (1966) |
| E. coli | E. coli | 1 | — | 4·3 | 7·7 | ±4 | 0·24 d | 2·5 d | 8·7–8·9 | Hafkenscheid and Bonting (1968, 1969) |
| Liver | Rat | 1 | 6 | 0·9 | 7·3 | 3·9 | 0·37 d | 3·6 d | 8·7 | Bakkeren and Bonting (1968a) |
| Stria vascularis | Guinea pig | 0·5–1 | 4·5 | 0·9 | 7·3 | 5·5 | 8·0 d | 5·6 d | 8·7 | Kuijpers and Bonting (1969) |
| Pancreas | Rabbit | 1·5 | 10 | 0·8 | 7·2 | 5·4 | 0·23 d | 2·7 d | 8·8 | Ridderstap and Bonting (1969b) |
| Kidney | Rat | 0·5 | 8 | 0·7 | 7·4 | 3·9 | 6·8 d | 9·3 d | 8·7 | v.d. Beek and Bonting unpublished |

Mg/ATP opt.: molar ratio of $Mg^{2+}$ to ATP at which maximal activity occurs.
Na+ $K_m$: half-maximal activation concentration for Na+ in meq/l at K+ concentration of about 5 meq/l.
K+ $K_m$: half-maximal activation concentration for K+ in meq/l at Na+ concentration of about 60 meq/l.
pH opt.: pH optimum.
ouabain $pI_{50}$: negative log molar concentration for 50% inhibition.
activity MKH: activity in moles/kg/hr, d = on dry wt. basis, w = on wet wt. basis.

nearly the same as for $K^+$. Thallium ions can also substitute for $K^+$ ions (Gehring and Hammond, 1967) and have an affinity 10 times greater than $K^+$ for the binding site (Britten and Blank, 1968). Both ions are actively accumulated by cells, apparently by the same mechanism as $K^+$. In the case of $Rb^+$ the uptake has been shown to be inhibitable by ouabain (lens: Becker and Cotlier, 1962; ciliary body: Bonting and Becker, 1964; liver: Bakkeren and Bonting 1968b). Anions have little effect on Na–K ATPase, except for $F^-$ which is strongly inhibitory at concentrations over 1 mM (Opit, Potter and Charnock, 1966). This inhibition is not overcome by raising the $Mg^{2+}$ concentration to 10 mM.

A characteristic property of the Na–K ATPase system is its inhibition by the cardiac glycosides. Of these, ouabain is most widely used because it is fairly soluble in water. Other cardiac glycosides have to be applied in alcoholic solution, which complicates the assay because of the inhibitory effect of ethanol on the enzyme system. Figure 6 shows the inhibition curve for Na–K ATPase from elasmobranch rectal gland (Bonting, 1966), obtained by adding graded amounts of ouabain ($10^{-9}$–$10^{-4}$ M) to the complete medium A.

Complete inhibition is caused by $10^{-4}$ M ouabain, half inhibition by about $2 \times 10^{-7}$ M ouabain and a slight stimulation by about $10^{-8}$–$10^{-9}$ M ouabain. Table 3 lists for a number of tissues the $pI_{50}$ value, i.e. the negative log of the molar concentration, causing 50% inhibition. The biphasic effect, stimulation

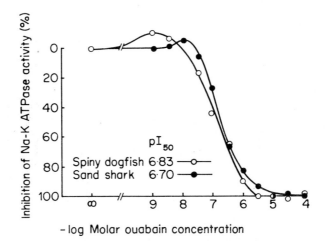

**Figure 6.** Effect of ouabain on Na–K ATPase activity in elasmobranch rectal gland homogenates; $pI_{50}$ is the negative logarithm of the molar ouabain concentration, causing 50% inhibition. (From Bonting, 1966).

at very low concentrations and inhibition at higher concentrations, has been observed in several other tissues (heart muscle: Lee and Yu, 1963; brain: Bonting, Hawkins and Canady, 1964; choroid plexus: Vates, Bonting and Oppelt, 1964; kidney: Palmer and Nechay, 1964; brain, kidney: Palmer, Lasseter and Melvin, 1966; pancreas: Ridderstap and Bonting, 1968, 1969b). Both inhibition and stimulation are confined to the Na–K ATPase system; the Mg ATPase activity is not affected by cardiac glycosides. $K^+$ ions have an antagonising effect on the inhibition of Na–K ATPase by ouabain (erythrocytes: Dunham and Glynn, 1961; brain: Bonting, Hawkins and Canady, 1964; liver: Ahmed and Judah, 1964), as is illustrated in figure 7. This effect is of a non-competitive nature (Matsui and Schwartz, 1966b) and is shown by ouabain concentrations, which are only partially inhibitory. In view of this effect, it is necessary to study inhibition of Na–K ATPase activity at the minimal $K^+$ level consistent with optimal or near-optimal activity of the enzyme system.

An important question to be considered is whether the Na–K ATPase and the Mg ATPase are two different enzymes or represent two activities of a single enzyme system. The fact that they are always found together in

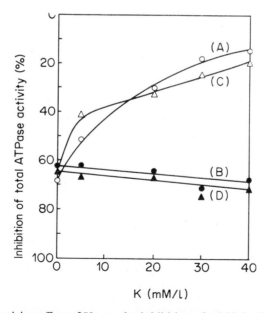

**Figure 7.** Antagonizing effect of $K^+$ on the inhibition of rabbit brain Na–K ATPase by erythrophleine (A, $5 \times 10^{-7}$M; B, $10^{-4}$M) and ouabain (C, $5 \times 10^{-7}$M; D, $10^{-4}$M) (From Bonting, Hawkins and Canady, 1964).

K

membrane preparations would favour the latter possibility. However, there are many arguments favouring the conclusion that they are two different enzymes: (i) the widely differing ratios for the two activities in different tissues, (ii) the different pH optima, (iii) the different effects of various agents, like ouabain, urea, NaI, sulphydryl reagents etc., the different effects of preincubation at 37° for 30–60 min (Hokin and Reasa, 1964), (iv) the different temperature-activity curves for the two activities (Bowler and Duncan, 1968; Bakkeren and Bonting, 1968a), (v) the different rate and extent of increase for the two activities in the salt gland of ducks given saline to drink (Ernst, Goertemiller and Ellis, 1967). The Mg ATPase in erythrocytes hydrolyses only ATP present outside, the Na–K ATPase only ATP present inside the cell (Hoffman, 1962). Exposure of erythrocyte membranes to ultrasound and detergent, followed by density gradient centrifugation, yielded two fractions, large and light fragments containing Na–K ATPase, and small and heavy fragments containing Mg ATPase (Tosteson, Cook and Blount, 1965). It has also been observed that removal of phospholipids from the membrane preparation by means of detergents or phospholipase leads to loss of Na–K ATPase activity, which is partially restored by addition of lecithin or other phospholipids (Tanaka and Abood, 1964b; Fenster and Copenhaver, 1967; Portela and associates, 1967), while Mg ATPase is not affected by phospholipid addition (Tanaka and Strickland, 1965). It thus appears that Na–K ATPase consists of a phospholipid–protein membrane complex, while Mg ATPase is primarily a protein present on the outside of the membrane. Molecular weights of 293,000 (Nakao and associates, 1967), 670,000 (Medzihradsky, Kline and Hokin, 1967; Uesugi and associates, 1969) and 250,000 (Kepner and Macey, 1968) have been reported for the Na–K ATPase complex, and a value of 250,000 for Mg ATPase and of 140,000 for the $K^+$-activated phosphatase thought to be part of the Na–K ATPase system (Kepner and Macey, 1968).

## C. Occurrence and Correlation with the Cation Pump

After the discovery of the Na–K ATPase system in crab nerve by Skou (1957) the enzyme was reported in erythrocyte membranes by Dunham and Glynn (1960, 1961) and Post and associates (1960). These authors demonstrated that the properties of the enzyme system and the transport system were qualitatively and quantitatively the same:

(i) both systems are located in the membrane;
(ii) both systems utilise ATP but not inosine triphosphate;
(iii) both systems require $Na^+$ and $K^+$ together, while either cation alone is ineffective; ammonium ion can substitute for $K^+$;

(iv) the half-maximal activation concentrations of the cations are the same in both systems;

(v) the inhibitory effects of various cardiac glycosides on the two systems are closely similar, give equal half-inhibition concentrations for both systems and require the unsaturated lactone group;

(vi) $K^+$ ions antagonise the cardiac glycoside effect in both systems.

Bonting, Simon and Hawkins (1961) studied the distribution of the enzyme system in 39 tissues of the cat and man, and found the enzyme present in all but 6 of these tissues. The tissues in which no significant activity could be demonstrated (corneal stroma, lens capsule, lens fibres, vitreous body, adipose tissue and serum) were all tissues in which few or no cells were present. The activities, expressed in mole/kg wet wt/hr at 37°, ranged widely from 1·52 in cat grey matter to 0·00027 in cat erythrocytes. Highest activities were noted in the nervous and secretory tissues. This wide distribution suggests that the enzyme system is a common feature of all mammalian cells.

Next a systematic study was made of the occurrence of the enzyme system in all tissues known at that time to possess a cardiac glycoside-sensitive cation transport system (Bonting, Caravaggio and Hawkins, 1962). The enzyme was found to be present in all 21 tissues from 10 species with activities ranging from 1·8 (guinea pig brain) to 0·00124 (human erythrocytes) mole/kg wet wt/hr at 37°. For 6 tissues (4 organs, 4 species) the ouabain inhibition curves were determined. The half-inhibition concentrations varied over a 17-fold range, depending both on species and organ.

Having established this qualitative correlation between Na–K ATPase and the cation transport system, an attempt was made to demonstrate a

**Table 4.** Comparison of cation fluxes and Na–K ATPase activities

| Tissue | Temp. (°C) | Cation flux ($10^{-14}$mole cm$^{-2}$ sec$^{-1}$) | Na–K ATPase activity* ($10^{-14}$mole cm$^{-2}$ sec$^{-1}$) | Ratio |
|---|---|---|---|---|
| Human erythrocytes | 37 | 3·87 | 1·38 ($\pm0·36$; 4) | 2·80 |
| Frog toe muscle | 17 | 985 | 530 ($\pm94$; 4) | 1·86 |
| Squid giant axon | 19 | 1200 | 400 ($\pm79$; 5) | 3·00 |
| Frog skin | 20 | 19,700 | 6640 ($\pm1100$; 4) | 2·97 |
| Toad bladder | 27 | 43,700 | 17,600 ($\pm1640$; 15) | 2·48 |
| Electric eel, non-innervated membrane, Sachs organ | 23 | 86,100 | 38,800 ($\pm4160$; 3) | 2·22 |
| | | $\tau = +0·866$ | $P = 0·017$ | 2·56 ($\pm0·19$) |

*In parentheses standard error of the mean and number of determinations. After Bonting and Caravaggio (1963).

quantitative correlation between active cation flux and Na–K ATPase activity. In six tissues, for which the active cation flux had been reported (human erythrocyte, frog toe muscle, squid giant axon, frog skin, toad bladder and noninnervated membrane of the Sachs organ of the electric eel), the Na–K ATPase activity per sq cm and per sec was determined at the same temperature as were the flux measurements (Bonting and Caravaggio, 1963). A significant correlation was found over a 25,000-fold range with an average molar cation/ATP ratio of $2·6 \pm 0·19$ (table 4). The Mg ATPase activity was not significantly correlated with the flux values. This finding further demonstrated the close relationship of the Na–K ATPase system to the active cation transport involved in maintenance of ion gradients in single cells, transport of salt and water across epithelial membranes and repolarisation processes in muscle, nerve and electric organ. A cation/ATP ratio of nearly 3 has now been found in a great number of tissues, mostly on a Na–K ATPase basis, in some, on the basis of the extra oxygen or glucose consumption or lactate production occurring when cation transport takes place (table 5).

**Table 5.** Ratios of cation transport to hydrolysis of ATP in various tissues

| Tissue | Cation/ATP (eq/mol) | References |
|---|---|---|
| *Based on Na–K ATPase assay* | | |
| Human erythrocytes | 2·8 (Na) | Bonting and Caravaggio, 1963 |
| Frog toe muscle | 1·9 (Na) | Bonting and Caravaggio, 1963 |
| Squid giant axon | 3·0 (K) | Bonting and Caravaggio, 1963 |
| Frog skin | 3·0 (Na) | Bonting and Caravaggio, 1963 |
| Toad bladder | 2·4 (Na) | Bonting and Canady, 1964 |
| Electric eel, non-inn. membrane electroplax | 2·2 (Na) | Bonting and Caravaggio, 1963 |
| Calf lens | 3·1 (Na) | Bonting, Caravaggio and Hawkins, 1963 |
| Rabbit lens | 2·5 (Na) | Bonting, Caravaggio and Hawkins, 1963 |
| Herring gull nasal gland | 2·0 (Na) | Bonting and coworkers, 1964 |
| Human leukocytes | 1·8 (Na) | Block and Bonting, 1964 |
| Rabbit ciliary epithelium | 3·0 (Na) | Bonting and Becker, 1964 |
| Spiny dogfish rectal gland | 2·2 (Na) | Bonting, 1966 |
| Frog sartorius muscle | 2·2 (K) | Corrie and Bonting, 1966 |
| *Escherichia coli* | 1·6–3·2 (K) | Hafkenscheid and Bonting, 1968 |
| Dog pancreas | 1·9 (Na) | Ridderstap and Bonting, 1969a |
| *Not based on Na–K ATPase assay* | | |
| Human erythrocytes | 2·7 (Na) | Sen and Post, 1964 |
|  | 3·0 (Na) | Glynn, 1962a |
| Frog muscle | 2·5–3·0 (Na) | Harris, 1967 |
| Frog skin | 2·7–3·3 (Na) | Zerahn, 1956 |
|  | 3·5 (Na) | Leaf and Renshaw, 1957 |
| Toad bladder | 2·8 (Na) | Leaf, Page and Anderson, 1959 |
| Calf lens | 3·0–4·5 (Na) | Kern, Roosa and Murray, 1962 |

Another indication of the close relationship between the Na–K ATPase system and the cation transport system is the inhibition of both by the erythrophleum alkaloids, which differ considerably in structure from the

Table 6. Occurrence of Na–K ATPase in animals and plants

| | | Animals |
|---|---|---|
| Vertebrates | | |
| Mammals | + | many species and tissues |
| Birds | | |
| Chicken brain | + | Bignami, Palladini and Venturini, 1966 |
| Chicken heart | + | Kelin, 1963 |
| Chicken kidney | + | Palmer and Nechay, 1964 |
| Gull salt gland | + | Hokin, 1963; Bonting and coworkers, 1964 |
| Duck salt gland | + | Fletcher, Stainer and Holmes, 1967 |
| Reptiles | | |
| Gecko gastric mucosa | + | Hansen and Bonting, unpublished results |
| Amphibia | | |
| Frog skin | + | Bonting and Caravaggio, 1963 |
| Frog retina | + | Bonting, Caravaggio and Canady, 1964 |
| Toad bladder | + | Bonting and Caravaggio, 1963; Bonting and Canady, 1964 |
| Toad heart | + | Kennedy and Nayler, 1965 |
| Fish | | |
| Carp retina | + | Ostrovskii and Trifonov, 1967 |
| Goldfish intestine | + | Smith, 1967. |
| Eel intestine | + | Oide, 1967 |
| Eel gills | + | Kamiya, 1968 |
| Killifish gills | + | Epstein, Katz and Pickford, 1967 |
| *Electrophorus electricus* electric organ | + | Glynn, 1962b; Albers and Koval, 1962; Bonting and Caravaggio, 1963 |
| Elasmobranch rectal gland | + | Bonting, 1967 |
| Invertebrates | | |
| Arthropods | | |
| Crab nerve | + | Skou, 1957, 1960 |
| Crab gills | + | Quinn and Lane, 1966 |
| Crayfish muscle | + | Bowler and Duncan, 1967a |
| Insects | | |
| Cockroach | + | Grasso, 1967 |
| Molluscs | | |
| Barnacle muscle | + | Beaugé and Sjodin, 1967 (ouabain-sensitive Na-efflux and K-influx) |
| Squid axon | + | Bonting and Caravaggio, 1962 |
| Squid retina | + | De Pont, Duncan and Bonting, to be published |
| Octopus gill | − | Schoffeniels, 1962 (Na-activated ATPase, insensitive to ouabain) |
| Echinodermata | | |
| Sea urchin egg | + | Ohnishi, 1963 |
| Worms | | |
| Earthworm axon | + | Mirsalikhova, Temoriv and Tashmukhamedov, 1968 |
| Protozoa | − | Conner, 1967 (ATPase and ion transport, insensitive to ouabain) |

**Table 6.** Occurrence of Na–K ATPase in animals and plants—*continued*

Plants

Cormophytes
  *Arachis hypogaea*　　　+　Brown and Altschul, 1964
  Barley root　　　　　　+　Brown, Chattopadhyay and Patel, 1967 (ouabain-
  　　　　　　　　　　　　　sensitive, Na– or K–activated ATPase
  Carrot, pea root　　　　–　Dodds and Ellis 1966
  　　　　　　　　　　　　　(K– or Na–activated ATPase insensitive to
  　　　　　　　　　　　　　ouabain)
Euthallophytes
  Algae　　　　　　　　　+　Janacek and Rybova, 1966 (ouabain-sensitive
  　　　　　　　　　　　　　cation transport); Mirsalikhova, 1968
  Valonia　　　　　　　　–　Gutknecht, 1967 (ion transport slightly sensitive
  　　　　　　　　　　　　　to ouabain)
  *Ulva lactua*　　　　　–　Bonting and Caravaggio, 1966
Schizophytes
  Mycoplasma　　　　　　–　Rottem and Razin, 1966
  *Escherichia coli*　　　+　Hafkenscheid and Bonting, 1968, 1969a
  *Staphylococcus aureus*　–　Gross and Coles, 1968 (Na– or K–activated
  　　　　　　　　　　　　　ATPase, insensitive to $10^{-3}$ M ouabain)

cardiac glycosides (figure 8), but are remarkably similar to them in pharmacological properties and in inhibitory effects on active cation transport
and Na–K ATPase (Bonting, Hawkins and Canady, 1964; Bonting, 1967).
The erythrophleum alkaloids erythrophleine and cassaine are slightly stimulatory at $10^{-8}$ M like ouabain, inhibitory at higher concentrations and their
inhibitory effect at intermediate concentrations is antagonised by $K^+$ (figure
7), while these compounds, like ouabain, do not affect the Mg ATPase
activity. The erythrophleum alkaloids have been shown to inhibit active
cation transport in erythrocytes (Kahn, 1962), toad bladder (Bonting and
Canady, 1964), cerebrospinal fluid formation (Vates, Bonting and Oppelt,
1964), frog sartorius muscle (Corrie and Bonting, 1966), cochlear function
(Kuijpers, van der Vleuten and Bonting, 1967), and exocrine pancreatic
secretion (Ridderstap and Bonting, 1969b). Kahn (1962) has, moreover,
observed a slight stimulation of the active $K^+$ influx in erythrocytes with very
low concentrations of erythrophleine alkaloids.

At this point it is worthwhile to consider the distribution of the Na–K
ATPase system in the animal and plant kingdom (table 6). In the earlier
part of this chapter we have already seen its occurrence in a wide variety of
mammalian species and tissues. It has also been demonstrated in various
species and tissues of birds, amphibia and fishes, and in one case in reptiles.
Among the invertebrates it has been found in arthropods, insects and
molluscs. In octopus gill a Na activated ATPase insensitive to ouabain has
been found. In protozoa the system has not been demonstrated and in plants

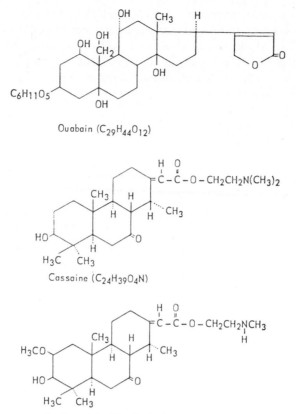

**Figure 8.** Chemical structure of ouabain and the erythrophleum alkaloids cassaine and erythrophleine. Position and configuration of the ring constituents in erythrophleine are still uncertain. (From Bonting, Hawkins and Canady, 1964).

the situation is still confusing. Positive results are reported for the soy bean (*Arachis hypogaea*), barley root, certain algae and *Escherichia coli*. Negative results have been recorded for carrot and pea roots, *Ulva lactuca*, *Mycoplasma* and *Staphylococcus aureus*. It may well be that further investigations will also reveal the existence of the system here, since it must be kept in mind that these organisms may have a very low sensitivity to ouabain, and also that detection of a small amount of Na–K ATPase activity in the presence of a high Mg ATPase activity is quite difficult. This was borne out in our own study of *Escherichia coli*, where only 5% of the total ATPase activity was ouabain-sensitive Na–K activated ATPase, which required $10^{-4}$–$10^{-3}$ M ouabain for inhibition (Hafkenscheid and Bonting, 1968, 1969).

In the absence of a reliable histochemical test for the Na–K ATPase system, fractionation techniques have had to be used to establish its location in the cell. For erythrocytes, where it is fairly easy to obtain a relatively pure membrane preparation, the occurrence in the cell membrane was easily established (Post and associates, 1960; Dunham and Glynn, 1961). Centrifugal fractionation of brain and liver homogenates gave a distribution of enzyme activity (in 'nuclear' fraction of liver; in all three particulate fractions of brain), which was consistent with membrane localisation in both tissues (Bonting, Caravaggio and Hawkins, 1962). Isolation of plasma membrane preparations with Na–K ATPase activity has been accomplished in some cases, e.g. liver (Emmelot and coworkers, 1964) and kidney (Landon and Norris, 1963). For intestinal epithelium a predominant localisation in the brushborder region of the cell has been demonstrated (Berg and Chapman, 1965). Gradient centrifugation of brain homogenates has shown the presence of the enzyme in external synaptic membranes of nerve (Hosie, 1965; Whittaker, 1966). The absence of Na–K ATPase from mitochondria has been demonstrated for brain (Tanaka and Abood, 1964a), muscle (Marcus and Manery, 1966), kidney (Whittam and Blond, 1964), and liver (Ulrich, 1963; Drabikowski and Rafalowska, 1968). It now appears that the enzyme is restricted to the plasma membrane and the smooth endoplasmatic reticulum, which is often considered to be an intracellular extension of the plasma membrane.

In concluding this section, mention may be made of the problems concerning a histochemical staining test for the Na–K ATPase system. The high substrate specificity of the enzyme makes the use of a chromogenic substrate impossible. The widely used ATPase staining method of Wachstein and Meisel (1957) uses ATP as substrate and incorporates 3 mM $Pb^{2+}$ in the substrate medium to precipitate inorganic phosphate released by the enzyme. The lead phosphate precipitate is visualised directly in electronmicroscopic observation or is first transformed with ammonium sulphide into the black, insoluble lead sulphide for light microscopic observation. Unfortunately, $Pb^{2+}$ and all other divalent cations ($Ca^{2+}$, $Co^{2+}$, $Zn^{2+}$, $Hg^{2+}$) are quite inhibitory to Na–K ATPase, more so than to Mg ATPase (Bonting, Caravaggio and Hawkins, 1962). In addition, the usual aldehyde fixatives greatly inhibit the former enzyme, again more than the latter (Gordon and Torack, 1967). Novikoff and coworkers (1961) could not observe any effect of the absence of $Na^+$ or presence of ouabain ($10^{-3}$ M) in cell membranes of kidney, brain and muscle. In order to avoid the complication of an appreciable Mg ATPase activity we made Wachstein–Meisel stains of minimally fixed sections of electric cell electroplax (95 % of total ATPase Na–K activated and ouabain-sensitive) with adjacent muscle in the presence and absence of ouabain ($10^{-4}$M, fully inhibitory *in vitro*). In both cases the electroplax showed no activity, while the muscle was strongly positive. However, McClurkin (1964) claimed to be able to demonstrate Na–K ATPase activity with this technique. Similar claims have been made by Ohkawa (1966), Szmigielski (1966), Wollenberger and Schulze (1966) and Laurent, Dunel and Barets (1968). None of these authors, however, has explained why they would have succeeded where others failed. Critical studies by Tormey (1966), Farquhar and Palade (1966), Tice and Engel (1966), Marchesi and Palade (1967), Bosch (1967) and Jacobson and Jørgensen (1969) came to negative conclusions. Evidence has been presented that the lead phosphate precipitation in membranes is largely due to non-enzymatic hydrolysis of ATP, catalysed by lead ion, and adsorption of the resulting lead complex on to certain membrane components (Rosenthal and associates, 1966; Moses and associates, 1966; Moses and Rosenthal, 1968). It would, therefore, seem wise to discount the claims made so far for a histochemical demonstration of Na–K ATPase activity by the Wachstein–Meisel method or modifications of it.

## III.  SINGLE CELL SYSTEMS

### A.  Erythrocytes

In erythrocytes, as in other single cells, the primary role of the active cation transport system appears to be the prevention of swelling due to the large number of charged molecules which cannot pass through the cell membrane (Jacobs, 1931; Wilbrandt, 1941; Hoffman, 1958; Tosteson and Hoffman, 1960). In Section IIC we have described the evidence for the close relationship between the Na–K ATPase system and active cation transport in erythrocytes, the first case where this was established after the discovery of the Na–K ATPase enzyme system in crab nerve by Skou. The use of reconstituted erythrocyte ghosts with a controlled intracellular environment (Gardos, 1954) permitted proof that $K^+$ ions stimulated both systems externally and not internally, while $Na^+$ ions stimulated internally and not externally (Glynn, 1962a; Whittam, 1962a, 1962b, 1962c; Whittam and Ager, 1964; Garrahan and Glynn, 1967d). It was also shown that ATP has to be present inside the ghost to serve as substrate and that ouabain inhibits only when it is present externally (Hoffman, 1962). Phosphate is released inside the cell and can only leave the cell slowly by passive diffusion (Schatzmann, 1964). From the effects of very low glycoside concentrations on K uptake and measurement of the amount of glycoside bound, Glynn (1957) estimated that there are not more than 1000 potassium influx sites per erythrocyte. Measurements of binding of tritiated digoxin indicate the presence of only 200 sites per cell (Ellory and Keynes, 1969).

The ionic interrelations have been investigated in great detail in recent years (Glynn, 1968). External $Na^+$ interferes with the activation of the enzyme by external $K^+$ (Whittam and Ager, 1962), while it reduces the antagonistic effect of high external $K^+$ on ouabain inhibition (Schatzmann, 1965). These facts suggest an internal enzymic phosphorylation site, activated by $Na^+$, and an external dephosphorylation site, activated by $K^+$ and inhibited by ouabain. In the absence of external $K^+$ the pump system catalyses a 1-for-1 exchange of Na ions across the cell membrane (Garrahan and Glynn, 1967a; Levin, Rector and Seldin, 1968). It could be shown that in this case ATP is synthesised at the expense of energy derived from cation concentration gradients and thus the sodium pump is reversed (Garrahan and Glynn, 1967e). This requires simultaneous downhill movements of $Na^+$ into and $K^+$ out of the ghosts, while the ATP synthesis is blocked by ouabain or by addition of external $K^+$ (Lant and Whittam, 1968). The effect of external $K^+$ indicates that ATP synthesis and hydrolysis by the Na–K ATPase system cannot proceed simultaneously. A curious phenomenon is that in the absence of both $Na^+$ and $K^+$ externally there is still an ouabain-sensitive efflux of $Na^+$ ions from erythrocyte ghosts (Garrahan and Glynn, 1967b) as well as from crab nerve (Baker, 1964). This could possibly be explained by local accumulation at the outside of the membrane of $K^+$ released from cation channels close to the pumping sites.

Per molecule of ATP hydrolysed by the Na–K ATPase system 3 $Na^+$ ions are extruded, while 2–2·5 $K^+$ ions are accumulated (Sen and Post, 1964; Gardos, 1964; Whittam and Ager, 1965; Garrahan and Glynn, 1967d). These values agree very well with those obtained for a large number of tissues (table 5). Under certain conditions a correlation between pump activity and lactic acid production has been noticed (Whittam, Ager and Wiley, 1964). In the absence of pump activity there is a certain basal lactic acid production which can be stimulated to an extent proportional to the Na–K ATPase activity, maximally by 75% (Whittam and Ager, 1965). In other words, the cation transport system acts as a pacemaker for the cellular energy metabolism. Such a pacemaker effect has been demonstrated in brain (Whittam, 1962d), kidney cortex (Whittam, 1961), frog skin (Zerahn, 1956; Leaf and Renshaw, 1957) and toad bladder (Leaf, Anderson and Page, 1958). Further studies with erythrocyte ghosts demonstrated that the pump activity was directly related to the internal ATP concentration (Whittam and Wiley, 1967) and that ATP alone is the substrate for the Na–K ATPase system, but that any reaction generating ATP (e.g. ADP + phosphoenolpyruvate) will run the pump (Hoffman, 1960; 1962; see however Schoner, Beusch and Kramer, 1968).

The point at which the pump system influences the energy metabolism appears to be the phosphoglycerate kinase catalysed conversion: 1,3-diphosphoglycerate + ADP $\rightleftharpoons$ 3-phosphoglycerate + ATP, where the ADP would be a product of the pump reaction (Parker and Hoffman, 1967). This conversion, stimulated by ADP formed by the activity of the pump system, activates in turn the whole glycolytic cycle, thus producing more ATP, which can be used for active cation transport.

Various correlations between pump activity and Na–K ATPase activity have been found in erythrocytes. Potassium-rich erythrocytes of certain strains of sheep (Tosteson, Moulton and Blaustein, 1960) and opossum (Baker and Simmonds, 1966) have much higher Na–K ATPase activity than potassium-poor erythrocytes of other strains of these animals. Comparison of different species has also shown a qualitative correlation between K : Na ratio and Na–K ATPase activity (Nakamaru, 1964; Chan, Calabrese and Theil, 1964; Greeff, Grobecker and Piechowski, 1964; Lucaroni and Millo, 1964; Debenedetti and Lucaroni, 1965; Brewer and associates, 1968). An example are the inversed K : Na ratios for human erythrocytes (7·2) and cat erythrocytes (0·056) derived from table 1, while the Na–K ATPase activity in human erythrocytes is 8 times as high as in cat erythrocytes (Bonting, Simon and Hawkins, 1961). Only Duggan, Baer and Noll (1965) came to different conclusions. It should be noted that in some of these reports absence of Na–K ATPase activity is concluded where more sensitive assay techniques are able to demonstrate significant, but low activity.

Storage of erythrocytes at 4°, in media maintaining intracellular ATP levels as well as in media not maintaining these levels, produced marked increases in $Na^+$ content and decreases in $K^+$ content. This was found to be due to the very high temperature coefficient of the Na–K ATPase system (Wood and Beutler, 1967) particularly below 20° (Gruener and Avi-Dor, 1967), and not to irreversible loss of Na–K ATPase activity (Grobecker and Piechowski, 1966). Irradiation by X-rays caused parallel inhibition of active cation transport and Na–K ATPase activity (Bresciani, Auricchio and Fiore, 1964). The high rate of loss of $K^+$ upon incubation at 37° of erythrocytes of newborn has been correlated with a low Na–K ATPase activity in these cells as compared to erythrocytes of adults (Whaun and Oski, 1969). Correlations between clinical abnormalities and erythrocyte Na–K ATPase activity have also been sought. Brewer and associates (1968) noted an elevated activity in acquired haemolytic anaemia, normal activity in hereditary spherocytosis and in erythrocytes with an inherited elevated ATP level. Normal enzyme activity in hereditary spherocytosis was also found by Micrevoca, Brabec and Palek (1967), but an increased Na–K ATPase activity was observed in 7 out of 10 patients with this disease by Wiley (1969). Seeman and O'Brien (1963) claimed to have found twice the normal Na–K ATPase activity in erythrocytes of schizophrenic patients, but Parker and Hoffman (1964) found neither a difference in enzyme activity nor in cation levels. Evidence for an ethacrynic acid-sensitive, ouabain-insensitive Na efflux pump in erythrocytes, representing about 30% of total Na efflux capacity, has been adduced by Hoffman and Kregenow (1966).

### B. Lens

Though the mammalian lens (figure 9) is made up of a composite of cells, a frontal layer of epithelial cells and a mass of lens fibre cells surrounded by a collagenous capsule, it behaves in cation movements like a giant single cell. Duncan (1968) has shown this quite conclusively for the toad lens by simultaneous potential measurements with three microelectrodes, of which two were placed at various places inside the lens.

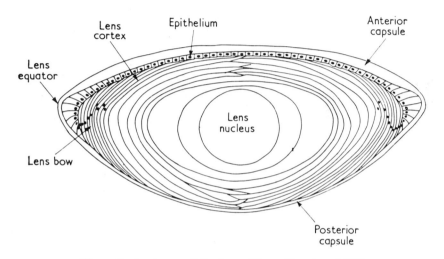

**Figure 9.** Anatomy of the lens. (From Bonting, 1965).

The lens, like most animal tissues, is characterised by a high potassium content and a low sodium content. Inhibition of lens metabolism *in vitro* by refrigeration, absence of glucose or presence of inhibitors leads to $Na^+$ influx and $K^+$ efflux, accompanied by swelling (Harris and Gehrsitz, 1951). The cation shift due to refrigeration can be reversed upon incubation at 37° in the presence of glucose (Harris, Gehrsitz and Nordquist, 1953; Harris, Hauschildt and Nordquist, 1954; Kinoshita, Kern and Merola, 1961). Inhibition of cation transport in the lens by ouabain was demonstrated by Kinoshita, Kern and Merola (1961) and by Becker (1962).

The presence of the Na–K ATPase system in cat, calf and rabbit lens was reported by Bonting, Caravaggio and Hawkins (1963). High activities were found in the epithelium, while the rest of the lens had no significant activity (table 7). The predominant localisation in the epithelium agrees with the finding that active uptake of [86]Rb and [42]K is ouabain-sensitive and occurs

**Table 7.** Distribution of total and Na–K ATPase activity in lens of cat, calf and rabbit

| Structure | Cat Total | Na–K | Calf Total | Na–K | Rabbit Total | Na–K |
|---|---|---|---|---|---|---|
| Anterior capsule | 0·2 | 0·0 | 1·0 | 0·1 | 0·2 | 0·1 |
| Epithelium + capsule | 57·2* | 41·5* | 136·0* | 90·0* | 207·0* | 149·0* |
| Cortex | 3·9* | 0·6 | 3·8 | 0·6 | 9·8* | 1·8 |
| Nucleus | 2·0 | 0·3 | 2·5 | 0·1 | 4·0 | 0·7 |
| Posterior capsule | 0·7 | 0·3 | 1·0 | 0·0 | 0·0 | 0·0 |

All values expressed in millimoles ATP per kg wet wt per hr hydrolysed at 37 °C.
*Significantly different from zero at the $P = 0.05$ level.
After Bonting, Caravaggio and Hawkins (1963).

mainly through the anterior face of the rabbit lens (Becker and Cotlier, 1962; Kinsey and Reddy, 1965). [42]K entry through the anterior face is much larger than through the posterior face, but in the presence of ouabain it decreases to the latter value (Kinsey and Reddy, 1965). These results indicate that the active cation transport is located in the epithelium and that movement across the capsule is by passive diffusion. This is not surprising if we bear in mind that the posterior surface of the lens is in contact with the more or less inert, gelatinous vitreous humour, while the aqueous humour constantly flows over the anterior surface, supplying materials and removing waste products.

The finding that $Rb^+$ may replace $K^+$ in the activation of the epithelial Na–K ATPase system (figure 5; Bonting, Caravaggio and Hawkins, 1963), agrees with the active transport of $Rb^+$ into the lens in competition with $K^+$ (Becker, 1962). There is also good agreement between the optimum pH for the epithelial Na–K ATPase system and the [86]Rb uptake in rabbit lens (figure 1), and between the $Q_{10}$ values for both activities which are 2·4 and 2, respectively (Bonting, 1965). Excellent agreement was also found for the $pI_{50}$ values of cation transport and Na–K ATPase system in calf and rabbit lens (Bonting, 1965). Calculation of the cation/ATP ratios for the lenses of both species gave values of 2·9 and 3·2 for calf and rabbit respectively (Bonting, 1965), in good agreement with those found in a large number of tissues (table 5).

The many common characteristics, both qualitative and quantitative, of the two systems summarized in table 8 form a strong argument for the primary and rate-limiting role of Na–K ATPase in the active cation transport of the lens. It is possible to assume that ATP supplies the energy for cation transport in view of the high ATP level in the epithelium (Frohman and Kinsey, 1952; Mandel and Klethi, 1958), the correlation between the high-energy phosphate level and the effectiveness of cation transport (Kinoshita,

**Table. 8.** Properties of cation transport and epithelial Ka–K ATPase systems in the lens

|  | Cation transport | Na–K ATPase |
|---|---|---|
| Location | Epithelium | Epithelium |
| Substrate | ATP (?) | ATP |
| Cation requirements | $Mg^{2+}$ (?), $Na^+$, $K^+$ | $Mg^{2+}$, $Na^+$, $K^+$ |
|  | $Rb^+$ competes with $K^+$ | $Rb^+$ can replace $K^+$ |
| pH optimum (rabbit) | 7·5 | 7·3 |
| Temperature coefficient $Q_{10}$ (rabbit) | 2 | 2·4 |
| Ouabain | Inhibitor | Inhibitor |
| $\quad$ pI$_{50}$, calf (5 mM $K^+$) | 7·0–7·4 | 7·1 |
| $\quad$ pI$_{50}$, rabbit (5 mM $K^+$) | 6·0 | 6·0 |
| Rates ($\mu$mole/lens/hr) |  |  |
| $\quad$ Calf | 2·6 ($K^+$), 3·1 ($Na^+$) | 0·78 (ATP) |
| $\quad$ Rabbit | 6·5 ($K^+$), 3·9 ($Na^+$) | 1·56 (ATP) |
|  | 3·9 ($Rb^+$) | 1·34 (ATP) |

After Bonting, Caravaggio and Hawkins (1963).

Kern and Merola, 1961) and the cation/ATP ratios of approximately 3 (Bonting, 1965). A pacemaker effect of cation transport in the lens appears from the decreased lactate production by intact cattle lens in the presence of ouabain (Kern, Roosa and Murray, 1962). The function of the active cation transport system in the lens could be threefold (Bonting, 1965): (i) prevention of osmotic swelling, which would result in opacity of the lens, (ii) a role in coupled transport of nutrients such as glucose and amino acids, (iii) requirement of a high intralenticular $K^+$ level for various enzymatic processes, such as protein synthesis. The importance of the first function has been stressed by Duncan (1968).

## C. Escherichia coli

In an attempt to establish the extent of the distribution of the Na–K ATPase cation transport system in the animal and plant kingdom it was important to investigate its possible occurrence in bacteria. Evidence for $K^+$ and $Na^+$ gradients across the membrane of *Escherichia coli* in the logarithmic growth phase was obtained by Schultz and Solomon (1961). In addition to a $K^+$–$Na^+$ exchange process, there is a faster and larger $K^+$–$H^+$ exchange (Schultz, Epstein and Solomon, 1962; 1963). Several investigators were unable to demonstrate ouabain-sensitive Na–K ATPase activity in bacteria (*E. coli*: Günther and Dorn, 1966; *Vibrio parahaemolyticus*: Hayashi and Uchida, 1965; *Staphylococcus aureus*: Gross and Coles, 1968; *Pseudomonas* and *Cytophaga*: Drapeau and MacLeod, 1963; *Mycoplasma*: Rottem and Razin, 1966). However, in these cases either the assay conditions were not quite suitable or the sensitivity of the assay was too low for the small activities to be detected.

Careful assay of the *E. coli* $Mg^{2+}$ activated ATPase activity indicated that about 5% of the total activity was Na–K activated and ouabain-sensitive (Hafkenscheid and Bonting, 1968). Urea treatment of the freeze-dried bacteria decreased Mg ATPase activity by more than 50%, thus permitting earlier determinations of the properties of the Na–K ATPase activity. Preparation of membrane fragments by removal of the cell wall through ultrasonic treatment, lysozyme–EDTA treatment or penicillin treatment led to disappearance of Na–K ATPase activity. Hence in all experiments freeze-dried bacteria, reconstituted with water, were used. An optimal Mg/ATP ratio of 1 : 1 and an indistinct pH optimum of 7·7 were found. The half-maximal activation concentration for $K^+$ was 4·3 mM, considerably higher than for vertebrate tissues (table 3), but understandable in view of the dominating $K^+$–$H^+$ transport. $Rb^+$ and $NH_4^+$ can replace $K^+$ ions for activation. Comparison of the $K^+$ flux data of Epstein and Schultz (1965) with our Na–K ATPase activity values gave a K/ATP ratio of 1·6–3·2, which agrees with the values for many vertebrate systems (table 5).

The relatively high Mg ATPase activity in *E. coli*, which is strongly activated by a single cation and has a pH optimum of about 9, necessitated a further study in which the presence of the Na–K ATPase system was reinvestigated (Hafkenscheid and Bonting, 1969). While the Na–K ATPase system had a high substrate specificity for ATP and ADP, the Mg ATPase

**Table 9.** Relative ATPase activities of *Escherichia coli* in various substrate media

| Code | Substrate medium $Mg^{2+}$ (2 mM) | $Na^+$ (60 mM) | $K^+$ (5 mM) | Ouabain (0·1 mM) | ATPase activity (%) |
|------|------|------|------|------|------|
| A | + | + | + | − | 100     (21) |
| B | + | + | − | − | 91·5 ± 0·9 (21) |
| C | + | − | + | − | 92·6 ± 0·8 ( 8) |
| D | + | + | + | + | 93·1 ± 1·0 (21) |
| E | + | + | − | + | 91·9 ± 0·5 (21) |
| F | + | − | + | + | 92·0 ± 0·5 ( 8) |
| G | + | − | − | − | 83·8 ± 0·7 (21) |
| H | + | − | − | + | 83·8 ± 0·7 (21) |

Relative activities expressed with standard error and, in parentheses, the number of determinations. Differences between medium A and each of the other media are significant ($P < 0·001$).
Differences between medium G and H and each of the other media are significant ($P < 0·001$).
Differences between media B and E, C and F, G and H are not significant ($P > 0·1$).
Differences between media B, C, D, E and F are not significant ($P > 0·1$).
After Hafkenscheid and Bonting (1969).

system readily hydrolysed all nucleotides except adenosinemonophosphate. There is also a different localisation of the two systems. Osmotic shock treatment followed by assay of bacterial sediment and supernatant showed most Na–K ATPase activity in the sediment (Hafkenscheid, 1968) and most Mg ATPase activity in the supernatant (Hafkenscheid and Bonting, 1969). Careful comparison of the ATPase activities in a variety of substrate media confirmed the presence of ouabain-sensitive Na–K ATPase activity in *E. coli* unequivocally (table 9).

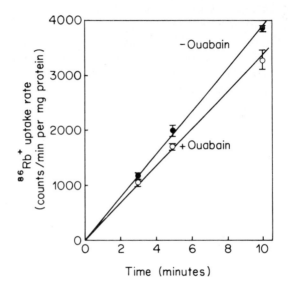

**Figure 10.** Inhibitory effect of ouabain on $^{86}$Rb uptake by *Escherichia coli*. (From Hafkenscheid and Bonting, to be published).

Correlation of this Na–K ATPase activity with ouabain-sensitive cation transport has also been possible (Hafkenscheid and Bonting, 1970). A significant increase in Na content and an equal decrease in K content was detected upon incubation of *E. coli* for $1\frac{1}{2}$ hr in the presence of $10^{-4}$ M ouabain. Active $^{86}$Rb$^+$ uptake in *E. coli* is competitively inhibited by K$^+$ and activates the Na–K ATPase system like K$^+$; it can therefore be used to study cation transport in this organism. No effect of ouabain ($0.8 \times 10^{-4}$ M) on $^{86}$Rb uptake at pH 7 was noticed. Bearing in mind that at acid pH the K$^+$–H$^+$ exchange is diminished, the experiment was repeated with $8 \times 10^{-5}$ M ouabain at pH 5.3. In this case a significant inhibition of about 10% was established (figure 10). Thus the functioning of a Na–K ATPase cation transport system in *E. coli*, though perhaps of minor importance for the total cation transport process, has been established. This is the first instance where this system has been demonstrated in bacteria.

## D. Various Cell Types

The presence and operation of a Na–K ATPase system in human leukocytes was demonstrated by Block and Bonting (1964). A reversible cation shift upon incubation at 2° C, an equivalent gain of Na and loss of K upon incubation in $2 \times 10^{-4}$ M ouabain and a K/ATP ratio of 2·0 (table 5) were observed. There was no effect of ouabain on the $H^+$ ion gradient across the leukocyte membrane. The pH optimum for the $Na^+–K^+$ exchange was 7·35, which is in good agreement with the optimum pH of the Na–K ATPase system in mammalian tissues (table 3). The Na–K ATPase activity in leukaemic leukocytes was higher than in normal leukocytes and increased parallel to the morphologic immaturity of the leukaemic leukocytes. Highest activity was found in acute and chronic lymphatic leukaemia. At the same time the Na content of the leukocytes was increased in the latter conditions. This suggests that the increased Na–K ATPase activity might be an adaptive response to an increased Na permeability. The pacemaker effect of the Na–K ATPase system could then explain, in view of the hexokinase deficiency of leukaemic leukocytes, the increased respiratory rate of these leukocytes.

The presence of the Na–K ATPase system has also been demonstrated in rabbit and cat marrow cells (Archdeacon, Rohrs and Marta, 1964), rabbit reticulocytes (Yunis and Arimura, 1966), boar epididymal spermatozoa (Uesugi and Yamazoe, 1966) rat epididymal adipose tissue (Modolell and Moore, 1967), the human cell strain HeLa (Mohri and colleagues, 1968) and Ehrlich ascites tumour cells (Bonting, Caravaggio and Hawkins, 1962; Wallach and Ullrey, 1964; Schoen and Menke, 1967). In reticulocytes there was a drop in Na–K ATPase activity upon maturation to erythrocytes (Yunis and Arimura, 1966) similar to that observed in leukocytes and concomitant with a decrease in active uptake of amino acids. There is also evidence for coupling of iron uptake by reticulocytes to the Na–K ATPase cation transport system (Wise and Archdeacon, 1965). In ghosts isolated from rat epididymal adipose cells a $Na^+$-activated, ouabain-inhibited active $K^+$ uptake has been demonstrated (Clausen, Rodbell and Dunand, 1969).

## IV.  EXCITATORY TISSUES

Under this heading we shall discuss those tissues which upon stimulation (electrical, chemical, or light stimulus) show a passive influx of sodium and a passive efflux of potassium: nerve, brain, muscle, retina and electric organ.

Common to all these systems is a cation pump mechanism serving to maintain the cation gradients required for the cation fluxes occurring during stimulation. The cochlea is included in this section, because it is a system, which is excited by sound stimuli.

## A. Nerve

The role of passive cation movements in nerve conduction has been elucidated very fully in the squid axon by Hodgkin and his colleagues (Hodgkin, 1964). Potential measurements with microelectrodes, measurement of currents flowing across the membrane in voltage-clamped axons, and tracer studies, were used in these investigations. In the resting state there is a membrane potential (inside of axon negative), largely determined by the $K^+$ gradient across the membrane. Upon stimulation there is a rapid, transient increase in $Na^+$ permeability, causing a reversal of the membrane potential (action potential), followed by reduction of the $Na^+$ permeability and a slow, maintained increase in $K^+$ efflux. When the resting membrane potential is re-established, the $K^+$ efflux decreases to its resting value. The existence of separate channels for $Na^+$ and $K^+$ ions is demonstrated by the action of tetrodotoxin, which blocks the increase in $Na^+$ ion permeability but not the delayed increase in $K^+$ ion permeability (Moore, Narahashi and Shaw, 1967), and is confirmed by various other observations (Baker, 1968). The number of $Na^+$ channels is estimated to be 10–13 per square micron of axon surface (Moore, Narahashi and Shaw, 1967).

For maintenance and restoration of the cation gradients across the nerve membrane a cation pump is required which actively extrudes $Na^+$ and moves $K^+$ inwardly. Inhibition of $Na^+$ efflux from the squid axon by ouabain was reported by Caldwell and Keynes (1959). Requirement of ATP for the $Na^+$ efflux was demonstrated by micro injection studies (Caldwell and co-workers, 1960; Mullins and Brinley, 1967; de Weer, 1968). ATP injection did not restore Na extrusion blocked by ouabain (de la Cruz, 1965), indicating that ouabain blocks the pump and not the substrate supply to the pump (Martin and Shaw, 1966). $Na^+$ efflux, sensitive to ouabain and requiring external $K^+$, was also demonstrated in the perfused squid axon freed of axoplasm (Baker and associates, 1968). $Rb^+$ influx in the squid axon was inhibited by ouabain ($10^{-7}$ M), and there was little discrimination between $Rb^+$ and $K^+$ ions (Celedon, 1967). After the original discovery of Na–K ATPase in crab nerve by Skou (1957, 1960), the enzyme system was demonstrated in the squid axon sheath by Bonting and Caravaggio (1962) and a K/ATP ratio of 3·0 was determined (Bonting and Caravaggio, 1963) in good agreement with the Na/ATP ratios of 2·8–3·0 derived from ATP injection studies (Baker and Shaw, 1965; Mullins and Brinley, 1967) and with the values for several other tissues (table 5). The Na–K ATPase system has also been demonstrated in the axon sheath of other species of squid (Rojas, 1965; Canessa–Fischer, Zambrano and Moreno, 1967). Electronmicroscopic observations of preparations treated by the Wachstein–Meisel technique showed highest ATPase activity, sensitive to ouabain, in the axolemma, located between axoplasm and Schwann cell (Sabatini, Dipolo and Villegas, 1968), although these results should be treated with caution in view of what is said about this technique in Section IIC. A correlation between the activation of Na extrusion from squid axon and of Na–K ATPase activity in crab nerve ($K^+ > Rb^+ > Cs^+$) has been noted by Sjodin and Beaugé (1968). These authors also found that Na and K transport are coupled, but in a ratio varying from 1 : 1 at low internal Na concentration to 2 : 1 at high internal Na concentration.

An extensive study of the crab nerve cation transport system has been carried out by Baker (1963, 1964, 1965), who could demonstrate an ouabain-sensitive enzymatic hydrolysis of energy-rich phosphate, activated by internal $Na^+$ and external $K^+$, and which is responsible for active $Na^+$ efflux and $K^+$ influx. In the absence of external ions there was an ouabain-sensitive loss of $Na^+$ and $K^+$, balanced by a net efflux of glutamate and aspartate (Baker, 1964). External $K^+$ could be replaced by $Th^+$, $Rb^+$, $Cs^+$ and $NH_4^+$ ions, and the Na/ATP ratio was 2·7–4 (Baker, 1965). Externally applied ATP could not serve as substrate for the pump. Oxidative phosphorylation supplies most of the ATP utilised by the pump, and the $K_m$ for external $K^+$ is 1 mM (Baker and Connelly, 1966), in good agreement with the $K_m$ values for $K^+$ activation of the Na–K ATPase system shown in table 3.

Information on other types of nerves is less complete, but what is known supports the role of the Na–K ATPase system demonstrated for the squid axon and crab nerve. Stimulation of rabbit demyelinated vagus nerve fibres increased the resting $K^+$ efflux by 1 % per impulse, and neither of these passive effluxes was affected by ouabain (Keynes and Ritchie, 1965). An ouabain-sensitive, electrogenic Na pump has been demonstrated in snail nerve cells (Kerkut and Thomas, 1965). Isolation of crustacean sensory neurons led to loss of $K^+$ and gain of $Na^+$, which upon incubation in a suitable physiological salt solution was reversed by an ouabain-sensitive cation pump, while tetrodotoxin had no effect on the cation levels of the resting neuron (Giacobini, Hovmark and Kometiani, 1967). In crayfish nerve cord an ouabain-sensitive ATPase activity has been demonstrated (Bowler and Duncan, 1967a). A nerve growth factor, isolated from mouse submaxillary glands, increased the Na–K ATPase activity of the superior cervical ganglion in the newborn mouse to the adult level after 5 daily subcutaneous injections (Grasso, Zanini and Sabatini, 1968). In degenerating cat sciatic nerve Na–K ATPase activity decreased sharply between 2 to 8 days after injury and increased to above control levels at 16–32 days after injury. The decreased activity in the centrifugal fractions could be partially restored by the addition of a phospholipid mixture during assay (Bachelard and Silva, 1966). In cockroach nerve an active, ouabain-sensitive $Na^+$ efflux, stimulated by external $K^+$ was demonstrated (Treherne, 1961, 1966), while the presence of ouabain-sensitive Na–K ATPase activity was also established (Grasso, 1967).

## B. Brain

The presence of Na–K stimulation of Mg activated ATPase in brain was first noted, without being correlated with active cation transport, by Hess and Pope (1957). Thereafter, it was described in rat brain microsomes (Järnefelt, 1960, 1961a), in cat grey and white matter, optic and sciatic nerve and superior cervical ganglion (Bonting, Simon and Hawkins, 1961), in cattle brain (Deul and McIlwain, 1961), in guinea pig and rabbit brain (Bonting, Caravaggio and Hawkins, 1962), and in human brain (Samaha, 1967). The appearance of Na–K ATPase activity in brain during development has also been studied. While Samson, Dick and Balfour (1964) could not detect activity in rat brain until 10 days after birth, Cote (1964) was able to demonstrate its presence in the 16-days old rat embryo and a rapid increase during the rest of embryonic life. Bignami, Palladini and Venturini (1966) found significant activity in the chick embryo brain at the 17th day of incubation, increasing parallel with the electrophysiological activity to maximal activity soon after hatching. Zaheer, Iqbal and Talwar (1968) point to the rapid rise in enzyme activity in chicken embryo brain between 10 and 12 days of incubation, immediately preceding the appearance of electrical activity in the 12-day old embryo. Microdissection studies indicated a high Na–K ATPase activity in membranes of neurons and low activities in glia and capillaries (Cummins and Hyden, 1962). The low activity in glial cells was confirmed by analysis of an astrocytoma (Hess and colleagues, 1963). The distribution of Na–K ATPase activity over various regions of the rabbit brain (Harmony, Urba-Holmgren and Urbay, 1967) and the monkey brain (Fahn and Cote, 1968) has been determined. Centrifugal fractionation studies have shown a high Na–K ATPase activity in nerve endings (Hosie, 1965; Whittaker, 1966; Albers, de Lores Arnaiz and de Robertis, 1965; Abdel-Latif, Brody and Ramahi, 1967). This activity represents a very substantial part of the activity in whole brain (Kurokawa, Sakamoto and Kato, 1965).

Correlation between cardiac glycoside sensitive cation transport and Na–K ATPase activity in brain was first reported by Bonting, Caravaggio and Hawkins (1962). Electrical stimulation of cerebral cortex slices *in vitro* for 10 min at a frequency of 100 per sec caused a $Na^+$ gain and a $K^+$ drop with reversal in about 10 min upon cessation of stimulation (Keesey, Wallgren and Mc-Ilwain, 1965). Reversal of the cation shift depended upon the presence of metabolic substrates and was prevented by the metabolic inhibitors dinitrophenol and fluoride (Joanny and Hillman, 1963). Addition of tetrodotoxin ($1$–$5 \times 10^{-6}$ M) during stimulation prevented this cation shift, while addition after stimulation did not prevent the reversal of the cation shift (Swanson, 1968). Preincubation with ouabain ($5 \times 10^{-6}$ M) for 30 min caused a cation shift, which was not further increased by stimulation and which would not reverse in the presence of ouabain. These experiments indicate that in brain, as in nerve, stimulation causes cation shift by increasing cation permeability, and that the Na–K ATPase system serves to maintain and reestablish the cation gradients. An ouabain-sensitive, active $Na^{+}$–$K^{+}$ exchange in isolated nerve endings with intact synaptosomal membranes, antagonised by high $K^{+}$ concentrations in the medium, has been reported (Ling and Abdel-Latif, 1968; Escueta and Appel, 1969). Thus it appears that in brain a substantial part of the Na–K ATPase activity functions in recovering cation gradients at neuronal synapses after synaptic impulse conduction.

Understandably, a variety of drugs with central nervous system activity has been tested for effects on cerebral Na–K ATPase activity. Inhibition of brain microsomal Na–K ATPase activity by ethanol might play an important, though not primary role in alcohol intoxication (Israel, Kalant and Laufer, 1965; Israel, Kalant and Le Blanc, 1966). For the general anaesthetics acetone, chloroform, ethanol, ethyl ether, *n*-octanol, and *n*-pentanol good agreement was found between narcotic concentrations and concentrations required for Na–K ATPase inhibition (Israel and Salazar, 1967). However, the general anaesthetics diethyl ether and halothane, bubbled through rabbit brain microsomal preparations, had inhibitory effects only at partial pressures clearly exceeding the clinical range (Ueda and Mietani, 1967). The antidepressant imipramine in $0.2$ mM concentration inhibits Na–K ATPase of guinea pig brain microsomes by $50\%$ (Tarve and Brechtlova, 1967). Possibly this effect would explain the inhibitory action of this compound on catecholamine uptake by neurons. No enzyme effects after *in vivo* administration of the anti-epileptic diphenylhydantoin, metrazole, pentobarbital and pyrithiamine and no *in vitro* effects of prochlorperazine, lysergic acid diethylamide, tetrodotoxin and pyrithiamine were noticeable (Pincus and Giarman, 1967). Diphenylhydantoin nearly doubled activity in isolated rat cortical nerve endings in the presence of only $0.2$ mM $K^{+}$, but inhibited activity at $10$ mM $K^{+}$ concentration (Festoff and Appel, 1968). Metrazole in dosages inducing convulsions in rats raised cerebral Na–K ATPase activity (De Robertis, Alberici and Rodriguez de Lores Arnaiz, 1969). An *in vitro* inhibitory effect was demonstrated for chlorpromazine and chlorprothixene at $0.25$ mM concentration on guinea pig brain microscomes (Okada, 1968), and for various phenothiazine derivatives in 0·5 mM concentration on rat brain microsomes (Davis and Brody, 1966). Subsequent work indicates that a semiquinone free radical of chlorpromazine, rather than chlorpromazine itself, may be responsible for inhibition of the enzyme (Akera and Brody, 1968). Both $d$-tubocurarine and benadryl (2·5 mM) inhibited bovine brain microsomal Na–K ATPase activity, while eserine (2·5 mM) activated the enzyme and antagonised the tubocurarine effect (Rossini and Valeri, 1966). No effects were found for atropine, alderlin, benzodioxan, ephedrine, $l$-methyllysergic acid butanolamide, caffeine, cocaine and procaine at 2·5 mM concentration. Convulsions caused by $Cu^{2+}$ in pigeons appear to be due to inhibition of cerebral Na–K ATPase activity (Peters, Shorthouse and Walshe, 1965). Inhibition of brain Na–K ATPase by $Be^{2+}$ (0·6 mM) has also been reported (Thomas and Aldridge, 1966). Strychnine, both *in vivo* and *in vitro*, increased the activity of brain microsomal Na–K ATPase without affecting Mg ATPase (Pal and Ghosh, 1968). This effect might be related to the convulsant effect of this drug. An adaptive increase in Na–K ATPase activity after induction of cortical lesions in rats by freezing was suggested by Lewin and McCrimmon (1967, 1968). No change in enzyme activity was noted in the rat spinal cord after exposure and injury of the cord (Fried, 1965).

## C. Muscle

Skeletal and cardiac muscle from various species resemble nerve in having a high K/Na ratio and a large membrane potential (70–110 mV) with negative cell interior. An active $Na^+$ efflux, dependent on the presence of extracellular $K^+$ and requiring 6–16% of total cellular energy metabolism was demonstrated (Keynes, 1954; Keynes and Maisel, 1954). Anoxia causes a net loss of $K^+$ and a slightly larger gain of $Na^+$ (Creese, 1954). All of the potassium and sodium appears to be exchangeable (Creese, 1954, Keynes, 1954). Stimulation causes, as in nerve, a rapid transient increase of $Na^+$-permeability with membrane depolarisation and reversal of potential (action potential), followed by contraction (Hodgkin, 1951; Hodgkin and Horowicz, 1959). Stretching of the muscle causes an increase in $Na^+$ efflux with a fall in internal sodium level and increased oxygen consumption (Harris, 1954). The cardiac glycoside strophanthin causes inhibition of active Na efflux and loss of $K^+$ in resting frog sartorius muscle (Witt and Schatzmann, 1954; Johnson, 1956; Edwards and Harris, 1957) and also in giant muscle fibres of the barnacle (Beaugé and Sjodin, 1967).

Na–K ATPase was first demonstrated in muscle by Bonting, Simon and Hawkins (1961). Correlation between cardiac glycoside-sensitive cation transport and occurrence of Na–K ATPase activity was established in cardiac muscle of frog, dog, rabbit, guinea pig and cat, in skeletal muscle of frog and in uterus muscle of rabbit (Bonting, Caravaggio and Hawkins, 1962). In frog toe muscle (Bonting and Caravaggio, 1963) and frog sartorius muscle (Corrie and Bonting, 1966) cation/ATP ratios of 1·9 and 2·2 were found (table 10), in fair agreement with the Na/ATP ratio of 2·5–3 calculated by

**Table 10.** Relative and absolute ATPase activity of frog muscle in various substrate media

| Medium* | Toe (%) | Sartorius (%) |
|---|---|---|
| A, complete | 100 | 100 |
| B, — $K^+$ | 88·7 | 74 |
| C, — $Na^+$ | 85·3 | 58 |
| D, + ouabain ($10^{-4}$ M) | 85·1 | 72 |
| E, — $K^+$, + ouabain | 80·3 | 81 |
| | | |
| Na–K ATPase activity | | |
| in % of total ATPase | 14·9 | 24·3† |
| in mmol/kg wet wt/hr | 67 | 29 |
| cation/ATP ratio | 1·9 | 2·2 |

*For composition of media see table 2.
†Omitting the result for the $Na^+$-free medium, which deviates because of Na-sensitivity of Mg ATPase activity.
After Bonting and Caravaggio (1963) and Corrie and Bonting (1966).

Harris (1967) from direct determinations of ATP consumption during ion movement in frog sartorius muscle, as well as with the values for a variety of tissues in table 5. The decrease in membrane potential of frog sartorius muscle after a 4 hr treatment with $10^{-4}$ M ouabain or erythrophleine was found to be equal to that calculated from the measured decrease in $K^+$ concentration

**Table 11.** Effects of ouabain and erythrophleine on membrane potential and chemical composition of frog sartorius muscle

| Agent | No. of expts. | $\Delta E_{obs}$ (mV) | $\Delta E_{calc.}$ (mV) | P | $\dfrac{\Delta Na^+ - \Delta K^+}{\Delta H_2O}$ (M) | $\dfrac{\Delta Cl^-}{\Delta H_2O}$ (M) | P |
|---|---|---|---|---|---|---|---|
| ouabain, $10^{-4}$ M, 4 hr | 13 | 19 ± 1·6 | 22 ± 1·8 | 0·28 | 0·12 ± 0·016 | 0·15 ± 0·018* | 0·28 |
| erythrophleine, $10^{-4}$ M, 4 hr | 4 | 21 ± 3·5 | 12 ± 3·7 | 0·42 | 0·16 ± 0·043 | | |

P is bilateral probability, calculated from t–test.
*6 experiments.
After Corrie and Bonting (1966).

(table 11; Corrie and Bonting, 1966). There was also a gain in $Na^+$, slightly larger than the loss of $K^+$, of $H_2O$ and $Cl^-$. Calculation showed that the cation and anion gains were equivalent and amounted to an isotonic uptake of NaCl (table 11). In frog sartorius muscle $10^{-5}$ M ouabain has a small inhibitory effect on $K^+$ influx and $Na^+$ efflux at low (5 mM) internal $Na^+$ concentration and a large effect at high (15–20 mM) $Na^+$ level (Sjodin and Beaugé, 1968). These observations fit very well with the $Na^+$ activation curve of the Na–K ATPase system (figure 3). Absence of external $K^+$ reversibly inhibits the active $Na^+$ extrusion by frog sartorius muscle (Noda, 1968), which agrees with the requirement of extracellular $K^+$ for activity of the Na–K ATPase system. These findings demonstrate that also in muscle the Na–K ATPase system plays a major role in maintaining the cation gradients, necessary for muscle function.

From the many reports on the Na–K ATPase system in muscle a few may be mentioned here. Isolation and properties of the cardiac muscle Na–K ATPase system have been described by Matsui and Schwartz (1966a) and by Portius and Repke (1967a, b). Digoxin binding in cardiac muscle required ATP and $Mg^{2+}$, was stimulated by $Na^+$ and depressed by $K^+$ (Matsui and Schwartz, 1968). Active cardiac glycosides could displace bound digoxin, while inactive ones could not. Binding of digoxin was also observed with acetyl phosphate instead of ATP; in this case $Na^+$ did not stimulate, while $K^+$ still depressed the binding. These results suggest that the cardiac glycoside binds to the phosphorylated form of the enzyme thereby inhibiting its hydrolysis. The therapeutic effect of low doses of cardiac glycosides on heart muscle and the toxic effects of high doses are both ascribed to their inhibitory effect on the Na–K ATPase system, whereby low doses would lead to intracellular $Ca^{2+}$ release through $Na^+$ influx (Repke and Portius, 1963; Repke, 1964). Further evidence for the role of Na–K ATPase in the therapeutic effect of inotropic agents follows from the observation that inotropically active azasteroids inhibit cardiac muscle Na–K ATPase, while inactive azosteroids do not (Brown, 1966). Ouabain ($10^{-3}$ M) inhibited rat uterus Na–K ATPase, decreased the $K^+$ concentration and caused marked contraction of the muscle (Allen and Daniel, 1964). Surprising is the stability of the Na–K ATPase activity in the sarcoplasmic reticulum of patients with progressive muscular dystrophy, while other ATPases and dehydrogenases are greatly decreased (Sugita and associates, 1966). In embryonic chick heart the appearance of Na–K ATPase activity was found to coincide with the reversal of the Na/K ratio (Klein, 1961, 1963). The activity of the enzyme system in toad heart was higher in summer than in winter (Kennedy and Naylor, 1965). The interesting suggestion was made by Bowler and Duncan (1967b) that the Mg ATPase of crayfish muscle could control passive cation permeability into these cells, since the activity *vs.* temperature curve for the enzyme is

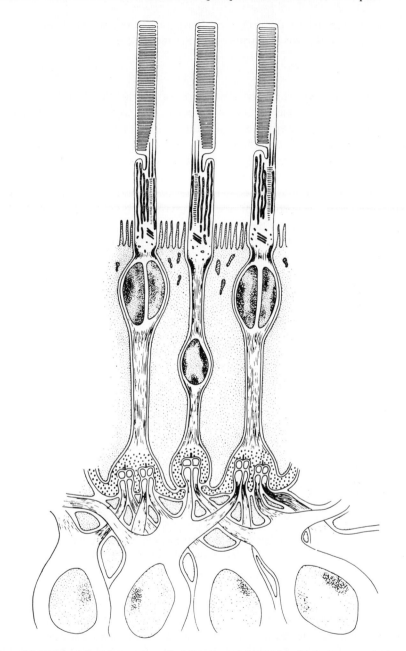

**Figure 11.** Schematic representation of the photoreceptor cell. From top to bottom: outer segment with rod sacs, inner segment with mitochondria, nucleus and synapse with bipolar cell. (From Sjöstrand, 1961).

remarkably parallel to the $LD_{50}$ *vs.* temperature curve and heat death is accompanied by $Na^+$ disappearance and $K^+$ increase in haemolymph.

## D. Retina

The mechanism by which the absorption of a light quantum by the visual pigment leads to nervous excitation appears to involve a cation exchange process in the photoreceptor cell, similar to that in nerve (Bonting, 1969). The rod photoreceptor cell (figure 11) consists of the outer segment with numerous rod sacs, the inner segment with mitochondria and nucleus, and the synapse connecting the photoreceptor with the bipolar cell. The visual pigment rhodopsin is located in the membranes of the rod sacs, of which it forms a substantial part. Absorption of one light quantum by a molecule of rhodopsin is sufficient to stimulate the synapse, thereby generating a graded receptor potential. Indications that the generation of this potential involves a passive influx of $Na^+$ ions have been obtained from its abolition in $Na^+$-free media (Hamasaki, 1963, 1964; Hanawa, Kuge and Matsumara, 1967), with ouabain (Frank and Goldsmith, 1965, 1967; Langham, Ryan and Kostelnik, 1967) and with tetrodotoxin (Benolken and Russell, 1967).

A high Na–K ATPase activity was found in whole cat retina (Bonting, Simon and Hawkins, 1961). Centrifugal fractionation studies demonstrated that a substantial part of the retinal Na–K ATPase activity in cattle, rabbit and frog (table 12) is located in the rod outer segments (Bonting, Caravaggio

**Table 12.** Distribution of Mg–ATPase and Na–K ATPase activities in retinal fractions

|                     | Rhodopsin | Cattle | | Rabbit | | Frog | |
|---------------------|-----------|--------|------|--------|------|------|------|
|                     |           | Mg     | Na–K | Mg     | Na–K | Mg   | Na–K |
| Whole retina        | 9·3       | 1·5    | 4·6  | 1·7    | 2·9  | 1·7  | 3·2  |
| Nuclear fraction    | 4·5       | 0·7    | 1·9  | 0·5    | 1·1  | 0·5  | 0·6  |
| Rod fraction        | 38·2      | 2·4    | 6·2  | 2·1    | 3·7  | 1·3  | 2·9  |
| Mitoch.-micros. frn.| 6·7       | 2·6    | 5·8  | 1·7    | 7·4  | 2·6  | 3·9  |

All ATPase activities expressed in moles ATP hydrolysed per hr per kg dry wt. of fraction. Rhodopsin expressed in extinction drop $\Delta E_{500}$, divided by dry wt. per $\mu l$ extract.
After Bonting, Caravaggio and Canady (1964).

and Canady, 1964). Calculations for the cattle rod have shown that the pump capacity is sufficient to restore the cation gradient up to a quantum incidence causing rod saturation (Bonting and Bangham, 1967), and that the necessary amount of ATP, which is not generated in the outer segments, can be supplied by diffusion from the inner segment (Bonting, 1969).

Studies with isolated rod outer segments have shown that illumination indeed causes a cation exchange. A cattle rod outer segment suspension in sucrose with low cation content gave, upon illumination, a reduced loss of $Na^+$ and an equivalent net loss of $K^+$ (Bonting and Bangham, 1967), and a frog rod outer segment suspension behaved similarly (Daemen, Lion and Bonting, unpublished results). Table 13 summarises these data. Retinalde-

**Table 13.** Net cation movements in rod outer segments upon illumination

| Species | Lux. sec. | K | Na |
|---------|-----------|-----|-----|
| Cattle | $5 \times 10^6$ | —2·4 | +2·9 |
|        | $8 \times 10^5$ | —2·5 | +3·0 |
|        | $2 \times 10^3$ | —1·0 | +1·0 |
| Frog   | $4 \times 10^3$ | —2·8 | +1·7 |

All values in meq/kg dry wt.
Cattle results: Bonting and Bangham (1967)
Frog results: Daemen, Lion and Bonting, unpublished results.

hyde, added to the suspension in $10^{-6}$–$10^{-4}$ M concentration, had the same effect as light, while retinol and retinoic acid had no effect (Daemen and Bonting, 1968).

Since these experiments had to be done in media with unphysiologically low cation concentrations, they have been repeated with tracers (Duncan, Daemen and Bonting, 1970). Full confirmation was obtained for frog rod outer segments suspended in a Ringer solution, fortified with glucose, ATP and $Mg^{2+}$. Light and retinaldehyde caused a $^{42}K$ efflux (figure 12a), and a $^{22}Na$ influx (figure 12b). Ouabain increased $^{22}Na$ influx in the dark and abolished eventually the light effect (figure 12c). Upon centrifugation of these rods in a choline–Ringer solution (3 mM $Na^+$) for 8–10 min the K/Na ratio increased from 1 to 3 owing to loss of $Na^+$, indicating that the intrasaccular space in fresh, dark adapted rods is high in $K^+$ and low in $Na^+$, while the extrasaccular space is low in $K^+$ and high in $Na^+$. The retinaldehyde effect could be explained from the results of model studies with phospholipid micelles, which indicate that binding of retinaldehyde to a membrane amino group causes loss of a positive charge on the membrane (De Pont, Daemen and Bonting, 1968; Daemen and Bonting, 1969). This charge effect leads to an increased cation permeability of the micelles (Bonting and Bangham, 1967). There is now evidence that *in vivo* illumination causes the retinaldehyde group of the visual pigment to transfer from its original bond in the pigment

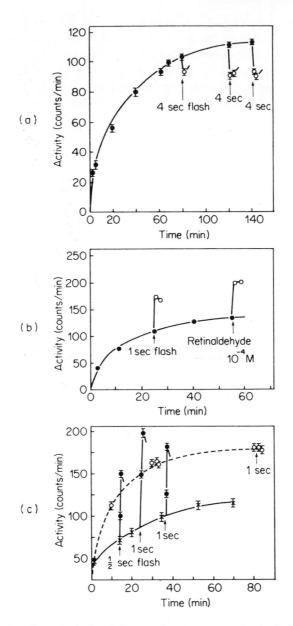

**Figure 12.** Ionic effects in isolated frog rod outer segments studied with tracers. a. $^{42}$K-efflux by light and retinaldehyde. b. $^{22}$Na-influx by light and retinaldehyde. c. Increased $^{22}$Na-influx and abolition of light effect by ouabain. ×contol uptake in dark; ● light stimulated samples; ○ rods incubated in 10$^{-4}$ ouabain. (From Duncan, Daemen and Bonting, to be published).

molecule to an amino group in the rod sac membrane. This could result, in a similar fashion as in the phospholipid micelles, in cation exchange and stimulation of the synapse with the bipolar cell (Bonting, 1969).

### E. Cochlea

The cochlea is the organ which is responsible for the transduction of acoustical stimuli into nervous excitation. Active cation transport plays an essential role in this process. The mammalian cochlea consists of a spirally wound three-fold tube. The middle tube is the scala media, filled with endolymph, an extracellular fluid with a cation composition (12–16 mM $Na^+$, 140–150 mM $K^+$) resembling an intracellular medium (Smith, Lowry and Wu, 1954; Rauch and Köstlin, 1958; Johnstone, Schmidt and Johnstone, 1963; Fernández, 1967). The scala media is surrounded on two sides with spaces filled with perilymph, the scala vestibuli and the scala tympani (figure 13). The cation composition of perilymph (130–150 mM $Na^+$, 4–5 mM $K^+$) agrees with that of extracellular fluids. The separation between scala media and scala tympani is formed by the thick fibrous membrana basilaris upon which rests, on the endolymphatic side, the organ of Corti. The scala vestibuli and scala media are separated by Reissner's membrane, consisting of two cell layers with microvilli on the endolymphatic side. The third side of the scala media is formed by an epithelial structure, called stria vascularis, which rests on a layer of connective tissue, the spiral ligament. The cells of the stria vascularis are characterised by a large number of mitochondria, a high activity of enzymes participating in oxidative metabolism (Vosteen, 1961; Nakai and Hilding, 1966), and deeply invaginated basal and lateral cell wall (Iurato, 1967).

Two characteristic electrical potentials have been detected in the cochlea (Davis, 1957): the endolymphatic resting potential (ERP) and the cochlear microphonic potential (CMP). The ERP is a steady potential of about 80 mV, with the endolymph positive relative to perilymph and blood plasma. The CMP is an alternating potential, which arises upon acoustical stimulation of the ear and which can be derived from an electrode placed anywhere in the cochlea. It follows exactly the frequency of the acoustical stimulus, without a true threshold, refractory period or adaptation, and is, like the endolymphatic resting potential, very sensitive to oxygen deprivation (Honrubia, Johnstone and Butler, 1965). Both the ERP and the CMP are strongly inhibited by metabolic poisons (Chambers and Lucchina, 1966; Konishi and Kelsey, 1968) and are depressed when the $K^+$ concentration in the tympanal perilymph is increased (Tasaki and Fernandez, 1952; Butler, 1965) or the $K^+$ concentration in the endolymph is decreased (Konishi, Kelsey and Singleton, 1966). These findings indicate that the cochlear potentials, and thus cochlear function,

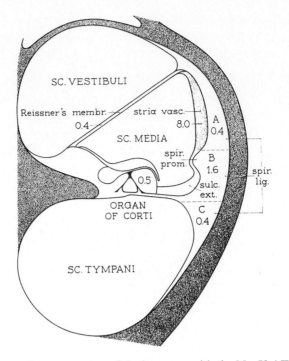

**Figure 13.** Schematic cross-section of the inner ear with the Na–K ATPase activities (in moles/kg dry wt/hr) of the various structures indicated by numbers. (From Kuijpers and Bonting, 1969).

depend on the existence of the cation gradients between endolymph and perilymph and that these gradients are maintained by an active transport process.

Questions to be answered were: what is the site of the pump and what is its chemical nature? Reissner's membrane, separating endolymph and peri-lymph is a possible site. Rauch (1964) found a rapid transport of $^{42}K$ from vestibular perilymph to endolymph, which was inhibited by cyanide, iodo-acetate and ouabain. No transport of $K^+$ and $Na^+$ could be demonstrated from scala tympani to endolymph through the membrana basilaris. On the other hand, the submicroscopic structure of the cells of Reissner's membrane does not resemble that of secretory cells. Structurally the cells of the stria vascularis, which is highly vascularised, have the characteristics of secretory cells: many mitochondria and deeply invaginated basal and lateral cell walls.

Therefore, we undertook a study of the occurrence of Na–K ATPase in the cochlea and its possible correlation with the CMP (Kuijpers, van der Vleuten and Bonting, 1967). The cochlea of the guinea pig and the 1-day old chicken were dissected at 0 °C under a binocular microscope, and either the whole membranous structure or the stria vascularis alone, in the case of the guinea pig, and the tegmentum vasculosum alone, in the case of the chick, were collected, freeze-dried, homogenised (20–200 $\mu$g in 100 $\mu$l) and assayed for ATPase activity. The guinea pig stria vascularis had an activity on a dry weight basis 10·5 times as high (8·2 mole/kg/hr) as the total membranous structure, whereas the tegmentum vasculosum (homologous with stria vascularis plus Reissner's membrane in mammalians) had an activity 3·6 times as high (7·0 mole/kg/hr) as the total membranous structure of the chick cochlea. It was also shown that perfusion of the guinea pig scala vestibuli with 10 $\mu$l/min Krebs–Ringer solution, containing ouabain or erythrophleine, inhibited the CMP fairly rapidly, reaching full effect in about 30 min. These findings proved the involvement of the cochlear cation pump system, and suggested the stria vascularis as the site of the pump.

In further studies, involving careful microdissection and ultramicro enzyme assays, the exact distribution of Na–K ATPase activity over the cochlear structures was determined (Kuijpers and Bonting, 1969). It is clear from the results presented in table 14 and figure 13 that the Na–K ATPase activity is

**Table 14.** ATPase activities in various cochlear structures

| Structure | Mg ATPase (mole/kg dry wt/hr) | Na–K ATPase | | No. of detns. |
|---|---|---|---|---|
| | | (mole/kg dry wt/hr) | (% of total ATPase) | |
| Stria vascularis | 5·6 ± 0·41 | 8·0 ± 0·43 | 59 ± 1 | 16 |
| Spiral ligament | | | | |
| A. behind stria vasc. | 0·36 ± 0·09 | 0·41 ± 0·09 | 55 ± 7 | 4 |
| B. with prominentia spiralis and sulcus externus | 1·3 ± 0·43 | 1·6 ± 0·39 | 58 ± 6 | 4 |
| C. bordering scala tympani | 0·62 ± 0·07 | 0·35 ± 0·07 | 36 ± 7 | 4 |
| Reissner's membrane | 0·91 ± 0·10 | 0·35 ± 0·06 | 28 ± 4 | 12 |
| Organ of Corti | 2·8 ± 0·79 | 0·47 ± 0·27 | 11 ± 6 | 3 |

Activities given as means with standard errors. From Kuijpers and Bonting (1969).

very predominantly located in the stria vascularis, where the activity is about 20 times as high as in all other structures except for that part of the spiral ligament comprising the prominentia spiralis and sulcus externus. These latter structures form the continuation of the stria vascularis. The activity in Reissner's membrane is low on a dry weight basis, and even more so on an absolute basis. Taking into account the dry weights of the two structures, the ratio of total Na–K ATPase activity in stria vascularis and Reissner's membrane is 68 : 1. And on a surface area basis, if the multiple infoldings of the

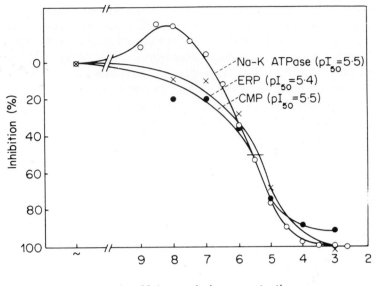

- log Molar ouabain concentration

**Figure 14.** Inhibitory effects of ouabain on Na–K ATPase activity of stria vascularis *in vitro* and on the endolymphatic resting potential (ERP) and cochlear microphonic potential (CMP) upon perfusion of the scala vestibuli for 45 min. (From Kuijpers and Bonting, to be published).

stria vascularis cell membranes are disregarded, the ratio is even 168:1. Hence Reissner's membrane cannot contribute significantly to the active cation transport in the cochlea. Additional arguments for this conclusion can be derived from the fact that this membrane is avascular and has a very high electrical resistance (Johnstone, Johnstone and Pugsley, 1966). The properties of the Na–K ATPase system of the stria vascularis were also determined and found to agree in all respects with those of the enzyme in other tissues (table 3). The ouabain inhibition curves for the CMP, the ERP, and the enzyme system show a nearly identical $pI_{50}$ (figure 14), while the time course for the inhibition of both potentials is also about the same (figure 15). The stimulating effect of very low concentrations of ouabain on the enzyme, which has been demonstrated in several other tissues (section IIB), could not as yet be obtained for the cochlear potentials.

It thus appears that the Na–K ATPase system of the stria vascularis plays a primary role in the maintenance of the cochlear cation gradients and hence in the generation of the two cochlear potentials. At the same time it appears highly unlikely that Reissner's membrane is the site of significant cation pump

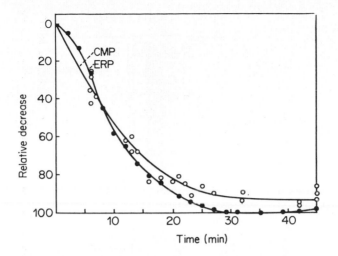

**Figure 15.** Time course of relative decrease of endolymphatic resting potential (ERP) and cochlear microphonic potential (CMP) upon perfusion of the scala vestibuli with $10^{-3}$M ouabain in modified Krebs–Ringer solution. (From Kuijpers and Bonting, to be published).

activity. The transport of $^{42}$K from the vestibular perilymph to the endolymph, observed by Rauch (1964), could take place through the spiral ligament to the stria vascularis, where it would be actively secreted into the endolymph. This supposition is favoured by the very high isotope content in spiral ligament and stria vascularis immediately after perilymphatic $^{42}$K injection, even after interruption of the blood circulation.

## F. Electric Organ

The electric organ of electrical fish, particularly that of the electric eel, *Electrophorus electricus*, is able upon nervous stimulation to generate discharges of considerable voltage and current (Keynes, 1957). The organ consists of thousands of identical units, called electroplax, the schematic structure of which is shown in figure 16. The active part of the electroplax is a syncytium consisting of an innervated and a non-innervated membrane, in close apposition. The fluid enclosed by the two membranes has a high K/Na ratio, while the exterior fluid has a low K/Na ratio (Schoffeniels, 1959). Thus, in the resting state the membrane potentials across the two membranes would oppose each other. Nervous stimulation of the electroplax causes a sharply increased Na$^+$ influx across the innervated membrane, followed by K$^+$ efflux, thus reversing the potential across this membrane and making the two membrane potentials additive. Since thousands of the electroplax units are

L

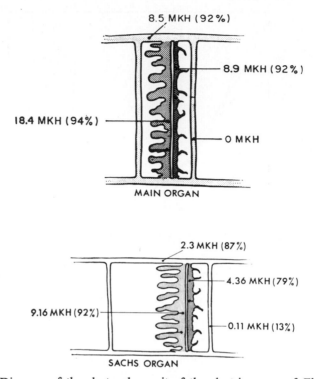

**Figure 16.** Diagram of the electroplax unit of the electric organ of *Electrophorus electricus.* Top main organ, bottom Sachs organ. In the syncytium the non-innervated membrane is on the left. The nerves, terminating in the innervated membrane, run through the horizontal walls of the electroplax unit. Numbers indicate Na–K ATPase activity (in mole/kg dry wt/hr) in the various components of the unit.

arranged in series, and can be stimulated simultaneously, very potent electrical spikes can be generated in this way. Maintenance of the cationic gradients in the electroplax and their restoration after stimulation requires an active cation transport system. The energy for the latter system is supplied by glycolysis (Aubert, Chance and Keynes, 1964; Chance, Lee and Oshino, 1964). Activation of glycolysis occurs through stimulation of phosphoglycerate kinase, glyceraldehyde-3-phosphate dehydrogenase and phosphofructokinase by increasing the ADP/ATP ratio (Maitra and associates, 1964), similar to the case of the erythrocyte (Section IIIA).

The occurrence of a high Na–K ATPase activity in the electric organ was established by Bonting, Simon and Caravaggio (1961). Its properties are similar to those of the system from other tissues. Remarkable was the high relative activity in terms of total ATPase activity in unfractionated homogenate of the main organ (95 %), higher than in any other tissue, while the Sachs organ had a slightly lower percentage (78 %). By means of microdissection of lyophilised, longitudinal sections of the main organ and the Sachs organ and ultramicro enzyme assay, the activities in the innervated and non-innervated membranes and other parts of the electroplax could be determined separately (figure 16). The major activity was present in the non-innervated membranes (18·4 and 9·2 mole/kg dry wt/hr), the highest noticed in any tissue so far. Since Schoffeniels (1959) found the active Na efflux across this membrane in the Sachs organ to be about 14 times as high as that across the innervated membrane, the Na–K ATPase activity determined in the innervated membrane (4·4 mole/kg/hr) appears rather high. This might be due to a contribution from the nerves ending in this membrane, because the wall of the electroplax unit through which the nerves run also had a much higher activity (2·3 mole/kg/hr) than the other walls (0·1 mole/kg/hr). The high Na–K ATPase activity in the non-innervated membrane also explains the fact that upon subcellular fractionation of the electric organ the Na–K ATPase-rich fraction contains numerous fragments derived from the non-innervated membrane (Sheridan, Whittaker and Israel, 1966). The decrease of the enzyme activity in the electric organ from head to tail (Fahn, 1968) is also understandable, because in that direction the units become wider and hence the number of electroplaxes per cm decreases.

The active Na flux across the non-innervated membrane of the Sachs organ, determined by Schoffeniels (1959), was compared with the Na–K ATPase activity in this membrane (Bonting and Caravaggio, 1963). The membrane surface area in the lyophilised sections was calculated as the product of the height of the membrane and the section thickness. Dividing this surface area by the dry weight of the dissected membrane (0·1–0·3 $\mu$g) yielded the surface area to dry weight ratio. The enzyme activity in moles/kg dry wt/hr could then be converted to moles/cm$^2$/sec by dividing through this ratio. Comparison of this figure, after conversion to the temperature at which the Na flux had been determined by Schoffeniels, with the active Na flux gave a Na/ATP ratio of 2·2 (table 4), in good agreement with the ratios found for other tissues (table 5). This is a strong indication that the Na–K ATPase system is involved in the active cation transport required for the maintenance of the cation gradient across the non-innervated membrane of the electroplax. Further support is given by our unpublished observation that after subcutaneous injection of 32 $\mu$g ouabain in a 65 cm electric eel the frequency of firing upon electrical stimulation of the animal was reduced by 44 % ($P = 0·03$) and the spike height by 8 % ($P = 0·0002$). It should, however, be kept in mind that the occurrence of nervous or circulatory effects of the injected ouabain cannot be ruled out in this experiment.

Na–K ATPase of the electric organ is 75% inhibited by $10^{-5}$ g/ml oligomycin, but it could be shown that this does not mean that oxidation–reduction reactions are involved in the transport mechanism (Glynn, 1962b; 1963). Rather it appears that the oligomycin affects a phosphorylation step in the Na–K ATPase mechanism, similar to the one it inhibits in the oxidative phosphorylation system. A high specificity of the Na–K ATPase system in electric organ for ATP among various nucleotides has also been demonstrated (Albers and Koval, 1962).

## V.  SECRETORY SYSTEMS

Of the many organs which secrete cations, we shall discuss the role of the Na–K ATPase system in kidney, frog skin, toad bladder, marine bird salt gland, elasmobranch rectal gland, liver, intestine, pancreas, formation of aqueous humour and cerebrospinal fluid, and in various other organs. Since several of these organs will be discussed more fully in later chapters, the treatment will be brief and will stress the role of Na–K ATPase system.

### A.  Kidney

One of the main functions of the kidney is the retention of $Na^+$ by means of tubular reabsorption, and thus it is logical to expect a role of the Na–K ATPase system in renal function. The occurrence of the enzyme in the kidney has been reported by several investigators (Bonting, Simon and Hawkins, 1961; Bonting, Caravaggio and Hawkins, 1962; Whittam and Wheeler, 1961; Wheeler and Whittam, 1962; Skou, 1962; Charnock and Post, 1963a; Landon and Norris, 1963; Rendi, 1964). Inhibition of renal tubular Na reabsorption by cardiac glycosides has repeatedly been demonstrated (Orloff and Burg, 1960; Wilde and Howard, 1960; Cade and associates, 1961; Strickler, Kessler and Knutson, 1961; Tanabe and associates, 1961; Vogel and Kluge, 1961). A parallel decrease of renal Na reabsorption and concentrating capacity and of the renal Na–K ATPase activity by digoxin infused into one renal artery of the dog has also been observed (Martinez-Maldonado and associates, 1969). It has even been possible to observe the biphasic effects of ouabain on tubular Na reabsorption *in vivo* and on the enzyme activity *in vitro* in the chicken kidney (Palmer and Nechay, 1964). Also, artificially raised plasma $K^+$ levels reduce the effect of cardiac glycosides (Heidenreich, Laaff and Fuelgraff, 1966). These findings constitute clear evidence of an important role of the Na–K ATPase system in tubular $Na^+$ reabsorption.

A pacemaker effect of renal cation transport on energy metabolism has been shown to exist in kidney cortex slices (Whittam and Willis, 1963) and to be mediated by the Na–K ATPase system (Blond and Whittam, 1964a). Homogenate studies have shown that this effect is due to the stimulating effect on phosphoglycerate kinase of ADP produced by the Na–K ATPase system (Blond and Whittam, 1964b, 1965; Landon, 1967). A parallel decrease in accumulation of $p$-aminohippurate and α-aminoisobutyric acid and of Na–K ATPase activity in kidney slices of rats with increasing ages has also been reported (Beauchene, Fanestil and Barrows, 1965). Upon greatly increasing the tubular reabsorptive load of Na by unilateral nephrectomy, feeding of a high protein diet or injection of methylprednisolone, an adaptive increase in renal Na–K ATPase activity (per mg protein) is found without changes in other microsomal enzyme activities, including Mg ATPase (Katz and Epstein, 1967). This adaptive increase is not really due to the increase of the tubular reabsorption load, but both compensatory renal enlargement and induction of Na–K ATPase appear to be related to the mass of functioning renal tissue and not to the mass of excreting renal tissue remaining in the animal. This was concluded from experiments in which one ureter was excised and urine allowed to drain in the peritoneal cavity (Fanestil, 1968).

In view of the effects of the hormones aldosterone, hydrocortisone, vasopressin and angiotensin on renal electrolyte excretion, it was logical to consider their possible effects on the Na–K ATPase system. No *in vitro* effects at well above normal physiological levels were found for aldosterone, hydrocortisone and vasopressin on brain and retinal Na–K ATPase (Bonting, Simon and Hawkins, 1961), for aldosterone and hydrocortisone on renal Na–K ATPase (Charnock and Post, 1963a), for vasopressin on toad bladder Na–K ATPase (Bonting and Canady, 1964), and for angiotensin on Na–K ATPase of rabbit kidney cortex and medulla, brain and ciliary body (Bonting, Canady and Hawkins, 1964). *In vitro* activation of mouse kidney Na–K ATPase by testosterone ($10^{-4}$ M) has been claimed (Akikusa and Yamamoto, 1968). Indirect effects of adrenal hormones on the renal Na–K ATPase activity have, however, been demonstrated. The activity of the renal enzymes was reduced by about half in 6–7 days after adrenalectomy, could be restored by corticosterone, but not by aldosterone in moderate dosage (Chignell and Titus, 1966a). With very high levels of aldosterone the lost activity could be restored (Landon, Jazab and Forte, 1966). It has been suggested that the aldosterone exerts its effect through changes in the $Na^+$ level or $Na^+/K^+$ ratio in plasma (Joergensen, 1968), but a specific renal receptor must be involved since adrenalectomy does not affect rat brain Na–K ATPase (Gallagher and Glaser, 1968). Micropuncture studies indicate that adrenalectomy primarily affects Na reabsorption in the proximal tubules (Hierholzer, Wiederholt and Stolte, 1966).

Since cardiac glycosides, like ouabain, have a diuretic effect on the kidney by inhibiting renal Na–K ATPase activity, it was also logical to study the effect of known diuretic drugs on the enzyme system. Contradictory results have been obtained. Organic mercury compounds (about 1 mM) inhibit ion transport in kidney cortex slices and autoradiography showed Hg to be located primarily in the basal membrane rather than the brush-border membrane of the proximal tubular cells and in lower concentration in the basal membrane of distal tubular cells (Kleinzeller and associates, 1963). These mercurials inhibited renal Na–K ATPase, rather than Mg ATPase, while sulphydryl reagents, such as iodoacetate and iodacetamide and the diuretics: caffeine, theobromine, theophylline and chlorothiazide did not inhibit Na–K ATPase (Taylor, 1963b). The latter finding and the fact that mercurials are known to affect citric acid cycle enzymes and to inhibit non-Na-stimulated respiration of kidney slices more than Na-stimulated respiration, suggest that the diuretic effect of mercurials could be due to inhibition of renal energy metabolism rather than of Na–K ATPase activity. On the other hand, diuretic organic mercurials have been found to inhibit renal Na–K ATPase activity *in vitro* as well as *in vivo* and to decrease the ability of renal membrane fragments to stimulate cytoplasmic glycolysis, while non-diuretic mercurials like *p*-chloromercuribenzoate did not have the latter effect (Jones, Lockett and Landon, 1965). Likewise the diuretic mercurial meralluride decreased intracellular K, passive K outflow and respiration in kidney slices, while the non-diuretic mercurial *p*-chloromercuribenzoate did not have these effects (Bowman and Landon, 1967). The diuretic ethacrynic acid inhibited renal Na–K ATPase activity more in the guinea pig than in the rat (Duggan and Noll, 1966). It was pointed out that here is another discrepancy, since ethacrynic acid has no diuretic effect in the rat, although it does inhibit the enzyme system *in vivo* (Hook and Williamson, 1965). In a critical study, comparing the effects of diuretic and non-diuretic mercurials and ethacrynic acid with those of ouabain, it is concluded that renal Na–K ATPase cannot be the site of diuretic action of mercurials and ethacrynic acid (Nechay and associates, 1967).

In concluding this section on the kidney, it is necessary to look at the more detailed picture we have nowadays of renal function. In figure 17 a schematic diagram of the nephron shows its components. According to current insights the descending and ascending limbs and the collecting tubule form a counter-current exchange system. In the ascending limb Na$^+$, unaccompanied by water, is actively extruded. As a result the osmotic strength of tubular and interstitial fluid increases from cortex to papilla. This leads to withdrawal of water from the collecting tubule, the permeability of which to water is controlled by the action of vasopressin. This constitutes the renal concentrating mechanism, maintaining the osmotic equilibrium in the body. In the proximal convoluted tubule isotonic Na$^+$ transport takes place, which withdraws some 80% of the Na$^+$ and fluid from the glomerular filtrate, and facilitates the reabsorption of valuable metabolites such as glucose and amino acids, as well as the excretion of unwanted products such as urea from

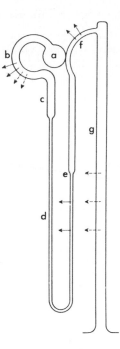

**Figure 17.** Diagram of the nephron, showing its components: (a) glomerulus, (b) proximal convoluted tubule, (c) pars recta, (d) descending limb, (e) ascending limb, (f) distal convoluted tubule, and (g) collecting tubule.

the plasma. In the distal convoluted tubule $Na^+$ reabsorption in exchange for $K^+$ takes place, which is controlled by aldosterone and serves to maintain the electrolyte balance in the body. By applying Lowry's quantitative histochemical technique of microdissection and ultramicro assay to lyophilised sections of rat kidney, the distribution of Na–K ATPase over the various parts of the nephron has recently been determined by Schmidt and Dubach (1969). The results in table 15 show that the highest activity is found in the lower part of the ascending limb, where $Na^+$ must be pumped against a considerable concentration gradient. The much lower activity in the proximal convoluted tubule would reflect the smaller energy demand of its isosmotic Na transport and perhaps also the larger surface area over which transport can take place. The lowest activities in glomerulus and pars recta would agree with the mainly passive ion movements in these structures.

The pathway of $Na^+$ and $Cl^-$ ions through the proximal tubular cells has been studied by means of precipitation as Na-antimonate and AgCl and

**Table 15.** ATPase activities in rat nephron

| Nephron segment | Mg–ATPase | Na–K ATPase |
|---|---|---|
| glomerulus (s.c.) | 9·3 | 1·6 |
| glomerulus (j.m.) | 10·4 | 0·3 n.s. |
| prox. conv. (c.s.) | 5·5 | 1·6 |
| prox. conv. (j.m.) | 4·7 | 2·8 |
| pars recta (s.c.) | 3·6 | 0·3 n.s. |
| pars recta (j.m.) | 4·1 | 1·3 |
| ascending, inner stripe | 5·5 | 12·1 |
| ascending, outer stripe | 6·5 | 6·1 |
| distal convol. | 6·9 | 7·9 |
| collecting tubule | (7.6) | (6.3) |

All values expressed as mole/kg dry wt/hr at 37°.
n.s. = Na–K ATPase value not significant at $P = 0.05$ level.
s.c. = subcapsular glomerulus and related parts of proximal tubule.
j.m. = juxtaglomerular glomerulus and related parts of proximal tubule.
From Schmidt and Dubach (1969).

electronmicroscopic observation (Nolte, 1966). Chloride was mainly localised in intercellular spaces, while sodium was distributed diffusely. From this the author draws the unlikely conclusion that $Na^+$ would be transported actively through the cell from brush-border to peritubular capillary, while $Cl^-$ would move passively and separately between the cells from lumen to capillary. More likely is that the relatively soluble Na–antimonate diffused during the treatment of the tissue, while the very insoluble AgCl did not diffuse and therefore indicates the real pathway of NaCl movement, *viz.* via the intercellular spaces to the peritubular capillary. This would agree with the observations of Schmidt-Nielsen and Davis (1968) on the proximal tubules of the reptile kidney in different functional states. At the lumen there were terminal bars between adjacent cells. In kidneys actively reabsorbing $Na^+$ there were wide intercellular spaces in the proximal tubules below the terminal bar down to the base of the cell. When no reabsorption was going on (lumen closed) or when $Na^+$ transport was inhibited by previous injection of ouabain (lumen open), the intercellular spaces disappeared. Apparently the isosmotic Na transport takes place by local osmosis as in the gall bladder (Diamond, 1962c; Kaye and associates, 1966), with $Na^+$ being transported actively across the lateral cell wall into the intercellular spaces. The resulting local hyperosmotic condition draws water into the space, which distends until isosmosis is reached. Hydrostatic pressure then forces the fluid into the capillary.

## B. Frog Skin and Toad Bladder

These two structures have served as simplified models of renal electrolyte transport and as such have been widely studied. Both structures, by means of an epithelial layer with a sodium pump and varying water permeability regulated by vasopressin, enable these amphibians to maintain their osmotic equilibrium on land and in water. When a piece of frog skin or toad bladder is clamped between two chambers, containing aerated physiological salt solution, there is net $Na^+$ transport from the mucosal to the serosal side. If the two chambers are short-circuited, a current flows which is equivalent to the net amount of $Na^+$ transported per second (Ussing and Zerahn, 1951; Leaf, Anderson and Page, 1958). Thus the short-circuit current is a measure of the active $Na^+$ transport rate.

In the case of frog skin, the presence of ouabain on the serosal side causes inhibition of active $Na^+$ transport (Koefoed-Johnsen, 1957; Nakajima, 1960) while at very low levels there is initially a stimulating effect (Marro and Pesente, 1962). Increasing the $K^+$ concentration in the medium reduces the inhibitory effect of ouabain (Whitney and Widdas, 1959; Marro, Pesente and Tiripicchio, 1962). The hydrolysis of ATP to ADP and inorganic phosphate appears to supply the energy required (Abdel-Wahab, 1961). Potassium, present in low concentration (0–2·5 mM) on the serosal side activates active $Na^+$ transport, but in higher concentrations it inhibits the system (Snell and Chowdhury, 1965). While ouabain inhibits $Na^+$ transport, it does not affect the increased skin permeability for water moving along an osmotic gradient in the presence of vasopressin, suggesting that $Na^+$ and water movement are conducted by different mechanisms (Natochin, 1966). These findings strongly suggest that the Na–K ATPase system is involved in the $Na^+$ transport through frog skin. Its presence was first demonstrated by Bonting, Caravaggio and Hawkins (1962), and the Na/ATP ratio was found to be 3·0 (Bonting and Caravaggio, 1963), which value agrees well with the values for other tissues in tables 4 and 5, and with the values of 2·7–3·3 (Zerahn, 1956) and 3·5 (Leaf and Renshaw, 1957) derived from the increased $O_2$ consumption upon stimulation of active $Na^+$ transport in frog skin. The occurrence of Na–K ATPase activity, evenly distributed across the whole skin, has been confirmed (Rotunno, Pouchan and Cerejido, 1966). An electron microscopic study with histochemical ATPase demonstration showed ATPase activity (Farquhar and Palade, 1966) on all inward facing membranes, but not on the outward facing membranes (figure 25). Between the cells there is a labyrynth of intercellular spaces, which appear to function in coupled transport of $Na^+$ and $H_2O$ (see section V E).

The presence of the Na–K ATPase system in toad bladder was first demonstrated by Bonting and Caravaggio (1963), and its activity expressed in moles/ $cm^2$/sec was compared to the active $Na^+$ flux determined from the short circuit

current by Leaf, Anderson and Page (1958). The calculated Na/ATP ratio
was 2·5, which again agreed with the values for other tissues shown in tables
4 and 5, and also with the values of 2·8 (Leaf, Page and Anderson, 1959)
and 3·2 (Leaf and Dempsey, 1960) derived from the increased $O_2$ consumption
upon stimulation of $Na^+$ transport. In a further study the properties of the
transport system and the enzyme system were compared in more detail
(Bonting and Canady, 1964). A Na/ATP ratio of 2·4 was found at a short
circuit current per mg dry wt within 14% of that reported by Leaf, Anderson
and Page (1958). Ouabain inhibited Na transport when added to the serosal
side, but not when added to the mucosal side, indicating that the Na pump
is located on the serosal side. The ouabain inhibition curves for $Na^+$ transport
and enzyme agree very well (figure 18) and give $pI_{50}$ values of 4·6 and 4·7,

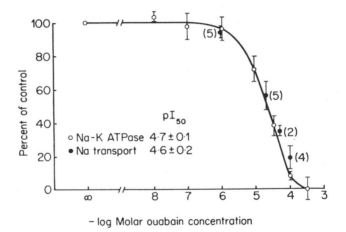

**Figure 18.** Effect of ouabain on Na–K ATPase activity and short circuit current
($Na^+$ transport). Enzyme activity (open circles) was determined in homogenates
(nine determinations per point). Short-circuit current (closed circles) was measured
in the Ussing chamber with ouabain added to the serosal side (numbers in parentheses indicate number of experiments). (From Bonting and Canady, 1964).

respectively. The ouabain effect was not changed when the animals were
depleted of endogenous aldosterone by maintaining them for 5–7 days in
110 mM NaCl, contrary to the statement of Sharp and Leaf (1963) that the
ouabain effect on $Na^+$ transport was only observed after prior stimulation by
aldosterone. Vasopressin, its mediator cyclic AMP (Orloff and Handler,
1967) and aldosterone (with or without preincubation) had no effect on the

Na–K ATPase activity, which agrees with the conclusion that these two hormones act on the passive permeability of the mucosal surface (Frazier, Dempsey and Leaf, 1962; Sharp and Leaf, 1963; Crabbé and De Weer, 1965; Civan, Kedem and Leaf, 1966). Erythrophleine also inhibited enzyme activity and Na transport with $pI_{50}$ values of 6·0 and 4·8 respectively. It was active on both the serosal and the mucosal surface, presumably because of its high permeability as a tertiary amine (figure 8). This may also explain its smaller effect on transport than on enzyme activity, which was also noted in the formation of cerebrospinal fluid (section V F; Vates, Bonting and Oppelt, 1964).

From electrophysiological measurements at varying ionic concentrations it is concluded that the transcellular transport of $Na^+$ by the toad bladder involves at least two steps (Frazier, Dempsey and Leaf, 1962; Gatzy and Clarkson, 1965). The first ·and rate-limiting step is entry into the cell across the mucosal membrane of the epithelial cells by a passive process. The second step is the active extrusion of $Na^+$ across the serosal membrane of these cells. The mucosal membrane is more permeable to $Na^+$ than to any other univalent cation, while the serosal membrane is preferentially permeable to $K^+$. By histochemical staining technique ATPase activity was found primarily at the serosal surface and the lateral interdigitating surface of the epithelial cells (Keller, 1963; Bartoszewicz and Barrnett, 1964). The rate of $Na^+$ transport is related to the intracellular $Na^+$ concentration (Frazier, Dempsey and Leaf, 1962) and is severely but reversibly depressed upon withdrawal of serosal $K^+$ (Bower, 1963), corresponding to the activation characteristics of the erythrocyte Na pump. The inhibition of $Na^+$ transport in the toad bladder is reversed by high serosal $K^+$ concentrations (Bower, 1964). A biphasic effect of ouabain (stimulation at a concentration of $10^{-9}$–$10^{-7}$ M) and inhibition at concentrations above $10^{-6}$ M has also been observed (McClane, 1965). Ouabain does not affect the passive $Cl^-$ movement across the bladder and the $Na^+$ flux from the serosal to the mucosal side into the cells, but rather the serosal pump (Herrera, 1966).

An interesting point is the low sensitivity of the toad bladder cation transport system to ouabain ($pI_{50} = 4·6$), which was also found for the toad heart (Chen, Hargreaves and Winchester, 1938). This is not due to rapid detoxication, absence of biotransformation into a truly active substance, low absorption or rapid excretion (Herrmann, Portius and Repke, 1964), but to low sensitivity of the Na–K ATPase system in the toad bladder ($pI_{50} = 4·7$). Most likely, this is related to the fact that the toad produces several bufagins in its venom, which are chemically and pharmacologically analogous to digitalis aglycones (Fieser and Fieser, 1959). This conclusion is supported by the much higher sensitivity of the enzyme system to the structurally dissimilar erythrophleine ($pI_{50} = 6·0$).

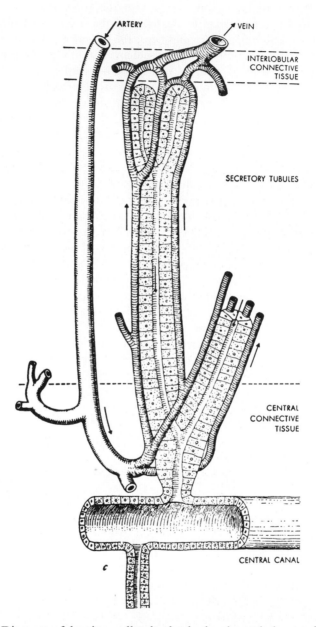

**Figure 19.** Diagram of herring gull salt gland, showing tubules, capillaries and central duct. (From Schmidt-Nielsen, 1959).

## C. Salt Gland and Rectal Gland

Various species of marine animals have developed structures, designed to assist their kidney in maintaining osmotic equilibrium: the salt gland or nasal gland of marine birds, the rectal gland of elasmobranch fishes and the gills of teleost fishes and molluscs.

The salt gland of the marine birds has been studied extensively by Schmidt-Nielsen (1959; 1960). These animals are able to excrete a highly hypertonic salt solution (600–800 mM $Na^+$, 45 mM $K^+$) through their nasal gland when receiving an oral or intravenous salt load, while in the absence of an osmotic load the gland was inactive. This explains how these animals can exist indefinitely in a marine environment, when their water supply has the high osmotic concentration of sea water while their kidneys are entirely unable to maintain the ionic plasma composition of vertebrates. The morphology of the salt gland (Fänge, Schmidt-Nielsen and Osaki, 1958; Komnick, 1963a, b,c,d) is similar to that of the kidney in having a great number of tubules, formed by epithelial cells, which converge to a collecting duct and which have blood capillaries running alongside in counter-current fashion (figure 19).

When inhibition by ouabain of the salt gland function was demonstrated (Thesleff and Schmidt-Nielsen, 1962), it seemed logical to look for an involvement of the Na–K ATPase system (Hokin, 1963; Bonting and associates, 1964). The enzyme was found to be present in high absolute and relative activity (table 16). Comparison of the Na-excretion rates after salt infusion (Nechay, Larimer and Maren, 1960) with the Na–K ATPase activity in the wild animals yields a Na/ATP ratio of 2·0, in good agreement with the values listed for a variety of tissues in tables 4 and 5. The properties of the enzyme were quite similar to those of the Na–K ATPase system in other tissues. The $Na^+$ activation and $K^+$ activation curves are shown in figures 3 and 4, respectively, and the kinetic data in table 3. Half-maximal inhibition with ouabain was at $5 \times 10^{-6}$ M, while at $10^{-8}$ M ouabain a slight stimulation was noticed. The results in table 16 show a marked difference between the salt glands of wild birds and birds held in captivity on fresh water. The much lower ATPase activities in the captive animals suggest that salt deprivation leads to significant loss of both ATPase activities over and beyond the weight loss of the gland. This loss is not quickly recovered, since giving four of these animals 3% NaCl solution instead of water led to their death within 10 days (Bonting and associates, 1964). Calculation showed that these animals were receiving twice as much $Na^+$ as their involuted salt glands could handle, while the remaining Na–K ATPase activity was just adequate to handle the salt ingestion on the fresh water regimen. The wild birds with their three times higher activity could have managed this high Na intake. This retrograde adaptation of the glandular Na–K ATPase activity is even more pronounced in the domestic duck, which has a salt gland with only 0·8 % of the Na–K ATPase activity of the herring gull gland on a body weight basis. Maintaining ducks on a NaCl solution increased the Na–K ATPase activity and the Na-excretory rate of their salt glands about 4-fold in 9 days (Fletcher, Stainer and Holmes, 1967; Ernst, Goertemiller and Ellis, 1967). The morphological and histochemical aspects of the salt-stimulated adaptation of the salt gland in duck has been reported by Ellis and coworkers (1963). The adaptive effect for the ATPase activities, esp. the Na–K dependent type, is rather specific, since carbonic anhydrase activity did not show such an effect (table 16).

**Table 16.** Enzyme activities in herring gull salt gland

|  | Wild gulls | Captive gulls | Ratio |
|---|---|---|---|
| Body wt (g) | 1052 ± 35 | 930 ± 77 | 0·88 |
| Gland wt (mg) | 920 ± 60 | 600 ± 80 | 0·65* |
| % dry wt | 29 ± 0·9 | 27 ± 0·4 | 0·93 |
| Na–K ATPase |  |  |  |
| mole/kg/hr | 2·68 ± 0·16 | 1·37 ± 0·15 | 0·51* |
| mmole/hr/gland | 2·46 ± 0·20 | 0·79 ± 0·09 | 0·32* |
| Mg ATPase |  |  |  |
| mole/kg/hr | 1·41 ± 0·10 | 0·90 ± 0·05 | 0·64* |
| mmole/hr/gland | 1·30 ± 0·11 | 0·55 ± 0·09 | 0·42* |
| Carbonic anhydrase |  |  |  |
| mole/kg/hr | 6380 ± 440 | 6020 ± 270 | 0·95 |
| mmole/hr/gland | 5340 ± 790 | 3560 ± 380 | 0·67 |

All data on wet weight basis, means with standard errors shown.
Wild gulls: 6 animals sacrificed within 30 hr after capture in coastal area.
Captive gulls: 5 animals sacrificed after 7 weeks on fresh water regimen.
*Significantly less than 1·0 at $P = 0·05$ level.
From Bonting and coworkers, 1964.

These findings clearly demonstrate the primary, rate-limiting role of the Na–K ATPase system in salt gland function. This conclusion permits extension of Schmidt-Nielsen's diagram (Schmidt-Nielsen, 1960) of the sequence of events in salt gland function (figure 20). The Na–K ATPase system is thought to be located at the luminal membrane of the tubular cells, because this is the site where $Na^+$ is excreted by the cells. $Na^+$ is expected to enter the tubular cell from the capillary by passive diffusion. In the absence of osmotic stimulation the capillaries are constricted, thus limiting the supply of $Na^+$ and oxygen to the tubular cells and consequently their pump activity. Stimulation of the osmoreceptors by a salt load (21 meq $Na^+$/l plasma or more; McFarland, 1964) leads through nerve stimulation by vasodilation in the gland and thus to an increased blood supply to the tubular cells. The resulting increased

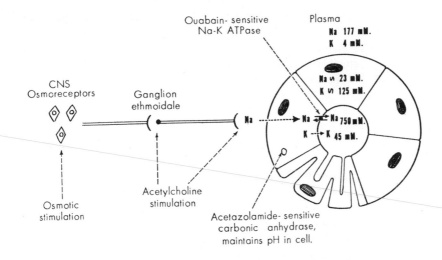

**Figure 20.** Sequence of events in the secretion of NaCl by the herring gull salt gland. Approximate concentrations of Na⁺ and K⁺ in plasma, tubule cell cytoplasm and secreted fluid are shown. Solid arrows for Na⁺ and K⁺ indicate active transport; broken arrows indicate diffusion. (From Bonting and coworkers, 1964), modified after Schmidt-Nielsen (1960).

intracellular Na level, which activates the Na–K ATPase system, and an increased oxygen supply, which may raise the cellular ATP level, make the Na pump operate at maximum rate. This mechanism allows the animal to dispose of sudden increases in salt intake.

Elasmobranch fishes have a compound tubular gland, opening mid-dorsally into the rectum by a duct (Sullivan, 1908). Between an outer layer with small arteries and a central region, consisting of ducts and veins arranged around a central canal terminating in the duct, there is a thick glandular layer consisting almost exclusively of tubules and capillaries. In sharks the tubules are arranged radially, while in skates and rays the glandular parenchyma is divided into lobules with tubules oriented mostly in a longitudinal direction (Bonting, 1966). The submicroscopic structure of the tubular cells of the rectal gland (Doyle, 1962) was quite similar to that of the tubular cells in the salt gland (Doyle, 1960). In spiny dogfish (*Squalus acanthias*) the rectal gland has been shown to secrete a fluid consisting essentially of NaCl, with a concentration twice that of the plasma and greater than that of sea water (Burger and Hess, 1960). The average glandular NaCl secretion was twice the urinary NaCl secretion. Ligation of the gland caused a rise in plasma and urine chloride levels, while injection of large amounts of NaCl caused a rapid increase of the glandular secretion to a maximum rate of 1·9 ml/hr/kg body weight after a lag period of 30–75 min and only a small increase in urinary Na-output, whereupon the glandular secretion returned to the normal level (Burger, 1962). It was concluded that this gland plays a significant role in the salt economy of this species, just as the nasal gland does in marine birds. Most of the Na$^+$ influx (0·52 mmole/kg/hr compared to a maximal excretion of 0·94 mmole/kg/hr by the rectal gland) into the animal occurs through the head, probably the gills (Horowicz and Burger, 1968).

In rectal glands from all 9 elasmobranch species studied, a fairly high (absolute and relative) Na–K ATPase activity was demonstrated (Bonting, 1966). The properties were quite similar to those of the enzyme in other species (table 3). The ouabain inhibition curve (figure 6) showed a pI$_{50}$ of 6·7–6·8 and a slight stimulation at $10^{-9}$–$10^{-8}$ M ouabain. The pH optimum of the Na–K ATPase system was between 6·3 and 7·0, slightly lower than in vertebrate tissues (table 3). The enzyme activities found in the 9 species studied (table 17) indicate a larger variation in Na–K ATPase activity (4·2

**Table. 17.** ATPase activities in rectal gland of nine elasmobranch species

| Species | No. of animals | Body wt in kg | Gland wt in g | Mg ATPase | | Na–K ATPase | | mmole/ hr/kg body wt |
|---|---|---|---|---|---|---|---|---|
| | | | | Wet | Dry | Wet | Dry | |
| Spiny dogfish | 32 | 2·27–4·54 | 0·97–2·09 | 0·69 | 2·90 | 1·32 | 5·68 | 0·60 |
| Sand shark | 10 | 9·1 –18·2 | 1·69–3·33 | 0·59 | 2·56 | 0·75 | 3·25 | 0·12 |
| Sickle-shaped shark | 1 | 6·80 | 1·29 | 0·79 | 3·08 | 0·54 | 2·13 | 0·10 |
| Smooth dogfish | 6 | 1·36–1·7 | 0·27–0·39 | 0·81 | 3·51 | 0·36 | 1·57 | 0·079 |
| Clear-nosed skate | 5 | 0·57–1·81 | 0·12–0·35 | 0·67 | 2·80 | 0·32 | 1·31 | 0·059 |
| Eagle ray | 6 | 0·79–6·80 | 0·11–1·28 | 0·67 | 2·62 | 0·33 | 1·31 | 0·051 |
| Butterfly ray | 1 | 36·3 | 3·13 | 0·67 | 2·61 | 0·43 | 1·65 | 0·037 |
| Sting ray | 4 | | 0·94–2·05 | 0·92 | 4·42 | 0·39 | 1·88 | — |
| Monkfish | 4 | 10·0–15·4 | 0·88–1·68 | 0·36 | 1·62 | 0·36 | 1·56 | 0·032 |

Enzyme activities in moles/hr/kg wet or dry wt of gland. In last column activities in mmole/hr/kg body wt given as averages of individual values, calculated with the appropriate values of body weight and gland weight.
From Bonting (1966).

fold) than for the Mg ATPase activity (2·7 fold) between the various species. Spiny dogfish had the highest Na–K ATPase activity, expressed in each of the three ways used in table 17. An interesting point is that the species with radially arranged tubules (4 shark species) had the highest Na–K ATPase activities per kg body weight (0·08–0·6 mmole/hr/kg), while the species with lobular arrangement (skates and rays) had the lowest values (0·03–0·06). In spiny dogfish, of which 32 animals were analysed, a significant decrease of Na–K ATPase activity with increasing body weight was observed, leveling off at body weights over 3·7 kg. Comparison of the maximal excretion rate for animals of 4·3 kg average body weight (Burger, 1962) with the Na–K ATPase activity for animals of 3·7–4·5 kg body weight (Bonting, 1966) yielded a Na/ATP ratio of 2·2, again in good agreement with the values for other species and tissues listed in tables 4 and 5. Arterial perfusion with $10^{-4}$ M ouabain of the functioning, isolated rectal gland of spiny dogfish gave, after a transient increase of secretion, complete inhibition of secretion in all three preparations tested (R.F. Palmer, personal communication). It thus appears that the Na–K ATPase system has also a primary, rate-limiting role in salt secretion by the elasmobranch rectal gland.

In teleost fishes the gills appear to play an important role in the salt economy. Marine teleosts and sea water adapted eels swallow sea water and absorb it intestinally, and the absorbed cations with excess water are excreted mainly by the gills (Parry, 1966). In eels adapted to sea water the Na efflux is 13 mmoles/hr/kg fish, while in fresh water eels it is only 0·04 mmole/hr/kg fish (Maetz, Mayer and Chartier-Baraduc, 1967). Upon incubation of gills of fresh water eels in sea water there is a large uptake of $Na^+$, while in isolated gills of sea water adapted eels the $Na^+$ content does not increase during incubation in sea water (Bellamy, 1961; Kamiya, 1967). Injection of ouabain in the latter animals made their gills behave in the same way as those of fresh water eels (Kamiya, 1967). In the gills of the sea water adapted animals a Na–K ATPase activity 4·4 times that in fresh water animals was demonstrated (Kamiya, 1968). The adaptation of the enzyme activity took 7 days. The Na–K ATPase system of the gills had all the properties of the enzyme from other species and tissues. It was inhibited by ouabain and high $K^+$ concentrations antagonised this effect. In the intestine, which maintains an ouabain-sensitive transport of $Na^+$ from mucosa to serosa, there was a similar increase in Na–K ATPase activity (Oide, 1967). A 7-fold increase in Na–K ATPase, without change in glutaminase and succinic dehydrogenase, was found in a microsomal fraction of the gills of killifish (*Fundulus heteroclitus*) during adaptation to sea water (Epstein, Katz and Pickford, 1967). Ouabain-sensitive Na–K ATPase activity has also been demonstrated in the land crab (*Cardisoma guanhumi*) which maintains constant $Na^+$ and $K^+$ levels in blood, urine and stomach fluid when placed in distilled water as well as in 1·33 times concentrated sea water (Quinn and Lane, 1966). These findings suggest that the enzyme plays a key role in osmotic regulation in the gills of teleost fishes and invertebrates, as it does in the marine bird nasal gland and the elasmobranch rectal gland.

## D. Liver

Although the liver cannot be considered primarily a secretory organ, it is discussed here because of its role in bile secretion. An active cation transport in liver slices, dependent on oxidative metabolism and sensitive to ouabain was observed (Van Rossum, 1961, 1966; Elshove and Van Rossum, 1963; Cascarano and Seidman, 1965), while the occurrence of the Na–K ATPase system in moderate activity was first demonstrated in cat liver (Bonting, Simon and Hawkins, 1961), subsequently in rat liver (Bonting, Caravaggio and Hawkins, 1962), in rat liver microsomal fraction (Järnefelt, 1962; Schwartz, 1963; Morgan and Leon, 1963; Boernig and Giertler, 1965), in a rat liver lipoprotein preparation (Ahmed and Judah, 1964) and in isolated plasma membranes of rat liver (Emmelot and Bos, 1962; Emmelot and associates, 1964; Emmelot and Bos 1966). The findings with plasma membranes demonstrated the occurrence of the enzyme system in the cell membrane, perhaps extended to the endoplasmatic reticulum in view of the microsomal activity. A fair agreement was found between the active uptake rate for $K^+$ (Van Rossum, 1961, 1963) and the Na–K ATPase activity in rat liver during pre- and post-natal development (Bakkeren, 1968).

An extensive study of the properties of the Na–K ATPase system was made in rat liver homogenates (Bakkeren and Bonting, 1968a). Whole liver homogenates were used, since the same $pI_{50}$ value for ouabain inhibition and the same pH optimum were found as in a plasma membrane fraction prepared according to Emmelot and associates (1964) and the relative Na–K ATPase activity was not appreciably higher in the latter preparation or after deoxycholate ($0.1\%$) treatment according to Ahmed and Judah (1964). In view of the low relative activity, the homogenate was treated with urea, which increased the relative activity from 13 to $37\%$ and facilitated the determination of the Na–K ATPase properties. Figure 2 shows the $Mg^{2+}$ activation curve with an optimal $Mg^{2+} : ATP$ ratio of $1:1$. This and other characteristics, summarised in table 3, are very similar to those obtained for the enzyme system in a variety of other tissues. There was a high substrate specificity towards ATP for Na–K ATPase, but not for the Mg ATPase activity. The Na–K ATPase activity was inhibited by ouabain ($pI_{50} = 3.9$) and by erythrophleine ($pI_{50} = 5.1$). The low sensitivity in kidney ($pI_{50} = 3.9$, table 3), heart muscle ($pI_{50} = 4.3$; Dransfeld and associates, 1966), rat liver and other rat tissues for ouabain is in accordance with the low sensitivity of the rat for cardiac glycosides (Repke, Est and Portius, 1965). The rat brain, however, has a $pI_{50} = 5.2$ (Aldridge, 1962), which probably explains why the rat dies in convulsions, rather than from cardiac arrest, after a lethal dose of cardiac glycoside. A higher sensitivity for erythrophleine than for ouabain has been found in several tissues: rabbit brain (Bonting, Hawkins and Canady, 1964), cat choroid plexus (Vates, Bonting and Oppelt, 1964), toad bladder (Bonting and Canady, 1964), and rabbit pancreas (Ridderstap and Bonting, 1969b).

In view of the ability of the regenerating rat liver after partial hepatectomy to maintain a higher $K^+$ and a lower $Na^+$ concentration than normal liver (Humphrey, 1965), the activity of the cation pump in the regenerating rat liver was studied (Bakkeren and and Bonting, 1968b). It had occurred to us that an increased activity of the Na–K ATPase system would explain both cation changes, while a decrease in membrane permeabilities for the two cations, as suggested by Humphrey (1965), would involve two separate changes. Contrary to an earlier report by Morgan and Leon (1963), the activity of the Na–K ATPase system was indeed found to be increased by 57% maximally from 3 to 6 days of regeneration, and the ouabain-sensitive $^{86}Rb^+$ uptake rate in regenerating liver slices was increased by about 54% maximally from 2 to 6 days after hepatectomy (figure 21). The Mg ATPase activity was not significantly increased, except for an 18% increase at 6 days.

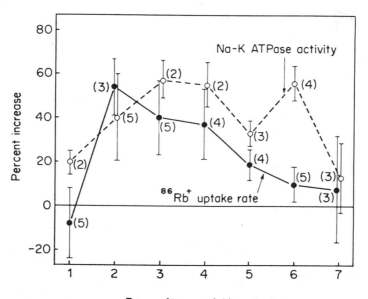

Days after partial hepatectomy

**Figure 21.** Increase of Na–K ATPase activity and ouabain-sensitive $^{86}Rb^+$ uptake rate in regenerating rat liver. Open circles, percentage increase of enzyme activity on dry weight basis, relative to liver tissue removed from same animal at time of partial hepatectomy. Closed circles, percentage increase after partial hepatectomy in ouabain-sensitive part of $^{86}Rb$ uptake, relative to liver tissue removed from same animal. Each point represents mean of 3 determinations on number of animals indicated in parentheses. (From Bakkeren and Bonting, 1968b).

Sham-operated animals did not show any changes in enzyme activity. The increases in Na–K ATPase activity and active $^{86}Rb^+$ uptake coincided with the time course for the changes in cation levels found by Humphrey (1965). Although these results seemed to settle the matter, the passive permeabilities for $^{86}Rb^+$ and $^{22}Na^+$ were also determined in rat liver slices before, and 3 days after, partial hepatectomy. Somewhat to our surprise a significant decrease in the passive $^{86}Rb$ efflux was found, though not in $^{22}Na^+$ influx. An increased pump activity might be expected as an adaptive process in cells with increased cation permeability, as in the case of leukaemic leukocytes (Section III D). In the regenerating rat liver, however, an increase in Na–K ATPase activity was accompanied by a decrease in permeability for $Rb^+$. These two changes would tend to reinforce each other in their effect on the cation levels.

With regard to the physiological significance, it is unlikely that the increased Na–K ATPase activity is related to the compensatory growth of the residual liver or the increased need for metabolites to support the accelerated mitotic rate. The weight increase is maximal in the first 2 days and the mitotic rate at 28 hr after hepatectomy, while the Na–K ATPase activity was maximal from 3 to 6 days after operation. Moreover, we could not demonstrate ouabain-sensitive active uptake of $^{14}C$-labelled amino acids and sugars. A more likely correlation would be with bile secretion, which would have to increase if the remaining one third of the liver is to produce sufficient bile for digestion. Leong, Pessotti and Brauer (1959) found in rats, after partial hepatectomy, a transient increase in bile flow rate per gram liver which reached a maximum of 53% 3 days after operation and returned to normal after 14 days. This finding agrees well with the 54% maximal increase in Na–K ATPase activity 3 days postoperatively. Moreover, a clear inhibition of the bile flow rate in rats after intravenous injection of ouabain could be established (Bakkeren, 1968).

A small, though significant diurnal variation in rat liver Na–K ATPase activity was observed (Bakkeren, 1968). The Na–K ATPase activity was 22% higher (P = 0·0033) during the period from 4.0 a.m. until 2.0 p.m. and the Mg ATPase activity was 15% higher (P = 0·013) during this period, compared with the other half of the 24-hour period. A similar increase (21–25%) in mitochondrial Mg ATPase activity in hamster liver during the dark period has been reported (Nishiitsutsuji-Uwo and associates, 1967). In general, diurnal changes are much smaller for membrane-bound enzymes than for cytoplasmic enzymes, where changes from 150–400% are found (Wurtman, Shoemaker and Larin, 1968). This diurnal effect on liver Na–K ATPase may be related to the increased physical activity (Browman, 1937) and food intake (Fuller and Snoddy, 1968) by rats during the nightly hours or to diurnal variations in adrenocortical and pituitary hormone levels (Critchlow and associates, 1963; Retiene and associates, 1968).

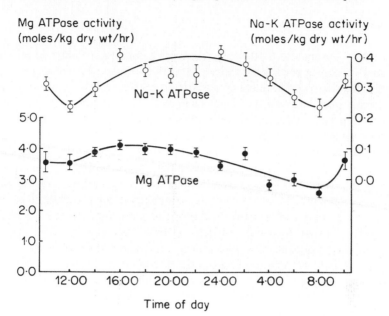

**Figure 22.** Diurnal variations in Na–K ATPase and Mg–ATPase activities of rat liver. Each point represents mean activity (in mole/kg dry wt/hr) for 10 animals with standard error. (From Bakkeren, 1968).

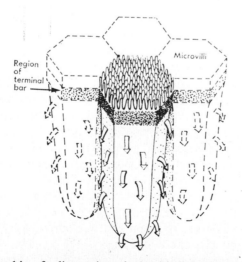

**Figure 23.** Relationship of adjacent intestinal epithelial cells. During fluid absorption the lateral membranes below the terminal bar region become separated, forming intercellular channels. (From Parsons, 1967a).

## E. Intestine

The Na$^+$ absorption by the intestine from lumen to serosal side resembles in many ways the processes of transcellular Na$^+$ transport occurring in kidney tubules, frog skin and toad bladder (Parsons, 1967a,b). It is an active process, which can occur against an electrochemical gradient and is dependent on metabolic processes taking place inside the absorbing cells. Coupled to the Na$^+$ transport there is absorption of fluid isotonic with the lumen content, and also of certain sugars and amino acids. It appears that the intestinal Na$^+$ absorption is of crucial importance in recovering the large amounts of water and salt entering the intestine from gastrointestinal secretions (saliva, gastric juice, bile, pancreatic juice) and in absorbing dietary metabolites. Morphological studies indicate that the mucosal cells are laterally attached to each other at the terminal bar region immediately below the microvilli (figure 23). During fluid absorption the lateral membranes below the terminal bar region become separated, forming intercellular channels. This points to the same system for salt and water absorption as observed in the proximal tubule of the kidney (Section V A). We shall return to this at the end of this section.

The presence of the Na–K ATPase system in intestinal mucosal cells was demonstrated for guinea pig (Taylor, 1962), rat (Taylor, 1963a; Berg and Chapman, 1965), rabbit (Richardson, 1968) and goldfish (Smith, 1967; Smith, Colombo and Munn, 1968). The properties of the guinea pig enzyme system, listed in table 3, agree closely with those of the enzyme in other tissues. Some differences in properties between the enzyme from rat intestine and rat erythrocytes in divalent cation stimulation and substrate specificity were noted (Berg and Szckerczes, 1966), and similarly between intestine from goldfish acclimatized to 8° and 30 °C (Smith, 1967; Smith, Colombo and Munn, 1968) and between colon from aldosterone-injected and saline-adapted toads (Ferreira and Smith, 1968). From centrifugal fractionation studies it was concluded that the epithelial membrane facing the serosa would have at least as much and perhaps more Na–K ATPase activity than the luminal side of the cell membrane (Berg and Chapman, 1965), and that the brush border fraction has only a small activity (Quigley and Gotterer, 1969a,b).

Various correlations between intestinal cation transport and Na–K ATPase activities were noted. In isolated rabbit ileum a short-circuit current was measured, which could be attributed to the net, active transport of Na$^+$ from the mucosal side of the ileum and was inhibited by ouabain present on the serosal side (Schultz and Zalusky, 1964a). This Na$^+$ transport is accompanied by a passive movement of Cl$^-$ and H$_2$O (Schultz, Zalusky and Gass, 1964) and is markedly stimulated by the addition of an actively transported, non-metabolised sugar on the mucosal side (Schultz and Zalusky,

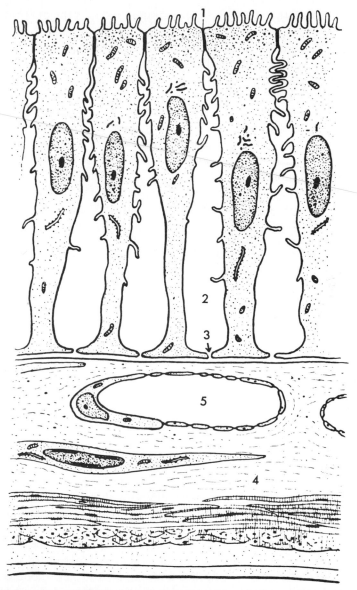

**Figure 24.** Structure of gall bladder wall showing junctional complex or desmosome (1), intercellular spaces (2), with narrow opening (3), towards lamina propria (4), containing capillaries (5). In local osmosis $Na^+$ enters cell passively from lumen, is transported actively across lateral cell membranes into intercellular spaces. The local hypertonicity draws water through the cell into the spaces, with dilution to isotonicity of the fluid and distension of the spaces. The resulting hydrostatic pressure expels the fluid into the lamina propria, where it is carried away by the capillaries. (After **Kaye** and coworkers, 1966).

1964b). Increasing the serosal $K^+$ concentration reduced the ouabain inhibition of sugar transport in rat small intestine (Newey, Sanford and Smyth, 1968), as it does the Na–K ATPase inhibition by ouabain (Section II B). In rat small intestine bile acids inhibited Na–K ATPase activity, Na transport, oxygen uptake and transport of urea and sorbose; the enzyme inhibition was considered the primary effect (Parkinson and Olson, 1964). Cetyltrimethylammonium bromide inhibited water and glucose transport and Na–K ATPase activity in rat small intestine about equally with a $pI_{50}$ of approximately 4·4 (Taylor, 1963a). There has also been found a simultaneous inhibition of intestinal $Na^+$ transport and Na–K ATPase activity by purgative drugs e.g. phenolphthaleine (Chignell, 1968; Richardson, 1968). The decrease in human intestinal Na–K ATPase activity in cholera (Hirschhorn and Rosenberg, 1968) appears to be due to the inactivating effect on the enzyme by mucinase produced intestinally by *Vibrio cholerae* (Richardson, 1968).

The gall bladder concentrates the bile by reabsorption of NaCl and water from its content. The mechanism of the coupling between $Na^+$ transport and water transport has been clarified by Diamond (1962a,b,c; 1964a,b). He could prove that the primary process is active $Na^+$ transport, and that water follows passively and isosmotically by local osmosis, and not by pressure filtration, classical osmosis, electro-osmosis, co-diffusion or the double-membrane effect (Curran and MacIntosh, 1962). Subsequent morphological studies (Diamond and Tormey, 1966; Kaye and associates, 1966) have further elucidated the mechanism of this process. The epithelial cells of the gall bladder are joined by a junctional complex or terminal bar region close to the lumen (figure 24, 1). When reabsorption was blocked by cooling or by bathing the gall bladder in a medium containing ouabain or in a $Na^+$- or $Cl^-$-free medium, the lateral surfaces of adjacent cells were in close apposition and the intercellular spaces were only 150–200 Å in width. In the reabsorbing gall bladder the intercellular spaces were distended up to about 20,000 Å (figure 24, 2), while the opening (figure 24, 3) towards the subepithelial lamina propria (figure 24, 4) with the blood capillaries (figure 24, 5) remained of the order of 200 Å. After fixation with osmium tetroxide–pyroantimonate, high $Na^+$ concentrations were detected in the distended intercellular spaces. Histochemically, ATPase activity was demonstrated along the lateral cell membranes. It was concluded that NaCl is actively transported out of the cells into the intercellular spaces, thus building up a local osmotic gradient. This gradient draws water into the spaces, leading to their distension and to dilution of the intercellular fluid to a concentration isotonic with the luminal fluid. The narrow opening towards the lamina propria retards the diffusion of solute from the intercellular space, thus aiding the osmotic equilibration. Then the increasing hydrostatic pressure in this space expels the fluid into the lamina propria, where it is carried away by the capillaries. The finding that

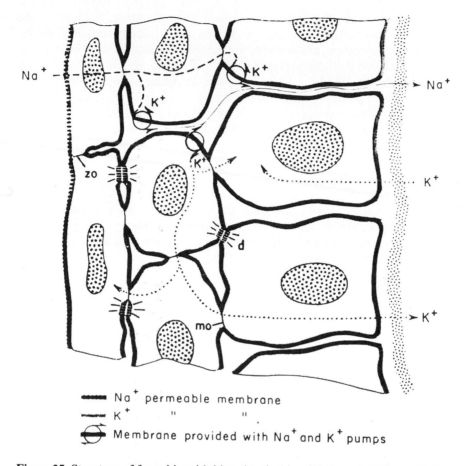

**Figure 25.** Structure of frog skin with histochemical localization of ATPase activity on lateral membranes of cells. Cell layers represent, from left to right: stratum corneum, s. spinosum, s. germinativum. Abbreviations: zo, zonula occludens; o, macula occludens; d, desmosome. The labryinth of intercellular spaces would function in coupled transport of $Na^+$ and $H_2O$ (from left to right) by local osmosis. (From Farquhar and Palade, 1966).

there is as much or more Na–K ATPase activity in the membrane on the serosal side as on the luminal side of intestinal epithelial cells (Berg and Chapman, 1965) suggests that the unidirectional $Na^+$ transport is not caused by a one-sided location of the enzyme, but by the asymmetric distribution of the cell surface through the excentric location of the terminal bar region (figures 23 and 24) and the resulting excess pump capacity on the lateral cell surfaces.

The situation in the frog skin, though more complex, appears to be basically similar (figure 25). $Na^+$ ions enter the stratum corneum (left) by passive permeability, are actively transported across inward facing cell membranes into the labyrynth of intercellular spaces, and finally emerge on the serosal side. Water follows passively by local osmosis. Strategically located terminal bars or desmosomes ensure that the net transport is from left to right. This process appears to explain isotonic water transport, coupled to active Na transport, in kidney, gall bladder, frog skin and possibly other structures as well.

## F. Pancreas

The exocrine secretion of the pancreas consists of a salt solution, isotonic with the blood plasma and containing a number of digestive enzymes. The secretion process takes place in the pancreatic acini with the associated ductuli. These structures consist of two types of cells: the acinar cells, richly endowed with rough endoplasmatic reticulum, which synthesise the enzymes and store them in the zymogen granules, and the acino-ductular cells, poor in endoplasmatic reticulum, which appear to be responsible for salt and fluid secretion. The hormone pancreozymin, released by the upper intestinal mucosa after food intake, stimulates the secretion of enzymes through 'reversed pinocytosis' in which fusion of zymogen granules with the luminal cell membrane is followed by opening of the membrane and release of enzymes into the acinar lumen. The hormone secretin, also released by the upper intestinal mucosa, stimulates the fluid and electrolyte secretion by the acino-ductular cells.

A role of the Na–K ATPase system in exocrine pancreatic secretion of fluid and electrolytes, but not of enzymes, has been demonstrated (Ridderstap and Bonting, 1968, 1969a,b). The presence of the enzyme system in dog and rabbit pancreas was established and its properties were determined. The properties were similar to those of the enzyme system in other tissues (table 3). The activities in whole homogenates were fairly low, both absolutely (0·33 and 0·23 mole/kg dry wt/hr for dog and rabbit, respectively) and relatively (26 and 8%, respectively). Pretreatment with urea was necessary to lower the Mg ATPase activity sufficiently for the determination of the enzyme properties.

In the dog the pancreatic fluid was collected *in vivo* after cannulating the pancreatic duct (Ridderstap and Bonting, 1969a). When ouabain (4·5 $\mu$g/kg body wt) was injected into the arterial blood supply of the pancreas, thus avoiding as much as possible systemic effects of the ouabain, the fluid secretion was inhibited (by 70%) without changes in $Na^+$ and $HCO_3^-$ concentrations. Comparison of the secretin-stimulated fluid secretion and the total pancreatic Na–K ATPase activity gave a Na/ATP ratio of 1·8. This ratio is somewhat low compared to the values in table 6 probably because not all the Na–K ATPase activity measured in whole pancreas homogenate is present in the acino-ductular cells, which are responsible for fluid secretion.

In view of the technical limitations of the *in vivo* approach, further studies were carried out with the isolated rabbit pancreas (Ridderstap and Bonting, 1969b). The rabbit pancreas is ideally suited for this purpose because of its thin sheetlike configuration (Rothman, 1964). A sustained high flow rate was obtained for at least 5 hr. The flow rate was constant during the 3rd and 4th hours after explantation, and therefore these periods were chosen as control and experimental periods respectively. Various cardiac glycosides and erythrophleine inhibited both the Na–K ATPase activity and the flow rate (table 18). The sequence in inhibitory effect was the same in both cases,

**Table 18.** Effect of transport inhibitors on Na–K ATPase activity and pancreatic flow rate of rabbit pancreas *in vitro*

| Inhibitor | Concn. (M) | Na–K ATPase inhibition (%) | Concn. (M) | Flow rate inhibition (%) |
|---|---|---|---|---|
| Erythrophleine | $10^{-4}$ | 100 (3) | $10^{-6}$ | 44 ± 0·2 (2) |
| Ouabain | $10^{-4}$ | 74 ± 2·8 (3) | $10^{-5}$ | 65 ± 1·3 (4) |
| Scillaren A | $10^{-4}$ | 64 ± 1·9 (3) | $10^{-5}$ | 60 ± 13·7 (2) |
| Digoxin | $10^{-4}$ | 52 ± 3·2 (3) | $10^{-5}$ | 35 ± 0·2 (2) |
| Hexahydroscillaren A | $10^{-4}$ | 0 (3) | $10^{-5}$ | 2 ± 3·7 (2) |

Means with standard errors and, in parentheses, the number of experiments.
From Ridderstap and Bonting (1969b).

erythrophleine having the greatest effect, ouabain less and hexahydroscillaren A almost none. Ouabain at various concentrations had similar biphasic effects on enzyme and fluid secretion with identical $pI_{50}$ of 5·4 for the two inhibitory effects (figure 26). The lack of inhibition of either activity by hexahydroscillaren A, a cardiac glycoside with a saturated lactone group, is in agreement with the previously established requirement of an unsaturated lactone group for the inhibitory effect of cardiac glycosides on the Na–K ATPase system and active $Na^+$ transport (Dunham and Glynn, 1961). In all of these experiments the $Na^+$ concentration of the secreted fluid remained unchanged. When $^{22}Na^+$ was added to the bathing medium, its secretion was inhibited to the same degree as the flow rate by ouabain ($5 \times 10^{-6}$ M; 57% and 58%, respectively). Lowering the $Na^+$ concentration in the bathing medium by 85% (from 170 to 25 mM) gave a proportional decrease in the $Na^+$

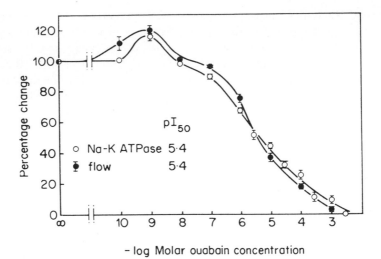

**Figure 26.** Effect of ouabain on fluid secretion by isolated rabbit pancreas and Na–K ATPase activity in the organ. (From Ridderstap and Bonting, 1969b).

secretion rate (84 %), which was reversible. From these results it was concluded that active Na$^+$ transport by means of the Na–K ATPase system is the primary and rate-limiting event in exocrine pancreatic secretion, with water following passively the actively secreted Na$^+$ by local osmosis. Morphological evidence for the latter process was obtained: in dog pancreas biopsies obtained after injection of secretin there was distension of the intercellular spaces between the acino-ductular cells, while biopsies obtained before injection showed no distension (Ridderstap, 1969). Since the terminal bar region was found to be on the luminal side, the process of local osmosis must function here in the opposite way to that in the intestinal epithelium. The sodium pump must be located mainly on the luminal side, drawing Na$^+$ passively across the lateral cell membrane from the intercellular space. The resulting hypotonicity in this space makes water flow with the Na$^+$ ions (Diamond and Bossert, 1968).

The enzyme secretion did not appear to be coupled to the Na–K ATPase system (Ridderstap and Bonting, 1969c). Ouabain, acetazolamide and Na-azide inhibited flow severely without affecting output of protein and α-amylase. NaF had the reverse effect, while anaerobiosis and lowering the Na$^+$ concentration in the bathing fluid depressed the flow rate significantly more than the protein and α-amylase secretion. Instead, cyclic AMP appears to play a role as a mediator of the pancreozymin effect on enzyme secretion (Kulka and Sternlicht, 1968; Ridderstap and Bonting 1969d).

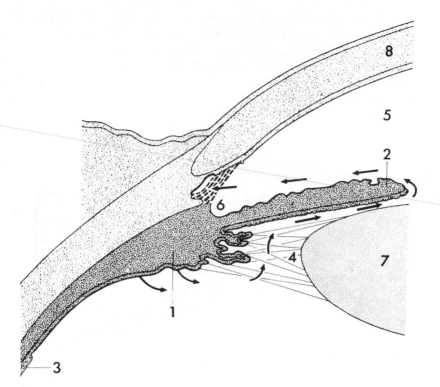

**Figure 27.** Flow of aqueous humor in the eye. 1. ciliary body, 2. iris, 3. retina, 4. posterior chamber, 5. anterior chamber, 6. chamber angle, 7. lens, 8. cornea.

## G. Aqueous Humour Formation

The flow of the aqueous humour in the eye is illustrated in figure 27. The ciliary body (1), located between the base of the iris (2) and the periphery of the retina (3) and extending in a full circle around the optical axis, secretes a fluid which flows through the posterior (4) and anterior (5) chambers and leaves the eye by the trabecular meshwork in the chamber angle (6). It has three functions: maintaining the shape of the eye by its pressure, acting as a refractory medium and transporting metabolites to and from the lens and perhaps the corneal endothelium. Its chemical composition approximates to that of a plasma ultrafiltrate. However, upon dialysis against plasma of the same species there is a slight loss of $Na^+$ and $Cl^-$ (Davson, Duke-Elder and Maurice, 1949), indicating that it cannot be formed by ultrafiltration from plasma. Studies of the accumulation of $^{24}Na^+$, $^{36}Cl^-$ and $H^{14}CO_3^-$ from the plasma in the posterior and anterior eye chambers showed that 2/3 of the

Na$^+$ is actively secreted by the ciliary body (Kinsey and Reddy, 1959). After occlusion of the chamber angle in the eyes of rabbits and collection of the effluent by a needle piercing the cornea, it was possible to show 75% inhibition of Na$^+$ secretion by dinitrophenol and fluoroacetamide and 62% inhibition by ouabain (Cole, 1960a,b; 1961). In the hypothermic rabbit there was an equal inhibition of Na$^+$ influx and aqueous humour flow (Becker, 1961). These findings suggested a role of active Na$^+$ transport in the formation of aqueous humour.

Substantial Na–K ATPase activity was demonstrated in cat and human ciliary body (Bonting, Simon and Hawkins, 1961) and in rabbit ciliary body (Bonting, Caravaggio and Hawkins, 1962). Ouabain inhibition curves for cat and rabbit ciliary body (pI$_{50}$ 6·6 and 6·0 respectively) were determined (Bonting, Caravaggio and Hawkins, 1962). Intravenous injection of ouabain (0·065–0·11 mg/kg) in the cat gave a significant decrease in ciliary body Na–K ATPase when compared to the activity in the other eye, removed before injection (Simon, Bonting and Hawkins, 1962). There was also a significant decrease in aqueous inflow, measured as the recovery rate of the intraocular pressure after draining a small amount of aqueous humour from the eye (table 19; first column). The smaller decrease in enzyme activity is

**Table 19.** Effects of cardiac glycosides on aqueous humour formation and ciliary body Na–K ATPase activity

|  | Cat (ouabain i.v. 0·065–0·11 mg/kg) (% change) | P | Man (digoxin orally, 0·5 mg twice daily) (% change) | P |
|---|---|---|---|---|
| Na–K ATPase activity | −41 ± 9·7 (7) | 0.02 | — |  |
| Intraocular pressure | −19 ± 3·9 (7) | 0·004 | −15 ± 3·8 (16) | 0·001 |
| Recovery rate IOP | −70 ± 4·7 (6) | 0·007 | −46 ± 4·0 (10) | <0·001 |
| Outflow rate | — | — | −7 ± 5·9 (16) | 0·25 |
| Scleral rigidity | — | — | −2 ± 3·9 (13) | 0·55 |

Means with standard error and, in parentheses, number of animals or patients.
$P$ = bilateral probability, calculated from $t$-test.
From Simon, Bonting and Hawkins (1962).

due to the dilution of the tissue during assay, since ouabain is not irreversibly bound. Similar measurements of aqueous humour formation in man by means of aplanatic tonometry after bulbar compression showed a significant decrease of flow after oral administration of digoxin (0·5 mg twice daily) without changes in outflow facility and scleral rigidity (table 19; second column). This proves that the cardiac glycoside did not lower the recovery rate of intraocular pressure through increasing the aqueous humour outflow or decreasing the rigidity of the eye ball. These therapeutic doses of digoxin lowered intraocular pressure in glaucoma patients as much as acetazolamide did, while a combination of these drugs had a significantly greater effect than either alone (Simon and Bonting, 1962), indicating a possible usefulness of cardiac glycosides in the treatment of glaucoma.

A very pronounced effect of ouabain on aqueous humour inflow was found after the injection of ouabain (0·1–0·5 μg) into the vitreous of the rabbit eye *in vivo* (Becker, 1963). This finding led to a further study of the relationship between the ciliary body Na–K ATPase system and aqueous humour flow (Bonting and Becker, 1964). A single injection of 0·5 μg ouabain inhibited flow by 78% and *in vitro* enzyme activity by 27% corresponding to an *in vivo* enzyme inhibition of 70% (table 20). There was a significant positive correla-

Table 20. Effect of intravitreal ouabain injection on ciliary body ATPase activities and aqueous humour formation in the rabbit

| Dose | Mg ATPase activity (% change) | Na–K ATPase activity (% change) | Aqueous humour formation (% change) |
|---|---|---|---|
| 0·5 μg | —4·1 ± 1·8 $P = 0·06$ | —26·6 ± 4·6 $P < 0·001$ | —78 ± 7·3 $P < 0·001$ |
| 0·2 μg | —3·8 ± 4·0 $P = 0·37$ | —16·2 ± 2·6 $P < 0·001$ | —51 ± 7·6 $P < 0·001$ |
| 0·1 μg | —2·3 ± 3·0 $P = 0·47$ | —23·2 ± 3·7 $P < 0·001$ | —47 ± 6·5 $P < 0·001$ |

Tonography was performed 4 days after injection, ciliary body obtained on 5th day. Percentage change between injected eye and control eye, and averaged for the 8 animals per dose group, given with standard error and probabilities calculated by the *t*-test.
From Bonting and Becker (1964).

tion between flow inhibition and enzyme inhibition at each dose of ouabain (table 21). The time course for the enzyme inhibition closely paralleled that for flow inhibition (figure 28). Maximum effects were reached 4–5 days after injection, while return to normal conditions occurred after 20 days. [86]Rb accumulation in the isolated ciliary body was inhibited to the same extent as the Na–K ATPase activity with $pI_{50}$ values of 5·9 and 6·0 respectively (figure 29). From the normal flow rate and the total ciliary body Na–K ATPase activity a Na/ATP ratio of 3·0 was calculated, in good agreement with the values listed for a variety of tissues in table 5. Another correlation is that erythrophleum alkaloids inhibit both aqueous humour inflow and the ciliary body Na–K ATPase system (Bonting, Hawkins and Canady, 1964;

**Table 21.** Correlation between inhibition of ciliary body Na–K ATPase inhibition and aqueous humour formation after intravitreal ouabain injection in the rabbit eye

| Animal no. | Dose 0·5 g | | Dose 0·2 g | | Dose 0·1 g | |
|---|---|---|---|---|---|---|
| | enzyme activity (% change) | aqu. humour formation (% change) | enzyme activity (% change) | aqu. humour formation (% change) | enzyme activity (% change) | aqu. humour formation (% change) |
| 1 | 42·1 | 90 | 14·8 | 33 | 11·7 | 25 |
| 2 | 7·5 | 40 | 20·0 | 70 | 14·0 | 45 |
| 3 | 19·3 | 80 | 11·0 | 50 | 32·2 | 65 |
| 4 | 40·7 | 90 | 11·5 | 67 | 34·8 | 65 |
| 5 | 14·4 | 50 | 18·8 | 60 | 10·7 | 35 |
| 6 | 37·4 | 90 | 32·2 | 80 | 35·6 | 75 |
| 7 | 26·8 | 90 | 10·5 | 20 | 26·5 | 50 |
| 8 | 24·3 | 90 | 10·6 | 30 | 23·1 | 30 |
| aver. | 26·6 | 78 | 16·2 | 51 | 23·6 | 49 |
| | $r = +0.83$ | | $r = +0.72$ | | $r = +0.86$ | |
| | $P = 0.011$ | | $P = 0.047$ | | $P = 0.006$ | |

Correlation coefficient $r$ and probability $P$ calculated according to Snedecor (1961). From Bonting and Becker (1964).

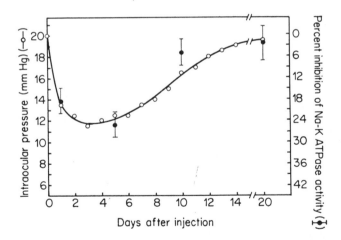

**Figure 28.** Time course of intraocular pressure decrease due to inhibition of flow and of ciliary body Na–K ATPase inhibition after intravitreal injection of 0·5 μg ouabain into the rabbit eye. (From Bonting and Becker, 1964).

Brown, Jackson and Waitzman, 1967). These results strongly suggest a primary and rate-limiting role of the ciliary body Na–K ATPase activity in aqueous humour secretion.

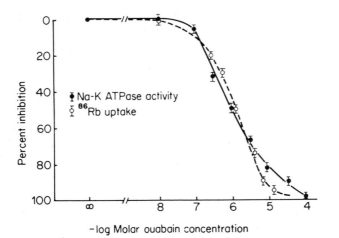

**Figure 29.** *In vitro* inhibition of ciliary body Na–K ATPase activity ($pI_{50} = 6\cdot0$) and $^{86}$Rb uptake in ciliary body ($pI_{50} = 5\cdot9$) by varying concentrations of ouabain. (From Bonting and Becker, 1964).

Some contradictory results have been obtained for the effects of ouabain on aqueous humour formation by a few investigators. The partially negative results of Langham and Eakins (1964) for intravenous and intra-arterial ouabain may have been due to unsuitable techniques, and their complex effects with intravitreal ouabain to excessive dosage. The suggestion that the inhibition of aqueous inflow by intravenous ouabain in cats might actually be due to pentobarbital anaesthesia (Warner and Drance, 1966) is made unlikely by the results obtained with a perfusion and inulin-dilution technique (Oppelt, 1967). Direct observation of aqueous humour secretion by an isolated ciliary body preparation confirmed *in vitro* inhibition by ouabain as well as by cyanide, azide and dinitrophenol (Berggren, 1964a,b). Subconjunctival as well as topical administration of ouabain led to inhibition of aqueous humour flow (Waitzman and Jackson, 1964, 1965). Measurement of flow from the dilution of intracamerally administered inulin showed inhibition by intravenous ouabain (Macri, Dixon and Rall, 1966). This was confirmed (Oppelt and White, 1968) with the more refined method of perfusion and inulin dilution measurement (Oppelt, 1967), but intracamerally ouabain had no effect, possibly because too much of it is carried away and perhaps bound by lens or cornea (Oppelt and White, 1968).

The Na–K ATPase activity of the ciliary epithelial cells, which are the site of aqueous humour secretion, is more than 20 times as high as that in the ciliary stroma (Riley, 1964a). Some evidence for a localisation of the Na–K

M

ATPase system in the non-pigmented epithelial cells (adjacent to the posterior chamber) of rabbit ciliary body, rather than in the pigmented epithelial cells (between the former and the ciliary stroma), has been obtained by a histochemical staining method with only 0·9 mM $Pb^{2+}$ as phosphate-precipitating reagent (Cole, 1964). Electronmicroscopic studies show high Mg activated ATPase activity in the infoldings of the plasma membrane at the basal side (adjacent to the posterior chamber) of the non-pigmented cells (Shiose and Sears, 1965, 1966; Kaye and Pappas, 1965).

A pacemaker effect of the ciliary body Na–K ATPase system is indicated by the observation that absence of $Na^+$ or presence of ouabain inhibits glycolysis *in vitro* by 61 % and oxygen uptake by 64 % (Riley, 1964b). The main energy source for $Na^+$ transport in the ciliary body is oxidative phosphorylation. The latter process and glycolysis are regulated by the ADP production from the Na–K ATPase system (Cole, 1965; Riley, 1966). Definitive proof of an active $Na^+$ transport from stroma to chamber follows from the observation that the isolated ciliary body of the cat gives in an Ussing chamber a short-circuit current, which is equivalent to the net $Na^+$ flux determined by means of a double isotope technique (Holland and Stockwell, 1967).

## H. Cerebrospinal Fluid Formation

The cerebrospinal fluid (CSF) is formed in the cerebral ventricles and flows around the brain through the subarachnoid space (figure 30). It protects the brain against mechanical shock and pressure changes, provides exchange of metabolites with the nerve cells and affords thermal insulation of the brain (Davson, 1967). The formation of the CSF takes place in the choroid plexus, lining a substantial part of the cerebral ventricles. There are several findings indicating that CSF is formed mainly by active secretion. The $Na^+$, $K^+$ and $Cl^-$ levels of CSF are significantly different from dialysed homologous plasma (Davson, 1955). Acetazolamide, both systematically and intraventricularly, inhibits the production rate as well as the entry of $^{24}Na$ from the blood into the CSF (Davson and Luck, 1957; Fishman, 1959), and vasopressin increases the $^{24}Na$ entry (Fishman, 1959). Newly formed CSF, collected directly from the choroid plexus *in vivo*, showed a $Na^+$ excess relative to plasma dialysate (De Rougemont and associates, 1960; Ames, Sakanoue and Endo, 1964). There is also a positive potential of 1–5 mV in CSF relative to blood (Held, Fencl and Pappenheimer, 1963; Patlak, 1964), similar to that found in aqueous humour (Cole, 1961), and nasal salt secretion of marine birds (Thesleff and Schmidt-Nielsen, 1962). These potentials are sensitive to ouabain, and presumably reflect the active secretion of $Na^+$ with water passively following.

The presence of Na–K ATPase activity in choroid plexus was first demonstrated in cat by Bonting, Simon and Hawkins (1961). It is inhibited by both

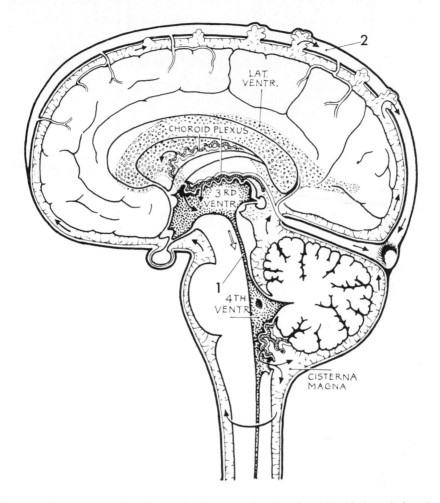

**Figure 30.** Cerebrospinal fluid circulation, showing lateral, third and fourth
ventricles, aqueduct (1), cisterna magna and subarachnoid space (2).

ouabain and erythrophleum alkaloids (Bonting, Hawkins and Canady,
1964). Its involvement in CSF formation in the cat was proven in various
ways (Vates, Bonting and Oppelt, 1964). The CSF formation rate, measured
by collection from a catheter placed in the aqueduct, was inhibited 18% by
intravenous desacetyl lanatoside C (0·2 mg/kg). Ouabain, injected into the
ventricle, caused flow inhibition, varying from 100% by $10^{-6}$ mole, 76% by
$5 \times 10^{-8}$ mole, 27% by $5 \times 10^{-9}$ mole to 0% by $5 \times 10^{-10}$ mole. There

**Table 22.** Effects of cardiac glycosides on CSF flow rate and enzyme activities in choroid plexus of lateral ventricle in cat

|  | Flow rate ($\mu$l/min) | Na–K ATPase (mole/kg/hr) | Mg ATP (mole/kg/hr) | Carbonic anhydrase ($10^3$ mole/kg/hr) |
|---|---|---|---|---|
| Controls | $8.9 \pm 0.86$ (6) | $0.98 \pm 0.069$ (10) | $2.56 \pm 0.18$ (10) | $29.3 \pm 3.09$ (8) |
| Desacetyl lanatoside C 0.2 mg/kg i.v. | $7.3 \pm 0.80$ (6) $P < 0.05$ | $1.07 \pm 0.11$ (4) | $2.79 \pm 0.38$ (4) | $31.0 \pm 3.83$ (5) |
| Ouabain intraventricularly, $10^{-6}$ mole | $0$ (4) $P < 0.001$ | $0.28 \pm 0.057$ (4) $P < 0.001$ | $2.60 \pm 0.19$ (4) | $36.1 \pm 6.15$ (4) |

Means with standard error and, in parentheses, number of animals.
Significant changes from control values are indicated with the corresponding probability values.
From Vates, Bonting and Oppelt (1964).

was also some evidence of reversal of ouabain inhibition by KCl injected intraventricularly. Table 22 compares the effects on flow and choroid plexus enzyme activities of intravenous and intraventricular cardiac glycoside administration. Mg ATPase and carbonic anhydrase activities were not affected, while *in vitro* Na–K ATPase activity was inhibited by 69% at full flow inhibition and not inhibited at 18% flow inhibition. Partial reversal of Na–K ATPase inhibition upon dilution (700 × !) of the tissue during the assay procedure explains why only 69% enzyme inhibition at 100% flow inhibition and no enzyme inhibition at 18% flow inhibition were found.

A better technique for quantitative comparison of flow and enzyme activity is the ventriculo-cisternal perfusion method with flow determination by the inulin-dilution technique (Pappenheimer and associates, 1962). The flow rate determined in this way was stable for the 5-hr duration of the experiment. After a 1-hr control period the inhibitor was added to the perfusion fluid in the desired concentration and a $1\frac{1}{2}$–3 hr experimental period was maintained. The flow inhibitions obtained at four different ouabain concentrations ($10^{-8}$, $10^{-7}$, $10^{-6}$ and $10^{-5}$ M) showed fair agreement with the *in vitro* enzyme inhibition curve (figure 31). Five other substances were tested at a single concentration in the perfusion fluid, and the concentrations giving equal enzyme inhibition were determined from their enzyme inhibition curves (table 23). For the three cardiac glycosides there was agreement between the concentrations required for equal inhibition of flow and Na–K ATPase activity. Erythrophleine and cassaine required much higher concentrations for flow inhibition than for the same degree of *in vitro* enzyme inhibition. A similar discrepancy was found for the toad bladder (Section V B),

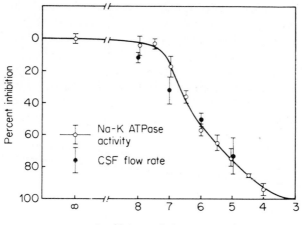

-log Molar ouabain concentration

**Figure 31.** Comparison of inhibition of choroid plexus Na–K ATPase activity *in vitro* and of CSF flow *in vivo* by intraventricular ouabian perfusion in cat. (From Vates, Bonting and Oppelt, 1964).

**Table 23.** Effects of various compounds on CSF flow and choroid plexus Na–K ATPase activity in cat

|  | Flow inhibition | Neg. log molar concn. intra- ventricular | Neg. log molar concn. for equal enzyme inhibn. | Difference neg. log molar concn. |
|---|---|---|---|---|
| Ouabain | 51 | 6·0 | 6·1 | 0·1 |
| Scillaren A | 51 | 6·7 | 6·8 | 0·1 |
| Hexahydro- scillaren A | 23 | 4·7 | 5·3 | 0·6 |
| Erythrophleine | 25 | 6·3 | 7·9 | 1·6* |
| Cassaine | 32 | 5·0 | 6·3 | 1·3* |
| Acetazolamide | 41 | 3·1† | 7·3‡ | 4·2* |

*Significantly different from zero at $P = 0·05$ level.
†From Pollay and Davson (1963).
‡Calculated from kinetic data published by Maren (1963).
From Vates, Bonting and Oppelt (1964).

and is probably due to the ease with which these tertiary amines penetrate cell membranes. The results for acetazolamide show that a 16,000 times higher concentration is needed for flow inhibition than for carbonic anhydrase inhibition. Furthermore, comparing the carbonic anhydrase activity present in choroid plexus with the $Na^+$ transport, gives a ratio of 1/3000 mole $Na^+$

transported per mole $CO_2$ hydrated. This means that carbonic anhydrase cannot be a limiting factor in CSF formation. On the other hand, the choroid plexus Na–K ATPase system appears to have a primary, rate-limiting role in the secretion of CSF in the cat.

This conclusion is further supported by the observation of stimulation of CSF formation as well as choroid plexus Na–K ATPase activity in the cat by low concentrations ($10^{-10}$ M) of ouabain (Oppelt and Palmer, 1966). In the choroid plexus of the spiny dogfish ouabain-sensitive Na–K ATPase activity has been found, while intraventricular administration of $10^{-5}$–$10^{-3}$ M ouabain decreased the rate of formation of ventricular fluid (Oppelt and associates, 1966).

## I. Various Systems

In this section the evidence for a role of the Na–K ATPase system in various other systems with a secretory function will be briefly reviewed. The following systems will be treated: salivary gland, sweat gland, cornea, thyroid gland and adrenal.

There is a great similarity between the salivary gland secretion and the exocrine pancreatic secretion. An aqueous solution with $Na^+$ as the predominant cation and mainly isotonic with plasma, is excreted. An active cation transport system, with ATP as an energy donor, appears to function (Burgen and Emmelin, 1961; Schneyer and Schneyer, 1967). In experiments with salivary gland slices an active $Na^+$ efflux and $K^+$ influx against electrochemical gradients, dependent on oxidative energy metabolism, was demonstrated (Schneyer and Schneyer, 1962). Ouabain inhibited this cation transport, and high $K^+$ levels in the medium reversed the ouabain effect (Schneyer and Schneyer, 1965). The presence of a Na–K ATPase system has been demonstrated in cat submaxillary gland (Bonting, Simon and Hawkins, 1961), dog parotid gland (Schwartz, Laseter and Kraintz, 1963), sheep parotid gland (Filsell and Jarrett, 1965) and rat submaxillary gland (Schwartz and Moore, 1968). In the sheep parotid gland the enzyme activity was low in the immature gland of lambs and high in the mature secreting gland. Micropuncture studies have shown that the primary secretion in the acini-intercalated duct region is isotonic with plasma and has an ionic composition close to that of plasma, both in the resting state and after stimulation by pilocarpine (Young and Schögel, 1966; Martinez, Holzgreve and Frick, 1966). Retrograde injection of ouabain into the glandular duct system up to the acini did not affect flow rate, but increased the $Na^+$ concentration and total osmolarity at all flow rates (Mangos and Braun, 1966). This would suggest that the primary secretion mechanism is ouabain-insensitive, and that ouabain-sensitive $Na^+$-reabsorption in excess of water absorption occurs in the striated duct region, which makes the isotonic secretion hypotonic. In the distal region of the duct re-equilibration with the plasma would take place.

The sweat gland shows great similarity to the salivary gland in structure and function. The primary secretion occurs in the secretory coil, which thus resembles the acini-intercalated duct region of the salivary gland. Reabsorptive processes take place in the coiled portion of the duct, resembling the striated duct region of the salivary gland (Cage and Dobson, 1965; Schulz and associates, 1965). The primary secretion is isotonic, but through active Na$^+$-reabsorption becomes gradually hypotonic while moving along the duct. At near-maximal rates of sweating the final product becomes isotonic, presumably because the contact-time with the ductular cells becomes too short to permit significant Na$^+$-reabsorption (Slegers, 1964, 1966). Ouabain placed in the duct abolished the potential between lumen and plasma, and increased the Na$^+$ concentration of the sweat (Schulz and associates, 1965), suggesting that the reabsorptive mechanism is ouabain-sensitive, as it is in the salivary gland. The presence of Na–K ATPase activity in the sweat gland of monkey and man has been demonstrated after microdissection (Adachi and Yamasawa, 1966; Gibbs, Griffin and Reimer, 1967). Its activity, like that of some dehydrogenases, was not greatly altered in sweat glands of patients with cystic fibrosis (Gibbs, 1967). In these determinations it was not possible to distinguish between the secretory coil and the coiled duct because of their close association. Electronmicroscopic observation after pyroantimonate treatment (Ochi, 1968) demonstrated Na$^+$ precipitates in the basal infoldings and lateral interdigitations, and to a lesser extent in the microvilli at the luminal side of the cells in the secretory coil, but not between these cells. In the ducts the precipitate was mainly found in the lateral, intercellular spaces.

The cornea possesses an outside epithelial layer and inside endothelial layer, between which is the stroma, consisting mostly of collagen fibres and mucopolysaccharides with few cells. Corneal transparency depends on the state of hydration of the stroma, which is determined primarily by the metabolic activity of the endothelial layer (Harris, 1957). There is a net influx of Na$^+$ from epithelium to endothelium (Donn, Maurice and Mills, 1959; Green, 1965), which is inhibited by ouabain (Lambert and Donn, 1964). Ouabain also causes swelling of the cornea when applied *in vivo* on the epithelial surface and prevents dehydration of the swollen cornea after cooling (Brown and Hedbys, 1965). Presentation of ouabain to the endothelial surface *in vitro* produced swelling without changing the permeability to sucrose and urea (Trenberth and Mishima, 1968). Removal of the epithelium had no effect on corneal hydration in the presence or absence of ouabain, indicating that the endothelium plays the major role in controlling corneal hydration. The presence of the Na–K ATPase system was first demonstrated in cat corneal epithelium and endothelium (Bonting, Simon and Hawkins, 1961), later in cattle and rabbit corneal epithelium (Langham and Kostelnik, 1965) and in cattle and cat corneal epithelium and endothelium (Rogers, 1968). The endothelial activity, on a protein basis, is 6·4 times as high as the epithelial activity in cattle cornea. Histochemical ATPase assays showed a localisation of Pb-phosphate on the lateral cell membranes and the intercellular spaces of the endothelium and epithelium (Maeda and Sakaguchi, 1965; Ehlers, 1965; Kaye and Tice, 1966). Treatment with pyroantimonate indicated high $Na^+$ concentrations within the endothelial cells along the lateral cell membranes, more so when ouabain was present in the aqueous humour prior to corneal fixation (Kaye, Cole and Donn, 1965). Although there remain some unexplained aspects of corneal electrolyte and water transport, these findings clearly demonstrate an important role of the Na–K ATPase system in the control of corneal hydration.

In thyroid gland function the active uptake of iodide is of crucial importance (Wolff, 1964). The presence of Na–K ATPase in the thyroid and evidence for its role in iodide accumulation could be demonstrated (Wolff and Halmi, 1963). The iodide uptake increased linearly with the $Na^+$ concentration of the medium up to 170 mM; above this concentration it decreased again. $Na^+$ did not reverse the ouabain inhibition of iodide uptake, but $K^+$ did (Alexander and Wolff, 1964). Cysteamine or cystamine administration to rats increased both the iodide accumulation by the thyroid and its Na–K ATPase activity (Wolff and Rall, 1965). A claim that thyrotropin *in vitro* stimulates thyroidal Na–K ATPase activity (Turkington, 1962, 1963) was shown to be false (Wolff and Halmi, 1963; Takagi, 1968). *In vivo* inhibition of thyroidal iodide uptake by near-lethal doses of ouabain has been demonstrated in the dog (Zakrzewska-Henisz, 1966). Oestradiol and other oestrogens, which stimulate thyroidal iodide uptake, increased thyroidal Na–K ATPase *in vitro* at low concentrations ($10^{-7}$ M), while goitrogens like propylthiouracil depressed the enzyme slightly (Takagi, 1968).

A role of the Na–K ATPase system in adrenal function is also suggested. The presence of the enzyme in cat adrenal was demonstrated by Bonting, Simon and Hawkins (1961). Simultaneous inhibition by ouabain with equal $pI_{50}$ of 7·0 was found for active cation transport and corticosterone biosynthesis by calf adrenal cortex slices, while ouabain had no effect on corticosteroid formation by adrenal mitochondria (Wellen and Benraad, 1969). A similar effect of ouabain on aldosterone secretion has been found in dog adrenal (Cushman, 1969). Ouabain, perfused through bovine adrenal gland, also increased the rate of release of catecholamines simultaneously with the inhibition of cation transport, while it did not affect the catecholamine release from isolated chromaffin granules (Banks, 1967). This effect may be due to an increase of $Ca^{2+}$ influx by ouabain, since in the absence of $Ca^{2+}$ ouabain does not affect catecholamine release.

## VI. MECHANISM OF THE Na$^+$–K$^+$ ACTIVATED ADENOSINETRI-PHOSPHATASE SYSTEM

### A. Reaction Sequence

In all previous sections of this chapter we have purposely referred to the "Na–K ATPase system", rather than to the "Na–K ATPase", because from the beginning it has seemed unlikely that the hydrolysis of ATP occurs in a single step. Since the discovery of the Na–K ATPase system by Skou, considerable effort has been spent in elucidating the reaction mechanism.

In addition to the hydrolysis of ATP to ADP and inorganic phosphate, the Na–K ATPase system of crab peripheral nerve catalyses an ATP–ADP exchange reaction (Skou, 1960). In electric organ preparations, after treatment with *N*-ethylmaleimide, an ATP–ADP exchange reaction was demonstrated, which is specific for ATP, is activated by Na$^+$ + Mg$^{2+}$ and inhibited by ouabain (Fahn, Albers and Koval, 1964, 1966; Fahn and associates, 1966). This reaction has also been demonstrated in guinea pig cerebral cortex (Swanson, 1968). This suggests that the first step in the reaction mechanism is the formation of an intermediary phosphorylated compound due to the transfer of an energy-rich phosphate bond from ATP to the Na–K ATPase system:

$$E + ATP \underset{Na^+}{\overset{Mg^{2+}}{\rightleftharpoons}} E \sim P + ADP \tag{1}$$

Besides treatment with *N*-ethylmaleimide, addition of arsenite with 2,3-dimercaptopropanol also enhanced the Na$^+$ activated ATP–ADP transphosphorylation, while simultaneously inhibiting the Na–K activated hydrolysis of ATP (Siegel and Albers, 1967). A non-Na$^+$ activated ATP–ADP kinase activity, as found in brain, is not associated with the Na–K ATPase system (Stahl, Sattin and McIlwain, 1966).

### B. The Phosphorylated Intermediate

The phosphorylated intermediate has been demonstrated by exposing Na–K ATPase preparations to terminally $^{32}$P-labelled ATP under various conditions and then localizing the incorporated radioactivity. The earliest report, which was for the electric organ Na–K ATPase system, indicated that in the presence of Mg$^{2+}$ and Na$^+$ maximal incorporation of $^{32}$P occurred in 30 sec (Albers, Fahn and Koval, 1963). Addition of K$^+$ released the bound $^{32}$P as inorganic phosphate, and this K$^+$ activated dephosphorylation was inhibited by *N*-ethylmaleimide and oligomycin (Fahn, Koval and Albers, 1968) and by ouabain (Post, Sen and Rosenthal, 1965; Gibbs, Roddy and Titus, 1965; Ahmed and Judah, 1965). The bound radioactivity was most stable at pH 2–3 and could be precipitated with trichloroacetic acid. After brief incubation of the sediment with pepsin at pH 2 several radioactive acid-soluble fragments were released. These peptic fragments showed the same acid stability of the $^{32}P$ as found in the sediment. In the absence of $Na^+$ there were three major peaks in addition to orthophosphate, while an additional peak appeared in the presence of both $Mg^{2+}$ and $Na^+$. These findings suggest that the phosphorylated intermediate is a phosphoprotein.

Formation of such an intermediary phosphorylated protein has also been shown in kidney cortex (Charnock and Post, 1963b; Charnock, Rosenthal and Post, 1963) and brain (Gibbs, Roddy and Titus, 1965; Lisovskaya and Tashmukhamedov, 1965). In erythrocyte ghosts Heinz and Hoffman (1965) reported $^{32}P$ incorporation into an acid-precipitable fraction, which did not release $^{32}P$ under conditions which reactivated ATPase. However, after brief incubation at 0 °C addition of excess non-radioactive Mg-ATP gave a rapid release of $^{32}P$, indicating that the formation of the Na–K ATPase intermediate is obscured by other modes of labelling (Blostein, 1966), simply because erythrocytes have a low Na–K ATPase activity compared to most tissues (Blake, Leader and Whittam, 1967). In a comparative study with six different tissues from eleven species $Na^+$ activated phosphorylation and $K^+$ stimulated dephosphorylation were detected in all cases (Bader, Post and Bond, 1968). Although the Na–K ATPase activity in these tissues varies over a 400-fold range, the ratio of phosphorylation and of dephosphorylation to Na–K ATPase activity varied only over a 2-fold range. The phosphorylated enzyme preparations yielded identical $^{32}P$-labelled peptides upon electrophoresis after peptic digestion. In all cases hydroxylamine released $^{32}P$ from these peptides. The phosphorylation step and the overall reaction show the same relative reaction rates towards different nucleotides: ATP : ITP : GTP = 27 : 2 : 1 under similar conditions (Schoner, Beusch and Kramer, 1968). These findings constitute strong evidence that the phosphorylated protein is a functional intermediate in active transport and that the mechanism is similar in different species and tissues.

It is unlikely that any phospholipid could be the phosphorylated intermediate in active cation transport, although this had been suggested (Hokin, and Hokin, 1960, 1963a,b; Hokin, Hokin and Mathison, 1963; Hokin, Yoda and Sandhu, 1966) The $^{32}P$ incorporation from ATP into the total phospholipid fraction of brain microsomes is too slow to represent the formation of an intermediate of the Na–K ATPase reaction (Järnefelt, 1962), while 95% of the label is incorporated into a phosphoprotein fraction (Ahmed and Judah, 1965). Exposure of electric organ preparations to

$^{32}$P-ATP and isolation of the various phospholipids showed no significant $^{32}$P-incorporation (Glynn and associates, 1965). This does not mean of course that phospholipids cannot be essential structural components of the total Na–K ATPase system (Section II B).

The nature of the phosphate link appears to be nearly settled. Its hydrolysis by hydroxylamine (Nagano and associates, 1965, 1967; Hokin and associates 1965; Bader, Sen and Post, 1966), the change in stability at varying pH (Nagano and associates, 1965; Hokin and associates, 1965; Bader, Sen and Post, 1966), its hydrolysis by molybdate (Bader, Sen and Post, 1966) and by acylphosphatase (Hokin and associates, 1965) suggest that it is an acyl phosphate. Formation of a phosphorylserine (Ahmed and Judah, 1965) was ruled out (Bader, Sen and Post, 1966) and also phosphorylation of a histidine or thiol group (Nagano and associates, 1965), or an arginyl group (Bader, Post and Jean, 1967). Study of the phosphorylated pronase fragment showed the presence of at least one dibasic (arginine or lysine) and one dicarboxylic amino acid, and a phosphate bond to the ω-carboxyl group was suggested (Jean and Bader, 1967). Incubation of a kidney Na–K ATPase preparation with $^{32}$P-labelled di-isopropylfluorophosphate (DPF) and hydrolysis of the phosphorylated product, permitted isolation of a phosphoserine, implicating the serine as the ATP substrate site (Hokin and Yoda, 1964). This conclusion was supported by the protective effect of ATP against inhibition by DFP (Hokin and Yoda, 1965). This was interpreted to mean that the active site of Na–K ATPase would be formed by serine and the dicarboxylic amino acid adjacent to each other, with the former binding the substrate ATP and the latter forming the acylphosphate intermediate (Hokin, Yoda and Sandhu, 1966). Increase of inhibition by DFP in the presence of K$^+$ or strophanthidin is thought to be due to an increase in the accessibility of the active site to DFP, in other words these substances bring about a conformational change in the Na–K ATPase system. In view of the lability of the phosphorylated intermediate, Kahlenberg, Galsworthy and Hokin (1967, 1968) converted the peptide fraction into a radioactive hydroxamate derivative with tritium-labelled *N*-(*n*-propyl)hydroxylamine, which was positively identified by comparison with synthetic derivatives. In this way the acylphosphate intermediate was identified as an *L*-glutamyl-γ-phosphate residue. Doubt has been expressed about the role of hydroxylamine. It might act as a K$^+$-like cation in releasing the phosphate (Charnock, Opit and Potter, 1967). Moreover it does not inhibit Na–K ATPase (Schoner, Kramer and Scheubert, 1966; Chignell and Titus, 1966b, 1968), and thus the acylphosphate group could not be part of the Na–K ATPase reaction. Against this criticism Kahlenberg, Galsworthy and Hokin (1968) point out that addition of Ca$^{2+}$ permits inhibition of Na–K ATPase by NH$_2$OH, apparently by making the acylphosphate accessible (Bader and Broom, 1967).

## C. The Dephosphorylation Step

The dephosphorylation step has also been clarified to some extent. It is stimulated by $K^+$ and inhibited by ouabain, and can be formulated:

$$E \sim P + H_2O \underset{\text{ouabain}}{\overset{K^+}{\rightleftharpoons}} E + P_i \tag{2}$$

Clear evidence for this step has been reported by Post (1968). Combination of (1) and (2) gives ATP hydrolysis as the overall reaction:

$$ATP + H_2O \longrightarrow ADP + P_i \tag{3}$$

By using *p*-nitrophenylphosphate, carbamylphosphate or acetylphosphate as substrates, a $K^+$ activated, ouabain-sensitive phosphatase activity has been detected in Na–K ATPase preparations of brain (Cooper and McIlwain, 1967; Sachs, Rose and Hirschowitz, 1967; Formby and Clausen, 1968), erythrocyte membranes (Rega, Garrahan and Pouchan, 1968), intestinal epithelial brush borders (Boyd, Parsons and Thomas, 1968), kidney cortex (Bader and Sen, 1966), electric organ (Albers and Koval, 1966), parotid gland (Takiguchi and associates, 1968), and liver (Nagai, Izumi and Yoshida, 1966; Bakkeren, 1968). In a comparative study the activities were determined in nine vertebrate species and eleven tissues (Hashiuchi, 1967). Similarities in heat inactivation (Yoshida, Izumi and Nagai, 1966) and cross-inhibition by the substrates (Bader and Sen, 1966; Formby and Clausen, 1968; Bakkeren, 1968) emphasise the close relationship between the $K^+$ activated phosphatase activity and the Na–K ATPase system. Stimulation, rather than inhibition of the $K^+$-dependent phosphatase activity was observed when $Na^+$ was added simultaneously with ATP, i.e. the properties of this enzyme activity came closer to those of the Na–K ATPase system upon addition of ATP and $Na^+$ (Yoshida and associates, 1969). The activity in gastric mucosa was not inhibited by ouabain, but the authors did not demonstrate an ouabain-sensitive Na–K ATPase activity in their preparation (Forte, Forte and Salt-man, 1967). In one report differences in degree of inhibition were cited as evidence against a connection of the two systems even though the authors found cross-inhibition by the substrates (Israel and Titus, 1967). This is not sufficient reason for concluding that no connection exists, as shown by the effects of extraction of a kidney Na–K ATPase preparation by an acetone–chloroform mixture (Rendi, 1966). The activation by $Na^+ + K^+$ and the inhibition by strophanthidin were abolished, but not the $^{32}P$-incorporation, while the $K^+$ activated dephosphorylation was abolished (Rendi, 1966). This is a clear demonstration of the two separate steps in the Na–K ATPase system, which may differ in degree in some properties, but nonetheless are connected in the total functioning system. Such differences as are found between the Na–K ATPase system and the $K^+$ activated phosphatase activity could be due to the existence of two different active sites (Fujita and associates, 1966).

At this point the reaction sequence of the Na–K ATPase system should be considered in more detail. The reaction sequence (1) and (2) is not adequate to explain some observations. No labelling of ATP was found on incubation of the system with $^{32}$P-acetylphosphate, and it has, therefore, been suggested that the phosphorylated intermediate formed by the latter substance is of lower energy than the intermediate formed from ATP (Sachs, Rose and Hirschowitz, 1967). This would agree with the conclusions reached by Siegel and Albers (1967), who formulate the reaction sequence (1) and (2) as follows:

$$E_1 + ATP \overset{Mg^{2+}}{\underset{Na^+}{\rightleftharpoons}} E_1 \sim P + ADP \tag{1a}$$

$$E_1 \sim P \rightleftharpoons E_2—P \tag{1b}$$

$$E_2—P + H_2O \overset{K^+}{\longrightarrow} E_2 + P_i \tag{2a}$$

$$E_2 \rightleftharpoons E_1 \tag{2b}$$

In this sequence $E_1$ and $E_1 \sim P$ would be inwardly oriented cation sites with high affinity for $Na^+$, while $E_2$ and $E_2—P$ would be outwardly oriented cation sites with low $Na^+$ affinity, permitting the release of $Na^+$ on the external side of the membrane. The reverse would be true for the $K^+$ affinity. In reaction (1b) exchange of $Na^+$ for $K^+$ on the binding site and in reaction (2b) the reverse exchange would take place. The release of free energy, associated with the transport process, would occur during reaction (1b).

## D. Models

While the kinetics of the system are thus known in considerable detail, we are still in almost complete ignorance of the structure of the system. This is due to its insolubility, making the application of modern techniques for the determination of protein structure impossible. It is, therefore, somewhat premature to design models visualising the mechanism of the Na–K ATPase system. Not only do we know very little of its macromolecular structure, but we cannot at present even decide whether $Na^+$ and $K^+$ are bound at one or two sites (Green and Taylor, 1964; Masiak and Green, 1968). Chemical and electronmicroscopic evidence of cation binding in Na–K ATPase preparations

(Järnefelt, 1961b; Järnefelt and Von Stedingk, 1963; Charnock, Opit and Casely-Smith, 1966) and at the inside of the cell membrane in corneal endothelium (Kaye, Cole and Donn, 1965) and some kinetic evidence for conformational changes occurring in the system during activity (Robinson, 1967)

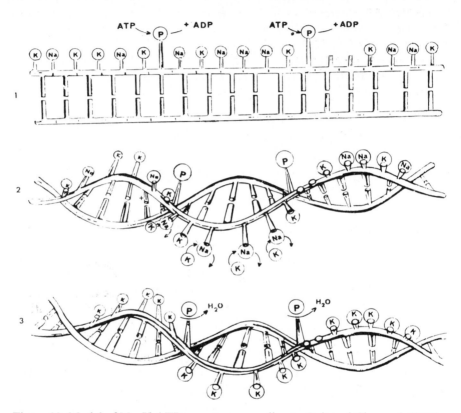

**Figure 32.** Model of Na–K ATPase system according to Opit and Charnock (1965). Resting state (1) shows protein–lipid bilayer with anionic sites on interior side carrying $Na^+$ and $K^+$ ions. Phosphorylation leads to elongation and rotation of peptide chain (2) and replacement of $Na^+$ from outward pointing anionic sites by $K^+$. This leads to dephosphorylation, back-rotation of the peptide chain and exchange of $K^+$ for $Na^+$ (3).

are the meagre aids for model builders. Nonetheless, some attempts have been made. An early model of Albers, Fahn and Koval (1963), now abandoned by the authors, placed the phosphorylation at the internal end of a pore in the membrane and the dephosphorylation at the external end. The phosphate group was thought to be translated from site to site on the pore wall, carrying Na$^+$ with it. The unsolved problem was how, after release, the phosphate group is returned to the cell interior, together with K$^+$ ions.

Opit and Charnock (1965) proposed a model based on the Danielli–Davson protein–lipid bilayer model (figure 32). In the resting state 1 the inner protein layer has anionic sites, binding both Na$^+$ and K$^+$. If many Na$^+$ ions are bound, ATP reacts with the phosphorylation sites. This would elongate the inner peptide chain, leading to rotation of the protein chain about the centre of the lipid core and with the phosphorylated sites resisting rotation (state 2). The externally oriented anionic sites have lost their high Na$^+$ affinity and will exchange Na$^+$ for K$^+$. This leads to dephosphorylation, back-rotation of the peptide chain and exchange of K$^+$ for Na$^+$ (state 3). Only a 5% elongation of the peptide chain would be necessary for this rotation.

A somewhat similar model has been proposed by Müller (1967), who noted that the viscosity of the polyanions dextran sulfate and polyvinylsulphonate depends on whether Na$^+$ or K$^+$ is the counterion. He assumes that a polyanion such as mucopolysaccharide, would link the phosphorylation sites. In the phosphorylated state the ATPase complexes would be large and consequently the polyanion chain contracted. This would make it selective for K$^+$ ions. Dephosphorylation would make the ATPase complexes smaller, the polyanion chain would extend, turn over and change its selectivity to Na$^+$ ions. The effect of inhibitors, such as ouabain and protamin (Yoshida, Fujisawa and Ohi, 1965), would be on the polyanion chain, and not on the ATPase complex. These last two models do not appear to take into account the very small number of Na–K ATPase sites on the cell membrane. There are probably only 1000 sites per erythrocyte (Glynn, 1957), which implies an average distance of 3600 Å between neighbouring sites. Moreover, the occurrence of mucopolysaccharide chains inside the cell membrane is highly unlikely. Jardetzky (1966) suggests that eversion of a pore, containing one or more cation binding sites, would occur. He notes that a displacement of a peptide chain over only 3 Å might correspond to such an eversion and that such displacements can easily occur during an allosteric conversion.

A model, which restricts itself to the known kinetics of the system, has been proposed by Albers, Koval and Siegel (1968). In figure 33 the circle is the cell membrane, on which six membrane sites in different stages of activity have been pictured. At 1 Na$^+$ activated phosphorylation takes place with the enzyme in the '*cis*'-form, i.e. with the cation binding sites pointing to the cytoplasm. The phosphorylated enzyme is rapidly converted (2) into the "*trans*"-form, with the cation binding sites pointing extracellularly. In the presence of K$^+$, reaction 3 activates the hydrolytic step (reaction 4), which makes the

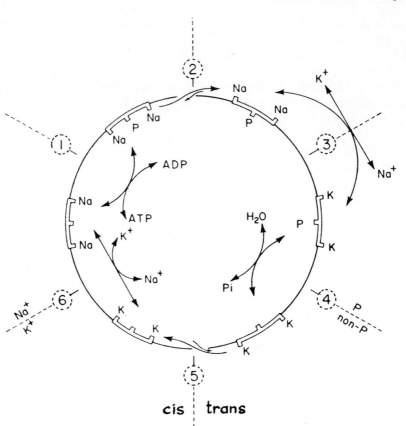

**Figure 33.** Model of Na–K ATPase mechanism according to Albers, Koval and Siegel (1968). Circle represents cell membrane. In the *cis*-form the enzyme system has its anionic sites turned inwardly, while in the *trans*-form they are turned outwardly. Reaction 1 is the $Na^+$-dependent ATP–ADP exchange. Phosphorylation converts the *cis*-form to the *trans*-form (2). Exchange of $Na^+$ for $K^+$ takes place (3). In the presence of $K^+$ dephosphorylation occurs (4), and the enzyme reconverts to the *cis*-form (5) with exchange of $K^+$ for $Na^+$ (6).

'*trans*'-enzyme less stable. It then reverts to the '*cis*'-form (reaction 5). The cycle is completed when $Na^+$ displaces $K^+$ from the *cis*-enzyme (reaction 6).

### E. Mechanism of Inhibition

In concluding this section on the mechanism of the Na–K ATPase reaction it is germane to consider our present views of the mode of action of some inhibitors of the system. The cardiac glycosides are powerful and specific inhibitors of Na–K ATPase activity (Section II B). It has been concluded (Charnock, Rosenthal and Post, 1963; Portius and Repke, 1964; Matsui and Schwartz, 1968) that they combine with the phosphorylated enzyme to inhibit the $K^+$-dependent dephosphorylation (figure 33, reaction 4). The antagonism between $K^+$ and the glycosides has been described as a reversible competition (Ahmed, Judah and Scholefield, 1966), but the reversal of glycoside inhibition is only demonstrable at low glycoside concentrations, and the calculated $K_i$ depends on the Na/K ratio (Matsui and Schwartz, 1966b). The binding of tritiated digoxin to Na–K ATPase requires ATP and $Mg^{2+}$, is stimulated by $Na^+$ and depressed by $K^+$, which suggests that digoxin binds to the phosphorylated intermediate (Matsui and Schwartz, 1968). Since binding is also observed with acetylphosphate present instead of ATP, it would appear that digoxin binds to the $E_2$—P intermediate, rather than to the $E_1 \sim P$ intermediate. The resulting complex must be rather stable to explain the inhibition of Na–K ATPase activity. Surprisingly, binding has also been found under conditions where the phosphorylated intermediate is not formed (Schwartz, Matsui and Laughter, 1968). This is interpreted to mean that a certain conformational change of the ATPase system, whether brought about by phosphorylation or other means, is essential for glycoside binding. This interpretation is further supported by the finding that in the presence of ouabain the enzyme can be phosphorylated by inorganic phosphate, which normally it cannot (Lindenmayer, Laughter and Schwartz, 1968; Albers, Koval and Siegel, 1968). Cardiac glycosides thus appear to inhibit the Na–K ATPase system by reducing the difference between the conformational energies of the phosphorylated and non-phosphorylated forms of the enzyme.

Oligomycin inhibits oxidative phosphorylation by blocking incorporation of inorganic phosphate into a high-energy phosphorylated intermediate, which in turn reacts with ADP to form ATP. It also inhibits active cation transport and Na–K ATPase activity (Glynn, 1963; Van Groningen and Slater, 1963; Blake, Leader and Whittam, 1967), even in erythrocytes, which do not have an oxidative phosphorylation system. Oligomycin does not, as might perhaps have been expected, inhibit the phosphorylation step (1a) (Whittam, Wheeler and Blake, 1964; Israel and Titus, 1967), neither does it inhibit the $K^+$ activated phosphatase activity (Israel and Titus, 1967; Askari and Koyal, 1968). However, it does inhibit the $K^+$ activated breakdown of the phosphorylated intermediate (Whittam, Wheeler and Blake, 1964; Fahn, Koval and Albers, 1968). This discrepancy can be explained by assuming that binding of ATP + $Na^+$, rather than the occurrence of $E \sim P$, creates a conformational change in the Na–K ATPase, which is favourable to the $K^+$-stimulated hydrolysis, and that oligomycin acts on this effect (Askari and Koyal, 1968; Inturrisi and Titus, 1968), perhaps in reaction (1b), (Stahl, 1968).

Another inhibitor of energy metabolism, 2,4-dinitrophenol, "uncouples" oxidative phosphorylation by hydrolysing the high-energy phosphorylated intermediate without forming ATP. In turtle bladder this compound has been found to inhibit anaerobic $Na^+$ transport and stimulate glycolysis, while at the same time reducing ATP levels slightly and nearly doubling the ADP/ATP ratio, but it did not affect Na–K ATPase activity (Klahr, Bourgoignie and Bricker, 1968). Simultaneous addition of ouabain or ethacrynic acid normalised the ADP/ATP ratio and glycolysis. This indicates that 2,4-dinitrophenol hydrolyses E $\sim$ P with dissipation of its energy ($E_1 \sim P \longrightarrow E_1 + P_i$), thus driving the Na–K ATPase reaction to the right increasing the ADP/ATP ratio and consequently the rate of glycolysis.

This concludes our description of the properties, occurrence, function and mechanism of the Na–K ATPase system in active cation transport across cell membranes. It is a field in which the disciplines of physiology, biophysics, biochemistry, morphology and pharmacology all interact to contribute to what might be called the molecular biology of active cation transport. This interaction between disciplines lends a fascination to this field and is without doubt largely responsible for the rapid progress in the thirteen years since the discovery of the Na–K ATPase system. Yet, much still remains to be learned!

## REFERENCES

*Reviews*
Albers, R. W. (1967) *Ann. Rev. Biochem.*, **36**, 727
Bonting, S. L. (1964) *West-Eur. Symp. Clin. Chem.*, **3**, 35
Burgen, A. S. V. (1967) *Proc. Roy. Soc. Med.*, **60**, 329
Glynn, I. M. (1964) *Pharmacol. Rev.*, **16**, 381
Glynn, I. M. (1966) *Sci. Basis Med., Ann. Rev.*, 217
Glynn, I. M. (1968) *Brit. Med. Bull.*, **24**, 165
Heinz, E. (1967) *Ann. Rev. Physiol.*, **29**, 21
Katz, A. I. and F. H. Epstein (1968) *N. Engl. J. Med.*, **278**, 253
Katz, A. I. and F. H. Epstein (1967) *Israel J. Med. Sci.*, **3**, 155
Langer, G. A. (1968) *Physiol. Rev.*, **48**, 708
Post, R. L. and A. K. Sen (1965) *J. Histochem. Cytochem.*, **13**, 105
Repke, K. (1964) *Klin. Wochenschr.*, **42**, 157
Rothstein, A. (1968) *Ann. Rev. Physiol.*, **30**, 15
Schatzmann, H. J. (1967) *Bull. Schweiz. Akad. Med. Wiss.*, **23**, 260
Skou, J. C. (1964) *Progr. Biophys.*, **14**, 131
Skou, J. C. (1965) *Physiol. Rev.*, **45**, 596
Smyth, D. H. and R. Whittam (1967) *Brit. Med. Bull.*, **23**, 231
Tosteson, D. C. (1963) *Fed. Proc.*, **22**, no. 1, 19

*Original Papers*
Abdel-Latif, A. A., J. Brody and H. Ramahi (1967) *J. Neurochem.*, **14**, 1133
Abdel-Wahab, M. F. (1961) *J. Chem. U.A.R.*, **4**, 247
Adachi, K. and S. Yamasawa (1966) *J. Invest. Dermatol.*, **46**, 510

Ahmed, K. and J. D. Judah (1964) *Biochim. Biophys. Acta.*, **93**, 603
Ahmed, K. and J. D. Judah (1965) *Biochim. Biophys. Acta*, **104**, 112
Ahmed, K., J. D. Judah and P. G. Scholefield (1966) *Biochim. Biophys. Acta*, **120**, 351
Akera, T. and T. M. Brody (1968) *Mol. Pharmacol.*, **4**, 600
Akikusa, Y. and K. Yamamoto (1968) *Nippon Seirigaku Zasshi*, **30**, 873
Albers, R. W. and G. J. Koval (1962) *Life Sci.*, **1**, 219
Albers, R. W. and G. J. Koval (1966) *J. Biol. Chem.*, **241**, 1896
Albers, R. W., S. Fahn and G. J. Koval (1963) *Proc. Natl. Acad. Sci. U.S.*, **50**, 474
Albers, R. W., G. J. Koval and G. J. Siegel (1968) *Mol. Pharmacol.*, **4**, 324
Albers, R. W., G. R. de Lores Arnaiz and E. de Robertis (1965) *Proc. Natl. Acad. Sci. U.S.*, **53**, 557
Aldridge, W. N. (1962) *Biochem. J.*, **83**, 527
Alexander, W. D. and J. Wolff (1964) *Arch. Biochem. Biophys.*, **106**, 525
Allen, J. C. and E. E. Daniel (1964) *Life Sci.*, **3**, 943
Ames, A., M. Sakanoue and S. Endo (1964) *J. Neurophysiol.*, **27**, 672
Archdeacon, J. W., H. C. Rohrs and H. Marta (1964) *Biochim. Biophys. Acta*, **82**, 647
Askari, A. and D. Koyal (1968) *Biochem. Biophys. Res. Commun.*, **32**, 227
Atkinson, A., S. Hunt and A. G. Lowe (1968) *Biochim. Biophys. Acta*, **167**, 469
Aubert, X., B. Chance and R. D. Keynes (1964) *Proc. Roy. Soc. (London) B.*, **160**, 211
Bachelard, H. S. and G. D. Silva (1966) *Arch. Biochem. Biophys.*, **117**, 98
Bader, H. and A. H. Broom (1967) *Biochim. Biophys. Acta*, **139**, 517
Bader, H. and A. K. Sen (1966) *Biochim. Biophys. Acta*, **118**, 116
Bader, H., R. L. Post and G. H. Bond (1968) *Biochim. Biophys. Acta*, **150**, 41
Bader, H., R. L. Post and D. H. Jean (1967) *Biochim. Biophys. Acta*, **143**, 229
Bader, H., A. K. Sen and R. L. Post (1966) *Biochim. Biophys. Acta*, **118**, 106
Baker, E. and W. J. Simmonds (1966) *Biochim. Biophys. Acta*, **126**, 492
Baker, P. F. (1963) *Biochim. Biophys. Acta*, **75**, 287
Baker, P. F. (1964) *Biochim. Biophys. Acta*, **88**, 458
Baker, P. F. (1965) *J. Physiol.*, **180**, 383
Baker, P. F. (1968) *Brit. Med. Bull.*, **24**, 179
Baker, P. F. and C. M. Connelly (1966) *J. Physiol.*, **185**, 270
Baker, P. F. and T. I. Shaw (1965) *J. Physiol.*, **170**, 424
Baker, P. F., R. F. Foster, D. S. Gilbert, and T. I. Shaw (1968) *Biochim. Biophys. Acta*, **163**, 560
Bakkeren, J. A. J. M. (1968) Ph.D. Thesis, Univ. of Nijmegen, Nijmegen, The Netherlands, p. 75
Bakkeren, J. A. J. M. and S. L. Bonting (1968a) *Biochim. Biophys. Acta*, **150**, 460
Bakkeren, J. A. J. M. and S. L. Bonting (1968b) *Biochim. Biophys. Acta*, **150**, 467
Banks, P. (1967) *J. Physiol.*, **193**, 631
Bartoszewicz, W. and R. J. Barrnett (1964) *J. Ultrastruct. Res.*, **10**, 599
Beauchene, R. E., D. D. Fanestil and C. Barrows, Jr. (1965) *J. Gerontol.*, **20**, 306
Beaugé, L. A. and R. A. Sjodin (1967) *Nature*, **215**, 1307
Becker, B. (1961) *Amer. J. Ophthalmol.*, **51**, 1032
Becker, B. (1962) *Invest. Ophthalmol.*, **1**, 502
Becker, B. (1963) *Invest. Ophthalmol.*, **2**, 325
Becker, B. and E. Cotlier (1962) *Invest. Ophthalmol.*, **1**, 642
Bellamy, D. (1961) *Comp. Biochem. Physiol.*, **3**, 125
Benolken, R. M. and C. J. Russell (1967) *Science*, **155**, 1576
Berg, G. G. and B. Chapman (1965) *J. Cell. Comp. Physiol.*, **65**, 361
Berg, G. G. and J. Szekerczes (1966) *J. Cell. Physiol.*, **67**, 487
Berggren, L. (1964a) *Invest. Ophthalmol.*, **3**, 266
Berggren, L. (1964b) *Invest. Ophthalmol.*, **4**, 83
Bignami, A., G. Palladini and G. Venturini (1966) *Brain Res.*, **3**, 207
Blake, A., D. P. Leader and R. Whittam (1967) *J. Physiol.*, **193**, 467
Block, J. B. and S. L. Bonting (1964) *Enzymol. Biol. Clin.*, **4**, 183
Blond, D. M. and R. Whittam (1964a) *Biochem. J.*, **92**, 158

Blond, D. M. and R. Whittam (1964b) *Biochem. Biophys. Res. Commun.*, **17**, 120
Blond, D. M. and R. Whittam (1965) *Biochem. J.*, **97**, 523
Blostein, R. (1966) *Biochem. Biophys. Res. Commun.*, **24**, 598
Boernig, H. and R. Giertler (1965) *Biochim. Biophys. Acta*, **100**, 603
Bojesen, E. and P. P. Leyssac (1965) *Acta Physiol. Scand.*, **65**, 20
Bonting, S. L. (1965) *Invest. Ophthalmol.*, **4**, 723
Bonting, S. L. (1966) *Comp. Biochem. Physiol.*, **17**, 953
Bonting, S. L. (1967) *Acta Physiol. Pharmacol. Neerl.*, **14**, 372
Bonting, S. L. (1969) in *Curr. Topics in Bioenergetics* (Ed. R. D. Sanadi), Academic Press, New York, **3**, 351
Bonting, S. L. and A. D. Bangham (1967) *Exptl. Eye Res.*, **6**, 400
Bonting, S. L. and B. Becker (1964) *Invest. Ophthalmol.*, **3**, 523
Bonting, S. L. and M. R. Canady (1964) *Amer. J. Physiol.*, **207**, 1005
Bonting, S. L. and L. L. Caravaggio (1962) *Nature*, **194**, 1180
Bonting, S. L. and L. L. Caravaggio (1963) *Arch. Biochem. Biophys.*, **101**, 37
Bonting, S. L. and L. L. Caravaggio (1966) *Biochim. Biophys. Acta,* **112**, 519
Bonting, S. L. and I. M. Rosenthal (1960) *Nature*, **185**, 686
Bonting, S. L., M. R. Canady and N. M. Hawkins (1964) *Biochim. Biophys. Acta*, **82**, 427
Bonting, S. L., L. L. Caravaggio and M. R. Canady (1964) *Exptl. Eye Res.*, **3**, 47
Bonting, S. L., L. L. Caravaggio and N. M. Hawkins (1962) *Arch. Biochem. Biophys.*, **98**, 413
Bonting, S. L., L. L. Caravaggio and N. M. Hawkins (1963) *Arch. Biochem. Biophys.*, **101**, 47
Bonting, S. L., N. M. Hawkins and M. R. Canady (1964) *Biochem. Pharmacol.*, **13**, 13
Bonting, S. L., K. A. Simon and L. L. Caravaggio (1961) *Ann. Mtg. Amer. Soc. Cell Biol.*, *Chicago*, p. 19.
Bonting, S. L., K. A. Simon and N. M. Hawkins (1961) *Arch. Biochem. Biophys.*, **95**, 416
Bonting, S. L., L. L. Caravaggio, M. R. Canady and N. M. Hawkins (1964) *Arch. Biochem. Biophys.*, **106**, 49
Bosch, R., A. Salibian and F. G. Romeu (1967) *J. Histochem. Cytochem.*, **15**, 114
Bower, B. F. (1963) *Fed. Proc.*, **22**, 445
Bower, B. F. (1964) *Nature*, **204**, 786
Bowler, K. and C. J. Duncan (1967a) *Comp. Biochem. Physiol.*, **20**, 543
Bowler, K. and C. J. Duncan (1967b) *J. Cell. Physiol.*, **70**, 121
Bowler, K. and C. J. Duncan (1968) *Comp. Biochem. Physiol.*, **24**, 1043
Bowman, F. J. and E. J. Landon (1967) *Amer. J. Physiol.*, **213**, 1209
Boyd, C. A. R., D. S. Parsons and A. V. Thomas (1968) *Biochim. Biophys. Acta*, **150**, 723
Bresciani, F., F. Auricchio and C. Fiore (1964) *Radiation Res.*, **22**, 463
Brewer, G. J., J. W. Eaton, C. C. Beck, L. Feitler and D. C. Shreffler (1968) *J. Lab. Clin. Med.*, **71**, 744
Britten, J. S. and M. Blank (1968) *Biochim. Biophys. Acta*, **159**, 160
Browman, L. G. (1937) *J. Exptl. Zool.*, **75**, 375
Brown, H. D. (1966) *Biochem. Pharmacol.*, **15**, 2007
Brown, H. D. and A. M. Altschul (1964) *Biochem. Biophys. Res. Commun.*, **15**, 479
Brown, H. D., S. K. Chattopadhyay and A. Patel (1967) *Enzymologia*, **32**, 205
Brown, H. D., R. T. Jackson and M. B. Waitzman (1967) *Life Sci.*, **6**, 1519
Brown, S. I. and B. O. Hedbys (1965) *Invest. Ophthalmol.*, **4**, 216
Burgen, A. S. V. and N. G. Emmelin (1961) Physiology of the Salivary Glands. Edward Arnold Ltd., London
Burger, J. W. (1962) *Physiol. Zool.*, **35**, 205
Burger, J. W. and W. N. Hess (1960) *Science*, **131**, 670
Butler, R. A. (1965) *J. Acoust. Soc. Amer.*, **37**, 429
Cade, J. R., R. J. Shaloub, M. Canessa-Fischer and R. F. Pitts (1961) *Amer. J. Physiol.*, **200**, 373
Cage, G. W. and R. I. Dobson (1965) *J. Clin. Invest.*, **44**, 1270
Caldwell, P. C. and R. D. Keynes (1959) *J. Physiol.*, **148**, 8P

Caldwell, P. C., A. L. Hodgkin, R. D. Keynes and T. I. Shaw (1960) *J. Physiol.*, **152**, 561
Cannessa-Fischer, M., F. Zambrano and V. R. Moreno (1967) *Arch. Biochem. Biophys.*, **122**, 658
Cascarano, J. and I. Seidman (1965) *Biochim. Biophys. Acta*, **100**, 301
Celedon, H. (1967) *An. Fac. Quim. Farm., Univ. Chile*, **19**, 21
Chambers, A. H. and G. G. Lucchina (1966) *J. Audit. Res.*, **6**, 13
Chan, P. C., V. Calabrese and L. S. Theil (1964) *Biochim. Biophys. Acta*, **79**, 424
Chance, B., C. P. Lee and R. Oshino (1964) *Biochim. Biophys. Acta*, **88**, 105
Charnock, J. S. and R. L. Post (1963a) *Aust. J. Exptl. Biol. Med. Sci.*, **41**, 547
Charnock, J. S. and R. L. Post (1963b) *Nature*, **199**, 910
Charnock, J. S., L. J. Opit and J. R. Caseley-Smith (1966) *Biochim. Biophys. Acta*, **126**, 350
Charnock, J. S., L. J. Opit and H. A. Potter (1967) *Biochem. J.*, **104**, 17C
Charnock, J. S., A. S. Rosenthal and R. L. Post (1963) *Austr. J. Exptl. Biol. Med. Sci.*, **41**, 675
Chen, K. K., C. C. Hargreaves and W. T. Winchester (1938) *J. Amer. Pharm. Assoc.*, **27**, 307
Chignell, C. F. (1968) *Biochem. Pharmacol.*, **17**, 1207
Chignell, C. F. and E. Titus (1966a) *J. Biol. Chem.*, **241**, 5083
Chignell, C. F. and E. Titus (1966b) *Proc. Natl. Acad. Sci. U.S.*, **56**, 1620
Chignell, C. F. and E. Titus (1968) *Biochim. Biophys. Acta*, **159**, 345
Civan, M. M., O. Kedem and A. Leaf (1966) *Amer. J. Physiol.*, **211**, 569
Clausen, T., M. Rodbell and P. Dunand (1969) *J. Biol. Chem.*, **244**, 1252
Cohn, W. E. and E. R. Cohn (1939) *Proc. Soc. Exptl. Biol. Med.*, **41**, 445
Cole, D. F. (1960a) *Brit. J. Ophthalmol.*, **44**, 225
Cole, D. F. (1960b) *Brit. J. Ophthalmol.*, **44**, 739
Cole, D. F. (1961) *Brit. J. Ophthalmol.*, **45**, 202
Cole, D. F. (1964) *Exptl. Eye Res.*, **3**, 72
Cole, D. F. (1965) *Exptl. Eye Res.*, **4**, 211
Conner, R. L. (1967) *Chem. Zool.*, **1**, 319
Cooper, J. R. and H. McIlwain (1967) *Biochem. J.*, **102**, 675
Corrie, W. S. and S. L. Bonting (1966) *Biochim. Biophys. Acta*, **120**, 91
Cote, L. J. (1964) *Life Sci.*, **3**, 899
Crabbé, J. and P. De Weer (1965) *J. Physiol.*, **180**, 560
Creese, R. (1954) *Proc. Roy. Soc., B*, **142**, 497
Critchlow, V., R. A. Liebelt, M. Bar-Sela, W. Montcastle and H. S. Lipscomb (1963) *Amer. J. Physiol.*, **205**, 807
de la Cruz, R. L. (1965) *Anales Fac. Quim. Farm., Univ. Chile*, **17**, 22
Cummins, J. and H. Hyden (1962) *Biochim. Biophys. Acta*, **60**, 271
Curran, P. F. and J. R. MacIntosh (1962) *Nature*, **193**, 347
Cushman, P. (1969) *Endocrinology*, **84**, 808
Daemen, F. J. M. and S. L. Bonting (1968) *Biochim. Biophys. Acta*, **163**, 212
Daemen, F. J. M. and S. L. Bonting (1969) *Biochim. Biophys. Acta*, **183**, 90
Dainty, J. and K. Krnjević (1955) *J. Physiol.*, **128**, 489
Danowski, T. S. (1941) *J. Biol. Chem.*, **139**, 693
Davis, H. (1957) *Physiol. Rev.*, **37**, 1
Davis, P. W. and T. M. Brody (1966) *Biochem. Pharmacol.*, **15**, 703
Davson, H. (1955) *J. Physiol.*, **129**, 111
Davson, H. (1967) *Physiology of the Cerebrospinal Fluid*, Churchill, London
Davson, H. and C. P. Luck (1957) *J. Physiol.*, **137**, 279
Davson, H., S. Duke-Elder and D. M. Maurice (1949) *J. Physiol.*, **109**, 32
Debenedetti, A. and A. Lucaroni (1965) *Boll. Soc. Ital. Biol. Sper.*, **41**, 999
De Robertis, E., M. Alberici and G. Rodriguez de Lores Arnaiz (1969) *Brain Res.*, **12**, 461
Deul, D. H. and H. McIlwain (1961) *J. Neurochem.*, **8**, 246
Diamond, J. M. (1962a) *J. Physiol.*, **161**, 442
Diamond, J. M. (1962b) *J. Physiol.*, **161**, 474
Diamond, J. M. (1962c) *J. Physiol.*, **161**, 503

Diamond, J. M. (1964a) *J. Gen. Physiol.*, **48**, 1
Diamond, J. M. (1964b) *J. Gen. Physiol.*, **48**, 15
Diamond, J. M. and W. H. Bossert (1968) *J. Cell Biol.*, **37**, 694
Diamond, J. M. and J. M. Tormey (1966) *Nature*, **210**, 817
Dodds, J. J. A. and R. J. Ellis (1966) *Biochem. J.*, **101**, 31P
Donn, A., D. M. Maurice and N. L. Mills (1959) *Arch. Ophthalmol.*, **62**, 741
Doyle, W. L. (1960) *Exptl. Cell Res.*, **21**, 386
Doyle, W. L. (1962) *Amer. J. Anat.*, **111**, 223
Drabikowski, W. and U. Rafalowska (1968) *Acta Biochim. Pol.*, **15**, 45
Dransfeld, H., K. Greeff, H. Berger and V. Cautius (1966) *Arch. Pharmak. Exptl. Path.*, **254**, 225
Drapeau, G. R. and R. A. MacLeod (1963) *J. Bact.*, **85**, 1413
Duggan, D. E. and R. M. Noll (1966) *Biochim. Biophys. Acta*, **121**, 162
Duggan, D. E., J. E. Baer and R. M. Noll (1965) *Naturwissenschaften*, **52**, 264
Duncan, G. (1968) Ph.D. Thesis, Univ. of East Anglia, Norwich
Duncan, G. (1969) *Exptl. Eye Res.*, **8**, 315
Duncan, G., F. J. M. Daemen and S. L. Bonting (1970); *Vision Res.*, in press
Dunham, E. T. and I. M. Glynn (1960) *J. Physiol.*, **152**, 61P
Dunham, E. T. and I. M. Glynn (1961) *J. Physiol.*, **156**, 274
Edwards, C. and E. J. Harris (1957) *J. Physiol.*, **135**, 567
Ehlers, N. (1965) *Exptl. Eye Res.*, **4**, 48
Ellis, R. A., C. C. Goertemiller, Jr., R. A. De Lellis and Y. H. Kablotsky (1963) *Develop. Biol.*, **8**, 286
Ellory, J. C. and R. D. Keynes (1969) *Nature*, **221**, 776
Elshove, A. and G. D. V. Van Rossum (1963) *J. Physiol.*, **168**, 531
Emmelot, P. and C. J. Bos (1962) *Biochim. Biophys. Acta*, **58**, 374
Emmelot, P. and C. J. Bos (1966) *Biochim. Biophys. Acta*, **120**, 369
Emmelot, P., C. J. Bos, E. L. Benedetti and P. Rümke (1964) *Biochim. Biophys. Acta*, **90**, 126
Epstein, W. and S. G. Schultz (1965) *J. Gen. Physiol.*, **49**, 221
Epstein, F. H., A. I. Katz and G. E. Pickford (1967) *Science*, **156**, 1245
Ernst, S. E., C. C. Goertemiller, Jr. and R. A. Ellis (1967) *Biochim. Biophys. Acta*, **135**, 682
Escueta, A. V. and S. H. Appel (1969) *Biochemistry*, **8**, 725
Fänge, R., K. Schmidt-Nielsen and H. Osaki (1958) *Biol. Bull.*, **115**, 162
Fahn, S. (1968) *Experientia*, **24**, 544
Fahn, S. and L. J. Cote (1968) *J. Neurochem.*, **15**, 433
Fahn, S., R. W. Albers and G. J. Koval (1964) *Science*, **145**, 283
Fahn, S., R. W. Albers and G. J. Koval (1966) *J. Biol. Chem.*, **241**, 1882
Fahn, S., G. J. Koval and R. W. Albers (1968) *J. Biol. Chem.*, **243**, 1993
Fahn, S., M. R. Hurley, G. J. Koval and R. W. Albers (1966) *J. Biol. Chem.*, **241**, 1890
Fanestil, D. D. (1968) *Nature*, **218**, 176
Farquhar, M. G. and G. E. Palade (1966) *J. Cell Biol.*, **30**, 359
Fenster, L. J. and J. H. Copenhaver, Jr. (1967) *Biochim. Biophys. Acta*, **137**, 406
Fernández, C. (1967) *Acta Otolaryngol.*, **86**, 222
Ferreira, H. G. and M. W. Smith (1968) *J. Physiol.*, **198**, 329
Festoff, B. W. and S. H. Appel (1968) *J. Clin. Invest.*, **47**, 2752
Fieser, L. J. and M. Fieser (1959) *Steroids*, Reinhold, New York, pp. 787–808
Filsell, O. H. and I. G. Jarrett (1965) *Biochem. J.*, **97**, 479
Fishman, R. A. (1959) *J. Clin. Invest.*, **38**, 1698
Fletcher, G. L., I. M. Stainer and W. N. Holmes (1967) *J. Exptl. Biol.*, **47**, 375
Formby, B. and J. Clausen (1968) *Hoppe-Seyler's Z. Physiol. Chem.*, **349**, 909
Forte, J. G., G. M. Forte and R. F. Bils (1966) *Exptl. Cell Res.*, **42**, 662
Forte, J. G., G. M. Forte and P. Saltman (1967) *J. Cell. Physiol.*, **69**, 293
Frank, R. N. and T. H. Goldsmith (1965) *Arch. Biochem. Biophys.*, **110**, 517
Frank, R. N. and T. H. Goldsmith (1967) *J. Gen. Physiol.*, **50**, 1585
Frazier, H., E. F. Dempsey and A. Leaf (1962) *J. Gen. Physiol.*, **45**, 529

Fried, R. (1965) *J. Neurochem.*, **12**, 815
Frohman, C. E. and V. E. Kinsey (1952) *Arch. Ophthalmol.*, **48**, 12
Fujita, M., T. Nakao, Y. Tashima, N. Mizuno, K. Nagano and M. Nakao (1966) *Biochim. Biophys. Acta*, **117**, 42
Fuller, R. W. and H. D. Snoddy (1968) *Science*, **159**, 738
Gallagher, B. B. and G. H. Glaser (1968) *J. Neurochem.*, **15**, 525
Gardos, G. (1954) *Acta Physiol. Acad. Sci. Hung.*, **6**, 191
Gardos, G. (1964) *Experientia*, **20**, 387
Garrahan, P. J. and I. M. Glynn (1967a) *J. Physiol.* **192**, 159
Garrahan, P. J. and I. M. Glynn (1967b) *J. Physiol.*, **192**, 175
Garrahan, P. J. and I. M. Glynn (1967c) *J. Physiol.*, **192**, 189
Garrahan, P. J. and I. M. Glynn (1967d) *J. Physiol.*, **192**, 217
Garrahan, P. J. and I. M. Glynn (1967e) *J. Physiol.*, **192**, 237
Gatzy, J. T. and T. W. Clarkson (1965) *J. Gen. Physiol.*, **48**, 647
Gehring, P. J. and P. B. Hammond (1967) *J. Pharmacol. Exptl. Ther.*, **155**, 187
Giacobini, E., S. Hovmark and Z. Kometiani (1967) *Acta. Physiol. Scand.*, **71**, 391
Gibbs, G. E. (1967) *Bibl. Paediat.*, **86** (Pt. 1), 95
Gibbs, G. E., G. Griffin and K. Reimer (1967) *Pediat. Res.*, **1**, 24
Gibbs, R., P. M. Roddy and E. Titus (1965) *J. Biol. Chem.*, **240**, 2181
Glynn, I. M. (1957) *J. Physiol.*, **136**, 148
Glynn, I. M. (1962a) *J. Physiol.*, **160**, 18P
Glynn, I. M. (1962b) *Biochem. J.*, **84**, 75P
Glynn, I. M. (1963) *J. Physiol.*, **169**, 452
Glynn, I. M. (1968) *Brit. Med. Bull.*, **24**, 165
Glynn, I. M., C. W. Slayman, J. Eichberg and R. M. C. Dawson (1965) *Biochem. J.*, **94**, 692
Gordon, J. S. and R. M. Torack (1967) *J. Neurochem.*, **14**, 1155
Grasso, A. (1967) *Life Sci.*, **6**, 1911
Grasso, A., A. Zanini and M. T. Sabatini (1968) *Brain Res.*, **7**, 320
Greeff, K., H. Grobecker and U. Piechowski (1964) *Naturwissenschaften*, **51**, 42
Green, A. L. and C. B. Taylor (1964) *Biochem. Biophys. Res. Commun.*, **14**, 118
Green, K. (1965) *Amer. J. Physiol.*, **209**, 1311
Grobecker, H. and U. Piechowski (1966) *Z. Klin. Chem.*, **4**, 126
Van Groningen, H. E. M. and E. C. Slater (1963) *Biochim. Biophys. Acta*, **73**, 527
Gross, R. and N. W. Coles (1968) *J. Bacteriol.*, **95**, 1322
Gruener, N. and Y. Avi-Dor (1967) *Israel J. Med. Sci.*, **3**, 143
Günther, T. and F. Dorn (1966) *Z. Naturforsch.*, **21b**, 1076
Gutknecht, J. (1967) *J. Gen. Physiol.*, **50**, 1821
Hafkenscheid, J. C. M. (1968) *Biochim. Biophys. Acta*, **167**, 582
Hafkenscheid, J. C. M. and S. L. Bonting (1968) *Biochim. Biophys. Acta*, **151**, 204
Hafkenscheid, J. C. M. and S. L. Bonting (1969) *Biochim. Biophys. Acta*, **178**, 128
Hafkenscheid, J. C. M. and S. L. Bonting (1970), to be published.
Hamasaki, D. I. (1963) *J. Physiol.*, **167**, 156
Hamasaki, D. I. (1964) *J. Physiol.*, **173**, 449
Hanawa, I., K. Kuge and K. Matsumara (1967) *Jap. J. Physiol.*, **17**, 1
Harmony, T., R. Urba-Holmgren and C. M. Urbay (1967) *Brain Res.*, **5**, 109
Harris, E. J. (1954) *J. Physiol.*, **124**, 242
Harris, E. J. (1956) *Transport and Accumulation in Biological Systems*, Butterworths, London
Harris, E. J. (1967) *J. Physiol.*, **193**, 455
Harris, J. E. (1941) *J. Biol. Chem.*, **141**, 579
Harris, J. E. (1957) *Amer. J. Ophthalmol.*, **44**, 262
Harris, J. E. and L. B. Gehrsitz (1951) *Amer. J. Ophthalmol.*, **34**, 131
Harris, J. E., L. B. Gehrsitz and L. T. Nordquist (1953) *Amer. J. Ophthalmol.*, **36**, 39
Harris, J. E., J. D. Hauschildt and L. T. Nordquist (1954) *Amer. J. Opthalmol.*, **38**, 141
Hashiuchi, Y. (1967) *Nara Igaku Zasshi*, **18**, 469
Hayashi, M. and R. Uchida (1965) *Biochim. Biophys. Acta*, **110**, 207

Heidenreich, O., H. Laaff and G. Fuelgraff (1966) *Arch. Pharmakol. Exptl. Pathol.*, **255**, 317
Heinz, E. and J. F. Hoffman (1965) *J. Cell Comp. Physiol.*, **65**, 31
Held, D., V. Fencl and J. Pappenheimer (1963) *Fed. Proc.*, **22**, 332
Herrera, F. C. (1966) *Amer. J. Physiol.*, **210**, 980
Herrmann, I., H. J. Portius and K. Repke (1964) *Arch. Exptl. Pathol. Pharmakol.*, **247**, 1
Hess, H. H. and A. Pope (1957) *Fed. Proc.*, **16**, 196
Hess, H. H., G. Schneider, M. Warnock and A. Pope (1963) *Fed. Proc.*, **22**, 333
Hierholzer, K., M. Wiederholt and H. Stolte (1966) *Pflügers Arch. Ges. Physiol.*, **291**, 43
Hirschhorn, N. and I. H. Rosenberg (1968) *J. Lab. Clin. Med.*, **72**, 28
Hodgkin, A. L. (1951) *Biol. Rev.*, **26**, 339
Hodgkin, A. L. (1964) *The Conduction of the Nervous Impulse*, University Press, Liverpool
Hodgkin, A. L. and P. Horowicz (1959) *J. Physiol.* **148**, 127
Hodgkin, A. L. and R. D. Keynes (1955) *J. Physiol.*, **128**, 28
Hoffman, J. F. (1958) *J. Gen. Physiol.*, **42**, 9
Hoffman, J. F. (1960) *Fed. Proc.*, **19**, 127
Hoffman, J. F. (1962) *Circulation*, **26**, 1201
Hoffman, J. F. and F. M. Kregenow (1966) *Ann. N.Y. Acad. Sci.*, **137**, 566
Hoffman, J. F., D. C. Tosteson and R. Whittam (1960) *Nature*, **185**, 186
Hokin, M. R. (1963) *Biochim. Biophys. Acta*, **77**, 108
Hokin, L. E. and M. R. Hokin (1960) *J. Gen. Physiol.*, **44**, 61
Hokin, L. E. and M. R. Hokin (1963a) *Fed. Proc.*, **22**, 8
Hokin, L. E. and M. R. Hokin (1963b) *Biochim. Biophys. Acta*, **67**, 470
Hokin, L. E. and D. Reasa (1964) *Biochim. Biophys. Acta*, **90**, 176
Hokin, L. E. and A. Yoda (1964) *Proc. Natl. Acad. Sci. U.S.*, **52**, 454
Hokin, L. E. and A. Yoda (1965) *Biochim. Biophys. Acta*, **97**, 594
Hokin, L. E., M. R. Hokin and D. Mathison (1963) *Biochim. Biophys. Acta*, **67**, 485
Hokin, L. E., A. Yoda and R. Sandhu (1966) *Biochim. Biophys. Acta*, **126**, 100
Hokin, L. E., P. S. Sastry, P. R. Galsworthy and A. Yoda (1965) *Proc. Natl. Acad. Sci. U.S.*, **54**, 177
Holland, M. G. and M. Stockwell (1967) *Invest. Ophthalmol.*, **6**, 401
Honrubia, V., B. M. Johnstone and R. A. Butler (1965) *Acta Otolaryngol.*, **60**, 105
Hook, J. B. and H. E. Williamson (1965) *Proc. Soc. Exptl. Biol. Med.*, **120**, 358
Horowicz, P. and J. W. Burger (1968) *Amer. J. Physiol.*, **214**, 635
Hosie, R. J. A. (1965) *Biochem. J.*, **96**, 404
Humphrey, E. W. (1965) *J. Cell Biol.*, **27**, 47A
Inturrisi, C. E. and E. Titus (1968) *Mol. Pharmacol.*, **4**, 591
Israel, Y. and I. Salazar (1967) *Arch. Biochem. Biophys.*, **122**, 310
Israel, Y. and E. Titus (1967) *Biochim. Biophys. Acta*, **139**, 450
Israel, Y., H. Kalant and I. Laufer (1965) *Biochem. Pharmacol.*, **14**, 1803
Israel, Y., H. Kalant and A. E. LeBlanc (1966) *Biochem. J.*, **100**, 27
Iurato, S. (1967) *Submicroscopic Structure of the Inner Ear*, Pergamon, Oxford
Jacobs, M. H. (1931) *Ergebn. Biol.*, **7**, 1
Jacobsen, N. O. and P. L. Jørgensen (1969) *J. Histochem. Cytochem.*, **17**, 443
Järnefelt, J. (1960) *Exptl. Cell Res.*, **21**, 214
Järnefelt, J. (1961a) *Biochim. Biophys. Acta*, **48**, 104
Järnefelt, J. (1961b) *Biochem. Biophys. Res. Commun.*, **6**, 285
Järnefelt, J. (1962) *Biochim. Biophys. Acta*, **59**, 655
Järnefelt, J. and L. V. von Stedingk (1963) *Acta Physiol. Scand.*, **57**, 328
Janacek, K. and R. Rybova (1966) *Cytologia (Tokyo)*, **31**, 199
Jardetzky, O. (1966) *Nature*, **211**, 969
Jean, D. H. and H. Bader (1967) *Biochem. Biophys. Res. Commun.*, **27**, 650
Joanny, P. and H. H. Hillman (1963) *J. Neurochem.*, **10**, 655
Joergensen, P. L. (1968) *Acta Physiol. Scand.*, **73**, 21A
Johnson, J. A. (1956) *Amer. J. Physiol.*, **187**, 328
Johnstone, B. M., J. R. Johnstone and I. D. Pugsley (1966) *J. Acoust. Soc. Amer.*, **40**, 1398
Johnstone, C. G., R. S. Schmidt and B. M. Johnstone (1963) *Comp. Biochem. Physiol.*, **9**, 335

Jones, V. D., G. Lockett and E. J. Landon (1965) *J. Pharmacol. Exptl. Therap.*, **147**, 23

Kahlenberg, A., P. R. Galsworthy and L. E. Hokin (1967) *Science*, **157**, 434

Kahlenberg, A., P. R. Galsworthy and L. E. Hokin (1968) *Arch. Biochem. Biophys.*, **126**, 331

Kahlenberg, A., N. C. Dulak, J. F. Dixon, P. R. Galsworthy and L. E. Hokin (1969) *Arch. Biochem. Biophys.*, **131**, 253

Kahn, Jr., J. B. (1962) *Proc. Soc. Exptl. Biol. Med.*, **110**, 412

Kamiya, M. (1967) *Annot. Zool. Jap.*, **40**, 123

Kamiya, M. (1968) *Comp. Biochem. Physiol.*, **26**, 675

Katz, A. I. and F. H. Epstein (1967) *J. Clin. Invest.*, **46**, 1999

Kaye, G. I. and G. D. Pappas (1965) *J. Microscopie*, **4**, 497

Kaye, G. I. and L. W. Tice (1966) *Invest. Ophthalmol.*, **5**, 22

Kaye, G. I., J. D. Cole and A. Donn (1965) *Science*, **150**, 1167

Kaye, G. I., H. O. Wheeler, R. T. Whitlock and N. Lane (1966) *J. Cell Biol.*, **30**, 237

Keesey, J. C., H. Wallgren and H. McIlwain (1965) *Biochem. J.*, **95**, 289

Keller, A. R. (1963) *Anat. Record*, **147**, 367

Kennedy, K. G. and W. G. Nayler (1965) *Comp. Biochem. Physiol.*, **16**, 175

Kepner, G. R. and R. I. Macey (1968) *Biochim. Biophys. Acta*, **163**, 188

Kerkut, G. A. and R. C. Thomas (1965) *Comp. Biochem. Physiol.*, **14**, 167

Kern, H. L., P. Roosa and S. Murray (1962) *Exptl. Eye Res.*, **1**, 385

Keynes, R. D. (1954) *Proc. Roy. Soc.*, B, **142**, 359

Keynes, R. D. (1957) in *The Physiology of Fishes* (Ed. M. E. Brown), Acad. Press, New York, vol. II, pp. 323–43

Keynes, R. D. and G. W. Maisel (1954) *Proc. Roy. Soc.*, B, **142**, 383

Keynes, R. D. and J. M. Ritchie (1965) *J. Physiol.*, **179**, 333

Kinoshita, J. H., H. L. Kern and L. O. Merola (1961) *Biochim. Biophys. Acta*, **47**, 458

Kinsey, V. E. and D. V. N. Reddy (1959) *Docum. Invest. Ophthalmol.*, **13**, 7

Kinsey, V. E. and D. V. N. Reddy (1965) *Invest. Ophthalmol.*, **4**, 104

Klahr, S., J. Bourgoignie and N. S. Bricker (1968) *Nature*, **218**, 769

Klein, R. L. (1961) *Amer. J. Physiol.*, **201**, 858

Klein, R. L. (1963) *Biochim. Biophys. Acta*, **73**, 488

Kleinzeller, A., A. Knothova, Z. Vacek and Z. Lodin (1963) *Acta Biol. Med. Ger.*, **11**, 816

Koefoed-Johnsen, V. (1957) *Acta Physiol. Scand.*, **42**, Suppl. 145, 87

Komnick H. (1963a) *Protoplasma*, **56**, 274

Komnick, H. (1963b) *Protoplasma*, **56**, 385

Komnick, H. (1963c) *Protoplasma*, **56**, 605

Komnick, H. (1963d) *Protoplasma*, **58**, 96

Komnick, H. (1963e) *Zeitschr. für Zellforschung*, **60**, 163

Konishi, T. and E. Kelsey (1968) *Acta Otolaryngol.*, **65**, 381

Konishi, T., E. Kelsey and G. T. Singleton (1966) *Acta Otolaryngol.*, **62**, 393

Krnjević, K. (1955) *J. Physiol.*, **128**, 473

Kulka, R. G. and E. Sternlicht (1968) *Proc. Natl. Acad. Sci. U.S.*, **61**, 1123

Kurokawa, M., T. Sakamoto and M. Kato (1965) *Biochem. J.*, **97**, 833

Kuijpers, W. and S. L. Bonting (1969) *Biochim. Biophys. Acta*, **173**, 477

Kuijpers, W., A. C. van der Vleuten and S. L. Bonting (1967) *Science*, **157**, 949

Lambert, B. and A. Donn (1964) *Arch. Ophthalmol.*, **72**, 525

Landon, E. J. (1967) *Biochim. Biophys. Acta*, **143**, 518

Landon, E. J. and J. L. Norris (1963) *Biochim. Biophys. Acta*, **71**, 266

Landon, E. J., N. Jazab and L. Forte (1966) *Amer. J. Physiol.*, **211**, 1050

Langham, M. E. and K. E. Eakins (1964) *J. Pharmacol. Exptl. Therap.*, **144**, 421

Langham, M. E. and M. Kostelnik (1965) *J. Pharmacol. Exptl. Ther.*, **150**, 398

Langham, M. E., S. J. Ryan and M. Kostelnik (1967) *Life Sci.*, **6**, 2037

Lant, A. F. and R. Whittam (1968) *J. Physiol.*, **199**, 457

Laurent, P., S. Dunel and A. Barets (1968) *Histochemie*, **14** 308

Leaf, A. and E. Dempsey (1960) *J. Biol. Chem.*, **235**, 2160

Leaf, A. and A. Renshaw (1957) *Biochem. J.*, **65**, 90

Leaf, A., J. Anderson and L. B. Page (1958) *J. Gen. Physiol.*, **41**, 657
Leaf, A., L. B. Page and J. Anderson (1959) *J. Biol. Chem.*, **234**, 1625
Lee, K. S. and D. H. Yu (1963) *Biochem. Pharmacol.*, **12**, 1253
Leong, G. F., R. L. Pessotti and R. W. Brauer (1959) *Amer. J. Physiol.*, **197**, 880
Levin, M. L., F. C. Rector and D. W. Seldin (1968) *Amer. J. Physiol.*, **214**, 1328
Lewin, E. and A. McCrimmon (1967) *Arch. Neurol.*, **16**, 321
Lewin, E. and A. McCrimmon (1968) *Brain Res.*, **8**, 291
Lindenmayer, G. E., A. H. Laughter and A. Schwartz (1968) *Arch. Biochem. Biophys.*, **127**, 187
Ling, C. M. and A. A. Abdel-Latif (1968) *J. Neurochem.*, **15**, 721
Lisovskaya, N. P., B. A. Tashmukhamedov (1965) *Dokl. Akad. Nauk SSSR*, **163**, 1503
Lucaroni, A. and A. Millo (1964) *Arch. Sci. Biol.*, **48**, 409
Macri, F. J., R. Dixon and D. P. Rall (1966) *Invest. Ophthalmol.*, **5**, 386
Maeda, K. and K. Sakaguchi (1965) *Japan J. Opthalmol.*, **9**, 195
Maetz, J., N. Mayer and M. M. Chartier-Baraduc (1967) *Gen. Comp. Endocr.*, **8**, 177
Maitra, P. K., A. Ghosh, B. Schoener and B. Chance (1964) *Biochim. Biophys. Acta*, **88**, 112
Maizels, M. (1951) *J. Physiol.*, **112**, 59
Maizels, M. (1954) *J. Physiol.*, **125**, 263
Mandel, P. and J. Klethi (1958) *Biochim. Biophys. Acta*, **28**, 199
Mangos, J. A. and G. Braun (1966) *Arch. Ges. Physiol.*, **290**, 184
Marchesi, V. T. and G. E. Palade (1967) *Proc. Natl. Acad. Sci. U.S.*, **58**, 991
Marcus, G. J. and J. F. Manery (1966) *Can. J. Biochem.*, **44**, 1127
Maren, T. H. (1963) *J. Pharmacol. Exptl. Ther.*, **139**, 129
Marro, F. and L. Pesente (1962) *Boll. Soc. Ital. Biol. Sper.*, **38**, 541
Marro, F., L. Pesente and I. Tiripicchio (1962) *Boll. Soc. Ital. Biol. Sper.*, **38**, 544
Martin, K. and T. I. Shaw (1966) *J. Physiol.*, **184**, 25P
Martinez, J. R., H. Holzgreve and A. Frick (1966) *Arch. Ges. Physiol.*, **290**, 124
Martinez-Maldonado, M., J. C. Allen, G. Eknoyan, W. Suki and A. Schwartz (1969) *Science*, **165**, 807
Masiak, S. J. and J. W. Green (1968) *Biochim. Biophys. Acta*, **159**, 340
Matsui, H. and A. Schwartz (1966a) *Biochim. Biophys. Acta*, **128**, 380
Matsui, H. and A. Schwartz (1966b) *Biochem. Biophys. Res. Commun.*, **25**, 147
Matsui, H. and A. Schwartz (1968) *Biochim. Biophys. Acta*. **151**, 655
McClane, T. K. (1965) *J. Pharmacol. Exptl. Therap.*, **148**, 106
McClurkin, I. T. (1964) *J. Histochem. Cytochem.*, **12**, 654
McFarland, L. Z. (1964) *Nature*, **204**, 1202
Medzihradsky, F., M. H. Kline and L. E. Hokin (1967) *Arch. Biochem. Biophys.*, **121**, 311
Micrevoca, L., V. Brabec and J. Palek (1967) *Blut*, **15**, 149
Mirsalikhova, N. M. (1968) *Uzb. Biol. Zh.*, **12**, 6
Mirsalikhova, N. M., S. A. Temirov and B. A. Tashmukhamedov (1968) *Uzb. Biol. Zh.*, **12**, 63
Modolell, J. B. and R. O. Moore (1967) *Biochim. Biophys. Acta*, **135**, 319
Mohri, T., T. Oyashiki, I. Furuno, H. Kitagawa (1968) *Biochim. Biophys. Acta*, **150**, 537
Moore, J. W., T. Narahashi and T. I. Shaw (1967) *J. Physiol.*, **188**, 99
Morgan, W. S. and H. A. Leon (1963) *Exptl. Mol. Pathol.*, **2**, 297
Moses, H. L. and A. S. Rosenthal (1968) *J. Histochem. Cytochem.*, **16**, 530
Moses, H. L., A. S. Rosenthal, D. L. Beaver and S. S. Schuffman (1966) *J. Histochem. Cytochem.*, **14**, 702
Müller, H. E. (1967) *Pfluegers Arch. Ges. Physiol.*, **295**, 30
Mullins, L. J. and F. J. Brinley, Jr. (1967) *J. Gen. Physiol.*, **50**, 2333
Nagai, K., F. Izumi and H. Yoshida (1966) *J. Biochem. (Tokyo)*, **59**, 295
Nagano, K., T. Kanazawa, N. Mizuno, Y. Tashima, T. Nakao and M. Nakao (1965) *Biochem. Biophys. Res. Commun.*, **19**, 759
Nagano, K., N. Mizuno, M. Fujita, Y. Tashima, T. Nakao and M. Nakao (1967) *Biochim. Biophys. Acta*, **143**, 239

Nakai, Y. and D. A. Hilding (1966) *Acta Otolaryngol.*, **62**, 411

Nakajima, S. (1960) *Proc. Japan. Acad.*, **36**, 226

Nakamaru, Y. (1964) *Sci. Repts. Tohoku Univ.*, **30**, 11

Nakao, T., K. Nagano, K. Adachi and M. Nakao (1963) *Biochem. Biophys. Res. Commun.*, **13**, 444

Nakao, T., Y. Tashima, K. Nagano and M. Nakao (1965) *Biochem. Biophys. Res. Commun.*, **19**, 755

Nakao, M., K. Nagano, T. Nakao, N. Mizuno, Y. Tashima, M. Fujita, H. Maeda and H. Matsudaira (1967) *Biochem. Biophys. Res. Commun.*, **29**, 588

Natochin, Y. V. (1966) *Biofizika*, **11**, 626

Nechay, B. R., J. L. Larimer and T. H. Maren (1960) *J. Pharmacol. Exptl. Therap.*, **130**, 401

Nechay, B. R., R. F. Palmer, D. A. Chinoy and V. A. Posey (1967) *J. Pharmacol. Exptl. Ther.*, **157**, 599

Newey, H., P. A. Sanford and D. H. Smyth (1968) *J. Physiol.*, **194**, 237

Nishiitsutsuji-Uwo, J., R. N. Townsend, H. Nakajima and C. S. Pittendrigh (1967) *Comp. Biochem. Physiol.*, **22**, 319

Noda, K. (1968) *Kurume Med. J.*, **15**, 51

Nolte, A. (1966) *Z. Zellforsch.*, **72**, 562

Novikoff, A. B., J. Drucker, Woo-Yung Shin and S. Goldfischer (1961) *J. Histochem. Cytochem.*, **9**, 434

Ochi, J. (1968) *Histochemie*, **14**, 300

Ohkawa, K. (1966) *Igaku No Ayumi*, **59**, 113

Ohnishi, T. (1963) *J. Biochem. (Tokyo)*, **53**, 238

Oide, M. (1967) *Annot. Zool. Jap.*, **40**, 130

Okada, S. (1968) *Nara Igaku Zasshi*, **19**, 139

Opit, L. J. and J. S. Charnock (1965) *Nature*, **208**, 471

Opit, L. J., H. Potter and J. S. Charnock (1966) *Biochim. Biophys. Acta*, **120**, 159

Oppelt, W. W. (1967) *Invest. Ophthalmol.*, **6**, 76

Oppelt, W. W. and R. F. Palmer (1966) *J. Pharmacol. Exptl. Ther.*, **154**, 581

Oppelt, W. W. and E. D. White, Jr. (1968) *Invest. Ophthalmol.*, **7**, 328

Oppelt, W. W., R. H. Adamson, C. G. Zubrod and D. P. Rall (1966) *Comp. Biochem. Physiol.*, **17**, 857

Orloff, J. and M. Burg (1960) *Amer. J. Physiol.*, **199**, 49

Orloff, J. and J. S. Handler (1967) *Amer. J. Med.*, **42**, 757

Ostrowski, A. and Y. A. Trifonov (1967) *Biofizika*, **12**, 1037

Pal, B. K. and J. J. Ghosh (1968) *J. Neurochem.*, **15**, 1243.

Palmer, R. F. and B. R. Nechay (1964) *J. Pharmacol. Exptl. Therap.*, **146**, 92

Palmer, R. F., K. C. Lasseter and S. L. Melvin (1966) *Arch. Biochem. Biophys.*, **113**, 629

Pappenheimer, J. R., S. R. Heisey, E. F. Jordan and J. De C. Donner (1962) *Amer. J. Physiol.*, **203**, 763

Parker, J. C. and J. F. Hoffman (1964) *Nature*, **201**, 823

Parker, J. C. and J. F. Hoffman (1967) *J. Gen. Physiol.*, **50**, 893

Parkinson, T. M. and J. A. Olson (1964) *Life Sci.*, **3**, 107

Parry, G. (1966) *Biol. Rev.*, **41**, 392

Parsons, D. S. (1967a) *Brit. Med. Bull.*, **23**, 252

Parsons, D. S. (1967b) *Proc. Nutr. Soc.*, **26**, 46

Patlak, C. S. (1964) *Fed. Proc.*, **23**, 211

Peters, R. A., M. Shorthouse and J. M. Walshe (1965) *J. Physiol.*, **181**, 27P

Pincus, J. H. and N. J. Giarman (1967) *Biochem. Pharmacol.*, **16**, 600

Pollay, M. and H. Davson (1963) *Brain*, **86**, 137

De Pont, J. J. H. H. M., F. J. M. Daemen and S. L. Bonting (1968) *Biochim. Biophys. Acta*, **163**, 204

Portela, A., J. C. Perez, E. J. Urgoiti, P. A. Stewart, R. Nalar, B. L. de Vicente, R. J. Perez and J. A. Vicente (1967) *Acta Physiol. Lat. Amer.*, **17**, 292

Portius, H. J. and K. Repke (1964) *Arzneimittel Forsch.*, **14**, 1073

Portius, H. J. and K. Repke (1967a) *Acta Biol. Med. German.*, **19**, 879

Portius, H. J. and K. Repke (1967b) *Acta Biol. Med. German.*, **19**, 907
Post, R. L. (1968) *Regulatory Functions of Biological Membranes*, J. Järnefelt (Ed.), Elsevier, Amsterdam, 173
Post, R. L., A. K. Sen and A. S. Rosenthal (1965) *J. Biol. Chem.*, **240**, 1437
Post, R. L., C. R. Merritt, C. R. Kinsolving and C. D. Albright (1960) *J. Biol. Chem.*, **235**, 1796
Priestland, R. N. and R. Whittam (1968) *Biochem. J.*, **109**, 369
Quigley, J. P. and G. S. Gotterer (1969a) *Biochim. Biophys. Acta*, **173**, 456
Quigley, J. P. and G. S. Gotterer (1969b) *Biochim. Biophys. Acta*, **173**, 469
Quinn, D. J. and C. E. Lane (1966) *Comp. Biochem. Physiol.*, **19**, 533
Rauch, S. (1964) *Biochemie des Hörorgans*, Thieme, Stuttgart, pp. 221–3
Rauch, S. and A. Köstlin (1958) *Pract. Oto-Rhino-Laryngol.*, **20**, 287
Rega, A. F., P. J. Garrahan and M. I. Pouchan (1968) *Biochim. Biophys. Acta*, **150**, 742
Rendi, R. (1964) *Biochim. Biophys. Acta*, **89**, 520
Rendi, R. (1966) *Biochim. Biophys. Acta*, **128**, 394
Repke, K. (1964) *Klin. Wochenschr.*, **42**, 157
Repke, K. and H. J. Portius (1963) *Arch. Exptl. Pathol. Pharmakol.*, **245**, 59
Repke, K. H., M. Est and H. J. Portius (1965) *Biochem. Pharmacol.*, **14**, 1785
Retiene, K., E. Zimmerman, W. G. Schindler, J. Neuenschwander and H. S. Lipscomb (1968) *Acta Endocrinol.*, **57**, 615
Richardson, S. H. (1968) *Biochim. Biophys. Acta*, **150**, 572
Ridderstap, A. S. and S. L. Bonting (1968) *Fed. Proc.*, **27**, 834
Ridderstap, A. S. (1969) Ph.D. Thesis, Univ. of Nijmegen, Nijmegen, The Netherlands
Ridderstap, A. S. and S. L. Bonting (1969a) *Amer. J. Physiol.*, **216**, 547
Ridderstap, A. S. and S. L. Bonting (1969b) *Amer. J. Physiol.*, **217**, in press
Ridderstap, A. S. and S. L. Bonting (1969c) *Europ. J. Physiol.*, **313**, 53
Ridderstap, A. S. and S. L. Bonting (1969d) *Europ. J. Physiol.*, **313**, 62
Riley, M. V. (1964a) *Exptl. Eye Res.*, **3**, 76
Riley, M. V. (1964b) *Nature*, **204**, 380
Riley, M. V. (1966) *Biochem. J.*, **98**, 898
Robinson, J. D. (1967) *Biochemistry*, **6**, 3250
Rogers, K. T. (1968) *Biochim. Biophys. Acta*, **163**, 50
Rojas, V. R. (1965) *Anales Fac. Quim. Farm, Univ. Chile*, **17**, 74
Rosenthal, A. S., H. L. Moses, D. L. Beaver and S. S. Schuffman (1966) *J. Histochem. Cytochem.*, **14**, 698
Rossini, L. and P. Valeri (1966) *Boll. Soc. Ital. Biol. Sper.*, **42**, 1432
Rothman, S. S. (1964) *Nature*, **204**, 84
Rottem, S. and S. Razin (1966) *J. Bacteriol.*, **92**, 714
Rotunno, C. A., M. I. Pouchan and M. Cerejido (1966) *Nature*, **210**, 597
De Rougemont, J., A. Ames, F. B. Nesbett and H. F. Hofmann (1960) *J. Neurophysiol.*, **23**, 485
Sabatini, M. T., R. Dipolo and R. Villegas (1968) *J. Cell. Biol.*, **38**, 176
Sachs, G., J. D. Rose and B. I. Hirschowitz (1967) *Arch. Biochem. Biophys.*, **119**, 277
Samaha, F. J. (1967) *J. Neurochem.*, **14**, 333
Samson, F. E., H. C. Dick and W. M. Balfour (1964) *Life Sci.*, **3**, 511
Scarpelli, D. G. and E. L. Craig (1963) *J. Cell Biol.*, **17**, 279
Schatzmann, H. J. (1953) *Helv. Physiol. Pharmacol. Acta*, **11**, 346
Schatzmann, H. J. (1964) *Experientia*, **20**, 551
Schatzmann, H. J. (1965) *Biochim. Biophys. Acta*, **94**, 89
Schmidt, U. and U. C. Dubach (1969) *Eur. J. Physiol.*, **306**, 219
Schmidt-Nielsen, K. (1959) *Sci. Am.*, **200**, 109
Schmidt-Nielsen, K. (1960) *Circulation*, **21**, 955
Schmidt-Nielsen, B. and L. E. Davis (1968) *Science*, **159**, 1105
Schneyer, L. H. and C. A. Schneyer (1962) *Amer. J. Physiol.*, **203**, 567
Schneyer, L. H. and C. A. Schneyer (1965) *Amer. J. Physiol.*, **209**, 111
Schneyer, L. H. and C. A. Schneyer (Ed.) (1967) *Secretory Mechanisms of Salivary Glands*, Acad. Press, New York

Schoen, R. and K. H. Menke (1967) *Acta Biol. Med. Ger.*, **18**, 43
Schoffeniels, E. (1959) *Ann. N.Y. Acad. Sci.*, **81**, 285
Schoffeniels, E. (1962) *Life Sci.*, **1**, 437
Schoner, W., R. Beusch and R. Kramer (1968) *Eur. J. Biochem.*, **7**, 102
Schoner, W., R. Kramer and W. Seubert (1966) *Biochem. Biophys. Res. Commun.*, **23**, 403
Schoner, W., C. von Ilberg, R. Kramer and W. Seubert (1967) *Eur. J. Biochem.*, **1**, 334
Schultz, S. G. and A. K. Solomon (1961) *J. Gen. Physiol.*, **45**, 355
Schultz, S. G. and R. Zalusky (1964a) *J. Gen. Physiol.*, **47**, 567
Schultz, S. G. and R. Zalusky (1964b) *J. Gen. Physiol.*, **47**, 1043
Schultz, S. G., W. Epstein and A. K. Solomon (1962) *J. Gen. Physiol.*, **46**, 343
Schultz, S. G., W. Epstein and A. K. Solomon (1963) *J. Gen. Physiol.*, **47**, 329
Schultz, S. G., R. Zalusky and A. E. Gass (1964) *J. Gen. Physiol.*, **48**, 375
Schulz, I., K. J. Ullrich, E. Frömter, H. Holzgreve, A. Frick and U. Hegel (1965) *Arch. Ges. Physiol.*, **284**, 360
Schwartz, A. (1963) *Biochim. Biophys. Acta*, **67**, 329
Schwartz, A. and C. A. Moore (1968) *Amer. J. Physiol.*, **214**, 1163
Schwartz, A., A. H. Laseter and L. Kraintz (1963) *J. Cell. Comp. Physiol.*, **62**, 193
Schwartz, A., H. Matsui and A. H. Laughter (1968) *Science*, **160**, 323
Seeman, P. M. and E. O'Brien (1963) *Nature*, **200**, 263
Sen, A. K. and R. L. Post (1964) *J. Biol. Chem.*, **239**, 345
Shanes, A. M. (1951) *J. Gen. Physiol.*, **34**, 795
Sharp, G. W. G. and A. Leaf (1963) *J. Clin. Invest.*, **42**, 978
Sheridan, M. N., V. P. Whittaker and M. Israel (1966) *Z. Zellforsch. Mikroskop. Anat.*, **74**, 291
Shimizu, S., K. Horikawa and S. Yamazoep (1965) *Igaku To Seibutsugaku*, **70**, 332
Shiose, Y. and M. Sears (1965) *Invest. Ophthalmol.*, **4**, 64
Shiose, Y. and M. Sears (1966) *Invest. Ophthalmol.*, **5**, 152
Siegel, G. J. and R. W. Albers (1967) *J. Biol. Chem.*, **242**, 4972
Simon, K. A. and S. L. Bonting (1962) *Arch. Ophthalmol.*, **68**, 227
Simon, K. A., S. L. Bonting and N. M. Hawkins (1962) *Exptl. Eye Res.*, **1**, 253
Sjodin, R. A. and L. A. Beaugé (1968) *J. Gen. Physiol.*, **51**, 1528
Sjodin, R. A. and L. A. Beaugé (1968) *J. Gen. Physiol.* **52**, 389
Sjöstrand, F. S. (1961) *The Structure of the Eye*, (Ed. Smelser), Acad. Press, New York, p.1
Skou, J. C. (1957) *Biochim. Biophys. Acta*, **23**, 394
Skou, J. C. (1960) *Biochim. Biophys. Acta*, **42**, 6
Skou, J. C. (1962) *Biochim. Biophys. Acta*, **58**, 314
Skou, J. C. and C. Hilberg (1965) *Biochim. Biophys. Acta*, **110**, 359
Slegers, J. F. G. (1964) *Arch. Ges. Physiol.*, **279**, 265
Slegers, J. F. G. (1966) *Dermatologica*, **132**, 152
Smith, M. W. (1967) *Biochem. J.*, **105**, 65
Smith, M. W., V. E. Colombo and E. A. Munn (1968) *Biochem. J.*, **107**, 691
Smith, C. A., O. H. Lowry and M. L. Wu (1954) *Laryngoscope*, **64**, 141
Snedecor, G. W. (1961) *Statistical Methods*, 5th ed., Iowa State University Press, Ames, Iowa, p. 160
Snell, F. M. and T. K. Chowdhury (1965) *Nature*, **207**, 45
Stahl, W. L. (1968) *J. Neurochem.*, **15**, 511
Stahl, W. L., A. Sattin and H. McIlwain (1966) *Biochem. J.*, **99**, 404
Stein, H. H., A. J. Glasky and W. R. Roderick (1965) *Ann. N.Y. Acad. Sci.*, **130**, 751
Strickler, J. C., R. H. Kessler and B. A. Knutson (1961) *J. Clin. Invest.*, **40**, 311
Sugita, H., K. Okimoto, S. Ebashi and S. Okinaka (1966) *Explor. Conc. Muscular Dystrophy Relat. Disord., Proc. Int. Conf.*, Ed. A. T. Milhorat, Excerpta Med. Found., Amsterdam (pub. 1967), p. 321
Sullivan, M. X. (1908) *Bull. Bur. Fisheries*, **27**, 1
Swanson, P. D. (1968) *Biochem. Pharmacol.*, **17**, 129
Swanson, P. D. (1968) *J. Neurochem.*, **15**, 1159
Szmigielski, S. (1966) *Nature*, **209**, 412

Takagi, I. (1968) *Jap. J. Physiol.*, **18**, 723
Takiguchi, H., S. Furuyama, Y. Ogata and J. Kanno (1968) *Arch. Oral. Biol.*, **13**, 1285
Tanabe, T., I. Tsunemi, Y. Abiko and H. Dazai (1961) *Arch. Intern. Pharmacodyn.*, **133**, 452
Tanaka, R. and L. G. Abood (1964a) *Arch. Biochem. Biophys.*, **105**, 554
Tanaka, R. and L. G. Abood (1964b) *Arch. Biochem. Biophys.*, **108**, 47
Tanaka, R. and K. P. Strickland (1965) *Arch. Biochem. Biophys.*, **111**, 583
Tarve, U. and M. Brechtlova (1967) *J. Neurochem.*, **14**, 283
Tasaki, I. and C. Fernández (1952) *J. Neurophysiol.*, **15**, 497
Taylor, C. B. (1962) *Biochem. Biophys. Acta*, **60**, 437
Taylor, C. B. (1963a) *J. Physiol.*, **165**, 199
Taylor, C. B. (1963b) *Biochim. Pharmacol.*, **12**, 539
Thesleff, S. and K. Schmidt-Nielsen (1962) *Amer. J. Physiol.*, **202**, 597
Thomas, M. and W. N. Aldridge (1966) *Biochem. J.*, **98**, 94
Tice, L. W. and A. G. Engel (1966) *J. Cell Biol.*, **31**, 489
Tormey, J. McD. (1966) *Nature*, **210**, 820
Tosteson, D. C. and J. F. Hoffman (1960) *J. Gen. Physiol.*, **44**, 169
Tosteson, D. C., P. Cook and R. Blount (1965) *J. Gen. Physiol.*, **48**, 1125
Tosteson, D. C., R. H. Moulton and M. Blaustein (1960) *Fed. Proc.*, **19**, 128
Treherne, J. E. (1961) *J. Exptl. Biol.*, **38**, 629
Treherne, J. E. (1966) *J. Exptl. Biol.*, **44**, 355
Trenberth, S. M. and S. Mishima (1968) *Invest. Ophthalmol.*, **7**, 44
Turkington, R. W. (1962) *Biochim. Biophys. Acta*, **65**, 386
Turkington, R. W. (1963) *J. Biol. Chem.*, **238**, 3463
Ueda, I. and W. Mietani (1967) *Biochem. Pharmacol.*, **16**, 1370
Uesugi, S. and S. Yamazoe (1966) *Nature* **209** 403
Uesugi, S., A. Kahlenberg, F. Medzihradsky and L. E. Hokin (1969) *Arch. Biochem. Biophys.*, **130**, 156
Ulrich, F. (1963) *Biochem. J.*, **88**, 193
Ussing, H. H. and K. Zerahn (1951) *Acta. Physiol. Scand.*, **23**, 110
Van Rossum, G. D. V. (1961) *Biochim. Biophys. Acta*, **54**, 403
Van Rossum, G. D. V. (1963) *Biochim. Biophys. Acta*, **74**, 1
Van Rossum, G. D. V. (1966) *Biochim. Biophys. Acta*, **122**, 323
Vates Jr., T. S., S. L. Bonting and W. W. Oppelt (1964) *Amer. J. Physiol.*, **206**, 1165
Vogel, G. and E. Kluge (1961) *Arzneimittel Forsch.*, **11**, 848
Vosteen, K. H. (1961) *Arch. Ohren-Nasen-Kehlkopfheilk.*, **178**, 1
Wachstein, M. and E. Meisel (1957) *Amer. J. Clin. Pathol.*, **27**, 13
Waitzman, M. B. and R. T. Jackson (1964) *Exptl. Eye Res.*, **3**, 201
Waitzman, M. B. and R. T. Jackson (1965) *Exptl. Eye Res.*, **4**, 135
Wallach, D. F. H. and D. Ullrey (1964) *Biochim. Biophys. Acta*, **88**, 620
Warner, D. M. and S. M. Drance (1966) *Brit. J. Opthalmol.*, **50**, 701
de Weer, P. (1968) *Nature*, **219**, 730
Weinstein, S. M. and H. G. Hempling (1964) *Biochim. Biophys. Acta*, **79**, 329
Wellen, J. J. and Th. J. Benraad (1969) *Biochim. Biophys. Acta*, **183**, 110
Whaun, J. M. and F. A. Oski (1969) *Pediat. Res.*, **3**, 105
Wheeler, K. P. and R. Whittam (1962) *Biochem. J.*, **85**, 495
Whitney, B. P. and W. F. Widdas (1959) *J. Physiol.*, **146**, 35P
Whittaker, V. P. (1966) *Ann. N.Y. Acad. Sci.*, **137**, 982
Whittam, R. (1961) *Nature*, **191**, 603
Whittam, R. (1962a) *Biochem. J.*, **83**, 29P
Whittam, R. (1962b) *Nature*, **196**, 134
Whittam, R. (1962c) *Biochem. J.*, **84**, 110
Whittam, R. (1962d) *Biochem. J.*, **82**, 205
Whittam, R. and M. E. Ager (1962) *Biochim. Biophys. Acta*, **65**, 383
Whittam, R. and M. R. Ager (1964) *Biochem. J.*, **93**, 337
Whittam, R. and M. E. Ager (1965) *Biochem. J.*, **97**, 214

Whittam, R. and D. M. Blond (1964) *Biochem. Biophys. Res. Commun.*, **17**, 120
Whittam, R. and K. P. Wheeler (1961) *Biochim. Biophys. Acta*, **51**, 622
Whittam, R. and J. S. Wiley (1967) *J. Physiol.*, **191**, 633
Whittam, R. and J. S. Willis (1963) *J. Physiol.*, **168**, 158
Whittam, R., M. E. Ager and J. S. Wiley (1964) *Nature*, **202**, 1111
Whittam, R., K. P. Wheeler and A. Blake (1964) *Nature*, **203**, 720
Wilbrandt, W. (1941) *Arch. Ges. Physiol.*, **245**, 22
Wilde, W. S. and P. J. Howard (1960) *J. Pharmacol. Exptl. Therap.*, **130**, 232
Wiley, J. S. (1969) *Nature*, **221**, 1222
Wise, W. C. and J. W. Archdeacon (1965) *Proc. Soc. Exptl. Biol. Med.*, **118**, 653
Witt, P. and H. J Schatzmann (1954) *Helv. Physical. Acta*, **12**, C44
Wolff, J. (1964) *Physiol. Revs.*, **44**, 45
Wolff, J. and N. S. Halmi (1963) *J. Biol. Chem.*, **238**, 847
Wolff, J. and J. E. Rall (1965) *Endocrinology*, **76**, 949
Wollenberger, A. and W. Schulze (1966) *Naturwissenschaften*, **53**, 134
Wood, L. and E. Beutler (1967) *J. Lab. Clin. Med.*, **70**, 287
Wurtman, R. J., W. G. Shoemaker and F. Larin (1968) *Proc. Natl. Acad. Sci. U.S.A.*, **59**, 800
Yoshida, H. and H. Fujisawa (1962) *Biochim. Biophys. Acta*, **60**, 443
Yoshida, H., H. Fujisawa and Y. Ohi (1965) *Canad. J. Biochem.*, **43**, 841
Yoshida, H., F. Izumi and K. Nagai (1966) *Biochim. Biophys. Acta*, **120**, 183
Yoshida, H., K. Nagai, T. Ohashi and N. Yoshiko (1969) *Biochim. Biophys. Acta*, **171**, 178
Young, J. A. and E. Schögel (1966) *Arch. Ges. Physiol.*, **291**, 85
Yunis, A. A. and G. K. Arimura (1966) *Proc. Exptl. Biol. Med.*, **121**, 327
Zaheer, N., Z. Iqbal and G. P. Talwar (1968) *J. Neurochem.*, **15**, 1217
Zakrzewska-Henisz, A. (1966) *Bull. Acad. Polon, Sci., Ser. Sci. Biol.*, **14**, 149
Zerahn, K. (1956) *Acta Physiol. Scand.*, **36**, 300

CHAPTER 9

# Linked ion and amino acid transport

### Halvor N. Christensen

*Department of Biological Chemistry*
*The University of Michigan,*
*Ann Arbor, Michigan, U.S.A.*

## I. INTRODUCTION

The subject of this chapter needs to be considered principally for its possible relevance to the transport of $Na^+$ and $K^+$ across cellular membranes. If reactive sites exist in the plasma membrane which become specific binding points for $Na^+$ when an amino acid is simultaneously available for binding, this would deserve serious consideration by those who seek to account for specific cation transport. If the presence of a sufficient concentration of an amino acid can convert a binding site for $K^+$ into one for $Na^+$, as suggested by some of the evidence to be discussed, then that behaviour deserves all the more their attention, not because we suppose these sites serve directly for the alkali-metal pump, but because they show how such convertible sites can be designed.

365

N

Beyond the contribution that may thus be made to the eventual description of the transport of alkali-metal ions, I note other biological interests that will not I think be totally foreign to the reader. One of these interests centres on transport in general. Are organic metabolites transported uphill by a direct or a second-hand use of the energy inherent in the gradients of the alkali-metal ions? What is the meaning of the cosubstrate role of $Na^+$ in cases where direct energization by the $Na^+$ gradient may perhaps be excluded? What are the points of similarity and those of difference, between the action of transport mediators and those of enzymes? Another interest, perhaps a little less likely to appeal to the natural audience of this book, includes the question of what it means for the metabolism of the amino acids that their transport is carried out by a collection of $Na^+$-independent, $Na^+$-dependent and $Na^+$-inhibited transport systems. Are the flux relations of ions and amino acids significant to the clinical problems of the return of displaced alkali-metal ions to their appropriate body compartments, along with the return of organic nutrition and metabolism to their normal ranges? What does it mean that only alkali metal requiring transport systems for amino acids and sugars appear to be stimulated by insulin?

Well before the discovery in 1952 of an association between movements of the alkali-metal ions and movements of metabolites without net charge, amino acids had been recognized to function in some invertebrates to provide tissue osmotic pressure (see for example Camien and coworkers, 1951), as well as to serve as building stones for other molecules, large and small. It had also been recognized that cellular $K^+$ tends to be exchanged for extracellular $Na^+$, not only in $K^+$ deficiency, but also in the generalized disturbance of body fluids sometimes known as 'acute nutritional failure'. The subsequent restoration of the alkali metals to their normal compartments was recognized to depend on the restoration of normal nutrition with regard to the supplies of protein and calories as well as of inorganic ions.

Stern and coworkers noted in 1949 the ability of L glutamic acid included in the suspending medium to permit slices of mouse cerebral cortex to retain $K^+$. This effect can arise very simply from the necessity that a cation follow when glutamate is taken up, and the preference of cellular uptake processes for $K^+$. The extra $K^+$ would then be retained as long as the level of anionic metabolites were kept high by glutamate metabolism. An opposite effect was subsequently observed: the displacement of $K^+$ from cells and tissues during the uptake of a cationic amino acid with remarkable transport properties, namely $\alpha,\gamma$-diaminobutyric acid (Christensen and coworkers, 1952 a,b; Christensen and Liang, 1966b). The extravagant behaviour thereby produced appears to be a model for a more physiological increase in the tissue content of lysine and arginine observed during the development of $K^+$ deficiency (Eckel and coworkers, 1954; Iacobellis and coworkers, 1956).

These effects may appear not to deserve our primary concern under the present subject because they appear to be predictable 'electrogenic' consequences of the concentrative uptake of organic ions by cells. It appears, however, as we shall see in the final section of this chapter, that actual direct linkage between movements can also occur in such instances. At the beginning I shall want to consider the shifts in the positions and changes in the fluxes of inorganic ions that occur when amino acids and sugars without net charge are transported, since the necessity for linkage is so much more obvious in these cases.

As we shall see, our subject does not lend itself easily to the making of flat conclusions, or to an attempt at other than a historical account of how the present situation was reached, as I see it. It is therefore intended to divide the historical development of the subject into a sequence of episodes.

## II. OBSERVATIONS PRE-DATING RECOGNITION OF THE SEVERAL DISTINCT TRANSPORT SYSTEMS

For the studies I will consider under this heading, we can infer retrospectively which transport-mediating agency must have been responsible for the observed effects. The uptake of glycine or alanine by duck red blood cell, ordinarily 'uphill' events or *active* transport, was observed as early as 1952 by Christensen and coworkers to be strongly inhibited when part of the Na$^+$ in the medium was replaced by K$^+$ (figure 1). The same behaviour was seen

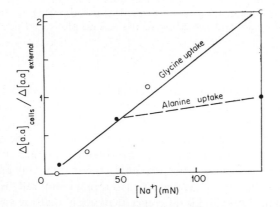

**Figure 1.** Inhibition of glycine and alanine uptake by the duck red blood cell when the external Na$^+$ is replaced by K$^+$. (Plotted from the data of table 6 in Christensen and Riggs, 1952).

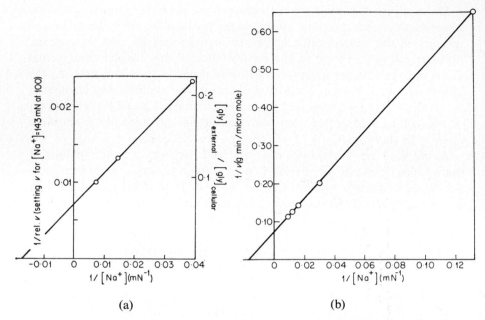

(a)                                                    (b)

**Figure 2.** Lineweaver–Burk plots of glycine uptake by the Ehrlich cell as a function of the Na$^+$ concentration, using choline to replace Na$^+$. a: plotted from the data of table 6, Christensen and Riggs, 1952. This plot treats the extent of uptake during 1 hr at 37° as a rate, under the provisional assumption that the opposing exodus is approximately described by a first-order rate constant. [gly]$_{external}$/[gly]$_{cellular}$, at a fairly constant [gly]$_{external}$, then represents the reciprocal of the entry rate. b: plotted from data from figure 1 of Kromphardt and coworkers, 1963. Uptake was measured during 2 min at 37°. (Similar data were published by Heinz in 1961). A half-maximal uptake of glycine should occur, according to the first plot at 71 mN Na$^+$, and according to the second plot at 60 mN Na$^+$.

in the Ehrlich cell by Christensen and Riggs (1952), and in this case the effect was also obtained when choline, rather than K$^+$, replaced Na$^+$ (figure 2a; Christensen and coworkers, 1952a). During the uptake of glycine (figure 3a) or of tryptophan (figure 3b) a shift of Na$^+$ from the medium and an outward shift of K$^+$ into the medium were observed. The action of tryptophan continued to be intensified after maximal gradients of this amino acid had developed, whereas that of glycine was associated temporally with the uptake of this amino acid itself. This difference may imply that the two amino acids influence Na$^+$ movements in two different ways, or it may arise from a continuing reaccumulation of tryptophan lost from the cell via the separate, Na$^+$-independent system.

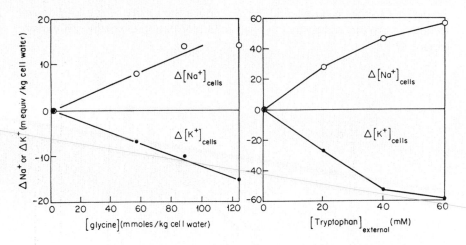

**Figure 3.** Loss of cellular $K^+$ and uptake of $Na^+$ by the Ehrlich cell associated with the uptake of glycine (a) and of tryptophan (b). The first plot shows the ion shifts as functions of the amount of glycine taken up by the cell. Although the observations illustrated were made at 3 hr, similar shifts were seen at 1 hr. b: shows the ion shifts as a function of the external tryptophan concentrations. These ion shifts may not be precisely of the same type as those seen in the presence of glycine, since they were observed after maximal gradients of tryptophan had been attained. Plotted from the data of table 6 of Christensen and Riggs (1952) and of table 8 of Riggs and coworkers, (1954).

These movements of $Na^+$ and $K^+$ had no obvious basis; they seemed unnecessary for maintaining electroneutrality. Similar results were obtained for alanine and three synthetic model amino acids. The effect was in fact observed for all amino acids whose uptake generated large concentration gradients between the interior and exterior of the cell, but not for those failing to establish such gradients. We know retrospectively that the former group includes the amino acids migrating primarily by the $Na^+$-dependent systems, with glycine almost completely specific to such systems. We also know that amino acids of the second group are taken up quickly, but principally by the $Na^+$-independent L system, which establishes much smaller concentration gradients because of its strong exchanging properties (Oxender and Christensen, 1963a, b).

A further investigation of the matter showed that a common factor accounting for the action of a number of toxic agents and conditions which inhibit the uptake of amino acids is their having first promoted a loss of cellular $K^+$, with $Na^+$ replacement (Riggs and coworkers, 1958). This disturbance in alkali-metal distribution appeared to be the basis on which cyanide and other metabolic inhibitors reduce the ability of the Ehrlich cell to take up glycine, rather than by their direct inhibition of energy metabolism (figure 4).

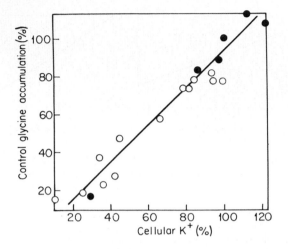

**Figure 4.** Association between the degree to which $K^+$ of the Ehrlich cell is displaced by $Na^+$, and the residual ability to concentrate glycine, after treatment with cyanide or 2,4-dinitrophenol. The cells were first incubated 30 min in the usual medium, to which either 5 mM NaCN ( ● ) or 0·15 to 0·5 mM dinitrophenol ( ○ ) had been added. The uptake of glycine (in mmoles/kg cell water for each millimolar unit of concentration in the medium) during an hour of incubation at 37° in medium containing 2 mM glycine was then determined, a subnormal or supranormal external $K^+$ concentration having been selected to obtain the indicated cellular $K^+$ content. The cell $K^+$ is expressed as a percentage of the level found in cells incubated in the normal medium in the absence of any inhibitor. The results were taken to be much the same whether the inhibitor continued to be present or not during the uptake phase of the experiment. The data for these two inhibitors were obtained concurrently, those for dinitrophenol having been selected previously for detailed documentation. (From Riggs and coworkers, 1958).

The lower line of figure 5 illustrates the progressive loss of the ability to accumulate glycine seen during the substitution of cellular $K^+$ by $Na^+$. When, however, $K^+$ was displaced through the spontaneous uptake of the organic cation $\alpha,\gamma$-diaminobutyric acid, rather than by $Na^+$, no handicap to subsequent glycine uptake was observed. The upper line of figure 5 shows that the uptake of glycine was instead somewhat enhanced, provided that we did not

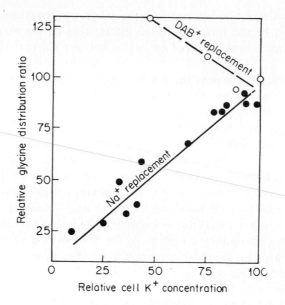

**Figure 5.** Correlation of the degree of K$^+$ displacement from the Ehrlich cell by Na$^+$ (lower line) or by $\alpha,\gamma$-diaminobutyric acid, with the subsequent ability of Ehrlich cells to take up glycine. Figure 4 from Riggs and coworkers (1958) has been traced to show again the plot of the lower line; data from figure 6 of the same paper have been replotted to show for contrast the upper line. The latter plot has been terminated at a point where the protocols show a serious degree of swelling, accompanied with substantial losses in the ability to accumulate glycine.

use so much diaminobutyrate as to cause severe swelling. Eddy (1968b) has recently made the very plausible suggestion that this enhancement may have arisen because diaminobutyric acid displaces not only cell K$^+$ but also some of the cell Na$^+$ (Christensen and coworkers, 1952b). In any event these results showed that K$^+$ *per se* does not need to be at its normal level for amino acid uptake to proceed normally. Either the organic cation diamino-butyric acid must be a superior substitute for K$^+$, or else the presence of abnormal levels of Na$^+$ in the cell rather than the depression of K$^+$, must be the factor that handicaps amino acid uptake when Na$^+$ displaces cellular K$^+$. The first alternative could mean that the K$^+$ gradient supplies the energy for glycine accumulation and the second that the Na$^+$ gradient drives the uptake process. Models based on these ideas were developed in that study; so too was the possibility that both gradients participate. The following description was given of the second model (Riggs and coworkers, 1958): 'A propelling function might instead be assigned to the Na$^+$ movements, which are reciprocal to the $K^+$ movements. In the same way consideration may be narrowed to a bound form of sodium influx; that is, the excess of sodium influx could occur in the form of a complex between the carrier, glycine, and sodium ion.'

Those who have followed the development of this area are aware of our odd preference as late as 1958 for the first of these models, and our tendency to de-emphasize aspects of our evidence that seem retrospectively to be overwhelmingly in favour of the second model. It has been gratifying to observe, however, that this preference has not greatly influenced the acceptance of the results or the accessibility of both models to those who followed us. The 1958 study also included the study of amino acid uptake during very brief intervals, introduced into our laboratory by Heinz (1954). The results confirmed that the handicap to amino acid uptake portrayed by the lower line of figure 5 arose from the abnormal cellular levels of $Na^+$ and $K^+$, and not from a secondary distortion of the composition of the suspending medium consequent to those abnormalities. (See figure 9 in Riggs and coworkers, 1958).

Heinz reported on two occasions in 1961, and again in 1963 in collaboration with his associates at Frankfurt (Kromphardt and coworkers, 1963), that over most of the range of replacement of extracellular $Na^+$ by $K^+$, the decreased $Na^+$ concentration of the extracellular fluid (figure 2b) had a larger influence on amino acid uptake than the increased external $K^+$ concentration. This conclusion was reached by replacing $Na^+$ not with $K^+$ but with choline, as we also had done earlier (see figure 2a). The later experiments appeared to have a great advantage over ours, however, in that uptake was observed during an interval as short as two minutes, so that initial rates were perhaps approximated. In the same study, Heinz and his associates found no significant influence of the replacement of about 60% of the internal $K^+$ by $Na^+$ on the apparent rate of glycine uptake by the Ehrlich cell.

Convincing evidence that the position of the $Na^+$ was important to sugar transport was provided for the small intestine of the toad by Csáky and Thale (1960) and for that of the rat by Csáky and Zollicoffer (1960). Thus the problem was extended to include the monosaccharides. Crane and coworkers (1965) have suggested that the observations reported by Kromphardt and coworkers in 1963 were derivative to those by Crane's group (1961), but this seems unlikely, since essentially the same evidence from the laboratory of Heinz was already reported in his monograph in early 1961, and in a symposium in May 1961. Subsequent study showed that the position of $Na^+$ in the system is similarly important to the intestinal absorption of amino acids (Csáky, 1961; Schultz and Zalusky, 1965), and to their uptake by thymus nuclei (Allfrey and coworkers, 1961), by kidney slices (Fox and coworkers, 1964), by the isolated, intact diaphragm (Kipnis and Parrish, 1965), and by brain slices (Margolis and Lajtha, 1968).

At this point in the history of the subject we begin to encounter what can be recognized retrospectively as a conceptual difficulty, namely an all-or-none expectation as to the energy source. Before accepting the proposal of a contribution of the gradients and fluxes of alkali-metal ions to the energy input for amino acids transport, the natural tendency was to presuppose that they provided all the energy required. For example, Hempling and Hare calculated in 1961 that insufficient energy is made available by the augmentation of $K^+$ efflux seen when glycine is added, to account for the extent to which glycine is concentrated by the Ehrlich cell. In the meantime, it became evident that the apparent $Na^+$ partition coefficient between the interior and exterior of the cell is also insufficient to account for the extent to which glycine or α-aminoisobutyric acid can be concentrated, assuming a 1:1 stoichiometry of coupling. These observations encouraged the viewpoint that $Na^+$ only permitted the carrier to complex with the organic substrate, without contributing to the energy of the subsequent translocation.

Nevertheless, the observation that the total omission of $K^+$ from the suspending medium handicaps glycine accumulation (Riggs and coworkers, 1958) has frequently been confirmed and extended to other cells, as have the effects of high external $K^+$ and of low external $Na^+$. The observations that $Na^+$ enters the cell and $K^+$ leaves the Ehrlich cell while amino acids are accumulated likewise have had only confirmation. The same is true of the continued glycine accumulation when energy metabolism is prevented, given that gradients of the alkali-metal ions are retained. It is questionable whether these associations can occur without providing an energy transfer. Yet, if the partition coefficient for glycine becomes higher than either the parallel one for $K^+$ or the opposite one for $Na^+$, one might expect the glycine gradient to drive $Na^+$ extrusion rather than the reverse (a view which bears some relevance to the subject of this book), unless of course the complex by which the $Na^+$ and amino acid joined receives some other help (a 'hand up', so to speak) in its movement across the membrane. Before we shall be prepared to consider the possibility that part of the energy in these cases might come from other sources, we will consider the various coexisting modes of amino acid transport found in various cells.

## III. DISCRIMINATION OF TRANSPORT SYSTEMS. DEMONSTRATION OF THE COMPLETENESS OF THEIR DEPENDENCE AND INDEPENDENCE OF THE Na⁺ CONCENTRATION

In 1963 Oxender and I reported extensive evidence for the operation of at least two transport systems for the neutral amino acids in the Ehrlich cell, one of them Na$^+$-dependent, and the other Na$^+$-independent (1963a,b). The question whether these systems were indeed independent needed to be considered carefully. The thesis was in fact not readily accepted that the component of transport of an amino acid which disappears on replacing the Na$^+$ of the medium with choline is kinetically and functionally unrelated to the component that remains. This question was answered in the affirmative by kinetic tests applied by Inui and Christensen (1966), with the conclusion that the so-called A system of the Ehrlich cell is totally dependent on Na$^+$, whereas the so-called L system has no significant dependence on Na$^+$. Figure 6 illustrates the results of such a test. The same conclusion was also

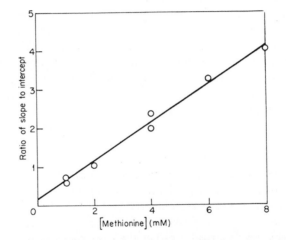

**Figure 6.** Plot to test whether the circumscribed inhibition of methionine uptake by MeAIB represents a simple competitive inhibition of one route of methionine uptake, without significant inhibition of another route, or a partial inhibition of an otherwise homogeneous mode of methionine uptake. The *inhibitor* is MeAIB, $K_i = 0.2$ mM. The mode of plotting, as shown in Inui and Christensen (1966) will yield a straight line with a positive slope only in the first case. This result is supplementary to that shown in figure 7 of the cited paper.

reached by systematically eliminating by N-methylation the Na$^+$-independent transport of amino acids migrating principally by the L system (Christensen and coworkers, 1965); also by the synthesis of a substrate specific to the

L system and the demonstration that its uptake is unaffected either by the $Na^+$ concentration or by the presence of substrates specific to the A system (Christensen and coworkers, 1969a). Obviously uphill transport systems for organic metabolites or even for amino acids do not always depend on $Na^+$, since they occur in microorganisms that do not require $Na^+$.

A number of other indications of independence among amino acid transport systems have been obtained. In bacteria one transport system may be lost by mutation without effects on another system contributing to the uptake of the same amino acid (Ames, 1964). During the maturation of the rabbit reticulocyte it has been possible to show the regression of the $Na^+$-dependent transport systems without any observable effect on the kinetic parameters of the $Na^+$-independent system. In that cell one of the $Na^+$-dependent systems, as well as the $Na^+$-independent system, show notable participation in uptake by exchange. Exchange attributable to the first can be eliminated by the removal of $Na^+$, whereas exchange by the second remains unaffected (Winter and Christensen, 1965; Wheeler and Christensen, 1967a). Indeed the augmentation of the $Na^+$ influx into the pigeon red blood cell caused by the presence of glycine could be shown to be additive to that caused by alanine, as would be expected if the $Na^+$-requiring systems function independently (Wheeler and Christensen, 1967b). The loss of an amino acid transport activity by *Escherichia coli* on osmotic shock was accompanied by the appearance in the medium of a protein that binds the corresponding substrate amino acids. No binding activity was found in the suspending solution for amino acids whose transport was unaffected by that treatment (Piperno and Oxender, 1966). The kinetic problems of the discrimination of transport systems have been summarized elsewhere (Christensen, 1969b).

The high reactivity of a number of amino acids with both the A and the L systems and the strong action of the latter for exchange, led us to suggest that such amino acids could serve as coins of exchange, by which energy could be transferred from one system to another. Because substrates of the A system are gradually accumulated to much higher gradients and because their uptake is sensitive to $Na^+$ and $K^+$ and more strongly slowed by addition of metabolic inhibitors, we suggested transfer of energy from system A to system L, still impressed by the prospect that the A system would prove to be energized, at least in part, by the pump for the alkali-metal ions. That is, if concentrative uptake by the A system is *secondary* to the pumping of the alkali-metal ions, then accumulation by the L system may be a *tertiary* process. Once the possible availability of energy from existing gradients of amino acids for the uptake of other amino acids is recognized, however, we also have to consider the possibility that supplementary energy may pass from that source to the uptake of a test amino acid presented alone.

## IV. STUDIES BY VIDAVER OF THE GLYCINE TRANSPORT SYSTEM OF THE PIGEON ERYTHROCYTE

We will for the moment leave unresolved the question of the meaning of the association of migrations of amino acids and the alkali metal ions in the Ehrlich cell, and turn to the same problem in the avian erythrocyte. The high transport activity already found for these cells in my earlier experiments with Riggs and Ray (1952), and their rather low fluxes of $Na^+$ and $K^+$ at the steady state, have made them particularly attractive. Having demonstrated that the concentrative uptake of glycine by this cell occurs by a transport system with which other amino acids are at most only weakly reactive (Vidaver and coworkers, 1964), Vidaver showed that the rate of glycine uptake by the pigeon red blood cell bears a sigmoid relationship to the $Na^+$ concentration. This relationship corresponded rather well to the cosubstrate reaction of two sodium ions for each glycine molecule (Vidaver, 1964a). The rate coefficient for the extrusion of glycine, however, proved to be independent of the external $Na^+$ concentration (figure 7a).

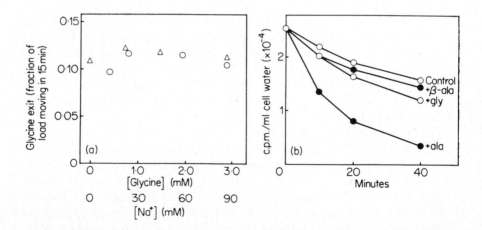

**Figure 7.** Effects of the external $Na^+$ concentration on glycine exodus (a), (○), as observed by Vidaver (1964a), contrasted with effects of external glycine concentration on $Na^+$ exodus (b), (○), as observed by Wheeler and Christensen (1967b). Vidaver's data are expressed as the fraction of the labelled glycine present in the cells lost in 15 min. Those at the right can readily be converted to the fraction of the labelled $Na^+$ lost in the indicated time interval by dividing readings from the ordinate by 2·55. The external concentration of the amino acids in this experiment was 2 mM. Other conditions in the experiment of Vidaver are mentioned in the text.

Vidaver showed further that the direction of the net movement of glycine into or out of the red cell ghost could be determined by the way in which the $Na^+$ gradient was disposed (table 1). If the internal and external $Na^+$ concentrations were set nearly equal, the two fluxes as estimated were nearly equal (Vidaver, 1964b). If the electrochemical gradient of $Na^+$ were reversed from the usual one, largely by manipulating the electric potential difference, the direction of the glycine pumping was reversed (Vidaver, 1964c). In this connection we wonder whether an electrical potential gradient may not have persisted to make a small contribution to the values in table 1. Vidaver noted that these nucleated red cells lose only about 45% of their dry weight on lysis and reconstitution; hence it seems likely that the Donnan effect might still be appreciable. Perhaps this factor is fully reflected in the flux ratio of 1·1 seen in three of the four experiments reported in table 1.

**Table 1.** Effect of producing apparent symmetry in $Na^+$ distribution on the inward and outward fluxes of glycine from lysed and reconstituted pigeon red blood cells (From Vidaver, 1964b)

| $Na^+$ outside 4 mmols/l water | $Na^+$ inside 4 mmols/l cell water | Glycine entry 4 mmols/l cell water · min | Glycine exodus 4 mmols/l cell water · min |
|---|---|---|---|
| 125 | 125 | 0·27 | 0·24 |
| 125 | 125 | 0·61 | 0·55 |
| 35 | 37 | 0·15 | 0·16 |
| 3·4 | 3·3 | 0·066 | 0·060 |

Desired internal $Na^+$ concentrations were obtained by adjustment of $Na^+$ content of the lysing and restoring solutions. One member of each pair containing [14]C-glycine inside and [12]C-glycine outside; for the other member the labelling was reversed. The external glycine was 0·9 mM in every case but the second, where it was 3·0 mM.

When Vidaver compared the increment in $Na^+$ entry caused by adding external glycine, with the increment in glycine entry caused by increasing the external $Na^+$ concentration, he obtained values for the ratio, $r$, in the range 1·5 to 2·2, implying that about two sodium ions enter the cell with each glycine molecule (table 2). In addition he found that the rate of glycine uptake was unrelated to the ATP content either of intact cells or of lysed and restored cells. On the basis of these and related observations Vidaver reached an unequivocal choice in favour of our second model, concluding that the cosubstrate action of $Na^+$ permitted the gradient of this ion to energize the uptake of glycine by the lysed and reconstituted pigeon red blood cell (1964c). This stoichiometry would of course make amino acid accumulation more expensive, but we note that it would dispose of the question of the thermodynamic adequacy of the $Na^+$ gradient to drive a steep accumulation of glycine into the pigeon red blood cell.

Let me comment at this point on a question that has received much attention, namely the question whether the sodium ion decreases the $K_m$ only, or influences both $K_m$ and $V_{max}$ for the transport of organic metabolites. Our own observations in various cells have generally shown that both are influenced in the direction favorable to transport stimulation, although the extent to which each is changed ranges widely (Wheeler and coworkers, 1965; Inui and Christensen, 1966; Wheeler and Christensen, 1967a). The expectation that only the $K_m$ would be changed is perhaps not well founded (Inui and Christensen, 1966); furthermore, the interest in showing which is the case may represent an as yet unjustified confidence that each parameter measures a distinct part of the given process.

## V.   LATER OBSERVATIONS ABOUT THE $Na^+$-DEPENDENT SYSTEMS OF RED BLOOD CELLS

Further work on the glycine transport system of the pigeon red blood cell, and on a similar system in the rabbit reticulocyte was then carried out in my laboratory. At the same time we compared that system with two other $Na^+$-dependent systems: one present in both cell types, for which alanine, serine, cysteine, threonine, asparagine and similar amino acids are substrates; and the other, reactive with β-alanine and taurine, but not found in the reticulocyte.

Studies carried out by Wheeler confirmed the sigmoid form of the relation observed by Vidaver for the rate of glycine uptake to the $Na^+$ concentration, and extended that observation to sarcosine (Wheeler and Christensen, 1967b), and also to the rabbit reticulocyte (Wheeler and Christensen, 1967a). For the pigeon red blood cell we confirmed Vidaver's conclusion that the augmentation ratio, $r$ (i.e. the ratio of the increase in the rate of $Na^+$ entry on adding glycine to the increase in the rate of glycine entry on adding $Na^+$) has a value greater than unity. Our average value for $r$ was 1·53.

The finding that the rates of alanine and serine uptake by the same cell showed an ordinary hyperbolic relation to the $Na^+$ concentration (i.e., a dependence on the first power of the $Na^+$ concentration), whereas that of β-alanine showed non-hyperbolic kinetics, permitted us to test the correlation between the kinetic indications as to whether one or more sodium ions are involved in the uptake process, with the stoichiometry indicated by the value of $r$ (table 2). The reader will observe that the results are about as inconsistent as they could be: β-alanine, despite its non-hyperbolic kinetics, caused the uptake of only one sodium ion per molecule of amino acid. Alanine, serine, and similar amino acids, despite the dependency of their rate of uptake on the first power of the sodium ion concentration, caused the uptake of two, three or more sodium ions per amino acid molecule. Furthermore, the value of $r$ was not the same for all amino acids transported by the system in question. For proline $r$ had a finite value well below one while for threonine and cysteine the value was four to five.

**Table 2.** Is the value of $r$, the ratio of the augmentation of $Na^+$ entry due to the presence of an amino acid, to the augmentation of the amino acid due to the presence of $Na^+$, correlated with the type of kinetics observed as to the dependency of amino acid uptake on $Na^+$?

|  | $\Delta v_{Na+}/\Delta v_{a.a.}$ | Nature of dependency of $\Delta v_s$ on $\Delta v_{Na+}$ |
|---|---|---|
| In the pigeon erythrocyte |  |  |
| glycine | $1 \cdot 8$[1] | sigmoid |
|  | $1 \cdot 5$[2] | sigmoid |
| alanine[2] | $2 \cdot 52 \pm 0 \cdot 34$ | hyperbolic |
| serine[3] | $3 \cdot 94 \pm 0 \cdot 52$ | hyperbolic |
| threonine[3] | $4 \cdot 50 \pm 0 \cdot 46$ | hyperbolic |
| proline[3] | $0 \cdot 22 \pm 0 \cdot 08$ | hyperbolic |
| hydroxyproline[3] | $3 \cdot 16 \pm 0 \cdot 20$ | hyperbolic |
| asparagine[3] | $1 \cdot 66 \pm 0 \cdot 28$ | hyperbolic |
| cysteine[3] | $4 \cdot 2 \pm 0 \cdot 52$ | hyperbolic |
| β-alanine[2] | $0 \cdot 96 \pm 0 \cdot 30$ | non-hyperbolic |
| In the rabbit reticulocyte |  |  |
| glycine | Not determined | sigmoid[4] |
| alanine | Not determined | hyperbolic[4] |
| In the Ehrlich cell |  |  |
| glycine | $0 \cdot 94 + 0 \cdot 11$[5] | hyperbolic |
| α-aminoisobutyric acid | $1$[6] | hyperbolic[7] |

Where the rate of amino acid uptake is stated to be a *hyperbolic* function of $Na^+$, a linear plot for $1/v$ against $1/[Na^+]$, or for $v$ against $v/[Na^+]$, was obtained; where the rate is described as a sigmoid or non-hyperbolic function, the plot was approximately linear for $1/v$ against $1/[Na^+]^2$ or for $v$ against $v/[Na^+]^2$

[1]Vidaver, 1964c; [2]Wheeler and Christensen, 1967b—for β-alanine the value of $K_m$ for $Na^+$ is too high to permit a sigmoid form to be verified; [3]Koser and Christensen, 1968; [4]Wheeler and Christensen, 1967a; [5]Eddy, 1968a, see figure 2; [6]Schafer and Jacquez, 1967; [7]Inui and Christensen, 1966; see also figure 2.

One may note in table 2 that the addition of a hydroxyl group to the amino acid sidechain has the effect in every case of increasing the value of $r$. This effect of the hydroxyl group increases in interest with the discovery of an enhancing effect of suitably positioned hydroxyl groups in the generation of a binding site for $Na^+$ in cationic and neutral amino acid transport systems (see below; Christensen and coworkers, 1969b). Koser showed that the value of $r$ was largely independent of several other variables (including the duration of the observations) and that the ratio of the maximal augmentations of the velocities of uptake of $Na^+$ and of the amino acid was not significantly different from the ratio obtained at lower concentrations of each augmenting substrate (Koser and Christensen, 1968).

The inconsistencies noted in table 2 do not support or disprove Vidaver's conclusion that the migration of two sodium ions may be linked with that of one glycine molecule but they introduce caution into our drawing any conclusions as to the number of sodium ions included in the presumed ternary complex, $E[Na^+]$[amino acid], where E is the molecule carrying the reactive site under study, either from the non-hyperbolic character of the kinetics or from the value of $r$. As Wheeler and I have suggested, the non-hyperbolic kinetics with respect to $Na^+$ could have an entirely different origin than the participation of two sodium ions in the reaction (Ferdinand, 1966; Wheeler and Christensen, 1967b).

One might wonder why such a cell as the pigeon erythrocyte does not become loaded with $Na^+$ in its natural state, if each entry of an amino acid molecule causes the entry of as many as four or five sodium ions. At first thought, the energy cost of driving amino acid uptake by $Na^+$ movements may

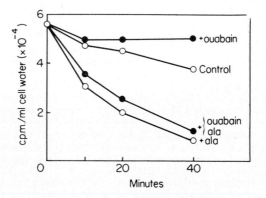

**Figure 8.** Effect of the presence of ouabain ($6.5 \times 10^{-5}M$) or of alanine ($2 \times 10^{-3}M$) or of both, on retention of $^{22}Na$ by the pigeon red blood cell. Although ouabain nearly abolished the extrusion of $Na^+$ occurring in the absence of alanine, it had very little effect on the incremental $Na^+$ exodus produced by the presence of external alanine. Parallel observations showed that the increases in cellular $Na^+$ caused by the ouabain were very nearly the same in the presence and in the absence of alanine, and that ouabain had no effect on the entry of alanine, either in the presence or the absence of $Na^+$. (From Wheeler and Christensen, 1967b).

appear to become prohibitive. In this connection another observation of Wheeler's experiments becomes highly significant: net shifts of alkali metals could not be demonstrated in the pigeon erythrocyte during the uptake of alanine (Wheeler and Christensen, 1967b). Both the inward and outward fluxes of $Na^+$ were observed to be stimulated by the presence of this amino acid. Furthermore, neither the inward nor the outward augmentations of $Na^+$ flux could be inhibited by ouabain (figure 8); the movement of alanine itself was likewise unaffected by this agent. In agreement with the proposed symmetry of these comigrations, no augmentation of either the inward or the outward flux of $^{42}K^+$ could be observed during the uptake of either alanine, serine, threonine, asparagine or proline.

Although inhibition of the uptake of neutral amino acids by ouabain has occasionally been observed, the circumstances seem to have been ones permitting substantial increases of the cellular $Na^+$ during the interval selected. Thus the two-minute observations reported by Bittner and Heinz (1963) may well have been associated with a doubling of cell $Na^+$. I emphasize this factor because the action of ouabain generally becomes much stronger as incubation is prolonged, as it was in that case.

The results of Wheeler and Christensen do not mean that equal inward and outward movements of $Na^+$ are caused by the movement of alanine only in the inward direction. Under the conditions of these experiments substantial decreases could be observed in the endogenous levels of several amino acids known to be substrates of the same transport system serving for alanine. This system appears to function largely for exchange; hence the observations of Wheeler and Christensen appear to show balanced inward and outward movements of $Na^+$ associated with inward movements of alanine that may very well have been balanced by outward movements of alanine and of analogues of alanine already present in the cell (1967b).

Studies by Koser (1969) of the linkage between amino acids and $Na^+$ in the pigeon red blood cell were done after loading the cells by holding them in 10 mM tritium-labeled alanine or threonine and $0.1\%$ glucose in a standard $^{22}NaCl$–Tris medium. Exodus was then observed during ten minutes into the usual medium containing 140 mM $Na^+$ and any of several unlabeled amino acids at concentrations two or four times their $K_m$ values for entry. The results showed that for 'homogenous exchange' (that is, the exit of tritiated alanine into an alanine solution, or the exit of tritiated threonine into a threonine solution), the exit ratio was not significantly different from the entry ratio for the same amino acid. When exodus was observed into a proline-containing solution the rates were low and the exit ratios characteristically lower than those for 'homogenous exchange'. The exit of alanine into a threonine solution was in contrast characterized by high exit ratios. Hence it appeared that both amino acids involved in the exchange could influence the value of the exit ratio $r$. As a more central matter, these experiments indicate that the linked fluxes occur in an outward as well as an inward direction.

The discovery that the system involved in alanine transport in the pigeon red blood cell (the ASC system described by myself, Liang and Archer, and by Eavenson and myself, both in 1967) permits $Na^+$ and an amino acid to be exchanged somewhat independently between the interior and the exterior of the cell when both are present, appears to deprive the measurement of the flux ratio of any value in revealing whether the net movement of $Na^+$ provides energy for the net movement of the amino acid. Under the experimental conditions used, the two fluxes of $Na^+$ appeared equally affected. Perhaps we must look to alkali metal metabolism to find biological meaning in the apparent service in this cell of such amino acids as alanine, serine or cysteine as 'exchange-diffusers' for $Na^+$ when they are joined with their transport mediator or 'carrier'. When Conway and Duggan observed in 1958 that various neutral amino acids accelerate the loss of $Na^+$ from $Na^+$-rich baker's yeast they probably were encountering the same phenomenon, rather than a direct competition for the alkali-metal pump.

We probably may continue to regard the flux ratios between $Na^+$ and glycine in the pigeon red blood cell as informative of the transfer of energy between the two. Vidaver found stimulation of transport between glycine and $Na^+$ only for the movement of each of these components *away from the phase in which the other component was present*. For example, although external $Na^+$ stimulated the uptake of glycine, it had no direct effect on glycine exodus. Figure 7a shows that when $Na^+$ and glycine were present together in the medium, modifications in the concentration of either component did not change the *exodus* rate for glycine; only the uptake rates were modified. In contrast, Wheeler and Christensen (1967b) found that glycine stimulated not only the uptake but also the exodus of $Na^+$ from the same cell. Several aspects of the inconsistency illustrated by figure 7 may be considered.

i) The stimulating effect of glycine on $Na^+$ movements was much weaker than that of alanine (figure 7b), so that it was harder to tell whether the two flux accelerations for glycine were really balanced. In at least one instance (table 1 in Wheeler and Christensen, 1967b), a net uptake of $Na^+$ accompanied that of glycine, although this behaviour was considered atypical. Furthermore, the possibility of stimulation of the $Na^+$ pump by the extra $Na^+$ entering with the glycine was not ruled out.

ii) A separate study has indicated that the glycine transport system of this cell is probably not an exchanger at all and that the exodus of other amino acids associated with the uptake of glycine probably arises from its appreciable reactivity with the alanine–serine–cysteine system (Eavenson and

Christensen, 1967). Glycine at 0·5 mM has a much weaker action than alanine in stimulating the loss of endogenous serine or threonine from the cell, although at high levels the effects of the two amino acids became similar (table 4 in Wheeler and Christensen, 1967b).

iii) Before the observations of figure 7a were obtained, the erythrocytes had been held for 107 minutes at 38·5° in a solution in which all $Na^+$ was replaced by $K^+$. In addition to other possible effects of this treatment, it may have largely eliminated the endogenous amino acids which can exchange for external amino acids via the ASC system. This system may, as suggested in ii, have provided the pathway for glycine stimulation of the exodus of various internal amino acids, along with $Na^+$. Conceivably then, the results in *a* may represent the glycine system operating alone and those in *b*, in conjunction with the exchanging ASC system.

In that connection, the demonstration by Wheeler and myself (1967b) of the additive nature of stimulation of $Na^+$ uptake produced by alanine on the one hand, and that produced by glycine, on the other, assumes extra importance. The experiments illustrated in table 3 were conducted at levels of the two amino acids nearly sufficient to 'saturate' or maximize their effects on $Na^+$ influx. The results so obtained show that even though glycine may have stimulated a small amount of amino acid exchange and $Na^+$ migration by the ASC system, most of its stimulation of $Na^+$ uptake occurred by an independent route. It would have been very significant by contrast if the action of external glycine in stimulating sodium exodus proved not additive to that of alanine, thereby supporting the view that the action of glycine on $Na^+$ exodus occurs by the same route as that of the alanine effect.

**Table 3.** Additive effects of alanine and glycine on $Na^+$ entry into the pigeon red blood cell

| Added to medium | $\Delta v_{gly}$ | $\Delta v_{ala}$ | $\Delta v_{Na+}$ | $\Delta v_{Na+}/\Delta v_s$ |
|---|---|---|---|---|
| 3·4 mM glycine | 0·157 | | 0·124 | 0·80 |
| 3·4 mM alanine | | 0·123 | 0·264 | 1·24 |
| 6·8 mM alanine | | 0·207 | 0·302 | 1·46 |
| Both at 3·4 mM | | | | |
| Found | 0·116 | 0·196 | 0·338 | |
| Predicted | | | 0·334 | |

Uptake of $^{22}Na$ and $^{14}C$-labelled amino acids measured during 5 min at 37° for a phosphate-buffered medium containing 8·3 mM glucose and a 135 mM level of either $Na^+$ or choline$^+$. $\Delta v_{Na+}$ represents the increment in $Na^+$ influx caused by addition of one or both amino acids. $\Delta v_{gly}$, $\Delta v_{ala}$ and $\Delta v_s$ represent the increments in uptake of a single amino acid due to the presence of external $Na^+$. The value marked *predicted* is the increment in $Na^+$ entry expected if the two systems operated independently, i.e., if the effects of glycine and alanine were completely additive. See Wheeler and Christensen (1967b) for details.

If glycine accumulated by the transport system largely specific to it, can be exchanged via the alanine–serine–cysteine system for other amino acids from the environment, then energy fed into accumulation by the glycine system could be retained in part in the accumulation of these amino acids. That is, the active transport of glycine may be secondary to that of $Na^+$, and the subsequent uphill transport of alanine, serine and similar amino acids by exchange for glycine may be a *tertiary* active transport. Or instead it remains possible that the ASC system is energized directly by net migrations of the alkali-metal ions, the linked net fluxes being heavily masked by the apparently independent, simultaneous exchanges of external $Na^+$ for internal $Na^+$ and of external amino acids for internal amino acids through this system. We may regard these large components of exchange, which tend to balance the gradients of the several amino acids and of $Na^+$, as a consequence of the high gradients already established endogenously.

Inherent in the preceding discussion is the significant feature that the exodus of amino acids from cells appears to occur to a large extent by the same mediating systems by which entry occurs. We have studied this question in considerable detail (Christensen and Handlogten, 1968) because we have encountered a persistent supposition that exodus occurs largely by diffusion or 'leaks' despite early evidence to the contrary (Riggs and co-workers, 1954; Heinz and Mariani, 1957). The portion of the exodus of the amino acids from the Ehrlich cell occurring by non-specific processes under ordinary physiological conditions is very small (Christensen and Liang, 1966a; Eavenson and Christensen, 1967; Christensen and Handlogten, 1968). Indeed, simple diffusion has not been formally proved rate limiting for any portion of normal uptake. We suggest that the leakiness of cells has frequently been greatly over-estimated, partly through under-estimates of the scope of the facilities present for the mediation of specific transport. Furthermore, much of the energy inherent in solute gradients probably is conserved during net migration in the direction of the gradient by linkages of the types under discussion.

## VI. RECENT OBSERVATIONS ON THE IONIC LINKAGE FOR GLYCINE AND α-AMINOISOBUTYRIC ACID TRANSPORT IN THE EHRLICH CELL

In the most thorough studies yet to appear of these complex relationships Eddy, Mulcahy and Thomson (1967) and Eddy (1968a,b) have combined studies of fluxes and the steady state. Generally one minute at $37°$ was used for the initial rate of glycine movement, whereas the steady state for the movements of glycine, $Na^+$ and $K^+$ was recognized by repeated analyses over an interval of thirty minutes, with the peak usually being reached by about ten minutes. An essential feature of the approach was to deplete the Ehrlich cells of ATP by incubating for twenty-five minutes in 2 mM NaCN. With that potential source of energy largely removed and the alkali-metal ion pump almost entirely stopped, the cellular and extracellular levels of $Na^+$ and $K^+$ remained the significant factors determining the rate and extent of glycine uptake and these concentrations were manipulated over a wide range.

Various metabolic inhibitors (fluoride, azide, 2,4-dinitrophenol and antimycin) were found to have no effect on glycine uptake or on the distribution of these alkali-metal ions in the cyanide-treated cells. The action of cyanide itself, as had been reported earlier on less complete evidence (figure 4), was almost entirely associated with its effects on $Na^+$ and $K^+$ distribution. Indeed, if the concentration of potassium ions in the incubation fluid was kept high enough and the sodium ion concentration low enough, for as long as the cyanide effect was present, cellular $K^+$ and $Na^+$ changes could be minimized and a nine-fold concentration of glycine observed subsequently, despite the presence of cyanide.

Furthermore, in the cyanide-treated cells, the uptake of glycine induced an accelerated concurrent entry of $Na^+$ and exodus of $K^+$, both in the direction of their downward gradients. A response of this type in ATP-depleted cells has not been shown in any other work cited. The stoichiometry $\Delta Na^+/\Delta gly$, observed under a variety of conditions, usually during three to five minutes, was $0.90 \pm 0.11$ and that for $\Delta K^+/\Delta gly$, $-0.62 \pm 0.11$. (Schafer and Jacquez had reported briefly in 1967 a similar ratio for $\Delta Na^+/\Delta\alpha$-aminoisobutyric acid, for one minute intervals at $37°$, and Schafer reported in 1968 a ratio of $\Delta K$ exodus/$\Delta\alpha$-aminoisobutyric acid uptake of $0.5-0.6$). Eddy showed that these actions were not predicted by any hypothesis for the energization of glycine transport, other than the proposal that the energy was derived from the previously generated gradients of the alkali metals (1968a).

A role for $K^+$ and $Na^+$ was included in an equation developed by Eddy and his colleagues (Eddy, Mulcahy and Thomson, 1967; Eddy, 1968b) on kinetic grounds under certain approximations:

$$\frac{[gly]_{in}}{[gly]_{out}} = \frac{[Na^+]_{out}}{[Na^+]_{in}} \cdot \frac{(1+\theta K^+_{in})}{(1+\theta[K^+]_{out})}$$

The experimental test of the predicted effect of $K^+$ indicated that the value of $\theta$ was not over $0.025$, which is to say, in agreement with all other work, that the effect of $K^+$ is small compared to that of $Na^+$. The results cannot be said to prove the equation. Nevertheless, when the $[Na^+]$ in both the extracellular and the cellular phases was set at 30mM (cyanide still present), the glycine gradient became somewhat larger when choline replaced

K+ in the suspending medium. On lowering the external [K+] from 128 to 10 mM, the maximal glycine uptake increased between 1·2–3·0-fold; these values represent the 95% confidence limits. This action may imply that a potassium complex of the carrier has a tendency to pass across the membrane without an amino acid present, whereas it is the Na+ complex that passes when an amino acid is included in the complex. Even though the effect of the K+ gradient appears relatively small, Eddy has calculated that it may contribute very significantly, so that the total energy made available may be sufficient for the glycine accumulation observed. Furthermore, as I observed in the Introduction, any addition of ubiquitous metabolites that can change a binding site for K+ into a binding site for Na+ deserves our close attention.

Figure 9 illustrates the relation found by Eddy between the steady-state distributions of glycine and Na+ in the cyanide-treated Ehrlich cell (a) and in the respiring Ehrlich cell (b). Both distribution ratios are plotted as their logarithms, a procedure justified by the thermodynamic argument that if a

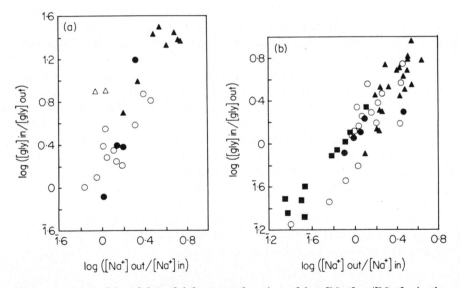

**Figure 9.** Plot of log [gly]$_{in}$/[gly]$_{out}$ as a function of log [Na+]$_{out}$/[Na+]$_{in}$ in the steady state reached when respiring (a) or cyanide-treated (b) Ehrlich cells were incubated with 0·9 mM glycine. The respiring cells were analysed after 40 min of incubation; the cyanide-treated cells had first been held for 25 min in a solution containing 2 mM NaCN at a selected set of concentrations of Na+ and K+, as keyed by the form of the symbol. Glycine uptake was then observed, again at selected levels of Na+ and K+ with NaCN still at 2 mM. The transient steady state prevailing when the ratio [gly]$_{in}$/[gly]$_{out}$ became maximal (usually at about 10 min) is recorded in the figure. The original article should be consulted for details.
(From Eddy, 1968).

constant fraction of the energy inherent in the Na⁺ gradient serves to concentrate glycine, a linear plot would be expected. The results obtained in the
absence of cyanide differ from those obtained in its presence by an apparently
greater scatter, and by their extension to somewhat higher values on both
coordinates. A third difference of greater importance is less readily apparent
on inspection of figure 9. In general, somewhat higher distribution ratios for
glycine for a given set of Na⁺ and K⁺ gradients were seen with the respiring
cells than with cyanide-treated cells. These results again show therefore that
most of the energy serving for glycine accumulation is retained by cyanide-
poisoned cells, given that favorable distributions of the alkali-metal ions
are obtained. The findings also suggest that somewhat greater accumulation
of glycine can occur when the cells are respiring, as though some energy
arises from a source not measured by the alkali-metal gradients.

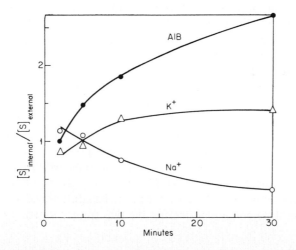

**Figure 10.** Accumulation of α-aminoisobutyric acid by 'cold-shocked' Ehrlich
cells, despite temporarily inverted gradients for both Na⁺ and K⁺—see text for
procedures used to obtain the indicated distributions of the alkali–metal ions.
Observations were made at pH 7·0, at 37°, in the presence of $2·5 \times 10^{-4}$ M ouabain
and 1 % alcohol. The curves show the changes in the distribution ratios for AIB,
K⁺ and Na⁺, as marked. (Drawn from data of table 14, Schafer, 1968).

This possibility is also stressed by the work of Schafer and Jacquez (1967,
1968), most of the documentation of which is available at this writing only
in the form of a doctoral dissertation (Schafer, 1968). This study was not
designed, as Eddy's was, to determine how much of the energy of amino acid

accumulation comes from the alkali-metal gradients, but to recognize whether other sources of energy must be considered. Ehrlich cells were depleted of K$^+$, with Na$^+$ replacement, first by chilling them, then by diluting the medium to half the usual osmolarity in the cold and finally by exposure of the cells to ouabain during their return to an environment of the usual osmotic pressure. The uptake of labelled $\alpha$-aminoisobutyric acid was then observed from a high-potassium medium containing ouabain (figure 10). The reader will note that in the interval two to five minutes the amino acid was accumulated into the cells to a modest extent before distinctly favorable gradients of either Na$^+$ or K$^+$ had developed. Similar experiments in which choline replaced part of the Na$^+$ of the medium gave more striking evidence that the test amino acid could be accumulated, in that case with the Na$^+$ gradient only inverted.

It needs to be re-emphasized that a failure of the alkali-metal ion gradients demonstrably to provide or to measure all the energy available for the accumulation of a solute does not mean that these ion gradients are excluded from providing an important part of the energy. The contrary conclusion that they do indeed provide a substantial part of that energy has been almost unavoidable for a decade in the case of glycine uptake by the Ehrlich cell; and this becomes, I think, inescapable with the results of Eddy. The supposition is not uncommon that the force applied to solute migration by a comigration with Na$^+$ can accomplish nothing unless the energy thereby provided is sufficient to do the whole task. This view is I think incorrect. Initially the question posed was, what force causes an amino acid to enter the cell against a large gradient of its chemical potential? Gradually it has become clear that this question needs to be repeated in a refined form. What force, if any, other than the K$^+$ gradient, causes the combination of amino acid molecule plus Na$^+$ to enter the cell against a now smaller gradient of the chemical potential, considering the two solutes together?

## VII. Na$^+$-DEPENDENT INHIBITION OF A TRANSPORT SYSTEM THAT DOES NOT REQUIRE EXTERNAL Na$^+$

A new relationship between Na$^+$ and amino acid transport has been recently uncovered in connection with the movements of the diamino acids. The uptake of lysine by the Ehrlich cell does not require the presence of Na$^+$ in the suspending fluid. On the contrary, a modest acceleration of lysine transport occurs when the Na$^+$ is replaced by choline; i.e. Na$^+$ appears to be slightly inhibitory (Christensen and Liang, 1966b). For homoarginine, the effect of Na$^+$ was also very small, although possibly stimulatory. (We prefer homoarginine as a test amino acid to lysine because a significant fraction of the latter is present as the $\alpha$-zwitterion, and is thus able to serve as substrate for systems L and, to a minor extent, A.) In rabbit reticulocyte we see instead a slowing of lysine uptake by about 10–15% upon the same replacement. Whether this effect arises from the uptake of a small part of the lysine by a Na⁺-dependent system is not yet known (Christensen and Antonioli, 1969).

When certain neutral amino acids are present, however, Na⁺ becomes strongly inhibitory to cationic amino acid uptake in both cell types. This is to say, Na⁺ is changed from an *accelerator* to an *inhibitor* of lysine uptake by the reticulocyte when certain neutral amino acids are added. Up to the present time, this effect has been seen in the reticulocyte for homoserine, cysteine, homocysteine and methionine, but not for valine and serine. This information already appears sufficient to indicate that the joint inhibitory

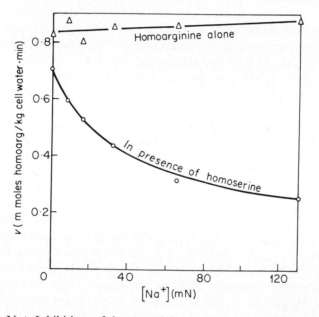

**Figure 11.** Na⁺ Inhibition of homoarginine uptake by the Ehrlich cell when homoserine is present, but not when it is absent. Uptake during 1 min at 37° from 0·2 mM homoarginine-¹⁴C in Krebs-Ringer's bicarbonate medium, pH 7·4, in which choline replaced Na⁺ to obtain the indicated Na⁺ levels. Homoserine was equal to 25 mM. The Na⁺-inhibitable rate of homoarginine uptake was estimated at 0·6 mmoles/kg cell water . min, i.e. six-sevenths of the total uptake rate, by plotting according to figure 3 in Inui and Christensen (1966). The curve has been drawn to correspond to the Michaelis–Menten equation for that Na⁺-inhibitable rate. By using a series of homoserine levels, a family of such curves was obtained, which showed that at somewhat higher levels of either Na⁺ or homoserine, inhibition became substantially complete. Similar results were obtained with phenylalanine, except that the $K_i$ values of *both* components were somewhat higher (From Christensen and Handlogten, 1969).

action of $Na^+$ and homoserine or other neutral amino acids on cationic amino acid uptake is not mediated by the A, ASC or L systems, or indeed by any known systems for the transport of neutral amino acids (Christensen and Antonioli, 1969).

In the Ehrlich cell, if one of several neutral amino acids is present, $Na^+$ also becomes intensely inhibitory (figure 11). Kinetic plots suggest that essentially all of the mediated uptake of lysine or homoarginine would be inhibited, were we able to secure sufficiently high levels of $Na^+$ and an appropriate amino acid. Phenylalanine, homoserine, cysteine, leucine and methionine show this effect (Christensen, 1969a). Hence in that cell also the group of amino acids so far recognized to be effective fails to correspond to any known transport system for neutral amino acids.

Knowing that $Na^+$ is necessary to the maximization of the inhibitory effect of neutral amino acids on the transport of cationic amino acids, the formerly puzzling problem of the apparently circumscribed nature of that inhibitory action can now be understood. Either $Na^+$ or the neutral amino acid alone shows inhibitory action on lysine uptake by the Ehrlich cell, but their action together is much greater, each increasing the reactivity of the other. The result is that no heterogeneity in uptake by the system for cationic amino acids needs to be proposed. This conclusion, along with the discovery that 3- and 4-carbon diamino acids serve as substrates for the A system for neutral amino acids (Christensen and Liang, 1966b) makes it so far unnecessary to propose more than one transport system for the uptake of cationic amino acids by the various cells under study.

To explain the $Na^+$-dependent inhibitory action of the homoserine, cysteine and other neutral amino acids in the rabbit reticulocyte, we have entertained the hypothesis that these amino acids serve as analogues for the diamino acids, provided that $Na^+$ occupies the position on the binding site normally taken by the positively-charged $\omega$-amino group of the diamino acid (Christensen and Antonioli, 1969; Christensen, 1969a; Christensen and coworkers, 1969). That is, the combination homoserine plus $Na^+$ appears to act as a 'surrogate amino acid', to occupy the diamino acid site. In the Ehrlich cell we find, however, another relevant action of the combination of neutral amino acid plus $Na^+$ (figure 12). In this case the external presence of any member of the same rather large group of neutral amino acids stimulates the exodus both of lysine and homoarginine much more strongly when $Na^+$ is present than when it is not. Under these conditions both $Na^+$ and the neutral amino acid appear to enter the cell in exchange for the cationic amino acids.

Figure 12 illustrates the $Na^+$-dependent stimulation of homoarginine exodus by external homoserine, along with the homologous stimulation by external homoarginine, which, however, we find to be decreased rather than increased by the presence of external $Na^+$.

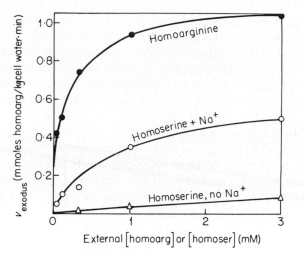

**Figure 12.** Concentration dependence of the action of external homoarginine (uppermost curve), of homoserine plus Na⁺ (middle curve) and of homoserine in the absence of Na⁺, to stimulate the exodus of homoarginine previously accumulated by the Ehrlich cell. The cells were first loaded to about 3 mM in homoarginine-14C by a 15 min incubation, and washed twice in KRB or in the same medium in which choline replaced Na⁺. Exodus was then observed by the decrease in cellular 14C during 1 min at 37° in about 100 volumes of KRB medium, pH 7·4, or the same medium with choline replacing Na⁺. The maximal stimulation of exodus of homoarginine in both cases occurred typically at a rate of 1·3 mmoles of homoarginine per kg cell water · min, and half this maximal stimulation was produced either by 0·09 mM homoarginine or by 5 mM homoserine, at a Na⁺ concentration of 116 mN.

Precedents can be cited for the concept that a specific site on a protein molecule can be occupied either by an intact substrate molecule, or alternatively, by cooperation between a smaller organic molecule, deficient by the absence of a charged group, and an ion, which appears to serve in the place of the missing part of the organic structure. I will select two examples that are familiar to me. Two pyridoxamine-oxaloacetate transaminases are known which do not act unless high concentrations of inorganic phosphate are present; these enzymes suffer inhibition by either pyridoxal phosphate or pyridoxamine phosphate (Wada and Snell, 1962). Also, trypsin splits α-acetyllysine ethyl ester rapidly but acetylglycine ethyl ester much more slowly, unless the ethylammonium ion is also supplied. For the hydrolysis of acetyllysine ethyl ester, the alkylammonium ions are instead competitive inhibitors (Inagami and Murachi, 1964). The present case is perhaps the first one from which we have evidence as to just what kind of a position the sodium ion occupies at a transport receptor site, and how that position is produced.

It seems likely that the amino acids present an important advantage over the subsequently introduced monosaccharides for study of the role of the alkali-metal ion in the transport complex. Amino acid molecules can readily be obtained or designed containing in their sidechains certain atoms (in particular N, O and S) which by their influence on the $Na^+$-facilitation of transport can show the position at which the alkali-metal binding site is generated.

We find we can also use the hydroxyl or mercapto group to signal the physical location of the $Na^+$ in the transport complex of the ASC system. The distance between the point at which $Na^+$ reacts and the points at which site ASC recognizes the $\alpha$-amino and the $\alpha$-carboxyl groups appears about two Angstrom units shorter than for the cationic amino acid system (Christensen and coworkers, 1969b).

## VIII.  CONCLUDING REMARKS

Some of the important unanswered questions suggested by the above account include the following:

i)  What is the nature of the energization of amino acid accumulation in cells in which $Na^+$ does not participate, and does that same mode also assist uphill transport by $Na^+$-dependent systems?

Even though no ATPase activated by neutral amino acids has as far as I know been described, nevertheless it appears likely that the uphill transport of some organic solutes is more or less directly energized by ATP. The occurrence of exchange seems likely, however, to cause the transport of other substances to be energized indirectly. A possibility that has on occasions appeared plausible to us is that the ATP-energized synthesis of glutamine could supply a sufficient cellular level of this abundant amino acid, so that it could serve in exchange for other neutral amino acids.

ii)  What is the chemical nature of the position occupied by $Na^+$ at the reactive site of systems where it accelerates a parallel amino acid flux?

iii)  What is the physical basis of the more rapid exchange of $Na^+$ than of amino acid observed for the so-called ASC system?

I am inclined to think that we have no right to suppose that $Na^+$ or $K^+$ ordinarily migrate across the plasma membrane without modifying the distribution of other substances. Furthermore such effects are reciprocal. Thus far discussions of the connections between the fluxes of $Na^+$ and organic metabolites appear to have reassured the student of cation transport that his subject is the fundamental one, and that the problems of metabolite transport are subordinate. My view, however, is that these intersections in the pathways followed by various transported solutes should for the present not be taken to establish which is the more basic transport process. I suggest that the ability of the sidechains of various neutral amino acids to complete a binding site for $Na^+$ where previously none could be found (perhaps at the same time eliminating one for $K^+$) may well be a model for a way in which the sidechains of amino acids contained in the polypeptide chain of proteins might generate a sequence of sites for alkali-metal transport through a region of low dielectric properties.

## Acknowledgements

This manuscript was largely prepared during the tenure of a Nobel Guest Professorship at the University of Uppsala, for which I want to express my appreciation to the Nobel Foundation and the Royal Swedish Academy of Science. Work by the author and his collaborators described here has been supported by the National Institutes of Health, U.S. Public Health Service, more specifically in recent years by grant HD01233 from the National Institute of Child Health and Human Development.

REFERENCES

Allfrey, V. G., R. Mendt, J. W. Hopkins and A. E. Mirsky (1961) *Proc. Natl. Acad. Sci. U.S.*, **47**, 907

Ames, G. F. (1964) *Arch. Biochem. Biophys.*, **104**, 1

Bittner, J. and E. Heinz (1963) *Biochim. Biophys. Acta.*, **74**, 392

Camien, M. N., H. Sarlet, G. Duchateau and M. Florkin (1951) *J. Biol. Chem.*, **193**, 881

Christensen, H. N. (1969a) In B. Kursonoglu and A. Perlmutter (Eds.), *Proceedings of a Symposium on Physical Principles of Biological Membranes*, Miami, Florida

Christensen, H. N. (1969b) *Advan. Enzymol.*, **32**, 1

Christensen, H. N. and J. A. Antonioli (1969a) *J. Biol. Chem.*, **244**, 1497

Christensen, H. N. and A. M. Cullen (1968) *Biochim. Biophys. Acta.*, **150**, 237

Christensen, H. N. and M. E. Handlogten (1968) *J. Biol. Chem.*, **243**, 5428

Christensen, H. N. and M. E. Handlogten (1969) *FEBS Letters*, **3**, 14

Christensen, H. N., M. E. Handlogten, I. Lam, H. S. Tager and R. Zand (1969a) *J. Biol. Chem.*, **244**, 1510

Christensen, H. N., M. E. Handlogten and E. L. Thomas (1969b) *Proc. Nat. Acad. Sci. U.S.*, **63**, 948

Christensen, H. N. and M. Liang (1966a) *Biochim. Biophys. Acta.*, **112**, 524

Christensen, H. N. and M. Liang (1966b) *J. Biol. Chem.*, **241**, 5542

Christensen, H. N., M. Liang and E. G. Archer (1967) *J. Biol. Chem.*, **242**, 5237

Christensen, H. N., D. L. Oxender, M. Liang and K. A. Vatz (1965) *J. Biol Chem.*, **240**, 3609

Christensen, H. N. and T. R. Riggs (1962) *J. Biol. Chem.*, **194**, 57

Christensen, H. N., T. R. Riggs, H. Fischer and I. M. Palatine (1952a) *J. Biol. Chem.*, **198**, 1

Christensen, H. N., T. R. Riggs, H. Fischer and I. M. Palatine (1952b) *J. Biol. Chem.*, **198**, 17

Christensen, H. N., T. R. Riggs and N. E. Ray (1952) *J. Biol. Chem.*, **194**, 41

Christensen, H. N., E. L. Thomas and M. E. Handlogten (1969) *Biochim. Biophys. Acta.* **193**, 228

Conway, E. J. and F. Duggan (1958) *Biochem. J.*, **69**, 265

Crane, R. K., D. Miller and I. Bihler (1961) In A. Kleinzeller and A. Kotyk (Eds.), *Membrane Transport and Metabolism*, Academic Press, New York p. 439

Crane, R. K., G. Forstner and A. Eichholz (1965) *Biochim. Biophys. Acta.*, **109**, 467

Csáky, T. Z. (1961) *Amer. J. Physiol.*, **201**, 999

Csáky, T. Z. and M. Thale (1960) *J. Physiol. (London)*, **151**, 59

Csáky, T. Z. and L. Zollicoffer (1960) *Amer. J. Physiol.*, **198**, 1056

Eavenson, E. and H. N. Christensen (1967) *J. Biol. Chem.*, **242**, 5386

Eckel, R. E., C. E. Pope, II and J. E. C. Norris (1954) *Arch. Biochem. Biophys.*, **52**, 293

Eddy, A. A. (1968a) *Biochem. J.*, **108**, 195

Eddy, A. A. (1968b) *Biochem. J.*, **108**, 489

Eddy, A. A., M. F. Mulcahy and P. J. Thomson (1967) *Biochem. J.*, **103**, 863

Ferdinand, W. (1966) *Biochem. J.*, **98**, 278

Fox, M., S. Thier, L. Rosenberg and S. Segal (1964) *Biochim. Biophys. Acta.*, **79**, 167

Heinz, E. (1954) *J. Biol. Chem.*, **211**, 781

Heinz, E. (1961a) *Biochemie des Actives Transports*. Springer-Verlag, Heidelberg p. 151

Heinz, E. (1961b) Presentation to Duarte Symposium, subsequently published in 1962 in J. T. Holden (Ed.), *Amino Acid Pools*. Elsevier, Amsterdam p. 539

Heinz, E. and H. Mariani (1957) *J. Biol. Chem.*, **228**, 97

Hempling, H. G. and D. Hare (1961) *J. Biol. Chem.*, **236**, 2498

Iacobellis, M., E. Muntwyler and C. L. Dodgen (1956) *Amer. J. Physiol.*, **183**, 275

Ingami, T. and T. Murachi (1964) *J. Biol. Chem.*, **239**, 1395

Inui, Y. and H. N. Christensen (1966) *J. Gen. Physiol.*, **50**, 203

Kipnis, D. M. and J. E. Parrish (1965) *Federation Proc.*, **24**, 1051

Koser, B. and H. N. Christensen (1968) *Federation Proc.*, **27**, 643

Koser, B. and H. N. Christensen (1969), unpublished results

Kromphardt, H., H. Grobecker, K. Ring and E. Heinz (1963) *Biochim. Biophys. Acta.* **74**, 549

Margolis, R. K. and A. Lajtha (1968) *Biochim. Biophys. Acta.*, **163**, 374

Oxender, D. L. and H. N. Christensen (1963a) *Nature*, **197**, 765

Oxender, D. L. and H. N. Christensen (1963b) *J. Biol. Chem.*, **238**, 3686

Piperno, J. and D. L. Oxender (1966) *J. Biol. Chem.*, **241**, 5732

Riggs, T. R., H. N. Christensen and I. M. Palatine (1952) *J. Biol. Chem.*, **194**, 53

Riggs, T. R., B. A. Coyne and H. N. Christensen (1954) *J. Biol. Chem.*, **209**, 395

Riggs, T. R., L. M. Walker and H. N. Christensen (1958) *J. Biol. Chem.*, **233**, 395

Schafer, J. A. (1968) Ph.D. thesis with J. A. Jacquez, The University of Michigan, Ann Arbor, Michigan

Schafer, J. A. and J. A. Jacquez (1967) *Biochim. Biophys. Acta.*, **135**, 1081

Schafer, J. A. and J. A. Jacquez (1968) *Federation Proc.*, **27**, 516

Schultz, S. G. and R. Zalusky (1965) *Nature*, **205**, 292

Stern, J. R., L. V. Eggleston, R. Hems and H. A. Krebs (1949) *Biochem. J.*, **44**, 410

Vidaver, G. A. (1964a) *Biochemistry*, **3**, 662

Vidaver, G. A. (1964b) *Biochemistry*, **3**, 795

Vidaver, G. A. (1964c) *Biochemistry*, **3**, 803

Vidaver, G. A., L. F. Romain and F. Haurowitz (1964) *Arch. Biochem. Biophys.*, **107**, 82

Wada, H. and E. E. Snell (1962) *J. Biol. Chem.*, **237**, 127

Wheeler, K. P. and H. N. Christensen (1967a) *J. Biol. Chem.*, **242**, 1450

Wheeler, K. P. and H. N. Christensen (1967b) *J. Biol. Chem.*, **242**, 3782

Wheeler, K. P., Y. Inui, P. F. Hollenberg, E. Eavenson and H. N. Christensen (1965) *Biochim. Biophys. Acta.*, **109**, 620

Winter, C. G., and H. N. Christensen (1965) *J. Biol. Chem.*, **240**, 3594

# CHAPTER 10

# Electrogenic or linked transport

R. P. Kernan

*Department of Physiology,*
*University College,*
*Dublin, Ireland*

## I. INTRODUCTION

It has been widely accepted for many years now that passive movement of ions down energy gradients may account adequately for both the resting membrane potential and for many of the transient potential changes observed during the activity of nerve and muscle. The latter include the action potential and also the non-propagated generator potentials of sensory receptors and electrical changes observed at muscle end-plates and at post-synaptic membranes. Here membrane depolarization was usually attributed to passive net movement of sodium ions into the cell, while membrane repolarization was believed to be due to net loss of potassium or gain of chloride ions by the cell. After prolonged membrane activity, as in nerve axons conducting a volley of impulses we might expect the eventual depletion of the energy, present as ionic concentration gradients between cell fluid and extracellular fluid, but for the presence at the cell membrane of a mechanism for restoring these gradients at the expense of metabolic energy. This mechanism is the ion pump. It is very likely that sodium ions are actively excreted from nerve, muscle and from most cells of the body, as they must be transported against both a chemical and electrical energy gradient. In the following pages the manner in which net movements of other ionic species are coupled to active excretion of sodium will be considered. It will be seen that in certain circumstances the active transport of ions may be responsible for significant changes of membrane potential.

## II.  GENERAL CONSIDERATIONS

During the spike or depolarization phase of the action potential in nerve axons about $3 \times 10^{-12}$ mole of sodium leak inward through each square centimeter of axon surface, with an equivalent loss of potassium during the phase of repolarization (Keynes and Lewis, 1951). Hodgkin (1958), in his Croonian Lecture to the Royal Society, proposed that the expulsion of sodium from nerve axon and reaccumulation of potassium following such activity, might be brought about by linked exchange of these ions on a carrier operating in a cyclic manner. Such a carrier would carry one potassium ion into the cell for each sodium ion excreted in a cycle of its movement across the cell membrane. If the ion pump operated in this manner it would contribute nothing directly to membrane potential but would be electrically neutral in its activity. Likewise an ion pump which excreted sodium accompanied by an equivalent amount of chloride would also be electrically neutral in its operation.

Coupled transport of sodium and potassium had been studied in squid giant axons by means of radioisotope flux measurements (Hodgkin and Keynes, 1955). In these experiments the interdependence of sodium and potassium fluxes during active transport provided the main evidence in favour of tightly linked exchanges of these ions on a single pump mechanism. Metabolic inhibitors including cyanide, azide and 2,4-dinitrophenol which decreased sodium excretion from axons, also greatly reduced the uptake of potassium from sea water bathing the axons. Hodgkin (1958) pointed out that the fact that inhibition of potassium uptake occurred here without appreciable change of membrane potential indicated that it was not mediated by electrical changes. Passive potassium uptake resulting from addition of potassium chloride to the bathing fluid was not inhibited by metabolic poisons. Evidence for coupling of sodium and potassium transport was also provided by the observation that in the absence of potassium from the external bathing fluid, sodium excretion as measured by radiosodium efflux was reduced to about one quarter of that in sea water. The dependence of radiosodium efflux on the presence of potassium in the bathing fluid did not hold true when the intracellular sodium concentration was about half the normal level (Frumento and Mullins, 1964). When potassium was absent from the bathing fluid the membrane potential increased by about 10mV, and this raised the electrical energy gradient against which sodium was excreted. If there was an upper limit to the electrochemical gradient against which the cells could excrete sodium, as seemed to be the case in muscle (Conway and coworkers, 1961), then perhaps this limit was exceeded in potassium-free conditions. However, when Hodgkin and Keynes (1954) increased the membrane potential of squid giant axon by up to 40mV through applied electrical current they found no reduction of sodium efflux. It is possible that here the inhibitory effect of hyperpolarizing current on sodium excretion may have been offset by potassium influx which must have taken place when the membrane potential, $E_m$, was made more negative than the potassium equilibrium potential defined by the Nernst equation,

$$E_K = -\frac{RT}{F} \ln \frac{[K]_i}{[K]_o} \tag{1}$$

where R, F and $T$ are the gas constant, Faraday constant and absolute temperature, respectively, and $[K]_i$ and $[K]_o$ the potassium concentrations in cell fluid and external bathing fluid.

As Harris (1954), Glynn (1956), Hodgkin (1958), Ussing (1960b) and others proposed models for tightly coupled or linked exchange of cations across the cell membrane on a single neutral pump, evidence in favour of chemically linked cation exchange was coming from studies of an enzyme system isolated from membrane fractions of cells and believed to be closely associated with the actual ion pump mechanism (Skou, 1960; Post and coworkers, 1960). This enzyme system was adenosine triphosphatase which catalyses the exergonic hydrolysis of adenosine triphosphate. As this system is described in detail in Chapter 9, I shall refer only to its relevant properties, namely that it requires the presence of both sodium and potassium ions for its activation and that in red cell ghosts (Glynn, 1962; Whittam, 1962) an asymmetrical activation of the enzyme system in the membrane has been demonstrated. In reconstituted red cell ghosts, sodium ions were required within the cell and potassium ions outside, before significant hydrolysis of adenosine triphosphate, as measured by formation of inorganic phosphate, took place. Since enzyme activation must necessarily involve some form of chemical combination between activating ion and enzyme, it was tempting to attribute to the adenosine triphosphatase system the role of ion carrier.

o

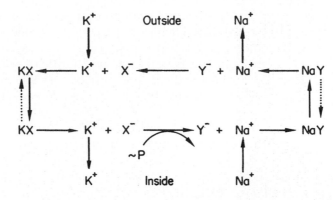

**Figure 1.** Scheme of a circulating carrier molecule for the active coupled exchange of sodium and potassium ions across the cell membrane in tight linkage. $Y^-$ is the phosphorylated form of the carrier which specifically binds sodium ions within the cell. At the outer face of the membrane this carrier is transformed into $X^-$ which has specificity for potassium ions.

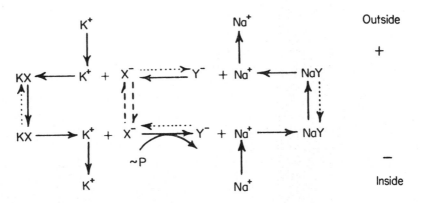

**Figure 2.** A circulating carrier differing from that shown in figure 1 in that form $X^-$ is capable of crossing the membrane without bound potassium. This would make it possible for the pump to become at least partly electrogenic. (From Cross and coworkers, 1965).

The first step in active excretion of sodium must be its specific binding to anionic sites on a carrier system at the inner surface of the cell membrane (figure 1) forming an electrically neutral complex. In this form the carrier with sodium attached may move to the outer surface of the membrane, where under activation by external potassium, energy-rich phosphate bonds are broken yielding energy for release of sodium ions into the external bathing fluid. It has been suggested that the carrier then transports potassium ions into the cell because it has acquired a greater specificity towards this cation as a result of the dephosphorylation reaction at the outer surface of the cell. Some modifications of this model have been proposed (figure 2) allowing for a less tightly coupled exchange of sodium and potassium with some degree of electrogenic pumping of sodium and electrical coupling of ion movement (Cross and coworkers, 1965). The wide acceptance of the hypothesis of chemically coupled sodium–potassium exchange during active transport must be viewed against the background of the considerable evidence for a role of adenosine triphosphatase as the carrier system, supplying both specific binding sites and the immediate source of energy. In his extensive review of the literature on cellular electrolyte metabolism (1960b) Ussing wrote, 'It seems, however, that the active transport processes, although being necessary for the maintenance of cellular ionic composition and thus ultimately for the potential development, do not otherwise contribute to the membrane potential.' This statement may at first sight seem strange coming as it does from one who has pioneered investigations into the role of the sodium pump in the electrogenesis of isolated frog skin. It will be seen presently, however, that the situation in epithelial tissues is complicated by the fact that we are dealing with transport across two membranes in series. This renders interpretation ambiguous.

The electrogenic pumping of ions may be recognized by a change of membrane potential which cannot be accounted for in terms of passive ion movement and which has some of the characteristics of a metabolic process, such as sensitivity to metabolic inhibitors and a high temperature coefficient. A change of membrane potential which is not accompanied by a significant change in membrane resistance might also be considered to indicate the presence of electrogenic pumping of ions.

In order to evaluate the evidence advanced for electrogenic pumping of ions in various tissues, it is worthwhile to consider how ionic gradients across the membrane and the relative permeability of the membrane to the more important ionic species may influence the membrane potential. Many resting cells have low permeability to sodium ions so that their membrane potential is determined mainly by the concentrations of potassium and chloride ions inside and outside the cell. In skeletal muscle and in other tissues where the Donnan equilibrium holds with respect to potassium and chloride ions the following relationship applies

$$[K]_i/[K]_o = [Cl]_o/[Cl]_i \qquad (2)$$

Here the membrane potential in the 'balanced state' (Conway, 1957) is defined by the Nernst equation, and we can choose to substitute either potassium or chloride ion concentrations in equation 1. As this equation is based on the activities of ions, the assumption is made that the activity coefficient of the various ionic species is similar in cell fluid and bathing fluid. The 'balanced state' has been found to apply in the case of isolated skeletal muscle bathed in plasma (Kernan, 1960, 1963), where membrane potential, $E_m$, measured by the microelectrode was found to be equal to the potassium equilibrium potential, $E_K$. However, this state is rarely achieved *in vitro* when isolated tissues are bathed in normal Ringer's fluid. For example, when $E_m$ was measured in freshly dissected frog sartorii bathed in Ringer's fluid with 2·5 mM K, it was found to be about $-92$mV (Kernan and Conway, 1955; Adrian, 1956), while $E_K$ calculated under the same conditions following potassium analysis of muscles and bathing fluid was $-100$mV. Adrian explained the difference between measured and calculated potentials by applying the more general constant field equation of Goldman to the situation, assuming that membrane permeability to sodium ions was sufficiently great to influence the membrane potential. In this equation (Goldman, 1943; Hodgkin and Katz, 1949), as used by Adrian,

$$E_m = -\frac{RT}{F} \ln \frac{P_K[K]_i + P_{Na}[Na]_i}{P_K[K]_o + P_{Na}[Na]_o} \tag{3}$$

it was assumed that chloride ions were in equilibrium with the membrane potential and permeability ratio $P_K : P_{Na}$ was about 100 : 1. The measured difference of 8mV between $E_m$ and $E_K$ might be accounted for in terms of the relative permeability of the membrane to potassium and sodium ions, taking into account the depolarizing action of net sodium influx. The hyperpolarizing effect of plasma, which has been shown to reduce net sodium influx in muscle (Carey and Conway, 1954; Creese and Northover, 1961; Creese, 1968) leads to an increase in the $P_K/P_{Na}$ ratio. This has the effect of making the second term in equation 3 insignificant, so that $E_m$ approaches $E_K$. In trying to determine whether the sodium pump is operating in an electrogenic manner it is important to know whether $E_m$ becomes more negative than $E_K$ and $E_{Cl}$. Unfortunately direct information on the concentrations of potassium, sodium and chloride within cells is not always available so that there may be some doubt of the values that can be ascribed to the equilibrium potentials of the various ionic species.

## III.  SPECIAL CONSIDERATIONS

### A. The Sodium Pump in the Neuron

#### 1. *Post-tetanic Hyperpolarization*

Perhaps the earliest indications that active excretion of sodium might contribute directly to membrane potential in nerve came from investigations into the nature of the long-lasting hyperpolarization of axon membrane which Hering (1884) first noted following repeated stimulation of peripheral nerve in frog. This post-tetanic hyperpolarization was found to be reduced under anoxic conditions (Gerard, 1930; Lorente de Nó, 1947) and following treatment of the nerve with metabolic inhibitors (Connelly, 1962; Greengard and Straub, 1962). It was augmented, on the other hand, when certain metabolites including glucose, alcohol, acetate or pyruvate were introduced into the fluid bathing the nerve (Greengard and Straub, 1962). There also appeared to be a close parallel between the time course of the hyperpolarization and of the extra oxygen consumption by the nerve following tetanic stimulation. On the basis of these observations, Connelly (1959) suggested that both these phenomena had their origins in the active transport of ions, triggered off by leakage of sodium into the axons during conduction of impulses. While Connelly appeared to favour the view that the sodium pump might be electrogenic here, that is to say the sodium excretion rate exceeded the rate of potassium uptake, Ritchie and Straub (1957) took a different view. They believed that the post-tetanic hyperpolarization (PTH) was due to depletion of potassium ions in the immediate proximity of the external surface of the axon membrane during the operation of an electrically neutral pump. They considered that $E_m$ was close to $E_K$ (equation 1), but that the latter increased because $[K]_o$ measured in the bulk of the bathing fluid was not representative of its lower concentration near the membrane following its uptake by the axons. Their main reason for believing this was that they were unable to detect PTH in potassium-free bathing fluid. Subsequently Gage and Hubbard (1966) were able to demonstrate PTH in rat phrenic nerve bathed in potassium-free fluid, but they attributed the phenomenon to an increase in the $P_K/P_{Na}$ ratio. Since the interpretation of their data seems questionable in some respects, it might be profitable to consider their experimental evidence and conclusions. They used depression of excitability in nerve as an index of membrane potential and found that reduction of the sodium concentration of the bathing fluid by 70% of normal increased PTH. They therefore argued that if PTH was produced by sodium pumping there should be less sodium in the axons following tetanic stimulation in this fluid and, therefore, PTH should decrease rather than increase. There remained however, the possibility that reduction of external sodium concentration might have a more pronounced effect on sodium excretion during the recovery phase than on sodium influx during the stimulation. There have been cases in which reduction of the sodium concentration in the bathing fluid has been found to greatly increase the active excretion of sodium from cells (Desmedt, 1953; Shaw and Simon, 1955; Carey and coworkers, 1959). While Gage and Hubbard observed PTH in potassium free fluid, they found that addition of 15 mM KCl to the bathing fluid reversibly abolished this hyperpolarization. They considered that the sodium pump could not function in the potassium-free bathing fluid and therefore could not contribute to the hyperpolarization and that addition of potassium chloride which should activate the pump, depressed rather than increased PTH. Since in potassium-free conditions the potential difference, $E_K - E_m$, was at a maximum, they suggested that this should augment a hyperpolarization produced by an increase in the permeability ratio, $P_K/P_{Na}$. In relation to these points it may be said that some excretion of sodium may take place in nerve and muscle even in the absence of external potassium (Ussing, 1960b). It seems likely that the potassium which leaks out of the axons and becomes trapped

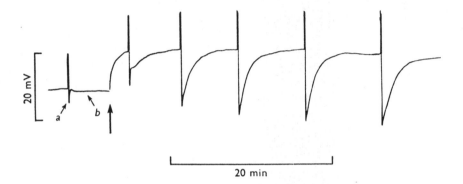

**Figure 3.** The effect of replacing chloride of Locke solution by isethionate on post-tetanic hyperpolarization of the non-myelinated fibres of a rabbit desheathed vagus nerve. The nerve was stimulated for 5 sec at a frequency of 30 stimuli/sec. Chloride of Locke solution was replaced by isethionate at the arrow. In this and subsequent records by these authors, the sharp upward deflection corresponds to the period of stimulation. Individual action potentials were not recorded. The fast and slow components of the post-tetanic response in unmodified Locke solution are indicated by *a* and *b* respectively. At least part of the general depolarization on switching from chloride- to isethionate-Locke is a result of change in the diffusion potential at the Locke/sucrose junction. (From Rang and Ritchie, 1968).

in the fluid between them, during repetitive stimulation in potassium-free bathing fluid, would be sufficient to maintain the activity of the sodium pump in this preparation. Furthermore, the addition of potassium chloride to the bathing fluid to the extent of 15 mM, may very well disturb the Donnan equilibrium, leading to a depolarizing passive movement of potassium into the axon, sufficient to mask PTH. Gage and Hubbard found that the cardiac glycoside ouabain, known to inhibit the sodium pump in many tissues (Schatzmann, 1953; Johnson, 1956) reduced PTH. However, they

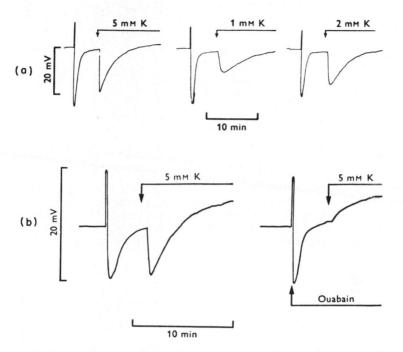

**Figure 4.** Activation by potassium of the recovery mechanism in non-myelinated fibres of rabbit desheathed vagus. a: the three responses show post-tetanic hyper-polarization obtained in potassium-free isethionate-Locke solution. Four minutes after each stimulation the bathing solution was changed (at the arrows) to one containing potassium in the concentration indicated by the number above each record. b: the left-hand record was obtained, as above, with the nerve in normal-Locke solution; the solution was changed to one containing 5mM potassium as indicated by the arrow. In the right-hand record the Locke solution also contained ouabain (1mM) from 30 sec after the end of the period of stimulation. (From Rang and Ritchie, 1968).

suggested that this was due to decrease in $[K]_i$ rather than to inhibition of an electrogenic sodium pump.

More recently, Rang and Ritchie (1968) used the sucrose gap technique to carry out an extensive investigation of PTH of C fibres of desheathed rabbit vagus nerves. They examined the effect of removal of chloride from the bathing fluid on PTH. The shunting effect of these permeable anions became evident when they found that in the presence of chloride PTH reached a value of 2–3mV and lasted for five seconds, while on replacement of chloride by the impermeant anion isethionate, PTH increased to a maximum of 35mV and lasted for five minutes (figure 3). Contrary to his

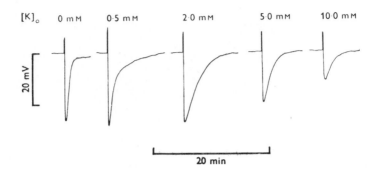

**Figure 5.** The effect of external potassium concentration on the post-tetanic hyperpolarization of the non-myelinated fibres of the rabbit desheathed vagus nerve. Post-tetanic responses to 5 periods of stimulation at 30/sec were elicited about every 10 min in isethionate-Locke solution whose potassium concentrations are indicated by numbers above each record. (From Rang and Ritchie, 1968).

earlier findings Ritchie now detected PTH even in the absence of external potassium, the main effect of its removal being a shortening of PTH to 1–2 minutes. A very significant observation of these workers was that

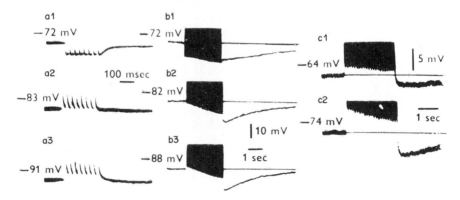

**Figure 6.** Effect of conditioning hyperpolarization on the after-potentials. a: short after-potential. b and c: post-tetanic hyperpolarization (PTH). a and b records from a same cell. c: another cell. a1, b1 and c1 show the after potentials at the membrane potential without conditioning current. With conditioning hyperpolarization, the polarity of the short after-potential was reversed (a2 and a3), but PTH became larger (b2, b3, c2). The membrane potentials are indicated over each base line. Tetanus was induced by intracellularly applied short current pulses of 31/sec in b and 50/sec in c. The spikes are off the trace at the amplification used (From Nakajima and Takahashi, 1966).

addition of 5 mM K to the bathing fluid after the PTH had declined in potassium-free conditions, (figure 4), produced a further hyperpolarization of about 20mV which decayed exponentially. This must surely have been due to reactivation of the sodium pump by the added potassium. Reactivation of PTH in potassium-free fluid could also be brought about by addition of rubidium, caesium, thallium or ammonium ions to the bathing fluid. The effect of the external potassium concentration on PTH in the experiments of Rang and Ritchie are shown in figure 5. Here it will be seen that PTH decreased significantly at $[K]_o$ values greater than 2 mM confirming the observation of Gage and Hubbard.

Nakajima and Takahashi (1966) examined the after-potentials in the stretch receptor neurons of crayfish and found two distinct types of potentials here. A short-term hyperpolarization, which followed a single action potential or a few action potentials, appeared to be produced by passive diffusion of potassium as it was dependent on the potential difference $E_K - E_m$. That is to say, the potential diminished and reversed in polarity when a hyper-polarizing current was applied across the nerve fibre membrane. The reversal of potential took place apparently when $E_m$ had been made equal in value to $E_K$ by the hyperpolarizing current. Furthermore, the potential at which reversal of polarity occurred depended on the external potassium concentration (figure 6a). The hyperpolarization which followed tetanic stimulation of the neuron differed in some important respects from this short-lasting increase in membrane potential. The former increased in amplitude when the membrane was hyperpolarized by applied current, (figure 6b,c) and was also increased by electrophoretic introduction of sodium ions into the cell from microcapillaries filled with sodium citrate. This procedure might be expected to stimulate the sodium pump into activity. Complete replacement of sodium of the bathing fluid by lithium depressed PTH. The fact that Rang and Ritchie also abolished PTH by replacement of sodium of the the external fluid with lithium raises doubts about the interpretation which Gage and Hubbard placed on the effects of partial sodium replacement in their experiments. Lithium ions are pumped out of cells very slowly (Keynes and Swan, 1959). Nakajima and Takahashi did not observe a significant change in membrane resistance during the PTH, which one would expect if there were a change in membrane permeability to a particular ionic species, as envisaged by Gage and Hubbard.

It is now evident from experiments by Holmes (1962) and Meves (1961) as well as from the ones mentioned above that PTH can occur under potassium-free conditions. It seems unlikely therefore that it is caused by potassium depletion at the axon membrane. Furthermore, its apparent independence of potassium equilibrium potential and the fact that it can occur without a significant membrane resistance change seems to rule out the possibility that it is produced by change in the ratio $P_K/P_{Na}$. There remains the possible electrogenic active transport of sodium ions out of the axon.

## 2. *Inhibitory Postsynaptic Potentials*

Eccles in his Ferrier Lecture to the Royal Society (1961) and in a later publication (1964) summarized the evidence of the nature of the inhibitory postsynaptic potential evoked in response to stimulation of presynaptic inhibitory interneurons. Considerable evidence has been collected by Brock and coworkers (1952) and by Coombs and coworkers (1955) suggesting that these hyperpolarizing potentials are produced by an increase in permeability of the postsynaptic membrane to potassium and chloride ions. More recently however, anomalous potential changes have been reported which suggest that in certain circumstances these inhibitory potentials may be produced by electrogenic pumping of sodium ions. In order that we may appreciate these anomalies we must consider briefly the methods employed by Eccles and his colleagues in their investigations. When a microelectrode was introduced into the cat spinal motoneuron and the membrane potential measured, it was found that electrical stimulation of a presynaptic inhibitory neuron increased this potential by about 2 mV and this transient potential change lasted about 2 msec. A second microelectrode was then introduced into the cell body near this electrode and electric current was passed through it, thereby changing the membrane potential of the cell to any desired value. It was possible with this voltage-clamp arrangement to test whether the membrane potential at which reversal of polarity of the IPSP took place corresponded with the equilibrium potential of potassium or chloride. The latter was altered by changing the composition of the fluid bathing the motoneuron or by electrophoretic introduction of various ions into the cell. One way in which $[K]_i$ and the associated $E_K$ was reduced was by introduction of sodium ions into the cell body. It was assumed that this procedure would tend to dilute the intracellular potassium by drawing chloride and water into the cell. In some cases instead of reducing the amplitude of the IPSP as expected, injection of sodium into the motoneurone resulted either in no change in the IPSP or even in an increase in its magnitude (Ito and Oshima, 1964). Kerkut and Thomas (1965) examined the IPSP in the abdominal ganglion of the common snail *Helix aspersa* in which spontaneous activity takes place. Using an experimental approach similar to that of Eccles they found that while injection of potassium acetate into the cells increased the membrane potential by about 7mV, the injection of sodium salts increased the potential by as much as 31mV (figure 7). The introduction of lithium salts into the cell increased $E_m$ by 20 mV but this hyperpolariza-

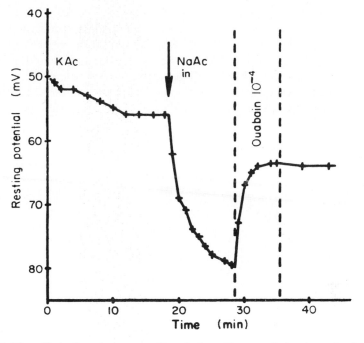

**Figure 7.** The effect of ouabain on sodium-induced hyperpolarization of cells of the abdominal ganglion of the common snail, *Helix aspersa*. The potential was recorded through a microcapillary electrode filled with potassium acetate, When sodium acetate was injected into the cell there was a hyperpolarization which was reduced on addition of ouabain. (From Kerkut and Thomas, 1965).

tion was maintained for a much shorter period than that produced by sodium injection. The hyperpolarization was reduced by addition of ouabain to, or absence of potassium from, the external bathing fluid. Thomas (1968) subsequently found that the introduction of 25 p-mole of sodium into the snail neuron increased $E_m$ by about 20 mV. He also employed the voltage-clamp technique to hold $E_m$ at a constant value during sodium injection and he found that for injections of less than 50 p-mole of sodium acetate, the voltage-clamp current was proportional to the amount of sodium injected (figure 8). This was a clear indication of a direct link between sodium excretion and membrane hyperpolarization.

Nishi and Koketsu (1967) used the sucrose gap method (figure 9) to examine the effect of changing $[K]_o$ and $[Cl]_o$ on the IPSP of bull-frog sympathetic ganglion cells in which normal orthodromic transmission was completely blocked by treatment with curare or nicotine. These cells, like those of the curarized superior cervical ganglion of turtle and rabbit (Laporte

and Lorente de Nó, 1950; Eccles, 1952), showed a long-lasting hyperpolarization (P-potential) after repeated preganglionic stimulation. In the
latter case, this potential was abolished by treatment with reserpine and may
therefore be produced by epinephrine. Epinephrine also appeared to produce
membrane hyperpolarization associated with metabolic changes in intestinal
smooth muscle (Axelsson and Bülbring, 1960). Nishi and Koketsu found
that the P-potential was independent of $E_K$ and $E_{Cl}$ but was very sensitive

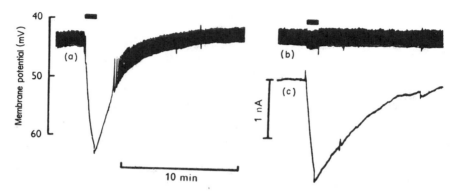

**Figure 8.** Response to sodium injection in cells of the abdominal ganglion of snail
before and after application of voltage clamp. a and b are records of membrane
potential change to injection of sodium acetate before and after application of the
clamp. The slow time constant of the pen recorder has reduced the action potentials
to only a few mV. The period of sodium injection is indicated by the solid bars
above the potential recordings. Record c is the clamp current recorded simultaneously with membrane potential b. All records are on the same time scale.
(From Thomas, 1968).

to inhibition by ouabain ($10^{-6}$M) and to reduction of temperature. Treatment of the preparation with cyanide or 2,4-dinitrophenol also decreased the
size of the potential. They concluded that it was generated by electrogenic
pumping of sodium ions. In the rabbit superior cervical ganglion the P-wave
appeared to be potassium sensitive (Kosterlitz and coworkers, 1968) and
may have been produced by a change in the $P_K/P_{Na}$ ratio, but when the
ganglion cell was first depolarized by treatment with acetylcholine, a hyperpolarization of the membrane was produced when this transmitter substance
was washed out of the preparation again. Unlike the P-wave this increase of
membrane potential was insensitive to changes of external potassium
concentration. The possibility that it was produced by diffusion of chloride

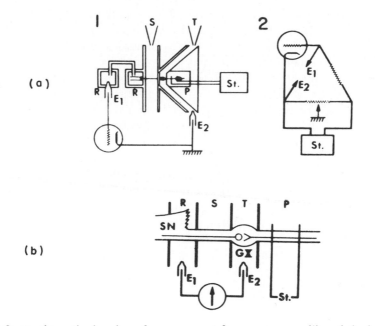

**Figure 9.** a: schematic drawing of arrangement for sucrose gap (1) and the bridge circuit (2) used for the application of conditioning currents to the bull-frog sympathetic postganglionic neurons. P, T, S and R represent the paraffin pool, test solution, sucrose solution and Ringer's solution pool respectively. $E_1$ and $E_2$ are calomel electrode and St is the stimulator. b: schematic illustration of the preparation. GX and SN represent the tenth sympathetic ganglion and spinal nerve respectively. (From Koketsu and Nishi, 1967).

ions was not excluded, so the conclusion of these workers that it might have been produced by electrogenic pumping of sodium was perhaps premature.

Temperature dependent resting potentials sensitive to addition of metabolic inhibitors and showing other properties consistent with electrogenic pumping of sodium ions have been observed in lobster axon (Senft, 1967) and in the visceral ganglion of the mollusk *Aplysia californica* (Carpenter and Alving, 1968).

## B. The Sodium Pump in Skeletal Muscle

Perhaps the best evidence for the operation of the sodium pump in an electrogenic manner came from studies with frog striated muscle. The author (1962a) reported that during net active transport of sodium from sodium-enriched sartorii into recovery fluid containing 10 mM K the mean membrane potential $E_m$, measured by the microcapillary electrode, was about 11 mV more negative than the calculated potassium equilibrium potential. The membrane potential was measured 10–20 minutes after immersion of the muscles in recovery fluid at room temperature and also after two hours when sodium excretion had been completed. Muscles were then analysed for potassium content after similar periods of immersion in this fluid and after measurements of the total water content and extracellular water volume of the tissue, the potassium concentration of the muscle fibre water and $E_K$ were accurately computed. It was found that after a two hour immersion $E_m$ and $E_K$ were almost identical. When the rate of sodium excretion was increased by immersing the sodium-rich muscles in sodium-free fluid (choline replacing sodium), the potential difference $E_m - E_K$ increased to about 33mV. When all the chloride of the bathing fluid was replaced by the impermeant anion sulphate, $E_m$ was still more negative than $E_K$, so chloride movements across the muscle fibre membrane were not responsible for the hyperpolarization. The author interpreted these results as indicating the electrogenic excretion of sodium by frog muscle and suggested that as muscle is normally quite permeable to potassium ions, these might be passively accumulated to a large extent under the influence of the potential generated by the sodium pump. In other words, he suggested that the coupling of sodium and potassium movements might be largely electrical. When $E_m$ was more negative than $E_K$ the potassium ions were not at equilibrium but tended to move down an electrochemical gradient into the muscle fibres. Subsequently other workers (Cross and coworkers, 1965; Frumento, 1965; Hashimoto, 1965; Mullins and Awad, 1965; Adrian and Slayman, 1966; Harris and Ochs, 1966) produced further evidence for electrogenic pumping of sodium ions in skeletal muscle.

Perhaps the most striking confirmation for this hypothesis came from the experiments of Adrian and Slayman, who eliminated the possibility that the hyperpolarization during active transport might be due to depletion of potassium at the muscle fibre membrane. They used rubidium instead of potassium in their recovery fluid and measured the effect of this substitution on $E_m$ during sodium excretion. The lower permeability of frog muscle to this cation compared with potassium was evident in the fact that its depolarizing effect on isolated muscle was significantly less than that of potassium when added to the bathing fluid at a concentration of up to 20 mM. About twice as much rubidium as potassium was required to produce an equal membrane depolarization (Feng and Liu, 1949; Sandow and Mandel, 1951). Permeability studies (Conway and Moore, 1945) based on relative osmotic activities of these two cations indicated a relative entrance rate of 100:38 in favour of potassium. Electrical measurements by Adrian

(1964) also confirmed greater membrane permeability to passive influx of potassium. If the coupling of sodium excretion with the inward movement of another cation were electrical, the sodium pump would generate a greater potential in attracting rubidium into the cell than in moving potassium in at a similar rate. Here the movement of cation into the cell in exchange for sodium would tend to shunt the potential generated by the sodium pump. If, on the other hand, membrane hyperpolarization was due to depletion of the external exchangeable cation and rubidium was taken up less rapidly than potassium, the membrane hyperpolarization should be slightly smaller where muscles were made to excrete sodium in exchange for rubidium. In their experiments, Adrian and Slayman made frog sartorii sodium-rich by immersing them for twenty-four hours at 1 °C in potassium-free fluid. Although sodium was excreted well from these muscles when they were reimmersed at room temperature in recovery fluid containing 10 mM Rb, this ion was taken up more slowly than potassium. Over the first half hour of active transport, intracellular rubidium increased at a rate of 50 mequiv./kg fibre . h compared with 70 mequiv./kg fibre . h for potassium uptake. For the measurement of $E_m$ during active transport, muscles were kept in recovery fluid at 1–3 °C for two hours to equilibrate before the temperature of the fluid was increased to activate the sodium pump. In the presence of 10 mM K, $E_m$ increased from about $-50$mV at low temperature to a maximum of $-85$mV after thirty minutes in Ringer's solution at 20 °C; whereas in the case of muscles recovering in 10 mM Rb, $E_m$ increased to about $-100$ mV. Furthermore, it was found that in freshly dissected muscles $E_m$ at a particular external rubidium concentration was never more than 7 mV negative to a muscle in the same concentration of potassium ions. This was much less than the difference in membrane potential between muscles actively transporting sodium in fluids containing these cations. It was concluded that if the pump exchanged sodium for potassium or rubidium ions on a one for one basis, the potentials of the two sets of muscles should be similar during active transport in the presence of the two cations.

Adrian and Slayman (1966) and also Harris and Ochs (1966) found that addition of the anaesthetic cocaine to recovery fluid containing 10 mM K, increased significantly both the membrane resistance and $E_m$ during active transport. The conditions then resembled those found where muscles were excreting sodium in exchange for rubidium ions. Some anaesthetics, including chloralose, paraldehyde and pentobarbital seem to decrease potassium conductance through the cell membrane (Guttman, 1939; Shanes, 1958). Harris and Ochs also found that the antihistaminic, mepyramine, increased membrane resistance from about 12 k$\Omega$/cm$^2$ to about 58 k$\Omega$/cm$^2$ while decreasing net uptake of potassium to about a fifth of the quantity of sodium excreted. In the process $E_m$ increased by about 10 mV over that found in the absence of the drug. These observations make it highly likely that a large part of the potassium uptake here may be electrically coupled to the membrane hyperpolarization generated by the sodium pump.

In trying to determine the possible contributions of electrical, as opposed to chemical, coupling to potassium accumulation during active transport, several approaches have been made. For example, some investigators have attempted to estimate the transmembrane current which might result from application of the potential difference $E_m - E_K$ across the membrane resistance, and compared this current in Cl-free fluid, presumably carried into the cell by $K^+$ ions, with the measured net influx of this cation. In the experiments of Harris and Ochs, the average potassium uptake in recovery fluid was 20 mequiv./kg . h, corresponding to a membrane current of $1 \cdot 3 \mu A/cm^2$. A membrane hyperpolarization of 15 mV would be required to drive this current through the measured membrane resistance of 12 k$\Omega$/cm$^2$. This was less than the value $E_m - E_K$ which they observed during sodium transport in chloride-free recovery fluid, so here it appeared that potassium uptake could be accounted for in electrical terms. In another calculation, Cross and his coworkers used a potassium conductance value of 480 $\mu$mho/cm$^2$ (Hodgkin and Horowicz, 1959) and a value of 10 mV for membrane hyperpolarization. The calculated inward potassium current was found to be $4 \cdot 8 \mu A/cm^2$ corresponding to 50 pmole/cm$^2$. This was about four times the value calculated by Harris and Ochs. The potassium accumulation per unit area of muscle fibre membrane, derived from the data of Desmedt (1953), was 30–60 pmole/cm$^2$ and this could therefore be accounted for in terms of passive inward movement under the influence of the potential difference $E_m - E_K$, provided that potassium conductance, $g_K$, was similar in sodium-rich and in freshly dissected muscles. However, the fact that the specific resistance of muscle fibre membrane increased to twice its normal value as intracellular potassium was replaced by sodium (Yamada and Yonemura, 1967) raises some doubt on this question. This may also explain to some extent the discrepancy between the two estimates of membrane current. Both these groups of investigators, and also Adrian and Slayman, agreed that the electrochemical potential difference for potassium ions between muscle fibre water and external fluid during active transport of sodium was sufficiently great to account for its uptake by passive electrical coupling. This is not to say that this ion can only enter the cell in this manner, since there is evidence for the presence of a potassium pump in the membrane. This evidence comes mainly from a consideration of rubidium uptake during active transport. In their investigation of this process Adrian and Slayman made use of the fact that membrane permeability to rubidium appeared to be independent of $E_m$, and the current carried across the membrane by these ions was a linear function of $E_m$. Using a form of the constant field equation, they calculated passive rubidium uptake by muscles at $E_m$ values measured in the presence and absence of active transport. They used ouabain to inhibit active transport and carried out their experiments at three different temperatures. They then compared the rubidium influx ratios derived in this way in the presence and absence of sodium excretion, with the influx ratios measured experimentally.

The calculated ratios of rubidium influx in the presence and absence of active transport, $M_a$ and $M_r$ respectively, were related to the membrane potential measured under these conditions, $E_a$ and $E_r$ respectively, as shown,

$$\frac{M_a}{M_r} = \frac{E_a}{E_r} \frac{\exp(E_r F/RT) - 1}{\exp(E_a F/RT) - 1} \tag{4}$$

Mean values of $E_a$ at 8 °C, 20 °C and 30 °C were found to be $-48$mV, $-70$mV and $-93$mV, respectively. The mean value of $E_r$, which was approximately proportional to the absolute temperature, was taken to be $-33$mV over the complete temperature range 8–30 °C. The values of $E_a/E_r$ in equation 4 were therefore 1·45, 2·12 and 2·82 at 8 °C, 20 °C and 30 °C, and the derived flux ratios, $M_a/M_r$, were 1·25, 1·66 and 2·08, respectively. At 30 °C the influx of rubidium measured where active transport was inhibited was 2·2 pmole/cm² of membrane, but where transport was not inhibited it was about 41 pmole/cm², giving a flux ratio of nearly 20. Since the measured influx ratio was about ten times that calculated for passive rubidium influx at 30 °C, it was evident that a large part of the rubidium influx, perhaps as much as 90% must have been accomplished through active accumulation. Unfortunately, the same type of calculation cannot be applied to potassium as its permeability is not independent of $E_m$. If, however, 90% of the accumulated rubidium ions must be actively transported into the cell, it seems reasonable to assume that the pump concerned with this transport is the physiological potassium carrier. Therefore the relative amount of potassium accumulated by electrostatic attraction to that chemically linked to sodium carriage may vary considerably depending on experimental conditions. Cross and co-workers, for example, reported finding cases in which $E_m$ was more positive than $E_K$ during active transport, where frogs had been stored in the cold for some time before use. Under these conditions potassium uptake would seem to have been against an electrochemical gradient and therefore active.

An interesting relevant observation (Lubin and Schneider, 1957) is that while most methods of measuring relative mobilities of rubidium and potassium ions through cell membranes have indicated the permeability of the former to be about half that of the latter, yet, where these ions are exchanging for sodium ions across the membrane, the $P_{Rb}/P_K$ ratio approaches unity. Indeed the following observations even suggest that the potassium carrier in the membrane may show a greater affinity for rubidium and caesium than for potassium. In one series of experiments, Relman and his colleagues (1956), over a period of 2–3 weeks, gave rats a mixture of potassium and rubidium (20 mequiv./l of each) in their drinking water. Then the animals were killed and the concentrations of these ions were measured in muscle fibre water and in plasma. When the distribution of both these cations between muscle and plasma was expressed as a ratio of concentration in the former over that in the latter, the ratio was three times greater for rubidium than for potassium. When the same type of experiment was carried out using caesium to which the cell membrane is less permeable, the ratio was five times greater for this ion than for potassium.

Mullins and Awad (1965) took sodium-rich muscles which had been soaked for twelve hours in potassium-free sulphate Ringer's solution at $4\,^\circ$C and placed them then in another aliquot of this fluid at $4\,^\circ$C for measurement of $E_m$. When the temperature of the fluid was raised to $20\,^\circ$C, the mean $E_m$ value increased from $-47$mV to $-76$mV. When this fluid was then replaced by a lithium sulphate Ringer's solution containing $2\cdot5$ mM K, $E_m$ increased still further to $-98$mV. When freshly dissected muscles were used for this experiment, instead of sodium-rich ones, the mean $E_m$ value in potassium-free Ringer's solution increased from $-104$mV to $-107$mV on warming the muscles from $4\,^\circ$C to $20\,^\circ$C. When the sodium sulphate was replaced by lithium sulphate in presence of $2\cdot5$ mM K, $E_m$ declined to $-88$mV instead of rising.

Two important points arise from these experiments. First, the $E_m$ of sodium-rich muscle, in contrast to freshly dissected ones markedly increases on warming in potassium-free fluid. At $4\,^\circ$C in potassium-free fluid there is a residual sodium excretion of about $4\cdot6$pmole/cm² sec and increase in bath temperature cannot but increase this. This strongly suggests electrogenic pumping of sodium. The second point of interest is the completely contrasting effects produced when sodium-rich muscles, on the one hand, and freshly dissected muscles, on the other, were transferred to lithium sulphate Ringer's solution. In the context of electrogenic pumping of sodium ions, the increased potential in the former is understandable, as replacement of extracellular sodium might be expected to increase its efflux rate. In the case of freshly dissected muscles there is not sufficient intracellular sodium to maintain the activity of the pump and the depolarization here is probably due to the combined effects of net lithium influx and addition of external potassium.

In equations (1) and (3) it can be seen that the membrane potential, based on passive movement of ions, is proportional to the absolute temperature. Therefore it should have a temperature coefficient of $1\cdot033$ in muscle, as found by Ling and Woodbury (1949). An increase in the temperature of the fluid bathing an isolated muscle preparation from $2\,^\circ$C to $21\,^\circ$C might be expected to increase $E_m$ by not more than 5mV unless a change in $P_K/P_{Na}$

occurs during the process. If, however, a metabolic process contributes directly to $E_m$, then we might expect to find a greater temperature coefficient for the membrane potential, as found by Harris and Ochs (1966), Adrian

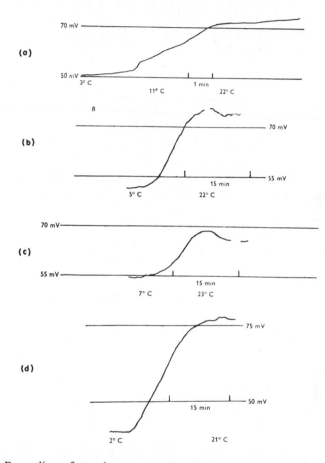

**Figure 10.** Recording of membrane potential changes of muscle fibres to warming from 4° to 20°C (or as indicated). The muscles had been depleted of potassium before being immersed in the fluid. a: in Cl⁻-mixture with 2·5 mM KCl present. Potassium uptake by muscle 11 mequiv./kg in 2·5h. b: In Cl⁻-mixture with 10 mM KCl added. The average potentials before warming and after were $54·1 \pm 0·5$ and $68·0 \pm 0·3$mV respectively. c: in Cl⁻-free mixture with 10 mM K⁺ methane sulphonate present. The average before warming and after were $51·2 \pm 3$ and $63·5 \pm 2$mV. d: in Cl⁻-free and Na⁺-free solution with 10 mM K⁺ methane sulphonate added. The average potentials before warming and after were $49·0 \pm 1·7$ and $73·8 \pm 1·4$ mV, respectively. (From Harris and Ochs, 1966).

and Slayman (1966) and Kernan (1968). Harris and Ochs found increases of $E_m$ of almost 25mV (figure 10) when the temperature of the bathing fluid was increased by about 20 °C, where sodium-rich muscles were reimmersed in recovery fluid composed mainly of choline–methane–sulphonate mixture with 10 mM K. This fluid was deficient in both sodium and chloride.

## C. The Sodium Pump in Epithelial Tissues

### 1. *Frog Skin*

As a fuller treatment of transepithelial transport of ions is found in Volume 3, only those properties of these tissues which are relevant to the present topic will be described here.

When isolated frog skin was mounted between two vessels and bathed on its inner and outer surfaces with identical fluids containing sodium or lithium with some potassium present, a potential difference of up to 100mV was measured between the vessels. The inner surface (chorion side) was usually positively charged with respect to the outer surface (epithelial side). Ussing and Zerahn (1951) short-circuited the potential found across the skin and measured the current required to maintain the skin potential at zero (figure 11). Using radioisotopes of sodium and chloride they then measured the unidirectional fluxes of these ions across the skin and from these measurements deduced the net fluxes in open and short-circuited conditions. In the short-circuited skin with similar solutions on both sides, the electrochemical potential gradients of the various ions across the tissue decreased to zero. Therefore in the case of the chloride ions, which appeared to pass through the skin passively, the measured isotopic influx and efflux were about equal, so no net flux of this ion took place. In open circuit conditions, the chloride flux ratio, influx:efflux, was consistent with passive distribution, as determined by its electrochemical gradient across the skin expressed by the equation.

$$RT \cdot \ln M_{in}/M_{out} = u_o - u_i \tag{5}$$

Assuming equal activity coefficients for chloride ion inside and outside the cell this approximates to,

$$\frac{M_{in}}{M_{out}} = \frac{[Cl]_o}{[Cl]_i} \cdot e^{EF/RT} \tag{6}$$

In open-circuit conditions therefore a net movement of chloride took place from the epithelial to the chorion side, under the influence of the potential difference across the skin. In contrast to chloride, the rate of influx of labelled sodium ($^{24}$Na) across short-circuited skin was as much as one

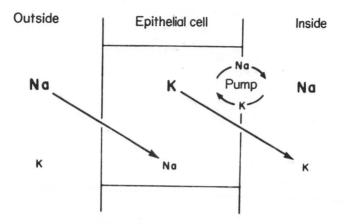

**Figure 11.** Schematic representation of the Koefoed-Johnsen–Ussing model of transepithelial transport of sodium ions. The sodium passes freely down an electrochemical gradient into the epithelial cells from the fluid bathing the outer surface. The potential across the outer membrane is therefore a sodium–equilibrium potential. The sodium is pumped out of the cell by an electrically neutral coupled sodium–potassium pump at the inner surface of the epithelial layer. Net efflux of potassium which has been carried into the cells by the pump occurs at this membrane thereby contributing to the transepithelial current.

hundred times greater than its efflux rate, and the resulting net flux towards the chorion side, expressed as gram equivalents, was approximately equal to the short-circuit current, expressed as faradays. In other words, it seemed that the flow of current through the skin, which was responsible for its potential under open-circuit conditions, was carried by sodium ions actively transported against an energy gradient proportional to this potential. Net transport of sodium through the skin did not alter significantly the composition of the bathing fluids during the course of the experiment.

The outer surface of the skin appeared to be freely permeable to sodium, but not to potassium, while the inner surface was permeable to potassium and impermeable to sodium (Ussing, 1960a). This became evident when chloride in the bathing fluid was replaced by impermeant sulphate anions and the effect of altering the sodium and potassium composition of the bathing fluids on the skin potential was examined. The skin potential was reduced when the sodium of the external bathing fluid, $[Na]_e$ was decreased or when the potassium concentration at the inner surface $[K]_e$ was increased. In other words, the outer surface of the skin acted as a sodium electrode and the inner surface as a potassium electrode. The total E.M.F. across the skin was then the sum of the two diffusion potentials,

$$E.M.F. = \frac{RT}{F} \cdot \ln\frac{[Na]_e}{[Na]_c} + \ln\frac{[K]_c}{[K]_i} \tag{7}$$

where c is cellular concentration, i is the concentration in chorion solution and e is the concentration in epithelial bathing fluid.

The Koefoed-Johnsen–Ussing model (1958) of epithelial transport was advanced to account for this paradoxical situation in which the skin potential could be represented as the sum of two diffusion potentials, while at the same time it clearly depended on the active transport of sodium. It was considered that sodium (figure 11) passed freely by passive diffusion from the external bathing fluid to the interior of the cells of the basal epithelium, making their interior positively charged with respect to the external fluid. This sodium was then pumped out of the cells once more, this time at the membrane of the chorion side of the epithelial layer. It was suggested that active transport of sodium at the basement membrane was brought about by a sodium-potassium linked pump operating in an electrically neutral manner. To operate in this manner, the pump had to transport potassium back into the cells at this membrane. It was believed that the potassium then diffused out of the cells towards the chorion side again, thereby contributing to transepithelial current. In this way Ussing and his colleagues were able to explain the generation of the skin potential by active transport, without the need for postulating an unlinked or electrogenic transport of sodium. If this model was valid it should have been possible to show the presence of at least two voltage steps with a microelectrode pushed gently through the skin. Engbaek and Hoshiko (1957), working in Ussing's laboratory, confirmed this prediction. They found the first stable potential step as the tip of the electrode penetrated the outer border of the cells of the basal epithelial layer and a second step of about the same magnitude as it passed through the chorion face of the basal epithelial cell. The sum of the two potential steps was equal to the skin potential. The electrical properties of the skin may be represented in an equivalent circuit as a sodium battery in series and in parallel with resistors. The parallel resistor is a shunt path provided by passive movement of chloride and of other ions through the skin under the influence of the skin potential. The contribution of chloride to this conduction became evident when chloride in the bathing fluids was replaced by sulphate, or where its membrane permeability was decreased by treatment of skin with copper sulphate solution. Here the skin potential increased from about 100mV to approximately 160mV. The nature of the series resistor, which might be regarded as associated with a rate-limiting step in the transfer of sodium across the skin, should also be mentioned here. Anything which facilitated the arrival of sodium ions at the carrier sites of the pump within the basement epithelial cells might be expected to increase its net transport provided that these sites were not already saturated by this ion. The rate-limiting step appeared to be diffusion of sodium chloride and water transfer at the epithelial side of the basement membrane, a process stimulated by addition of neurohypophyseal hormone (Fuhrman and Ussing, 1951).

Although Ussing (1960b) at one time appeared to accept the more fashionable concept of an electrically neutral 1:1 active exchange of sodium and potassium by the pump in frog skin, in his subsequent Harvey Lecture (1965) he seemed more inclined to consider that the sodium pump might be electrogenic. This change of attitude was mainly due, perhaps, to the outcome of experiments (Bricker and coworkers, 1963) in which the short-circuit current was measured while the chorion side of the skin was bathed in fluid containing a high concentration of potassium ions. When the electrochemical potential of potassium in the fluid bathing the chorion side of the skin had been raised in this way, above that in the cells, this cation could no longer leak out of the cell again after being transported in on a linked sodium–potassium pump at this membrane. According to the Koefoed-Johnsen–Ussing model the short-circuit current measured in the usual bathing fluids was carried across the outer membrane of the basal epithelial cell by sodium ions and across the inner membrane by potassium ions. The point of interest is what would happen to the short-circuit current when the latter flux was prevented as indicated. It was found that when potassium-rich fluid bathed the chorion side, both the skin potential and short-circuit current fell sharply, the former to 15mV, but the latter soon increased to 50–150% of its original value, declining again slowly over a period of hours. The net influx of sodium across the skin increased by about 30%; this movement was down an electrochemical gradient, but still appeared to be active as it was ouabain and cyanide sensitive. The pumping of sodium appeared to be uncoupled from potassium transport and was electrogenic. The membrane current was then carried entirely by sodium ions.

The electrogenic pumping of sodium across the body surface epithelium in larval salamander has also been reported by Dietz and coworkers (1967).

### 2. *Urinary Bladder*

Active transport of ions has been extensively investigated in another type of epithelial tissue, which is simpler histologically than frog skin. This is the toad bladder, which is a single layer of mucosal cells supported by a little connective tissue. Normally the serosal surface of this cell layer is positively charged with respect to the mucosal surface and this potential appears to be mainly due to the net transport of sodium across the bladder towards the serosal surface.

The net flux of sodium, derived from measurement of unidirectional fluxes of $^{22}$Na and $^{24}$Na (Leaf and coworkers, 1958), was found to equal the short-circuit current in forty-two experiments each lasting one hour under aerobic conditions. Frazier and Leaf (1963) used microelectrodes to measure the potential difference across the serosal membrane, where the sodium pump appeared to be located, while this surface was bathed in a potassium-rich solution containing little chloride. They found under these conditions that $E_m$ was more negative than $E_K$ or $E_{Cl}$. They therefore concluded that the sodium pump at this membrane could separate the charge or was electrogenic. The independence of net movement of sodium and potassium through the toad bladder was also demonstrated. The uptake of $^{42}$K from the serosal medium appeared to be in no way related to sodium transport across the bladder. Removal of sodium from the mucosal bathing fluid naturally greatly decreased active transport of this cation, but it had little effect on $^{42}$K uptake at the serosal surface, suggesting that a 1:1 exchange of sodium for potassium was not necessarily true of the pump here (Essig and Leaf, 1963). Furthermore, although potassium was required in the serosal medium for maintenance of sodium transport and short-circuit current, its absence from this fluid appeared to inhibit passive sodium movement across the mucosal membrane rather than the serosal pump directly. This was shown when $^{22}$Na was present in the fluid bathing the mucosal side of the bladder. When potassium was removed from the serosal fluid the $^{22}$Na labelling of the cells decreased, whereas if the absence of potassium had inhibited the serosal pump directly the $^{22}$Na labelling of the cells should have increased.

The following observations seem to emphasize the dependence of transepithelial potential and current on metabolism. The addition of insulin to the serosal fluid in the case of toad bladder (Herrera, 1965) and of isolated frog skin (Herrera and coworkers, 1963) and toad skin (Andre and Crabbé, 1966) increased both short-circuit current and net transport of sodium across the epithelial layer of cells. The author (1962b) also found that when insulin (100U/l) was added to recovery fluid in which sodium-rich frog sartorii were excreting this ion, the net excretion of sodium over a two hour recovery period was greatly increased. As this increased net transport of sodium was accompanied by membrane hyperpolarization, it was concluded (Kernan, 1965) that this hormone acted directly on electrogenic sodium excretion. The direct action of insulin on the sodium pump has been confirmed by measurements of sodium efflux in crab muscle (Bittar, 1967) and isolated rat diaphragm (Creese, 1968). This hormone also brought about membrane hyperpolarization of mammalian muscle when added to the bathing fluid (Zierler, 1959).

In the isolated bladder of the turtle *Pseudemys scripta* the short-circuit current which could be attributed to the net movement of sodium from mucosal to serosal side was greater than the measured short-circuit current by about 44 μA (Gonzalez and coworkers, 1967). The short-circuit current was then measured in a series of experiments in which sodium, chloride and bicarbonate, in turn, were omitted from the bathing fluid and their contributions to the transepithelial current measured. In this way it was revealed that the difference between net sodium flux and short-circuit current could be accounted for in terms of active transport of chloride and bicarbonate across the bladder. The sodium current was the same whether these anions were present or not (Gonzalez and coworkers, 1967), so it was concluded that all these ions were carried by parallel independent carriers.

## 3. Large Intestine

In gut the considerable reabsorption of water which takes place at the level of the colon reflects a very active transport of sodium ions from the lumen into the blood. When saline was introduced into the human colon it was found (Darrow, 1957) that its potassium concentration increased due to selective uptake of sodium chloride and water. While potassium salts are found in moderate amounts in faeces, sodium is virtually absent. Across the wall of the colon a potential difference has been found, such that the blood side may be as much as 60mV positive with respect to the lumen for many hours. This potential has been measured in the large intestine of the bullfrog (Cooperstein and coworkers, 1957) and of the dog (Cooperstein and Brockman, 1958). In a recent extensive investigation of this phenomenon Edmonds (1967) found that the magnitude of the potential across the colon wall could be increased by depletion of body sodium in rats. This depletion was induced by injection of a 5% glucose solution into the peritoneal cavity. This solution was then removed after a period of about two hours. Although this procedure appeared to have a negligible effect on plasma electrolyte concentrations it produced local reduction of sodium concentration in the gut tissue, leading to stimulation of active uptake of this cation. In rats on a diet of rice the mean potential difference across the caecum wall following sodium depletion was 40mV compared with 14mV in control rats. The E.M.F. across the wall of the descending colon was 38mV in the experimental rats compared with 13·6mV in controls. The importance of metabolism in generation of this potential was evident from the fact that following adrenalectomy of the animals it was absent (Edmonds and Marriott, 1968a) and it was also eliminated by anoxia, cyanide, iodoacetate and 2,4-dinitrophenol. It might be mentioned here that in toad colon, Crabbé (1967) found increased short-circuit current and transepithelial potential regularly, when he added insulin to the serosal bathing fluid, provided that the colon mucosa had been dissected free of underlying muscle layers.

Edmonds and Marriott (1968b) measured the short-circuit current in preparations of rat colon mucosa from both adrenalectomized and sodium-depleted rats. Removal of the musculature did not adversely affect trans-mucosal E.M.F., except when it was high, so it was considered that the preparation was satisfactory for the measurement of the short-circuit current, at least qualitatively. Although both sets of animals yielded mucosae with similar transepithelial resistance values, the latter had much greater short-circuit current in keeping with their greater E.M.F. and net sodium transport. Both E.M.F. and short-circuit current were greatly depressed when $10^{-3}M$ ouabain was added to the fluid bathing the serosal side. The following additional observations made it unlikely that the Koefoed-Johnsen–Ussing model could be applied to the electrogenesis of epithelial transport here.

i) The absence of potassium at the serosal surface had no effect either on potential or on short-circuit current. It will be recalled that sodium crosses the wall of the colon mainly in association with anions such as chloride rather than in exchange for potassium.

ii) Increasing the potassium concentration at the serosal surface from 5mM to 80mM raised the E.M.F. by about 3mV where one would have expected a much greater fall of E.M.F. from the model.

iii) When the transepithelial E.M.F. was high ($>30mV$) it was very dependent on luminal sodium concentration but only on that.

It was concluded from these and other observations that the active transport of sodium here did not depend on a coupling of sodium and potassium movement but was mainly electrogenic in nature.

### D. The Chloride Pump in Epithelial Tissues

Reference has already been made to the finding of anion pumps in isolated turtle bladder by Brodsky and his colleagues. Although chloride movement across the isolated skin of *Rana temporaria* appeared to be purely passive (Ussing, 1960a), in the case of intact frogs placed in dilute salt solutions the skin potential was not sufficiently large to account for its net influx, which suggested active uptake of this anion (Jørgensen and coworkers, 1954). It is possible that in preparing isolated skins, their ability to actively transport chloride may be lost, perhaps by denervation. However, in the case of the South American frog *Leptodactylus ocellatus* the chloride pump was evident, even in isolated skin. In this case it was found (Zadunaisky and coworkers, 1963) that the short-circuit current was less than the net sodium flux but the difference between the two disappeared when chloride ion in the fluids bathing the skin was replaced by impermeant methyl sulphate. Furthermore, the skin potential reversed in polarity when the skin was bathed on both sides with sodium-free solutions (Zadunaisky and de Fisch, 1964). The separate location of the sodium and chloride pumps was suggested by the fact that the antidiuretic hormone did not influence net chloride transport through the skin although it augmented sodium transport. In view of the suggested action of this hormone, its lack of action on chloride pumping could be explained if this pump were located at the outer membrane (Martin and Curran, 1966) and therefore readily accessible to chloride. Another possible explanation could be that the chloride pump was also situated at the inner membrane with the sodium pump, but that it was already saturated with chloride. The latter suggestion does not seem likely. One curious fact about the chloride pumping was that it was in the opposite direction to that induced in frog skin by treatment with epinephrine (Koefoed-Johnsen and coworkers, 1952). This was believed to be associated with secretion from the skin gland cells.

Although sodium pumping seems to be the most ubiquitous form of active transport in cells and tissues, active transport of chloride has also been widely found. It has been reported for example in the non-secreting gastric mucosa of cat (Kitahara, 1967), in squid giant axon (Keynes, 1963), in smooth muscle (Casteels, 1967) and in rumen epithelium of sheep (Stevens, 1964; Harrison and coworkers, 1968). Although the chloride pump in some cases appeared to function in an electrogenic manner, as in gastric mucosa (Hogben, 1955), in turtle bladder (Gonzalez and coworkers, 1967) and in toad bladder (Davies and coworkers, 1967), in other cases it appeared to operate in an electrically neutral manner, perhaps exchanging chloride for bicarbonate across the membrane (Dietz and coworkers, 1967).

## E. Electrogenic Transport of Potassium Ions

Active transport of ions in the small intestine of higher animals did not appear to give rise to the generation of significant levels of electrical activity (Curran and Solomon, 1957; Curran, 1960). However, the midgut of *Cecropia* the silkworm appeared to be endowed with a very effective potassium pump (Harvey and Nedergaard, 1964) which was electrogenic. The silkworm lives on a potassium-rich diet and although the luminal surface of the gut wall appeared to have a low permeability to this cation, potassium was pumped actively from the serosal side to the lumen. The E.M.F. across the gut wall was normally about 84mV (serosa negative) and this potential was very much dependent on the presence of potassium ions at this surface. Removal of potassium from the serosal bathing fluid caused the E.M.F. to fall to zero. Measurement of the unidirectional fluxes of potassium (Harvey and coworkers, 1968) revealed a net flux towards the lumen which was about thirty-two times that predicted for passive diffusion of this ion. About 83% of the short-circuit current of $400\mu A/cm^2$ could be accounted for by the net transport of potassium towards the lumen. This current was inhibited by anoxia and by other metabolic inhibitors including 2,4-dinitrophenol, but was very little influenced by ouabain (Haskell and Clemons, 1963).

Dockry and coworkers (1966) obtained evidence for the electrogenic pumping of potassium in the innervated toe muscle of the rat. When innervated and freshly denervated companion toe muscles were enriched with sodium by immersion for two hours in cold potassium-free Ringer's the former muscles lost much less potassium and had a lower initial $E_m$ than their companions. The observed change of $E_m$ and of net flux following denervation bore no relation to the change in sodium permeability associated with long-term denervation (Creese and coworkers, 1968a). It was possible that the innervated muscles had a lower membrane conductance for potassium than the denervated ones, which might have accounted for their more depolarized state and smaller potassium loss. However, when innervated companion muscles were made sodium-rich as described and re-immersed for two hours at 37°C in recovery fluid containing 10mM K, if one muscle of each pair was denervated just before reimmersion, the innervated muscles transported nearly twice as much sodium and potassium as the denervated muscles during recovery. Furthermore, the sodium pump appeared to be electrogenic in the denervated muscles. $E_m$, measured 10–20 minutes after reimmersion was more negative than $E_K$ by about 5mV, while in the innervated preparation $E_m$ and $E_K$ were not significantly different. The effects produced by innervation on both $E_m$ and net movement of potassium might once more be explained in terms of potassium permeability. This time, however, one must assume that the innervated muscles were more permeable to this cation, which is quite contrary to the earlier postulate. The effects of innervation of muscle on $E_m$ and net movement of ions in both potassium free and recovery conditions might be more consistently explained in terms of the electrogenic pumping of potassium into muscle fibres in the innervated preparation.

Investigations of the properties of Schwann cell membranes (Villegas and coworkers, 1968) have also revealed a possible electrogenic pumping of potassium. The resting potential of these cells was about −40mV and it appeared to be very dependent on external potassium concentration. However, when ouabain ($10^{-3}$M) was added to the fluid bathing this cell, $E_m$ increased by about 11mV. A hyperpolarization of 4mV was produced by $10^{-5}$M ouabain. Hyperpolarization was not produced if the external potassium concentration was below 1mM or above 30mM. It was concluded that the cardiac glycoside here inhibited an electrogenic inward pumping of potassium ions.

Finally, Salee and Vidrequin-Deliege (1957) found what appeared to be activation of an electrogenic potassium pump in the isolated nerve-skin preparation of the toad following electrical stimulation of the brachial plexus.

## F. Ion Movements in Plant Cells

Although the large coenocytic cells of algae, which may be several centimetres in diameter, might seem to be excellent material with which to investigate linkage of ion movement during active transport, they have a double membrane (figure 12) like the epithelial tissue. The internal medium is the vacuole and between it and the fluid bathing the cell is a layer of protoplasm. Techniques have been devised for measuring the ionic composition of vacuole and of protoplasm and also membrane potentials across the

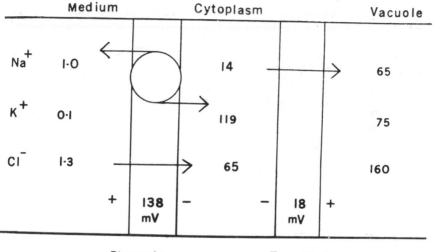

**Figure 12.** Ionic and electrical gradients across the tonoplast and plasmalemma of *Nitella translucens*. Active pumping of ions is indicated by arrows. The differences between $E_m$ and equilibrium potentials of sodium, potassium and chloride at the tonoplast were $+56$, $-5$ and $+4$ mV, respectively, and at the plasmalemma, $-72$, $+40$ and $-236$ mV, respectively (Spanswick and Williams, 1964).

plasmalemma and the tonoplast membranes. In *Nitella translucens* (Dainty, 1962; MacRobbie, 1962; Spanswick and Williams, 1964) the equilibrium potentials of potassium, sodium and chloride ions across these two membranes have been calculated and compared with the measured membrane potentials to determine whether their movements into the cell might be active or passive. These investigations indicated that active transport of sodium must take place at both membranes, and sodium was extruded from the cytoplasm into both the vacuole and external medium (figure 12).

In the plasmalemma its excretion was coupled to potassium uptake and also at this membrane there appeared to be an independent pumping of chloride into the cytoplasm. Both potassium and chloride ions appeared to be in electrochemical equilibrium across the tonoplast membrane so there seemed to be no need to postulate active transport of these ions.

Blount and Levendahl (1960) introduced a number of microcapillary tips into the large vacuole of *Halicystis ovalis*. These were used for internal perfusion, potential measurement and for measurement of the short-circuit current across the membranes. These workers were able to account for 97% of the total short-circuit current in terms of active transport of sodium (39·2%) and chloride ions (57·6%). This appeared to be in general agreement with findings in *Nitella*.

The mature hyphae of *Neurospora crassa* have an internal potential which may exceed −200mV. Slayman (1965) found that the measured potential was much greater than $E_K$ over the $[K]_o$ range of 0·2–100mM. When $[K]_o$, for example, was 10mM and the calculated $E_K$ was −73mV, the value of $E_m$ measured was −140mV. This high value of $E_m$ could not be accounted for in terms of passive diffusion of ions, and was subsequently shown to be very dependent on metabolism. Addition of metabolic inhibitors, including 2,4-dinitrophenol to the preparation caused $E_m$ to fall reversibly from −200mV to −30mV within one minute.

Poole (1966) found a similar large membrane potential in slices of beet tissue immersed in fluid containing $KHCO_3$. Both workers attributed the hyperpolarization to the electrogenic pumping of ions.

## G. Ion Transport in Red Blood Cells

Lassen and Sten-Knudsen (1968), using a very fine microelectrode mounted on a piezoelectric device, found that the mean membrane potential in the erythrocyte never exceeded −14mV and the chloride ions to which the membrane was permeable were in equilibrium with this potential. The concentration gradient of potassium ions across the cell membrane was about fifteen times greater than the chloride gradient and was consequently not at equilibrium with the potential. Therefore this cation must be actively transported into the cell against a large energy gradient and in this case, the potassium pump plays a more important role, perhaps, than in muscle and nerve where potassium ions are nearer to electrochemical equilibrium.

Several investigators have measured the coupling ratio of sodium–potassium transport across the red cell membrane. Post and Jolly (1957), using red cells which had been stored in the cold, found a flux ratio (sodium efflux to potassium influx) of 1·5 to 1. Later, McConaghey and Maizels (1962) found a 1:1 coupling ratio of these cations in lactose-treated erythrocytes. Finally, in a more recent study with reconstituted fresh red-cell ghosts, Garrahan and Glynn (1967a) measured the ratio of ouabain-sensitive [24]Na efflux to ouabain-sensitive [42]K influx in fluid containing 10 mM K and found a coupling ratio in favour of sodium of $1.35 \pm 0.01$. They had already established that ouabain-sensitive Na:Na exchange across the membrane did not take place in the presence of this concentration of potassium. When the measured coupling ratio was corrected for a possible K:K exchange, the resulting Na:K coupling ratio was as high as $1.86 \pm 0.06$. It seems likely then, that in red blood cells nearly twice as many sodium ions are excreted as there are potassium ions taken up, and the uneven exchange here is reflected in the swelling and shrinkage of cells accompanying sodium uptake and excretion (Harris, 1954). The uneven exchange of sodium and potassium across the membrane was in striking contrast to the ouabain-sensitive Na:Na exchange which Garrahan and Glynn (1967b) observed in the absence of external potassium and which presumably represented exchange on the sodium carrier. In this case one [24]Na ion exchanged for one inactive sodium ion moving in the opposite direction across the membrane and this exchange did not take place in the absence of ATP.

Glynn and Lüthi (1968a) and Garrahan and Glynn (1967c) have succeeded in showing a ouabain-sensitive incorporation of labelled inorganic ortho-phosphate into ATP by reversing the direction of operation of the sodium pump through net influx of sodium into red cell ghosts. The conditions required for this reversal were low $[ATP]_i$, high $[P]_i$, high $[K]_i$, zero $[K]_o$ and high $[Na]_o$. If net movements of sodium and potassium across the cell membrane are coupled on the same carrier system, then the step leading to dephosphorylation of the carrier and liberation of inorganic phosphate might also be reversed when there is a net efflux of potassium from the cell, and we might expect that the ouabain-sensitive efflux of potassium to be dependent on the presence of inorganic phosphate. When Glynn and Lüthi (1968b) reduced the intracellular concentration of inorganic phosphate by intro-ducing inositol into red cells, they found that the ouabain-sensitive potassium efflux was reduced or completely abolished. Although this observation strongly suggested carriage of potassium into red cells by a mechanism similar to that involved in sodium transport, the ouabain-sensitive potassium efflux from starved red cells did not require any external sodium but occurred in the presence of $5.3$ mM K. Where both sodium and potassium were absent from the external fluid this efflux was abolished.

## IV. CONCLUDING REMARKS

The many experiments cited in which observed changes of membrane potential could not be accounted for satisfactorily in terms of passive ion movement, make it clear that the electrogenic transport of ions, that is, the separation of charges across the cell membrane associated with active transport, may be more significant than previously realized. The potential generated by the sodium pump may be of physiological importance in nerve and in muscle and in this tissue may account for potassium uptake as a passive response to hyperpolarization. This phenomenon does not rule out the possibility of some degree of linked coupling of sodium–potassium transport on the same carrier system, but such coupling is unlikely to be an ion for ion exchange. In some respects the presence of several electrogenic pumps operating at the cell membrane would be a more attractive proposition than electrically neutral pumping of ions, particularly in view of the present interest in coupling of enzymatic reactions with electric current flow across membranes (Blumenthal and coworkers, 1967) and with natural fuel cells (Mitchell, 1967; Robertson, 1967). Although they are distinct, such pumps need not be independent because of energetic considerations. They might also be under nervous and hormonal control.

## REFERENCES

Adrian, R. H. (1956) *J. Physiol.* (*London*), **133**, 631
Adrian, R. H. (1964) *J. Physiol.* (*London*), **175**, 134
Adrian, R. H. and C. L. Slayman (1966) *J. Physiol.* (*London*), **184**, 970
Andre, R. and J. Crabbé (1966) *Arch. Intern. Physiol. Biochim.*, **74**, 538
Axelsson, J. and E. Bülbring (1960) *J. Physiol.* (*London*), **153**, 30P
Bittar, E. E. (1967) *Nature*, **214**, 726
Blount, R. W. and B. H. Levendahl (1960) *Acta Physiol. Scand.*, **49**, 1
Blumenthal, R., S. R. Caplan and O. Kedem (1967) *Biophys. J.*, **7**, 735
Bricker, N. S., T. Biber and H. H. Ussing (1963) *J. Clin. Invest.*, **42**, 88
Brock, L. G., J. S. Coombs and J. C. Eccles (1952) *J. Physiol.* (*London*), **117**, 431
Carey, M. J. and E. J. Conway (1954) *J. Physiol.* (*London*), **125**, 232
Carey, M. J., E. J. Conway and R. P. Kernan (1959) *J. Physiol.* (*London*), **148**, 51
Carpenter, D. O. and B. O. Alving (1968) *J. Gen. Physiol.* **52**, 1
Casteels, R. (1967) *Amer. J. Digest. Diseases*, **12**, 231
Connelly, C. M. (1959) *Rev. Mod. Phys.*, **31**, 475
Connelly, C. M. (1962) *Proc. Intern. Union Physiol. Sci. XXII Intern. Congr. Leiden*, 600
Conway, E. J. (1957) *Physiol. Rev.* **37**, 84
Conway, E. J. and P. T. Moore (1945) *Nature*, **156**, 170
Conway, E. J., R. P. Kernan and J. A. Zadunaisky (1961) *J. Physiol.* (*London*), **155**, 263
Coombs, J. S., J. C. Eccles and P. Fatt (1955) *J. Physiol.* (*London*), **130**, 326
Cooperstein, I. L., D. Chalfin and C. A. M. Hogben (1957) *Federation Proc.* **16**, 24
Cooperstein, I. L. and S. K. Brockman (1958) *J. Clin. Invest.*, **38**, 435
Crabbé, J. (1967) *Proc. Intern. Symp. on Polypeptide and Protein Hormones, Milan*, 377
Creese, R. (1968) *J. Physiol.* (*London*), **197**, 255
Creese, R. and J. Northover (1961) *J. Physiol.* (*London*), **155**, 343
Creese, R., A. L. El-Shafie and G. Vrbova (1968) *J. Physiol.* (*London*), **197**, 279
Cross, S. B., R. D. Keynes and R. Rybova (1965) *J. Physiol.* (*London*), **181**, 865
Curran, P. F. (1960) *J. Gen. Physiol.* **43**, 1137
Curran, P. F. and A. K. Solomon (1957) *J. Gen. Physiol.* **41**, 143

Dainty, J. (1962) *Ann. Rev. Plant Physiol.*, **13**, 379
Darrow, D. C. (1957) In Q. R. Murphy (Ed.), *Metabolic Aspects of Transport Across Cell Membranes*, University of Wisconsin Press, Madison, p. 23
Davies, H. E. F., D. G. Martin and G. W. D. Sharp (1968) *Biochim. Biophys. Acta*, **150**, 315
Desmedt, J. E. (1953) *J. Physiol. (London)*, **121**, 191
Dietz, T. H., L. B. Kirschner and D. Porter (1967) *J. Exp. Biol.*, **46**, 85
Dockry, M., R. P. Kernan and A. Tangney (1966) *J. Physiol. (London)*, **186**, 187
Eccles, R. M. (1952) *J. Physiol. (London)*, **117**, 196
Eccles, J. C. (1961) *Proc. Roy. Soc. London, Ser. B*, **153**, 445
Eccles, J. C. (1964) *The Physiology of Synapses*, Springer-Verlag, Berlin
Edmonds, C. J. (1967) *J. Physiol. (London)*, **193**, 603
Edmonds, C. J. and Jane Marriott (1968a) *J. Physiol. (London)*, **194**, 457
Edmonds, C. J. and Jane Marriott (1968b) *J. Physiol. (London)*, **194**, 479
Engbaek, L. and T. Hoshiko (1957) *Acta Physiol. Scand.*, **39**, 348
Essig, A. and A. Leaf (1963) *J. Gen. Physiol.*, **46**, 505
Feng, T. B. and Y. M. Liu (1949) *J. Cellular Comp. Physiol.*, **34**, 33
Frazier, H. S. and A. Leaf (1963) *J. Gen. Physiol.*, **46**, 491
Frumento, A. S. (1965) *Science*, **147**, 1442
Frumento, A. S. and L. J. Mullins (1964) *Nature*, **204**, 1312.
Fuhrman, F. A. and H. H. Ussing (1951) *J. Cellular Comp. Physiol.*, **38**, 109
Gage, P. W. and J. I. Hubbard (1966) *J. Physiol. (London)*, **184**, 335
Garrahan, P. J. and I. M. Glynn (1967a) *J. Physiol. (London)*, **192**, 159
Garrahan, P. J. and I. M. Glynn (1967b) *J. Physiol. (London)*, **192**, 175
Garrahan, P. J. and I. M. Glynn (1967c) *J. Physiol. (London)*, **192**, 237
Gerard, R. W. (1930) *Amer. J. Physiol.*, **92**, 498
Glynn, I. M. (1956) *J. Physiol. (London)*, **134**, 278
Glynn, I. M. (1962) *J. Physiol. (London)*, **160**, 18P
Glynn, I. M. and U. Lüthi (1968a) *J. Physiol. (London)*, **194**, 104P
Glynn, I. M. and U. Lüthi (1968b) *J. Gen. Physiol.*, **51**, 385s
Greengard, P. and R. W. Straub (1962) *J. Physiol. (London)*, **161**, 414
Goldman, D. E. (1943) *J. Gen. Physiol.*, **27**, 37
Gonzalez, C. F., Y. E. Shamoo and W. A. Brodsky (1967) *Amer. J. Physiol.*, **212**, 641
Gonzalez, C. F., Y. E. Shamoo, H. R. Wyssbrod, R. E. Sollinger and W. A. Brodsky (1967) *Amer. J. Physiol.*, **213**, 333
Guttman, R. (1939) *J. Gen. Physiol.*, **22**, 567
Harris, E. J. (1954) *Symp. Soc. Exp. Biol.*, **8**, 228
Harris, E. J. and S. Ochs (1966) *J. Physiol. (London)*, **187**, 5
Harrison, F. A., R. D. Keynes and L. Zurich (1968) *J. Physiol. (London)*, **194**, 48P
Harvey, W. R. and S. Nedergaard (1964) *Proc. Natl. Acad. Sci. U.S.*, **51**, 757
Harvey, W. R., J. A. Haskell and S. Nedergaard (1968) *J. Exp. Biol.*, **48**, 1
Hashimoto, Y. (1965) *Kumamoto Med. J.*, **18**, 23
Haskell, J. A. and R. D. Clemons (1963) *J. Cell. Biol.*, **19**, 32
Hering, E. (1884) *S.B. Akad. Wiss. Wein.*, **89**, 137
Herrera, F. C. (1965) *Amer. J. Physiol.*, **209**, 819
Herrera, F. C., G. Whittembury and A. Planchart (1963) *Biochim. Biophys. Acta*, **66**, 170
Hodgkin, A. L. (1958) *Proc. Roy. Soc. London Ser. B*, **148**, 1
Hodgkin, A. L. and B. Katz (1949) *J. Physiol. (London)* **108**, 37
Hodgkin, A. L. and R. D. Keynes (1954) *Symp. Soc. Exp. Biol.*, **8**, 423
Hodgkin, A. L. and R. D. Keynes (1955) *J. Physiol. (London)*, **128**, 28
Hodgkin, A. L. and P. Horowicz (1959) *J. Physiol. (London)*, **148**, 127
Hogben, C. A. M. (1955) *Amer. J. Physiol.*, **180**, 641
Holmes, O. (1962) *Arch. Intern. Physiol. Biochim.*, **70**, 211
Hoshiko, T. (1961) In *Biophysics of Physiological and Pharmacological Actions*, American Association for the Advancement of Science, Washington, D.C. p. 31
Ito, M. and T. Oshima (1964) *Proc. Roy. Soc. Ser. B*, **161**, 92
Johnson, J. A. (1956) *Amer. J. Physiol.*, **187**, 328

P

Jørgensen, C. B., H. Levi and K. Zerahn (1953) *Acta Physiol. Scand.*, **30**, 178
Kerkut, G. A. and R. C. Thomas (1965) *Comp. Biochem. Physiol.*, **14**, 167
Kernan, R. P. (1960) *Nature*, **185**, 471
Kernan, R. P. (1962a) *Nature*, **193**, 986
Kernan, R. P. (1962b) *J. Physiol.*, **162**, 129
Kernan, R. P. (1963) *Nature*, **200**, 474
Kernan, R. P. (1965) *Cell K*, Butterworths, London, p. 106
Kernan, R. P. (1968) *J. Gen. Physiol.*, **51**, 204s
Kernan, R. P. and E. J. Conway (1955) *Intern. Biochem. Congr. Brussels*, p.83 (Abstract)
Keynes, R. D. and P. R. Lewis (1951) *J. Physiol. (London)*, **114**, 151
Keynes, R. D. and R. C. Swan (1959) *J. Physiol. (London)*, **147**, 591
Keynes, R. D. (1963) *J. Physiol. (London)*, **169**, 690
Kitahara, S. (1967) *Amer. J. Physiol.*, **213**, 819
Koefoed-Johnsen, V., H. H. Ussing and K. Zerahn (1952) *Acta Physiol. Scand.*, **27**, 38
Koefoed-Johnsen, V. and H. H. Ussing (1958) *Acta Physiol. Scand.*, **42**, 298
Koketsu, K. and S. Nishi (1967) *Life Sci.*, **6**, 1827
Kosterlitz, H. W., G. M. Lees and D. I. Wallis (1968) *J. Physiol. (London)*, **195**, 39
Laporte, Y. and R. Lorente de Nó (1950) *J. Cellular Comp. Physiol.*, **35**, Suppl. 2, 61
Lassen, U. V. and O. Sten-Knudsen (1968) *J. Physiol. (London)*, **195**, 681
Leaf, A., J. Anderson and L. B. Page (1958) *J. Gen. Physiol.*, **41**, 657
Ling, G. and J. W. Woodbury (1949) *J. Cellular Comp. Physiol.*, **34**, 407
Lorente de Nó, R. (1947) In *Studies from the Rockefeller Institute for Medical Research*, **131**, 132
Lubin, M. and P. B. Schneider (1957) *J. Physiol. (London)*, **138**, 140
Martin, D. W. and P. F Curran (1966) *J. Cell. Physiol.*, **67**, 367
Meves, H. (1961) *Pflügers Arch. Ges. Physiol.*, **272**, 336
Mitchell, P. (1967) *Federation Proc.*, **26**, 1370
Mullin, L. J. and M. Z. Awad (1965) *J. Gen. Physiol.*, **48**, 761
McConaghy, P. D. and M. Maizels (1962) *J. Physiol. (London)*, **162**, 485
MacRobbie, E. A. C. (1962) *J. Gen. Physiol.*, **45**, 861
Nakajima, S. and K. Takahashi (1966) *J. Physiol. (London)*, **187**, 105
Nishi, S. and K. Koketsu (1967) *Life Sci.*, **6**, 2049
Poole, R. J. (1966) *J. Gen. Physiol.*, **49**, 551
Post, R. L. and P. C. Jolly (1957) *Biochim. Biophys. Acta*, **25**, 118
Post, R. L., C. R. Merritt, C. R. Kinsolving and C. D. Albright (1960) *J. Biol. Chem.*, **235**, 1796
Rang, H. P. and J. M. Ritchie (1968) *J. Physiol. (London)*, **196**, 183
Relman, A. S., A. T. Lambie, A. M. Roy and B. A. Burrows (1956) *Clin. Res.*, **4**, 150
Ritchie, J. M. and R. W. Straub (1957) *J. Physiol. (London)*, **136**, 80
Robertson, R. N. (1967) *Endeavour*, **26**, 134
Salle, M. and M. Vidrequin-Deliege (1967) *Comp. Biol. Physiol.*, **23**, 583
Sandow, A. and H. Mandel (1951) *J. Cellular Comp. Physiol.*, **38**, 271
Schatzman, H. J. (1953) *Helv. Physiol. Acta*, **11**, 346
Senft, J. P. (1967) *J. Gen. Physiol.*, **50**, 1835
Shanes, A. M. (1958) *Pharmacol. Rev.*, **10**, 59
Shaw, F. H. and S. E. Simon (1955) *Nature*, **176**, 1031
Skou, J. C. (1960) *Biochim. Biophys. Acta*, **42**, 6
Slayman, C. L. (1965) *J. Gen. Physiol.*, **49**, 69
Spanswick, R. M. and E. J. Williams (1964) *J. Exp. Botany*, **15**, 193
Stevens, C. E. (1964) *Amer. J. Physiol.*, **206**, 1099
Thomas, R. C. (1968) *J. Physiol. (London)*, **195**, 23P
Ussing, H. H. (1960a) *J. Gen. Physiol.*, **43**, 135
Ussing, H. H. (1960b) In *Handbuch der Experimentellen Pharmacology*, Springer-Verlag, Berlin pp. 1–195
Ussing, H. H. (1965) In *The Harvey Lecture Series 59*, Academic Press, New York–London pp. 1–30

Ussing, H. H. and K. Zerahn (1951) *Acta Physiol. Scand.*, **23**, 110
Villegas, J., R. Villegas and M. Gimenez (1966) *J. Gen. Physiol.*, **51**, 47
Whittam, R. (1962) *Biochem. J.*, **84**, 110
Yamada, K. and K. Yonemura (1967) *J. Physiol. Soc. Japan*, **29**, 268
Zadunaisky, J. A., O. A. Candia and D. J. Chiarandini (1963) *J. Gen. Physiol.*, **47**, 393
Zadunaisky, J. A. and F. W. deFisch (1964) *Amer. J. Physiol.*, **207**, 1010
Zierler, K. L. (1959) *Amer. J. Physiol.*, **197**, 524

# CHAPTER 11

# Models for sodium/potassium transport: a critique

## P. C. Caldwell

*Department of Zoology,*
*University of Bristol,*
*Bristol, England.*

## I. INTRODUCTION

Active transport and its explanation in molecular terms are at present very
live issues in biological research. In spite of the considerable amount of work

P*

which has been done, comparatively little is known about the precise molecular mechanisms involved. In only one case has a reasonably complete interpretation of an active transport process been put forward: this is by Shaw (1959) for the uptake of iodide into seaweed.

Shaw, as a result of experiments on the uptake of [131]I into fronds of *Laminaria digitata*, proposed the uptake mechanism shown in figure 1.

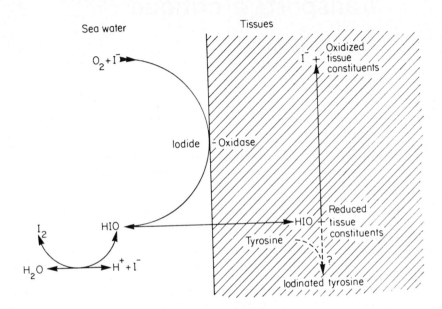

**Figure 1.** Diagram to illustrate the mechanism of iodine uptake proposed by Shaw (1959).

This involves first an oxidation of the iodide in the sea water to free iodine or hypoiodous acid. The iodide, or more likely the hypoiodous acid, penetrates into the tissues of the *Laminaria* where it is reduced to iodide. Since iodide is regarded as being unable to penetrate the cell walls, it remains in the tissue and is accumulated. In the mechanism shown in figure 1 the motive power for the active transport process is provided by the concentration gradient of hypoiodous acid which is set up between the tissue and its surroundings, this concentration gradient leading to a higher concentration of hypoiodous acid outside the tissue than inside it. One of the strongest pieces of evidence for the proposed mechanism is the existence of an iodide oxidase enzyme, apparently situated on the outside of the cells, which oxidizes iodide to free iodine, this oxidation being particularly evident if extra iodide is added to the sea water surrounding the weed. This process was first studied by Dangeard (1928a,b) and Kylin (1930), and was investigated further by Shaw. Evidence that the iodine penetrated into the tissue in an oxidized form, which could be hypoiodous acid, was provided by the observation that the reducing conditions provided by a combination of darkness and an atmosphere of nitrogen interrupt the uptake of $^{131}$I into the *Laminaria* fronds. If, however, elementary iodine, instead of iodide, was added to the sea water under these conditions, a rapid uptake of the labelled iodine took place. Shaw was aware that his mechanism might be oversimplified and that there were difficulties in that there have been suggestions (Kylin, 1929) that iodide oxidizing activity cannot be detected in certain brown algae. On the other hand, as has already been mentioned, the scheme shown in figure 1 represents the only suggested mechanism for an active transport process for which the experimental evidence is reasonably complete and clear-cut.

It has been possible to get a reasonably complete picture of the mechanism for the active transport of iodine into seaweed because most of the steps and intermediates involved are relatively simple from the chemical point of view, the actual transport process involving the diffusion of a small molecule. Complex molecules such as enzymes are only involved in the iodide oxidase system and in the reduction of hypoiodous acid in the tissue. Although a detailed knowledge of the mechanisms involved in these reactions and of the structure of the enzymes would be useful for a complete picture of the iodide transport mechanism, these details are not necessary for a basic understanding of the process. However, such a simple approach is not possible for most active transport processes, since complex molecular systems involving enzymes seem to be directly concerned with the movement of most substances. This is in fact the basic difficulty facing ion transport studies at the moment. It is comparatively easy to follow the movement of the small molecules which are being transported, and to suggest hypotheses to account for these movements, but it is extremely difficult to identify and characterize the more complex molecules constituting the machinery which carries out these transport processes. As a result, while there is a great deal of information available about ion movements across cell membranes and the ways in which these are affected by various ionic and metabolic conditions, there is comparatively little experimental information about the nature, structure, and functioning of the molecules in these membranes which form the actual transport mechanism.

One of the best examples of the present position in ion transport is provided by one of the most widely studied systems, namely the sodium/potassium transport system, and the rest of this article will be concerned with a discussion of this system and some of the present ideas about it.

The sodium pump system is in many ways the best to consider, since there are indications that certain concentration gradients across cell membranes may be maintained by linkage to the sodium gradient, the sodium pump thereby in effect being primarily responsible for the maintenance of the secondary concentration gradient. Particular examples of this appear to be the glycine gradient in *Ascites* cells (Eddy and Mulcahy, 1965) and the calcium gradient in squid nerve (Baker, Blaustein, Hodgkin and Steinhardt, 1969).

## II. ATTEMPTS TO ISOLATE AND IDENTIFY THE MEMBRANE SYSTEMS INVOLVED IN THE ACTIVE TRANSPORT OF SODIUM AND POTASSIUM

### A. Skou's Enzyme

Following the demonstration that the active transport of sodium in squid nerve depended on metabolism (Hodgkin and Keynes, 1955), and that this dependence was probably linked to the availability of high-energy phosphate compounds (Caldwell, 1956; Caldwell and Keynes, 1957), Skou (1957) isolated a particulate fraction from crab nerve which he examined for ATPase activity. A high ATPase activity had previously been demonstrated in the sheath remaining after the extrusion of the axoplasm from squid giant axons by Libet (1948), but Skou, in addition to demonstrating ATPase activity in the particles from crab nerve, also showed that this activity was enhanced by sodium ions and enhanced even further by potassium ions and sodium ions in combination. These properties are to be expected for a coupled sodium/potassium pump system if the energy for the transfer of sodium and potassium across the system is provided by the hydrolysis of ATP, this hydrolysis taking place only if sodium and potassium are available to move across the system. According to Skou (1965), the sodium/potassium activated ATPase has also been detected in membrane particles from red cells, brain, nerve, kidney, muscle, liver, intestine, electric tissue, parotid gland, frog skin, ciliary body, lens, retina, thyroid tissue and toad bladder.

Although the work done by Skou has not indicated the nature of the sodium and potassium carrying systems involved in the pump mechanism, it has provided some information about processes which may be involved in the interaction of the pump enzyme with ATP. Experiments in which $^{32}$P-labelled ADP was used (Skou, 1960) showed that some of the labelled ADP became converted into labelled ATP in the presence of unlabelled ATP, indicating that the terminal phosphate of the ATP can become attached to the enzyme to form a phosphorylated intermediate and can then be transferred back to ADP to reform ATP.

Further experiments with brain particles containing the sodium pump enzyme system support the idea that ATP interacts with the enzyme to form a phosphorylated intermediate which subsequently breaks down as it provides energy to drive the pump mechanism (Skou, 1965). The results of some of these experiments are shown in figure 2. In the presence of magnesium only the particles incorporated $^{32}$P from labelled ATP, the incorporation

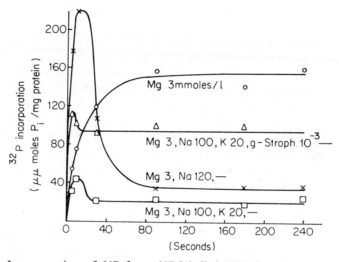

**Figure 2.** Incorporation of $^{32}$P from $^{32}$P-labelled ATP into the enzyme system isolated by Skou (from Skou, 1965). The ionic conditions under which each curve was obtained are indicated.

ceasing when the ATP had been broken down by the magnesium-activated ATPase present. This incorporation presumably represents the formation of a phosphorylated intermediate which may be related to the active transport of sodium and potassium. In the presence of sodium as well as magnesium the rate of incorporation of $^{32}$P from labelled ATP into the particles is faster. The $^{32}$P incorporated reaches a peak at the point at which all the ATP present has just broken down and then the amount of incorporated $^{32}$P decreases to a low level. In the absence of potassium the coupled pumping of sodium and potassium across the membrane fragments cannot presumably take place, so it seems possible that the $^{32}$P incorporation in the presence of sodium and magnesium represents the reversible build-up from ATP of a phosphorylated intermediate connected with the transport process, the formation of which depends on the movement of sodium across the membrane. However, this intermediate cannot break down again in the presence of ATP unless potassium as well as sodium is available for transport across the membrane fragments. This interpretation seems to be justified by the observation that much less $^{32}P$ is incorporated from labelled ATP if potassium is present in addition to sodium and magnesium. Under these conditions the intermediate would be expected to attain a much lower steady-state concentration on account of its rapid breakdown associated with the coupled movement of sodium and potassium across the fragments. A further confirmation of these ideas is provided by the observation that ouabain, which inhibits the sodium/potassium pump and its associated ATPase activity, increases the level of the intermediate in the presence of sodium, potassium and magnesium, presumably because it interferes with the breakdown of the intermediate. Results similar to those obtained by Skou have been obtained with kidney membrane fragments by Post, Sen and Rosenthal (1965).

## B. The Phosphatidic Acid Cycle

One of the earliest attempts to identify phosphorylated intermediates which might be involved in the transport of sodium was made by Hokin and Hokin (1960a,b) as a result of work on avian salt gland. The gland is stimulated to secrete sodium and chloride by acetylcholine, and it was shown that when slices of the gland were treated with acetylcholine in a medium containing $^{32}P$-labelled orthophosphate, there was a marked increase in the incorporation of $^{32}P$ in the phosphatidic acid in the gland. Hokin and Hokin postulated that phosphatidic acid was the sodium carrier, and produced the scheme for the mechanism of sodium transport shown in figure 3. ATP reacts with a membrane diglyceride on the inside of the membrane in the presence of a membrane diglyceride kinase to form phosphatidic acid which then complexes with two sodium ions. The sodium–phosphatidic acid complex then passes to the outside of the membrane where the phosphatidic acid is broken down by a membrane phosphatidic acid phosphatase to form a diglyceride and orthophosphate, the sodium being released into the external medium. The orthophosphate and the diglyceride then return to the inside of the membrane, and two chloride ions pass out passively to preserve electroneutrality.

Certain difficulties can be raised in connection with this scheme as a mechanism for the sodium pump, such as the return of the orthophosphate against an adverse electrical and concentration gradient, the lack of coupling of sodium exit with potassium entry and the fact that two rather than three sodium ions are transported. The most serious difficulties have however been raised by Glynn, Slayman, Eichberg and Dawson (1965), who have shown

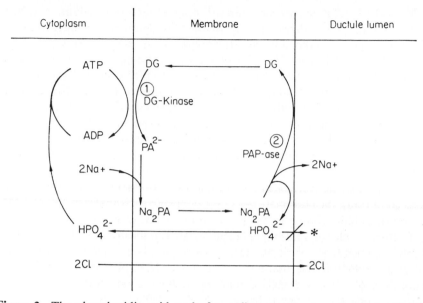

**Figure 3.** The phosphatidic acid cycle for sodium transport proposed by Hokin and Hokin (1960 a,b). DG = diglyceride; DG Kinase = diglyceride kinase; PA = phosphatidic acid; PAP-ase = phosphatidic acid phosphatase. The $HPO_4{}^{2-}$ does not leave the external surface of the membrane.

that, in a preparation of the Skou enzyme from *Electrophorus* electric organ, there was no labelling of phosphatidic acid by labelled ATP in a 30-second period during which 10–20% of the ATP was broken down. About 98% of the ATPase activity of the preparation was ouabain-sensitive, indicating that practically all this activity was due to sodium pump enzyme. These observations are clearly not consistent with the idea that phosphatidic acid is the main phosphorylated intermediate in the ATP splitting associated with sodium transport, and it now seems that Hokin and Hokin's phosphatidic acid cycle must be abandoned as a mechanism for the sodium pump.

### C. Phosphoproteins as Intermediates in the Mechanism of Sodium Transport

Certain authors (for example, Heald, 1962; Judah and Ahmed, 1964; Skou, 1964, 1965) have considered the possibility that the formation of a phosphoprotein may be a step in ion transport, and the evidence for this is stronger than that for the involvement of phosphatidic acid.

Albers, Fahn and Koval (1963) showed that, when a preparation of the Skou enzyme from the electric organ of *Electrophorus* was incubated with

[32]P-labelled ATP in the presence of magnesium and sodium and then subjected to peptic digestion, the main [32]P-containing product which is formed migrates cathodally on electrophoresis at pH 3·5. This product is probably derived from the phosphorylated intermediate investigated by Skou (1965) and by Post, Sen and Rosenthal (1965), since, like the intermediate, its amount is reduced if potassium is present as well as magnesium and sodium. Also, the sodium dependent formation of this product and the decrease in its amount induced by potassium have been shown by Hokin, Sastry, Galsworthy and Yoda (1965) to be inhibited by ouabain. These latter authors found two cathodally moving components in the product obtained after peptic digestion, and they have suggested that, since orthophosphate is liberated from the components more readily under alkaline conditions than acid conditions, the phosphate must be attached to a carboxyl group, probably in the side-chain of an aspartic acid or glutamic acid residue.

Further evidence that the phosphorylated intermediate is a phospho-protein comes from work on intact red cells. Thus Judah, Ahmed and McLean (1962) found that there was a build-up of radioactivity in the phosphoprotein of red cells containing a high concentration of internal sodium when they were incubated with [32]P-labelled orthophosphate in the absence of external potassium.

Addition of external potassium decreased the amount of labelled phospho-protein, and in this respect the phosphoprotein behaved similarly to the phosphorylated intermediate studied by Skou (1965) and Post, Sen and Rosenthal (1965), and to the components studied by Albers, Fahn and Koval (1963) and Hokin, Sastry, Galsworthy and Yoda (1965). Judah, Ahmed and McLean (1962) found that the effect of external potassium on the level of labelled phosphoprotein was abolished by ouabain, and in this respect too its behaviour is similar to that of the various intermediates and components which have been studied.

## D. Conclusions

It is apparent from the work which has been discussed in this section that the sodium pump system can be isolated and studied in the form of membrane fragments. While work on this type of preparation has not so far given any indication of the molecular mechanisms involved in the movement of sodium and potassium across the membrane, some indication has been obtained of the initial steps involved in the transfer of energy from ATP to the pump system. This appears to involve the transfer of the terminal phosphate of ATP to protein molecules which form part of the pump system. Since it has been found that ATP can only restore sodium transport in squid axons poisoned with metabolic inhibitors if applied internally and not externally

(Caldwell, Hodgkin, Keynes and Shaw, 1960a), this phosphate must be attached initially to protein at the inside surface of the membrane. The lability of this phosphorylated protein when isolated suggests that it may be an acyl phosphate, the phosphate probably being attached to an aspartic acid or glutamic acid side-chain (Hokin, Sastry, Galsworthy and Yoda, 1965). This intermediate accumulates when sodium and magnesium are present, but the amount present is reduced if potassium is also present. Whether the phosphorylated protein represents one of the actual ion carriers or whether it is only an intermediate in the transfer of energy to the pump system has not been finally decided.

## III.  CHARACTERISATION AND THEORETICAL INTERPRETATION OF SODIUM/POTASSIUM TRANSPORT FROM STUDIES OF THE ION FLUXES

### A. Introduction

A very great deal of information is available about the behaviour of sodium and potassium fluxes under a wide variety of conditions. Some of this work has been reviewed elsewhere (Caldwell, 1968) and it is not proposed to review it in detail here. One of the most complete studies of sodium and potassium movements so far carried out is that of Garrahan and Glynn (1967a,b,c,d,e) on red cells, and it seems that their work must be taken into account in any attempt to interpret these movements theoretically. Since there are other sodium and potassium movements in addition to those connected with the pump mechanism, some attempt must be made to assess the extent of the pump-mediated movements, and normally the fluxes inhibited by cardiac glycosides such as ouabain are regarded as those mediated by the pump.

### B. Coupling of Sodium Efflux and Potassium Influx

One of the earlier ideas to come out of studies on the effect of ionic changes in the external medium on sodium and potassium movements was that the outward movement of sodium ions is coupled to the inward movement of potassium. Evidence for this coupling is shown in figure 4 which is taken from work on *Sepia* axons by Hodgkin and Keynes (1955). When potassium was removed from the external medium there was a marked and reversible fall in the sodium efflux from the axons, and this can be interpreted in terms of a coupling of sodium efflux to potassium influx if it is assumed that these two processes form consecutive stages in the operation of the sodium pump mechanism. A coupling of sodium and potassium movements was also found in red cells (see for example, Glynn, 1956) and figure 5 shows a scheme

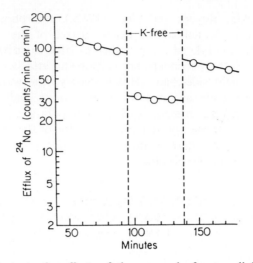

**Figure 4.** To illustrate the effects of the removal of extracellular potassium on the efflux of sodium from a *Sepia* axon (from Hodgkin and Keynes, 1955).

proposed by Shaw (quoted by Glynn, 1956) to account for this. This is based on interconvertible carriers for the sodium/potassium pump mechanism, and also depends on the presence of ATP which provides energy for the pumping mechanism by converting sodium carrier into potassium carrier on the inside of the membrane. The scheme shown in figure 5 can account for the fact that removal of external potassium reduces the rate of outward sodium

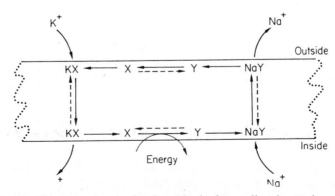

**Figure 5.** To illustrate the carrier hypothesis for sodium/potassium transport proposed by Shaw (from Glynn, 1956). X = potassium carrier. Y = sodium carrier.

transport, since the latter depends on the availability of sodium carrier on the inside of the membrane, and the amount of this will be reduced in the absence of external potassium on account of the accumulation of most of the carrier on the outside in the potassium carrier form.

## C. Sodium and Potassium Fluxes and Metabolism

The scheme proposed by Shaw also postulates the involvement of metabolic energy made available by the splitting of ATP. The likelihood of such an involvement had already become apparent when this scheme was put forward, as a result of studies on the effect of metabolic inhibitors on sodium and potassium movements (Harris, 1941; Maizels, 1951, 1954; Hodgkin and Keynes, 1955).

The involvement of ATP became more or less certain with the discovery of the Skou enzyme and the demonstration that the sodium efflux of squid axons poisoned with metabolic inhibitors could be restored by the injection of ATP or arginine phosphate (Caldwell and Keynes, 1957; Caldwell, Hodgkin, Keynes and Shaw, 1960a,b). Closer examination of the effects of metabolic inhibitors on sodium and potassium movements revealed an unexpected connection between the degree of inhibition of the metabolism and the coupling between the sodium efflux and potassium influx (Caldwell, Hodgkin, Keynes and Shaw, 1960b). An experiment which illustrates this is shown in figure 6. In this experiment the potassium was first removed from the external medium which resulted in a sharp fall in the sodium efflux, since at this stage the sodium efflux was tightly coupled to the influx of potassium and this could not take place. A very striking result was now obtained if a metabolic inhibitor such as cyanide was added while external potassium was absent. The sodium efflux first increased sharply, reached a peak after about half an hour and then fell again equally quickly. At the peak the sodium efflux was passing through a condition in which it was no longer tightly coupled to potassium influx, since addition of external potassium at this stage was found to make little difference to the efflux. This peak corresponded to a stage at which ATP was still present in the axon, whereas most of the arginine phosphate had broken down (Caldwell, 1960). On the other hand both of these compounds had been present before the addition of cyanide when the sodium efflux was tightly coupled to the potassium influx. The existence of this pattern was confirmed in experiments in which it was found that the injection of ATP into cyanide-poisoned axons brought about a restored sodium efflux which was not reduced by removal of external potassium, whereas the injection of arginine phosphate was able to bring about a restored sodium efflux which was reduced by removal of external potassium (Caldwell, Hodgkin, Keynes and Shaw, 1960a). In parallel with

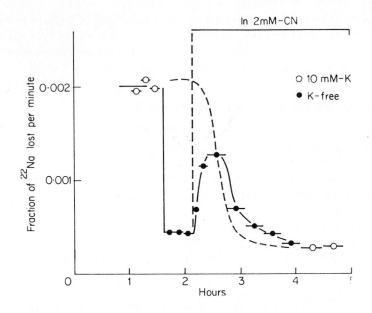

**Figure 6.** An experiment which illustrates a connection between the extent to which metabolism has been inhibited and the degree of coupling between sodium efflux and potassium influx (from Caldwell, Hodgkin, Keynes and Shaw, 1960b). The solid line shows the sodium efflux in the absence of external potassium and the broken line this efflux in its presence. As cyanide inhibition proceeds the reduction in sodium efflux brought about by the absence of external potassium becomes progressively less and less.

this the injection of ATP did not increase the potassium influx into cyanide-poisoned axons, whereas the injection of arginine phosphate did.

A scheme involving carriers for coupled sodium-potassium transport which attempted to take account of this variable degree of coupling of the two fluxes, and also for the observation that the specific sodium pump inhibitor, ouabain, appeared to act only on the outside of the membrane is shown in figure 7 (Caldwell, 1958).

Closer examination of a considerable variety of data suggested that there was a relationship between the degree of coupling of sodium efflux to potassium influx and the free energy available to the sodium pump (Caldwell and Schirmer, 1965), the degree of coupling being indicated by the ratio (Sodium efflux in the absence of external K$^+$)/(Sodium efflux in 10 mM external K$^+$).

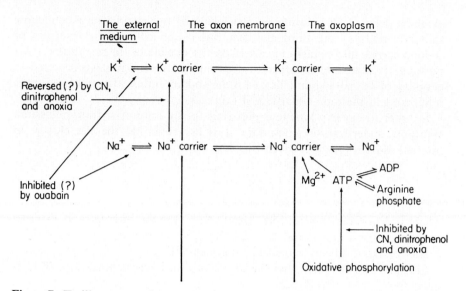

**Figure 7.** To illustrate a scheme for sodium transport proposed by Caldwell (1958).

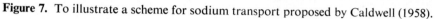
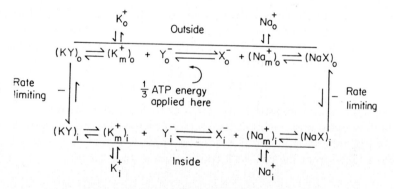

**Figure 8.** To illustrate a scheme for sodium/potassium transport used by Caldwell (1968) as the basis of a kinetic model. X = sodium carrier; Y = potassium carrier.

## D. A Kinetic Scheme to Account for Sodium and Potassium Movements

In an attempt to provide a basis for a kinetic scheme which would account for many of the features of sodium and potassium fluxes, the scheme involving carriers shown in figure 8 has recently been proposed (Caldwell, 1968). This is in many ways an extension of the schemes shown in figures 5 and 7, the main innovation being that it is necessary to postulate that the energy available from ATP splitting on the inside of the membrane is somehow split into units of one-third and that this is applied to the conversion of sodium carrier into potassium carrier on the outside of the membrane. This postulate is necessary in order to account for the progressive decrease in coupling of the potassium efflux with the sodium influx when metabolism is inhibited and the ratio (ADP)/(ATP) increases.

In order to develop kinetic expressions for the rate of sodium/potassium transport under various conditions it is assumed that the net change in internal sodium $[Na^+{}_i]$ is given by:

$$-\frac{d[Na^+]_i}{dt} = k_1[Na^+]_i X^-{}_i - k_1[Na^+]_o X^-{}_o \tag{1}$$

where $k_1$ is a rate constant, $[Na^+]_i$ and $[Na^+]_o$ are the internal and external sodium concentrations, and $X^-{}_i$ and $X^-{}_o$ are the amounts of free sodium carrier on the inside and outside of the membrane.

Similarly it is assumed that the net change in internal potassium $[K^+{}_i]$ is given by:

$$\frac{d[K^+]_i}{dt} = k_2[K^+]_o Y^-{}_o - k_2[K^+]_i Y^-{}_i \tag{2}$$

where $k_2$ is a rate constant, $[K^+]_o$ and $[K^+]_i$ are the external and internal potassium concentrations and $Y^-{}_o$ and $Y^-{}_i$ are the amounts of free potassium carrier outside and inside the membrane. The carriers are regarded as virtually unsaturated, so that:

$$X^-{}_i + X^-{}_o + Y^-{}_i + Y^-{}_o = \phi \tag{3}$$

where $\phi$ is the total amount of carrier. In order to account for the dependence of both coupled and uncoupled sodium effluxes on the presence of ATP in the cell (Caldwell, Hodgkin, Keynes and Shaw, 1960a), $\phi$ must be assumed to be a function of the cell ATP concentration. In addition:

$$\frac{X^-{}_i}{Y^-{}_i} = C_1 \tag{4}$$

where $C_1$ is an equilibrium constant, and

$$\frac{X^-{}_o}{Y^-{}_o} = C_2 \tag{5}$$

where $C_2$ is variable and related to the free energy available from ATP splitting $(\Delta G_{ATP})$ by

$$RT \log_e \frac{C_1}{C_2} = \frac{\Delta G_{ATP}}{3} \tag{6}$$

The net change in internal sodium is assumed to be equal to the net change in internal potassium, so the right-hand side of (1) can be equated with the right hand side of (2). The expression so obtained leads, in conjunction with equations (3), (4) and (5), to expressions for the amount of each carrier on each side of the membrane, and this enables calculations of the fluxes to be made under various conditions.

For example, the following expression is obtained for the amount of sodium carrier on the inside of the membrane $(X^-_i)$ as a fraction of the total carrier:

$$\frac{X^-_i}{\phi} = \frac{\dfrac{k_1[Na^+]_o}{(1+1/C_2)} + \dfrac{k_2[K^+]_o}{(C_2+1)}}{k_1[Na^+]_i + \dfrac{k_1[Na^+]_o(1+1/C_1)}{(1+1/C_2)} + \dfrac{k_2[K^+]_o(1+1/C_1)}{(C_2+1)} + \dfrac{k_2[K^+]_i}{C_1}} \quad (7)$$

If $X_i/\phi$ is multiplied by $[Na^+]_i$, then an expression is obtained which should describe the behaviour of the sodium efflux.

This has been used to provide a theoretical interpretation of the relationship already mentioned between the free energy available to the sodium pump from high-energy phosphate splitting and the extent to which the sodium efflux is cut down by removal of external potassium. Figure 9 (taken from Caldwell, 1968) shows the relation obtained from equation (7) when $k_2/k_1$ is taken as 5 and $C_1$ as 150, and compares it with the data collected by Caldwell and Schirmer (1965). It will be seen that equation (7) gives an adequate interpretation of the experimental data; it also gives an adequate interpretation of the variation of sodium efflux with $[K^+]_o$ for cephalopod giant axons, and in figure 10 the theoretical curve (taken from Caldwell, 1969) is compared with experimental data for *Sepia* obtained by Hodgkin and Keynes (1955).

Equation (7) and the other three equations which can be obtained for the calculation of the sodium influx, the potassium influx and the potassium efflux across the pump mechanism have recently been used (Caldwell, 1969) in an attempt to account for the very extensive data for ion movements across erythrocyte membranes obtained by Garrahan and Glynn (1967a,b,c,d,e).

These calculations cannot be discussed in detail here but two examples (taken from Caldwell, 1969) are shown in figures 11 and 12. With certain minor exceptions, such as the rise in sodium efflux at very low external sodium concentrations shown in figure 11, the calculated changes in the fluxes have been found to be in reasonable agreement with the changes observed by Garrahan and Glynn.

An interesting point arising from the kinetic treatment outlined in equations (1) to (7), and illustrated by the potassium influx in figure 12, is that although the carriers are postulated as being unsaturated at all times, the

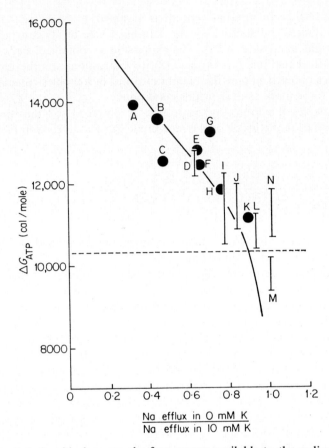

**Figure 9.** Relationship between the free energy available to the sodium pump of squid axons and the reduction in sodium efflux when potassium is removed from the bathing medium (from Caldwell and Schirmer, 1965). A–N refer to various metabolic conditions. The solid line has been calculated from equation (7).

kinetics lead to situations which look like the saturation of a carrier. In the particular example shown in figure 12 the potassium influx tends to increase as the external potassium increases. In order to keep the net outward movement of sodium equal to the net inward movement of potassium, the sodium efflux tends to increase, and this necessitates an increase in the amounts of sodium and potassium carrier on the inside of the membrane at the expense of the amounts of potassium and sodium carrier on the outside, the total amount of carrier remaining constant. This means that, as the external potassium is increased, the tendency for the potassium influx to increase

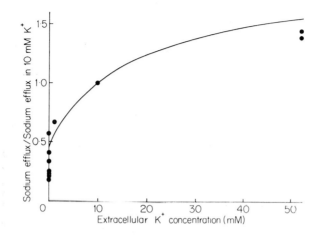

**Figure 10.** Variation of Na efflux from *Sepia* axons with extracellular potassium concentration. ● = experimental data taken from Hodgkin and Keynes (1955). Solid line = theoretical changes calculated from equation (7). (From Caldwell, 1969).

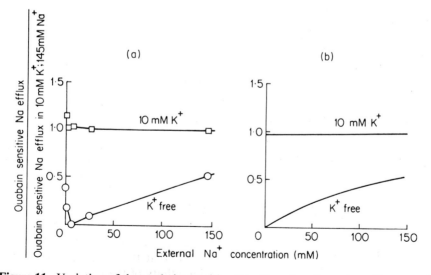

**Figure 11.** Variation of the ouabain sensitive Na efflux of erythrocytes with external Na⁺ concentration for external K⁺ concentrations of OmM and 10mM. (a) data taken from figure 3 of Garrahan and Glynn (1967a). (b) Theoretical changes calculated **from equation (7) (from** Caldwell, 1969).

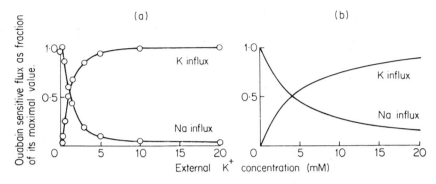

**Figure 12.** Variation of the ouabain-sensitive Na and K influxes into erythrocytes with external K+ concentration. (a) data taken from figure 3 of Garrahan and Glynn (1967c). (b) Theoretical changes calculated from equations (17) and (19) of Caldwell (1968). (from Caldwell, 1969).

gives rise to a drop in the external potassium carrier which counteracts the effect of the increase in external potassium on the influx. At high external potassium concentrations, the decrease in the amount of external potassium carrier almost exactly compensates the tendency of an increase in external potassium to increase the potassium influx, and so the influx process gives the appearance of becoming saturated. In the case of the sodium influx shown in figure 12, the concentration of external sodium is not changing, as increases in the external potassium produce decreases in the external sodium and potassium carriers. The sodium influx is therefore progressively decreased as the external potassium increases.

### E. The Kinetic Scheme proposed by Stone (1968)

An alternative kinetic scheme to explain many of the features of the kinetics of the sodium/potassium pump system has been proposed by Stone (1968), and the features of this scheme are illustrated in figure 13. It is rather more complicated than the scheme shown in figure 8, and it involves considerably more arbitrary constants, namely seven rate constants and twelve equilibrium constants. The reactions enclosed by broken lines in figure 13 are regarded as being at equilibrium. One of the main differences between this scheme and that outlined in figure 8 is that the carrier is regarded as existing in three forms instead of two, the affinity of the sodium form of the carrier changing as it passes across the membrane. The splitting of ATP is regarded as being geared to the interconversion of one form of the sodium carrier into the other as it crosses the membrane rather than to the conversion

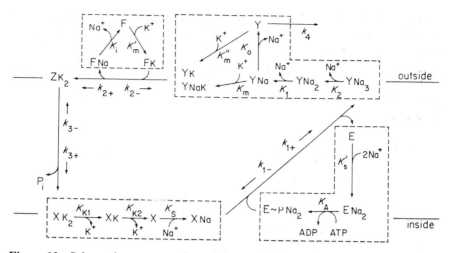

**Figure 13.** Schematic representation of the model for sodium transport proposed by Stone (from Stone, 1968). X, Y and Z refer to different forms of the carrier. The enzyme E and ATP phosphorylate X to give Y, Y and Z being phosphorylated. The various Ks are rate constants and equilibrium constants. The reactions enclosed by dotted lines are assumed to be at equilibrium.

of sodium carrier into potassium carrier on the outside. The interconversion is regarded as arising from a phosphorylation reaction brought about by ATP.

A consequence of the change in affinity of the sodium carrier as it crosses the membrane is that the rate constants for the movement of each carrier/ion complex in the two directions are different, and this is in contrast to the scheme shown in figure 8 where these constants are the same.

Stone states that computer calculations, based on the scheme shown in figure 13 with suitable values for the constants, enable an interpretation to be made of many features of the coupled sodium/potassium pump system, including the relationship shown in figure 9 and the variations found in the amount of phosphorylated intermediates under various conditions (e.g. Skou, 1965; Post, Sen and Rosenthal, 1965).

## IV. OTHER MODELS FOR THE SODIUM PUMP

### A. Introduction

Certain other models for the sodium pump mechanism have been proposed from time to time. The purpose of these models is not so much to interpret the kinetic features of the sodium pump as to give an explanation in molecular terms of the process of sodium and potassium transport. Some of these models will now be discussed, together with their relationship to the kinetic models discussed in the previous section. No mention will be made of models based on the coupling of ion movements to electron transfer processes since it seems unlikely that such mechanisms are involved in the operation of the sodium/potassium pump (see for example, Glynn, 1963). However, there is no doubt that electron transfer is involved in many other transport processes, and this aspect of ion transport has been excellently reviewed by Robertson (1968).

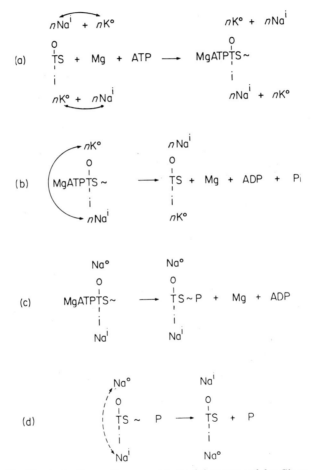

**Figure 14.** To illustrate the sodium pump model proposed by Skou (from Skou, 1965). (a) and (b) show the normal working of the pump. In (a) phosphorylation alters the affinities of the external and internal sites in favour of potassium and sodium respectively. In (b) transport and dephosphorylation takes place and the affinities of the external and internal sites reverse. (c) and (d) illustrate the build up in the system of a slowly decaying **phosphorylated** form in the absence of potassium.

## B. Skou's Model

The basic features of the model put forward by Skou (1964, 1965) are illustrated in figure 14. *n* is the number of ions transported per ATP split and is in the region of 2–3. In the inactive state depicted on the left-hand side of (a) in figure 14, sodium and potassium which have moved as a result of the operation of the previous transport cycle are regarded as being attached to sites on the transport system (TS) on the outside and inside of the membrane. On interaction of the transport system with magnesium and ATP, the affinities of the two sites are changed, the outer site becoming selective for potassium and the inner site selective for sodium. As a result the sodium on the outer site is replaced by potassium and the potassium on the inner site by sodium. Transport then takes place (figure 14b), the inner site transferring sodium as it moves outwards and the outer site transferring potassium as it moves inwards. Transport is immediately followed by a breakdown of the ATP to ADP and orthophosphate with the result that the system reverts to the inactive state once more. In order to account for the changes found in the phosphorylated intermediate, Skou postulated the sequence of events shown in figure 14c,d. In the absence of external potassium the system is converted into a phosphorylated form which can break down slowly to form inorganic phosphate.

Skou (1964) has also discussed the ways in which the changes in the affinity of the sites in his model may be related to the ideas of Eisenman and his colleagues (see for example, Eisenman, 1960) which relate ion selectivity to negative electrostatic field strength. He has extended these ideas in a scheme in which a role is assigned to an ATP-induced oxidation and reduction of SH and S–S groups in the sodium pump mechanism. In this latter scheme changes in the selectivity of the sites arise as a result of electron transfer mediated by SH and S–S groups which lead to changes in the field strength round the phosphate groups of phosphoprotein molecules at the sites.

## C. The Model proposed by Post, Sen and Rosenthal

Post, Sen and Rosenthal (1965) have proposed a model to account for the changes in the amount of the phosphorylated intermediate found in their work on kidney membrane fragments. Their scheme is shown in figure 15. It is in a sense an extension of the Shaw model, the sodium carrier being a phosphorylated protein which can carry three sodium ions, and the potassium carrier a phosphorylated protein which can carry two potassium ions. The potassium carrier is dephosphorylated on the inside of the membrane and, as in the Shaw model, is converted into the sodium carrier by a reaction involving the conversion of one molecule of ATP into ADP.

Q

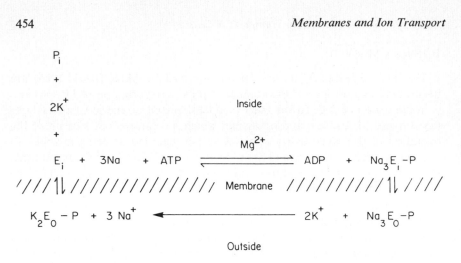

**Figure 15.** Scheme for sodium transport proposed by Post, Sen and Rosenthal. (from Post, Sen and Rosenthal, 1965). E–P is the phosphorylated intermediate. Ei and Eo indicate states of the system in communication with the inside and outside of the cell.

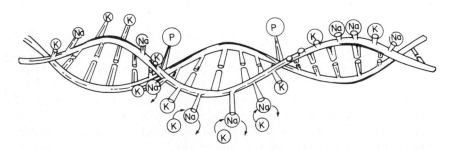

**Figure 16.** To illustrate the sodium/potassium pump model proposed by Opit and Charnock (from Opit and Charnock, 1965). This diagram shows the stage at which interaction with ATP has introduced a strain into the structure so that certain sites to which sodium is attached have been pushed to the outside, their affinity for potassium increasing in the process.

## D. The Model proposed by Opit and Charnock

This model, which is illustrated in figure 16 (see Opit and Charnock, 1965) aims to give a more pictorial idea of possible events in the bimolecular lipid layer structure of the membrane during sodium and potassium transport. In the initial state the sites of the transport system are localized inside the membrane and show a preference for sodium, although potassium is not completely excluded. ATP then phosphorylates an adjoining site on the inside of the membrane and the result of this phosphorylation is to introduce strain into the structure so that certain sites are pushed to the outside of the membrane and also change their affinity so that they now become selective for potassium. The phosphorylated sites are now dephosphorylated, and as a result of the removal of the strain the external sites now move back to the inside of the membrane taking potassium with them. On reaching the inside they become selective once more for sodium. This model is similar to those proposed by Stone (1968) and Jardetsky (1966) in that the sodium carrier changes its affinity as it crosses the membrane.

### E. Jardetsky's Model

Jardetsky (1966) has proposed an allosteric model for the sodium pump which is shown in figure 17. The mechanism consists of two proteins, and its

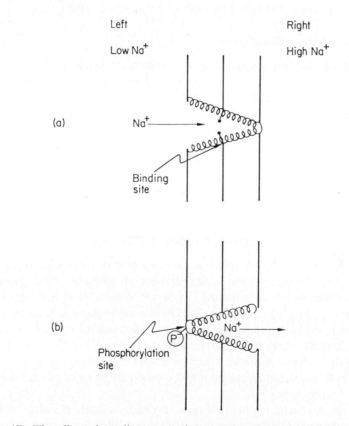

**Figure 17.** The allosteric sodium potassium pump proposed by Jardetzky (1966). (a) sodium form. (b) potassium form.

initial state is shown diagrammatically in figure 17A. Attached to the proteins is a site which has a high affinity for sodium, and which can interact with a sodium ion. When the site has interacted with sodium, the system is able to interact with an ATP molecule as a result of which it becomes phosphorylated and changes to the configuration shown in figure 17B. This new configuration has a low affinity for sodium arising from an allosteric effect of the phosphorylation on the properties of the site which previously had a high sodium affinity, this site now having a high potassium affinity. As a result sodium is lost to the outside and potassium is taken up. The system now dephosphorylates and reverts to configuration A, as a result of which potassium is lost to the inside and sodium is taken up in readiness for a repetition of the cycle. It should be noted that the change of affinity in the site as it changes from configuration A to configuration B is analogous to the changes in affinity as the sodium-carrying molecule crosses the membrane postulated by Stone (1968) and by Opit and Charnock (1965).

### F. The Model proposed by Lowe

Lowe (1968) has proposed the model shown in figure 18, and this is in some ways similar to Jardetsky's model. Three sodium ions, followed by ATP and magnesium, attach themselves to the bimolecular pump system in the manner indicated in the top of figure 18. The system can then go reversibly and spontaneously to the second stage shown in figure 18. If it loses the three sodium ions and gains two potassium ions it can then change to the third stage, and thence to the final stage with the breakdown of the ATP and the loss of the two potassium ions. The ADP, being at a low concentration inside the cell, moves out together with the inorganic phosphate, and the stage is then set for a repetition of the cycle.

## V.   CONCLUDING REMARKS

Many of the models for coupled sodium/potassium transport which have been discussed probably have some element of truth in them. Since the precise mechanism for this system is not yet known, it is not possible to evaluate the various models critically. On the other hand it is likely that a model which explains the largest number of features of sodium/potassium transport is more likely to be correct than one which explains only limited aspects of the process. A model which combines the features of several of the models which have been discussed is therefore likely to be of greater use than the individual models.

One of the intellectual difficulties raised by many models of active transport is the postulate of a carrier, since this implies the diffusion of what is perhaps a comparatively large ion-carrier complex across a membrane 50–100 Å

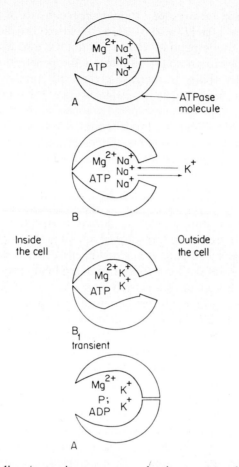

**Figure 18.** The sodium/potassium pump mechanism proposed by Lowe (1968).

thick. Models such as those of Jardetsky (1966) and Lowe (1968), in which the transport system is regarded as lying across the membrane and being capable of existing in different configurations with varying affinities for ions, go a long way towards resolving this difficulty. As Jardetsky has pointed out, a system of this type is not really a carrier system. Only the ions move across the membrane, not the molecules comprising the transport system.

On the other hand it would be unwise in the present state of knowledge to abandon the carrier idea completely. Peptides like valinomycin and the gramicidins might belong to a class of molecules which could act as carriers or possibly as components of the type of pump mechanism envisaged by

Jardetsky (1966) and Lowe (1968). Valinomycin consists of three units of the sequence—D-valine–L-lactic acid–L-valine–D-hydroxyisovalerate– joined together to form a cyclic peptide. It is an antibiotic and appears to owe its antibiotic properties to an ability to increase the penetration of membranes by certain ions. Its ability to increase greatly the uptake of potassium by mitochondria has been widely investigated (see for example, Pressman, 1965). It is possible that valinomycin and gramicidins A, B and C, which also increase the entry of sodium into mitochondria (Pressman, 1965) are modified components of ion transport systems, the modifications being such that they bring about disturbed permeability relationships in cells and hence have an antibiotic action. A further point is that valinomycin contains L-valine while gramicidins A, B and C contain glycine, L-alanine, L-valine, L-tryptophan, L-phenylalanine and L-tyrosine. If the molecules involved in sodium/ potassium transport were similar in structure to these antibiotics, they could then be involved in the transport of amino acids as well as ions, and a reason could be given for the apparent linkage of the transport of certain amino acids to the sodium ion concentration gradient across cell membranes (see for example, Eddy and Mulcahy, 1965).

This raises the question of whether hormones such as aldosterone and vasopressin, or molecules like them, might form part of the sodium transport mechanism since these substances stimulate sodium movement across the toad bladder (see for example, Sharp and Leaf, 1966), vasopressin being a cyclic peptide. While it is not certain whether vasopressin or something like it occurs in cephalopods, it might be noted that arginine vasopressin and the related vasotocin contain the amino acids glycine, arginine, aspartic acid, glutamic acid and tyrosine. Deffner (1961) has shown that considerable concentration gradients of these amino acids exist between the axoplasm of squid giant axons and the blood. If a vasopressin-like substance, which could occasionally be assembled or dissembled at various sites in the membrane, was involved in ion transport, it would be possible to see a way of accounting for these gradients. In addition, vasopressin and vasotocin contain cysteine, and such a mechanism could also set up a concentration gradient of this amino acid. Oxidation of cysteine in the axoplasm could then lead to the concentration gradients of cysteic acid amide, taurine and isethionic acid which have been found across the membrane of the squid giant axon (Deffner, 1961).

Any successful attempt to explain the mechanism of ion transport in molecular terms must also form the basis for the setting up of equations which will describe the kinetics of the transport satisfactorily. In this connection, it is of interest that the model proposed by Jardetsky involves a phosphorylation and change in the affinity of the sodium site as the sodium moves across the membrane. This is in many ways analogous to the change in the affinity of the sodium carrier as it crosses the membrane, which is one of the more controversial features of Stone's kinetic model (Stone, 1968). It is not possible at present to decide whether Stone's model or that discussed in connection with equations (1)–(7) is more likely to be true. The latter model has the virtue of greater simplicity, but this may be misleading. One test would be to see whether the relationship shown in figure 9 still holds if $\Delta G_{ATP}$ is decreased by an increase in the orthophosphate concentration in the cell rather than by a change in (ATP)/(ADP). From equations (6) and (7) the relationship should still hold under these conditions whereas according to Stone it should not. This point could be tested very simply by a study of the effects of the injection of orthophosphate on the ratio (Na efflux in 0 mM external $K^+$)/(Na efflux in 10 mM external $K^+$) in unpoisoned squid axons, but this experiment has not so far been carried out.

Baker and Stone (1966) have defined a quantity $\rho$ which they have called the rate reduction ratio.

$$\rho = v(x_{\frac{1}{2}} y_{\frac{1}{2}})/\tfrac{1}{4} v_1 \tag{8}$$

In the particular case of the coupled sodium/potassium pump, $v_1$ is the rate of operation of the pump at particular sodium and potassium concentrations, and $v(x_{\frac{1}{2}} \, y_{\frac{1}{2}})$ is its rate of operation at sodium and potassium concentrations which when used individually reduce the rate to $\frac{1}{2}$. Baker and Stone point out that $\rho$ can be determined experimentally and compared with values of $\rho$ which can be calculated for various suggested mechanisms for the sodium pump. They conclude, from a consideration of various experimental measurements, that $\rho$ probably lies between $\frac{2}{3}$ and 1. They point out that the models of Shaw (quoted by Glynn, 1956) and Opit and Charnock (1965), shown in figures 5 and 16, lead to values of $\rho$ which appear to exclude these models, at least in a simple form.

A calculation of $\rho$ can be made with the theory discussed in connection with equations (1)–(7). If the net rate of sodium pumping for a squid axon containing 100 mM sodium and in an extracellular potassium concentration of 100 mM is first calculated using the constants and the other ion concentrations used previously (Caldwell, 1968), then it can be calculated that the net pumping rate is reduced to about a half if the internal sodium is reduced to 45 mM or if the external potassium is reduced to 18 mM. Calculation of the net pumping rate in a nerve with an internal sodium concentration of 45 mM and exposed to an external potassium concentration of 18 mM gives a value of about 1·1 for $\rho$. This is reasonably close to the rough value of $\frac{2}{3}$–1 obtained by Baker and Stone (1966) from experimental data for crab nerve for ion concentrations similar to those used in these calculations. The calculated value of $\rho$ can in fact be regarded as reasonably close to that obtained from the experimental data in view of the lack of precision of the latter. Stone (1968) had calculated values of $\rho$ for his model, and states that satisfactory values of $\rho$ can be obtained by adjustment of one of the constants in the model, $K_1$-.

Finally mention must be made of the fact that the sodium pump can sometimes operate in an electrogenic way (see for example, Kerkut and Thomas, 1965), since this is not taken into account in the models which have been discussed. However, as has been pointed out by Cross, Keynes and Rybová (1965), an electrogenic component can be introduced into many of the models if the form of the transport system transporting potassium can move from the outside to the inside on its own as well as in association with potassium.

## REFERENCES

Albers, R. W., S. Fahn and G. J. Koval (1963) *Proc. Nat. Acad. Sci.*, **50**, 474

Baker, P. F., M. P. Blaustein, A. L. Hodgkin and R. A. Steinhardt (1969) *J. Physiol.*, **200**, 431

Baker, P. E. and A. J. Stone (1966) *Biochim. Biophys. Acta.*, **126**, 321

Caldwell, P. C. (1956) *J. Physiol.*, **132**, 35P

Caldwell, P. C. (1958) *Ier Colloque de Biologie de Saclay*, p. 88

Caldwell, P. C. (1960) *J. Physiol.*, **152**, 545

Caldwell, P. C. (1968) *Physiol. Rev.*, **48**, 1

Caldwell, P. C. (1969) *Current Topics in Bioenergetics*, Academic Press, New York, **3**, 251

Caldwell, P. C., A. L. Hodgkin, R. D. Keynes and T. I. Shaw (1960a) *J. Physiol.*, **152**, 561

Caldwell, P. C., A. L. Hodgkin, R. D. Keynes and T. I. Shaw (1960b) *J. Physiol.*, **152**, 591

Caldwell, P. C. and R. D. Keynes (1957) *J. Physiol.*, **137**, 12P

Caldwell, P. C. and H. Schirmer (1965) *J. Physiol.*, **181**, 25P

Cross, S. B., R. D. Keynes and R. Rybová (1965) *J. Physiol.*, **181**, 865

Dangeard, P. (1928a) *Compt. Rend.*, **186**, 892

Dangeard, P. (1928b) *Compt. Rend.*, **187**, 1156

Deffner, G. G. J. (1961) *Biochim. Biophys. Acta.*, **47**, 378

Eddy, A. A. and M. Mulcahy (1965) *Biochem. J.*, **96**, 76P

Eisenman, G. (1960) In A. Kleinzeller and A. Kotyk (Eds.) *Membrane Transport and Metabolism*, Academic Press, New York, p. 163

Garrahan, P. J. and I. M. Glynn (1967a) *J. Physiol.*, **192**, 159

Garrahan, P. J. and I. M. Glynn, (1967b) *J. Physiol.*, **192**, 175

Garrahan, P. J. and I. M. Glynn (1967c) *J. Physiol.*, **192**, 189

Garrahan, P. J. and I. M. Glynn (1967d) *J. Physiol.*, **192**, 217

Garrahan, P. J. and I. M. Glynn (1967e) *J. Physiol.*, **192**, 237

Glynn, I. M. (1956) *J. Physiol.*, **134**, 278

Glynn, I. M. (1963) *J. Physiol.*, **169**, 452

Glynn, I. M., C. W. Slayman, J. Eichberg and R. M. C. Dawson (1965) *Biochem. J.*, **94**, 692

Harris, J. E. (1941) *J. Biol. Chem.*, **141**, 579

Heald, P. J. (1962) *Nature*, **193**, 451

Hodgkin, A. L. and R. D. Keynes (1955) *J. Physiol.*, **128**, 28

Hokin, L. E. and M. R. Hokin (1960a) *J. Gen. Physiol.*, **44**, 61

Hokin, L. E. and M. R. Hokin (1960b) In A. Kleinzeller and A. Kotyk (Eds.), *Membrane Transport and Metabolism*, Academic Press, New York, p. 204

Hokin, L. E., P. S. Sastry, P. R. Galsworthy and A. Yoda (1965) *Proc. Nat. Acad. Sci.*, **54**, 177

Jardetzky, O. (1966) *Nature*, **211**, 969

Judah, J. D. and K. Ahmed (1964) *Biol. Rev.*, **39**, 160

Judah, J. D., K. Ahmed and A. E. M. McLean (1962) *Biochim. Biophys. Acta.*, **65**, 472

Kerkut, G. A. and R. C. Thomas (1965) *Comp. Biochem. Physiol.*, **14**, 167

Kylin, H. (1929) *Hoppe-Seyler z.*, **186**, 50

Kylin, H. (1930) *Hoppe-Seyler z.*, **191**, 200

Libet, B. (1948) *Fed. Proc.*, **7**, 72

Lowe, A. G. (1968) *Nature*, **219**, 934

Maizels, M. (1951) *J. Physiol.*, **112**, 59

Maizels, M. (1954) *J. Physiol.*, **125**, 263

Opit, L. J. and J. S. Charnock (1965) *Nature*, **208**, 471

Post, R. L., A. K. Sen and A. S. Rosenthal (1965) *J. Biol. Chem.*, **240**, 1437

Pressman, B. (1965) *Proc. Nat. Acad. Sci.*, **53**, 1076

Robertson, R. N. (1968) *Protons, Electrons, Phosphorylation and Active Transport*, Cambridge University Press, Cambridge

Sharp, G. W. and A. Leaf (1966) *Physiol. Rev.*, **46**, 593

Shaw, T. I. (1959) *Proc. Roy. Soc.*, **150**, 356

Skou, J. C. (1957) *Biochim. Biophys. Acta.*, **23**, 394

Skou, J. C. (1960) *Biochim. Biophys. Acta.*, **42**, 6

Skou, J. C. (1964) *Progr. Biophysics*, **14**, 131

Skou, J. C. (1965) *Physiol. Rev.*, **45**, 596

Stone, A. J. (1968) *Biochim. Biophys. Acta.*, **150**, 578

# Index